BOTANY
A BRIEF INTRODUCTION
TO PLANT BIOLOGY

2nd EDITION

BOTANY

A BRIEF INTRODUCTION TO PLANT BIOLOGY

THOMAS L. ROST
MICHAEL G. BARBOUR
ROBERT M. THORNTON
T. ELLIOT WEIER
C. RALPH STOCKING

University of California / Davis, California

JOHN WILEY & SONS

New York / Chichester / Brisbane / Toronto / Singapore

Production supervised by Ruth Greif Schild.
Book and cover designed by Madelyn Waxman Lesure.
    Cover photo: micrograph of Pine Wood/Michael
    Abbey/Photo Researchers, Inc.
Manuscript edited by Robyn Bem under the supervision of
    Deborah Herbert.

Library of Congress Cataloging in Publication Data

Main entry under title:

Botany: a brief introduction to plant biology.

    Includes indexes.
    1. Botany.   I. Rost, Thomas L.
QK47.B775   1984   581   83-21809
ISBN 0-471-87454-X

PREFACE

As science advances, it becomes increasingly more difficult to select topics for the introductory textbook. Texts tend to grow bulky and encyclopedic, far exceeding the material that can be covered in an academic quarter or semester. Some texts have been kept to a reasonable length through their authors emphasizing certain topics, such as human uses of plants, at the expense of other topics. Other texts have simply grown very large.

This text stands apart, offering a balanced yet compact introduction to all aspects of botany. Brevity was achieved by concentrating on the central facts and principles that are needed to understand the form, function, and evolution of plants and fungi. Illustrations, including both figures and examples, were carefully chosen for their educational value. The text covers all primary aspects of botany and the first edition served very well at many colleges and universities. We believe this revision preserves the virtues of the first edition, while offering significant improvements in several areas.

A major change is in the treatment of genetics and plant breeding. At users' request, the original genetics appendix has been expanded to a full chapter. Here are discussed reproduction and life cycles, principles of genetics with a slant toward plant improvement, and genetic engineering. Even for a course that omits genetics, the part of this chapter dealing with life cycles provides valuable background for the study of flowers and the survey of plant groups. The introduction to genetic engineering will also be of interest to many students.

Another change is in the use of color. We have included the most useful color plates from the sixth edition of *Botany* by T. E. Weier, C. R. Stocking, M. G. Barbour, and T. L. Rost (Wiley, New York, 1982). Most of the color photos deal with plant anatomy, a topic in which color is especially useful to the students.

There is more emphasis on evolution in this text than in the first edition. The adaptation to life on land is a unifying theme in the survey of plant groups (Chapters 16–20). As each division of land plants is introduced, the text singles out advanced traits and explains how they contribute to success in the terrestrial habitat. We think this unifying thread will make the survey chapters more interesting and easier to learn.

Much is happening in the fields of taxonomy and evolution; new methods and ideas have led to perspectives that were not available a few decades ago. Evolutionists argue whether evolution is gradual or whether it goes by jumps ("punctuated equilibria"); taxonomists armed with computers debate with more traditional workers as to how well we can define groups based on evolutionary connections. Chapter 14, "Classification and Evolution," has been almost totally rewritten to express these issues. It dwells on the mechanism of evolution and on the principles used to define plant groups. The survey of ancient plants, a mainstay of the original chapter, has been divided among the later survey chapters.

This text departs from the first edition in adopting a five-kingdom classification system. Such a move is consonant with current trends in taxonomy. Unfortunately, taxonomists differ in their criteria for defining the five kingdoms. The alternatives are discussed in Chapter 14. We chose the scheme described by L. Margulis and K. Schwartz in the book *Five Kingdoms* (Freeman, San Francisco, 1982) over the older Whittaker scheme, because the former reflects more clearly the evolutionary connections between groups. The change has an impact on lower levels of classification. Some groups previously ranked as classes are now divisions, and the fungi are now divided between two kingdoms. A table of kingdoms and divisions can be found inside the back cover of this book.

Lesser changes were made in the chapters on metabolism and photosynthesis. Chapter 2, "Metabolism," now has more illustrations, a slower pace, and less material. Chapter 8, "Photosynthesis," has also been lightened, but we responded to user requests by adding material on the C_4 pathway and photorespiration.

Those familiar with the first edition will find that the number of chapters has risen from 16 to 20. The two old chapters on vegetative and reproductive anatomy have been split into separate chapters on the stem, root, leaf, flower, and seed/fruit. The change is primarily one of packaging. Except for genetics, the new chapters retain the original depth of coverage. The text as a whole is slightly longer than before, but that is largely due to added illustrations.

The authors gratefully acknowledge the efforts of all those who have helped in developing the first edition and this revision. We are particularly grateful to Dr. James A. Doyle, University of California, Davis, for his meticulous review of Chapter 14 and for his advice in the areas of evolution and systematics. We are equally grateful to Professor Peter

Wejksnora, University of Wisconsin for his careful review of Chapter 11; to Dr. Norma J. Lang, University of California, Davis, for her advice on Chapter 16; and to Dr. Gerald C. Farr, Southwest Texas State University; Dr. William G. Lipke, Northern Arizona University; Professor Clifford M. Wetmore, University of Minnesota; Professor Jerry M. Baskin, University of Kentucky; Professor Harold Hopkins, St. Cloud State University; Professor William J. Pietraface, State University of New York at Oneonta; Dr. David Biesbore, University of Minnesota for their valuable review comments on the whole manuscript. Alice B. Addicott deserves our special thanks for providing many new drawings, in addition to those of the first edition. We also wish to thank Robyn Bem for her fine editorial work. Finally, we thank all the students in the botany courses at Davis, who made it clear where we could best improve the text.

THOMAS L. ROST
MICHAEL G. BARBOUR
ROBERT M. THORNTON
T. ELLIOTT WEIER
C. RALPH STOCKING

CONTENTS

LIST OF COLOR PLATES

Color plates appear following the page number listed below.

BOTANY

A BRIEF INTRODUCTION
TO PLANT BIOLOGY

1 INTRODUCTION

THE SCOPE OF BOTANY

Botany began with tribal lore about edible, medicinal, and poisonous plants. From this narrow focus on familiar leafy plants and mushrooms, curiosity spread to diverse forms until today more than 550,000 kinds, or **species,** of organisms are included in the realm of botany. New species are found continuously because there are still regions of the world that have not been thoroughly explored: the tropics, with their lush rain forests; the arctic; and the microscopic worlds of soils, oceans, and sediments.

In the earliest days of botany the field could easily be defined as the study of life forms that are rooted and essentially immobile. But the identification of additional species and more detailed studies have shown this to be an oversimplification. For example, the mosses **(Fig. 1.1, Color Plate 1)** have always been considered to be plants and appropriate subjects for botanists to study. But in early development the moss plant consists of green, threadlike filaments that resemble certain species of aquatic organisms, the **algae (Fig. 1.2, Color Plate 1).** Both the moss and the filamentous alga have a phase of the life cycle in which they produce free-living reproductive cells **(Fig. 1.3, Color Plate 1).** These cells swim by means of flagella that resemble those of animal sperm cells. Still other algae spend their whole lives as actively swimming, flagellated single cells. These discoveries confirm the fact that true natural boundaries between groups of organisms are difficult to find.

Is there any constant feature that is characteristic of all the organisms that botanists study and not of other forms of life? The answer is, "not quite." But two features—the presence of cell walls and the ability to photosynthesize—almost serve that purpose and are worth special comment.

Whenever large, complex forms of life are closely inspected, they are found to be composed of numerous microscopic units of living material called **cells.** In most kinds of organisms that botanists study, each cell is surrounded by a tough, fibrous **cell wall.** The walls of adjacent cells are cemented together, giving the plant as a whole a rigid shape and preventing individual cells from moving. But some of the unicellular algae do not have true walls. The slime molds **(Fig. 1.4, Color Plate 1)** also lack walls during most of their life cycle. This allows the slime mold to move about and pursue a predatory way of life reminiscent of animals (animal cells also have no walls).

Although cell walls are a characteristic of botanical life, there are major differences in wall structure and composition among organisms. In green plants the strength of the walls results from a network of **cellulose** fibers. In the fungi, **chitin** is usually found instead of cellulose, while the bacteria and blue-green algae have walls with a fishnet structure built from polymers of another, more complex set of subunits. These major differences in wall structure create a suspicion that the fungi, the bacteria, and the rooted green plants may be only remotely related.

Another property found only in botanical life is the ability to photosynthesize **(Fig. 1.5, Color Plate 1).** Photosynthesis uses the energy of sunlight to produce foods and other organic materials. The foods produced by photosynthetic organisms are essential not only for the organisms themselves but also for life forms such as animals (including human beings) that cannot trap sunlight. However, some of the "plants" discussed in this book do not photosynthesize. An example is "Indian pipe," a parasitic plant that has roots, stems, and flowers **(Fig. 1.6, Color Plate 1).** Most bacteria do not photosynthesize; nor do any of the 200,000 species of fungi **(Fig. 1.7, Color Plate 1).** We have no reason to suspect that the fungi ever had any photosynthetic ancestors. It is clear that botanists study these life forms because they have cell walls and some of the life cycle characteristics of the green, photosynthetic plants.

ANCESTRY AND CLASSIFICATION OF PLANTS

Because of the difficulties just described, not all scientists agree on the proper way to sort and classify the organisms included in botany. Nevertheless, classification is both a practical necessity and an important intellectual goal of botanists.

The highest goal in classification is to define natural groups of organisms that are related by common ancestry. To approach this goal we must reason from indirect evidence, since human observers did not trace events earlier than a few thousand years ago. The evidence includes fossils of ancient plants as well as similarities and differences between present-day plants (Chapter 14).

The most fundamental dividing line among life forms separates the **prokaryotes** from the **eukaryotes** (Fig. 1.8). The differences can be seen in the structure of the cells. One of several fundamental differences is that prokaryotes have their hereditary material (DNA) floating free in the same fluid mass as the rest of the cellular material, whereas the eukaryotes have their DNA separated from the rest of the cell contents by a surrounding membranous envelope. All

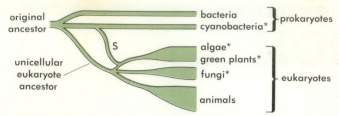

Figure 1.8. Ancestral relationships between modern organisms as deduced from protein structure and the structure and function of cells. **S** marks the symbiotic association between a cyanobacterium and a primitive eukaryote, by which chloroplasts are believed to have arisen. The starred groups are all studied in botany.

the common plants and fungi are eukaryotes, whereas the bacteria and the blue-green algae represent the world's only known prokaryotes. The entire animal kingdom consists of eukaryotes.

One of the most recent and promising tools for judging the hereditary relations between species depends on examining and comparing protein molecules. The hereditary information that is passed from generation to generation consists largely of instructions for building protein molecules. Since the number of possible different proteins is astronomical, a high degree of similarity between the proteins in two organisms indicates a common heredity. The proteins built by eukaryotes and prokaryotes are similar enough to indicate that they arose from a common ancestor, but they are different enough to suggest that the eukaryotes and prokaryotes diverged long before there were any organisms higher than unicells and before there were any distinctions between plants and animals.

Biologists divide the eukaryotes into kingdoms, but the number of kingdoms is subject to debate. One common approach is to define kingdoms on the basis of nutrition. The resulting groups are shown in Fig. 1.8. *Animals* typically ingest solid foods, a trait that is possible because their cells lack walls. *Fungi* are like animals in that they take in organic foods from the environment, but they have walls and can only absorb food molecules in dissolved form. *Plants*, in this classification, are eukaryotes that perform photosynthesis. Other points of view are presented in Chapter 14.

An interesting complication arises when we consider the origin of structures that occur within eukaryotic cells. For example, units called **chloroplasts** carry out photosynthesis in the cells of plants. Chloroplasts resemble certain bacteria in many ways, and it has been suggested that chloroplasts are descended from bacteria that entered early eukaryotic cells. In such a symbiotic association the bacterium might have been protected from harm and the eukaryote would have shared in the products of photosynthesis. With time, the association became permanent, so that modern chloroplasts cannot survive outside the eukaryotic cell. Comparable symbioses can be seen today between bacteria and the roots of legumes (Chapter 6), but in these, the partners are still separable. Certain other cell parts are also suspected of having arisen through symbioses between bacteria and eukaryotes. If these symbiotic origins really took place, the distinction between eukaryotes and prokaryotes may be less clear than once was thought. Such ideas leave us uncertain about how far back we must reach to find ancestral connections between the many forms of life that make up the subject matter of botany.

Figure 1.1. A common moss.

Figure 1.2. A filamentous green alga of the genus *Stigeoclonium*.

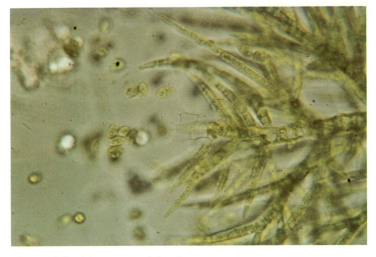

Figure 1.3. A later stage of development in the alga *Stigeoclonium*. The small free green cells are swimming reproductive cells.

Figure 1.4. Plasmodium of the slime mold *Physarum polycephalum.*

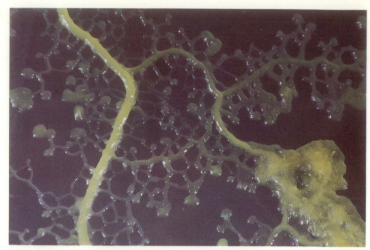

Figure 1.5. A species of *Mohavea,* a green flowering plant.

Figure 1.6. The nonphotosynthetic flowering plant *Monotropa uniflora.*

Figure 1.7. The fruiting body (basidiocarp) of *Amanita muscaria,* a true fungus.

2 METABOLISM

The visible signs of life in most plants are limited to slow changes such as the growth of organs. But if we look at the units of matter called molecules, of which plants are composed, the plant body proves to be a place of incessant, rapid activity. This chemical activity is called **metabolism.** The metabolic system within the plant generates thousands of chemical products, many of which are formed nowhere else in nature.

This chapter will show the principal molecules of life, how they are formed, and how they contribute to the life of the plant.[1]

RAW MATERIALS

Plants absorb raw materials from the environment and convert them into many complex products. Though the uptake of raw materials is discussed in detail in Chapter 7, it will be useful to introduce the raw materials now as a background for discussing the molecules manufactured by plants.

Water (H_2O) is the substance that plants take up in the greatest quantity. About 90% of the water that enters the plant is later evaporated away. Most of the retained water simply provides bulk and serves as a medium for transporting materials. However, water also provides atoms for building biological molecules. It is the chief source of the element hydrogen (H).

Carbon dioxide (CO_2) is taken up from the air (or from water, if the plant is aquatic). This compound is the plant's chief source of carbon (C) and is also a major source of oxygen.

Oxygen molecules (O_2) are also taken up. Molecular oxygen is needed to extract energy from the compounds that serve as fuels. During the day, the green parts of the plant make their own oxygen molecules from water. However, nongreen stems and roots must get oxygen from the air or water, and at night even leaves absorb oxygen from the air.

Mineral elements are absorbed from soil or water. They include nitrogen (N), phosphorus (P), sulfur (S), and many other elements. They occur as ions dissolved in water. Some ions, such as magnesium (Mg^{+2}) and potassium (K^+) are single charged atoms, but most are bound to hydrogen (as in ammonium, NH_4^+) or to oxygen (nitrate, NO_3^+, and sulfate, SO_4^{-2}) or both (phosphate, HPO_4^{-2}).

[1] Readers who have not previously studied chemistry may find it useful to read Appendix A along with this chapter.

MAJOR BIOLOGICAL MOLECULES

From the raw materials, plants make a great variety of compounds: at least 100,000 kinds have been described, and many more remain to be discovered. Fortunately, most biological compounds fall into just four categories: *carbohydrates, lipids, proteins,* and *nucleic acids.* The first three will be discussed immediately below, while nucleic acids are reserved for the discussion of hereditary information on page 11.

CARBOHYDRATES

The **carbohydrates** include sugars and compounds that are made by joining sugars into chains. They serve mainly as fuels and building materials.

Sugars have the general formula $(CH_2O)_n$, where n is an integer between 3 and about 10. Two of the most abundant sugars are **glucose** and **fructose,** shown in Fig. 2.1. They illustrate the common features of sugars as well as some ways in which sugars can differ from one another. First note that each sugar molecule can assume two forms: it can switch between being an open chain and a ring. This is possible because every open-chain sugar has a C$=$O group at the first or second carbon atom. The oxygen of C$=$O is reactive, and the rest of the molecule can twist around until this atom meets and joins another carbon atom to make the ring. (This cannot happen if the sugar has only three carbon atoms.)

Sugars are classified by the number of carbon atoms in their skeletons. Those with six carbons are called **hexoses;** *hex* means "six," and the ending *-ose* denotes a sugar. *Glucose* and *fructose* are hexoses. Sugars that contain five carbon atoms are called **pentoses.** An important example is **ribose,** which occurs in the molecules that carry hereditary information (see page 11). Ribose is also part of ATP, a compound that carries energy.

Most sugars have an OH group on every carbon except the one with the single oxygen (C$=$O). Many different sugars can be made by changing the orientation of the OH groups. As you see in Fig. 2.1*a*, glucose has one OH directed toward the left and the others directed toward the right. The sugar **galactose** (not shown) is exactly like glucose except that all the OH groups point to the right. These molecules are examples of **isomers,** or molecules with the same general structure but minor differences in organization.

Sugars can join into chains, which are classified by their length. The shortest chains have just two sugars and are

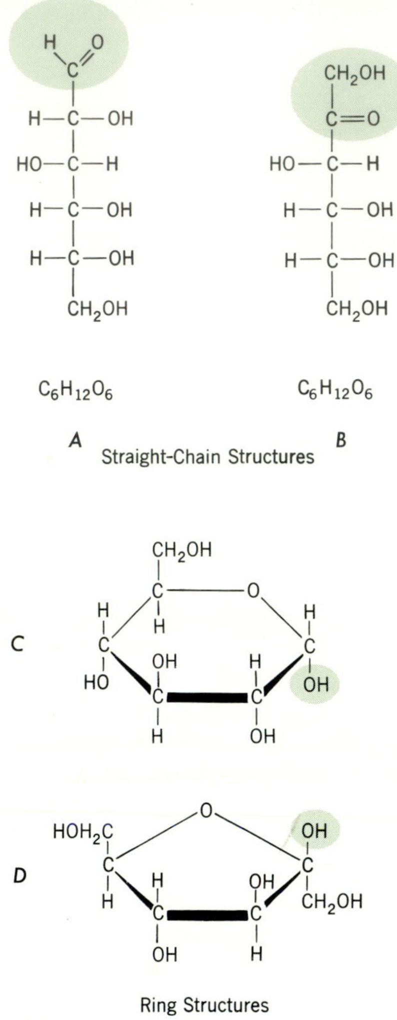

C₆H₁₂O₆   C₆H₁₂O₆

A   *B*

Straight-Chain Structures

Ring Structures

Figure 2.1. Glucose and fructose, two hexose sugars, have the same formula but different structures. Both can exist as a straight chain or ring. A and C, glucose; B and D, fructose.

called **disaccharides;** the longest have thousands of sugar units and are called **polysaccharides.** In these names the -*saccharide* portion means "sugar unit," while the prefix indicates the number. Thus *di-* means "two," and *poly-* means "many." A simple sugar such as glucose is a **monosaccharide;** *mono-* means "one."

The disaccharide **sucrose** (Fig. 2.2) is especially im-

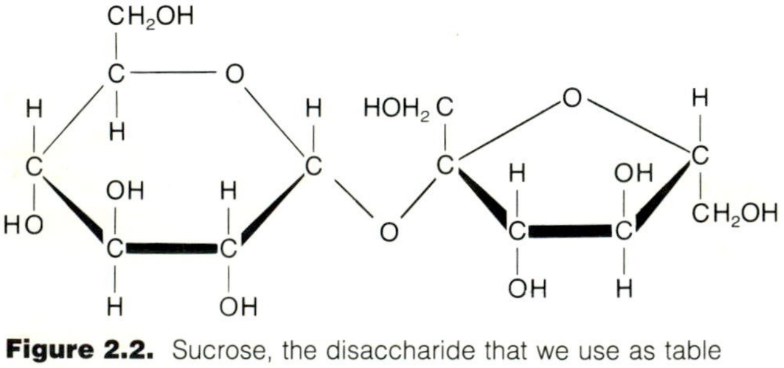

Figure 2.2. Sucrose, the disaccharide that we use as table sugar, is made of glucose and fructose.

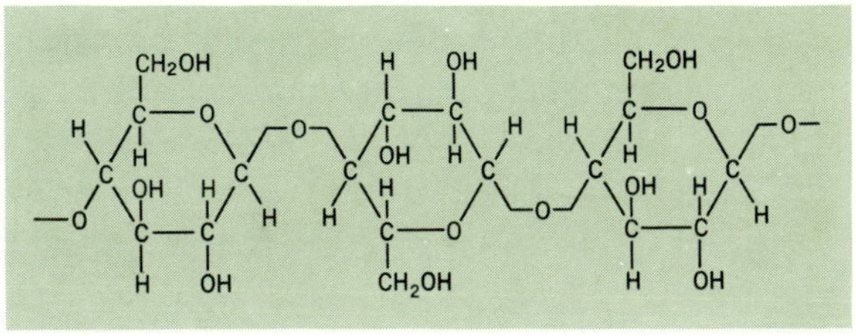

portant to plant and human life. Sucrose is composed of a glucose molecule linked to a fructose molecule. It is the main circulating energy storage compound in most plants; it is produced in leaves and transported to other parts of the plant. Sugar beets and sugarcane (the sources of table sugar) store large quantities of sucrose.

The polysaccharides illustrate the concept of **polymers:** that is, complex molecules constructed by linking many smaller repeating molecules into chains. Some polysaccharides contain just one kind of sugar, while others contain two or more kinds. *Starch* is composed entirely of glucose molecules. The starch molecule coils and may also branch because of the way the glucose units are linked together. Starch molecules clump together into grains that form a compact food reserve. *Cellulose* (Fig. 2.3) is also made entirely of glucose, but the sugars are linked in a way that leads to a straight, unbranched polymer. Cellulose molecules line up side-by-side to form tough, straight fibers. Cellulose contains as much energy as starch but is more resistant to attack. Appropriately, plants employ cellulose as a structural material rather than a food reserve. It makes up a large part of wood.

LIPIDS

Lipids are varied compounds with just one feature in common: they do not readily mix with water. Some lipids serve as energy reserves, some protect against water loss, and some serve structural roles in the plant.

Waxes such as **cutin** and **suberin** are solid lipids that coat the surfaces of plants and limit the loss of water. Some plant waxes (e.g., carnauba) are used in furniture and automobile polishing compounds.

Fats are excellent storage compounds because they have a high energy content and a tendency to accumulate in droplets. A fat molecule is composed of three **fatty acids** that are joined to a molecule of **glycerol** (Fig. 2.4). A fatty acid molecule has an acidic group at one end; the rest of the molecule is a long chain of carbon and hydrogen atoms. These **hydrocarbon chains** are insoluble in water.

Phospholipids are important structural compounds. To describe their function, it will be necessary to anticipate Chapter 3 and say a little about the organization of material in the plant. As shown in Fig. 2.5, the plant body consists of numerous units called *cells*, each of which contains functional units called *organelles*. Every cell is surrounded by a

Figure 2.3. Cellulose is an unbranched polymer composed of many thousands of glucose units. The diagram shows three of the units.

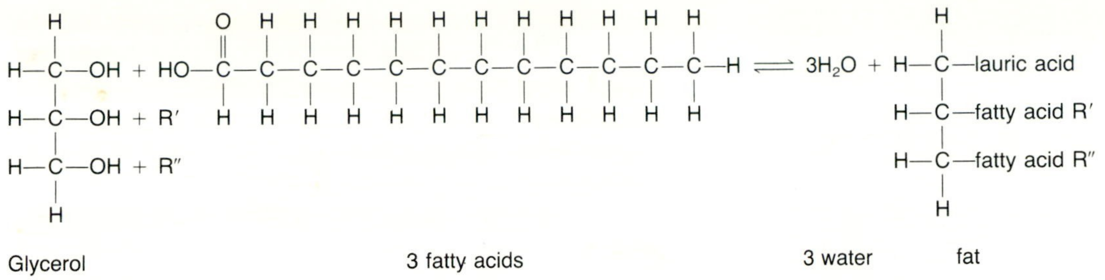

$$H-\underset{\underset{H}{\overset{H}{|}}}{\overset{H}{\overset{|}{C}}}-OH \;+\; HO-\underset{\overset{H}{|}}{\overset{O}{\overset{||}{C}}}-\underset{\overset{|}{H}}{\overset{H}{\overset{|}{C}}}-\underset{\overset{|}{H}}{\overset{H}{\overset{|}{C}}}-\underset{\overset{|}{H}}{\overset{H}{\overset{|}{C}}}-\underset{\overset{|}{H}}{\overset{H}{\overset{|}{C}}}-\underset{\overset{|}{H}}{\overset{H}{\overset{|}{C}}}-\underset{\overset{|}{H}}{\overset{H}{\overset{|}{C}}}-\underset{\overset{|}{H}}{\overset{H}{\overset{|}{C}}}-\underset{\overset{|}{H}}{\overset{H}{\overset{|}{C}}}-\underset{\overset{|}{H}}{\overset{H}{\overset{|}{C}}}-\underset{\overset{|}{H}}{\overset{H}{\overset{|}{C}}}-\underset{\overset{|}{H}}{\overset{H}{\overset{|}{C}}}-H \;\rightleftharpoons\; 3H_2O \;+\; H-\underset{\underset{H}{|}}{\overset{H}{\overset{|}{C}}}-\text{lauric acid}$$

Glycerol   3 fatty acids   3 water   fat

Figure 2.4. A fat is formed by joining three fatty acid molecules to a molecule of glycerol. One fatty acid (lauric acid, $C_{12}H_{25}O_2$) is shown in full; the others are symbolized as R′ and R″.

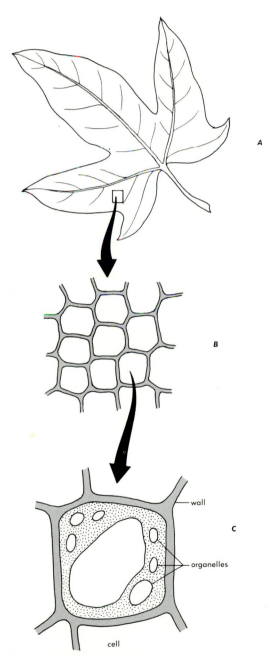

Figure 2.5. Levels of organization in the plant body. A, an organ, the leaf, B, on magnification, organs are found to contain cells within walls. C, further magnification shows that cells contain functional units, the organelles.

thin sheet of material called a **membrane,** and additional membranes surround most organelles. Phospholipids make up from one-fourth to two-thirds of the weight of each membrane. They combine into sheets because one end of the molecule is compatible with water and the other end is not. Therefore, when phospholipids are surrounded by water (as in the plant body), they line up in a double layer that keeps one end of each molecule away from the water. The arrangement is shown in Fig. 3.4, p. 22.

Phospholipids resemble fats, except that one of the fatty acids is replaced by a **phosphoryl group:**

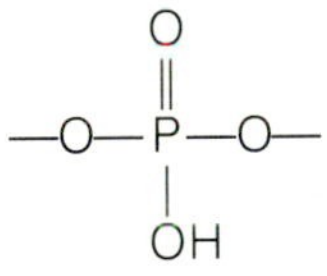

$$-O-\underset{\underset{OH}{|}}{\overset{\overset{O}{||}}{P}}-O-$$

One end of the phosphoryl group binds to the glycerol unit, while any of several organic groups may bind to the other end of the phosphoryl group.

PROTEINS

Proteins perform many functions in the plant. Some proteins regulate and control the chemical processes of life; some act as structural materials in membranes and elsewhere; and some are involved in generating motion.

Proteins are polymers made from subunits called **amino acids.** There are 20 common amino acids, 19 of which have the following structure[2]:

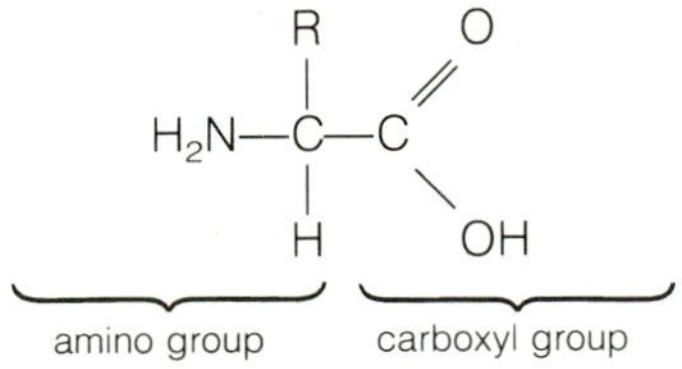

$$H_2N-\underset{\underset{H}{|}}{\overset{\overset{R}{|}}{C}}-\overset{\overset{O}{||}}{C}\begin{smallmatrix}\\OH\end{smallmatrix}$$

        amino group   carboxyl group

The symbol **R** signifies a group of atoms called a **side-chain.** The 20 amino acids have different side-chains; four of them are shown in Fig. 2.6.

[2]The 20th amino acid, **proline,** differs from the rest in that its R group bends over and binds to the N of the amino group.

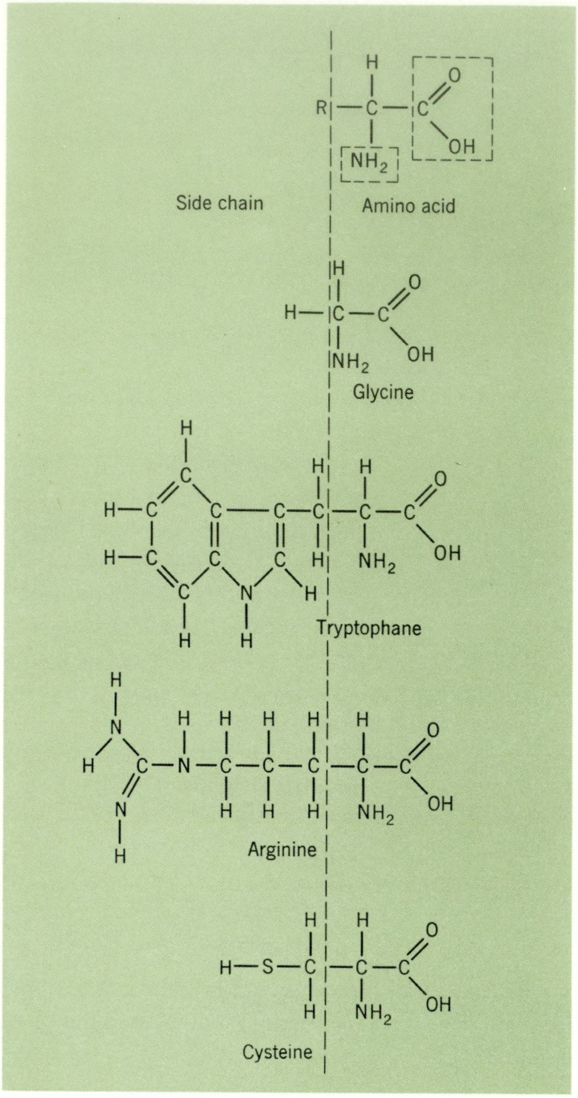

To form a protein, the amino acids are linked together as in Fig. 2.7. The link between two amino acids is a **peptide bond,** and for this reason proteins are often called **polypeptides.**

Proteins can perform many functions because they vary widely in structure. They range in size from below 100 to over 1000 amino acids. They combine the amino acids in different sequences, so countless different proteins are possible.

Proteins are flexible in shape, and each one can fold in many ways. This is important because each protein must fold into a specific shape to have a biological function. Figure 2.8 shows a folded protein. Unfortunately, the forces that stabilize the folding are quite weak. Proteins can be unfolded by mild heat and by many chemical agents. Unfolded proteins often become irreversibly tangled and cannot return to their original shapes. This process is called **denaturation.**

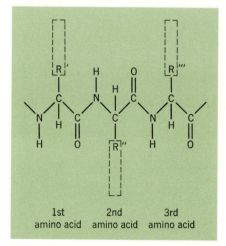

Figure 2.6. Some amino acids.

Figure 2.7. A polypeptide chain is a part of a protein molecule. The backbone of the protein molecule is formed by many amino acids joined by the union of the amino group (NH_2) of one amino acid to the acid group (COOH) of another amino acid by the removal of a water molecule. The R groups represent side-chains of the different amino acids.

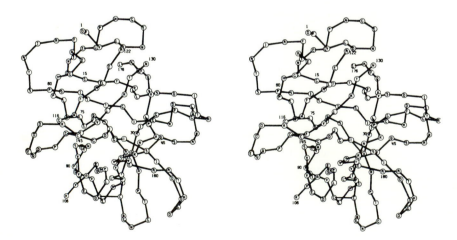

Figure 2.8. A protein that is found in soybeans. You can fuse the two drawings into a single image that looks three-dimensional if you stare at a point between the two pictures and let the images drift together. The drawing only shows the course of the protein's "backbone." In reality, side-chains fill the spaces between loops of the protein.

Several factors determine how a protein will fold. Most important is the influence of water, which completely surrounds most proteins in the living plant. Water molecules form hydrogen bonds with one another and resist being separated. Some of the side-chains of proteins are nonpolar and cannot break through the web of hydrogen bonds. Therefore, the surrounding water forces most proteins into a compact shape, with the nonpolar parts in the interior and the polar parts at the water surface. This is one reason why dehydration (excessive water loss) is fatal: it allows proteins to denature.

ENZYMES AND CATALYSIS

Life is based on chemical activity. However, most compounds are stable and will not react at useful rates without help from **catalysts.** A catalyst is an agent that speeds reactions without being consumed in the process. Catalysts in living plant cells are special proteins called **enzymes.** They have two important properties: they are selective in their action, and they work very rapidly.

Enzymes are selective in two ways. First, an enzyme reacts with only a few kinds of other molecules (the enzyme's **substrates**); secondly, each enzyme catalyzes only one kind of reaction.

Enzymes achieve their specificity by matching the shape of the substrate molecule. The enzyme has a pit or groove called the **active site** where the substrate must bind before the enzyme can react (Fig. 2.9).

With the substrate in place, an enzyme generates products as much as a billion times faster than an uncatalyzed reaction. Enzymes achieve their speed by varied mechanisms, and we know the full details in just a few cases. Here, we can only mention the most well-known aspect.

Uncatalyzed reactions tend to be slow because there are stabilizing forces within each molecule that resist change. To overcome these forces, a molecule must acquire an amount of energy called the **activation energy.** A collision between molecules can supply the energy; however, very few collisions provide enough energy to meet the activation requirement. In most collisions the molecules rebound without being changed.

How do enzymes speed reactions? The obvious answer, that they provide energy to the reactions, is wrong.

Enzymes work primarily by *reducing the activation energy.* This is done by providing a new pathway of change that involves weaker stabilizing forces and, therefore, smaller activation energies. To open up a new pathway, the enzyme itself reacts with the substrate. Rapid stepwise changes occur, and each step presents only a small energy barrier. These small energy demands can easily be met when molecules collide with the enzyme.

PHASES OF METABOLISM

Complex products are made by **metabolic pathways**—sequences of reactions in which one enzyme after another works on the product. Metabolism includes hundreds of pathways, some of which are shown in the following pages. As a framework for studying the pathways, let us point out that metabolism has a higher level of organization: the pathways can be grouped into several phases of chemical activity, and each phase plays a different part in the life of the plant. The phases of metabolism are *photosynthesis, anabolism,* and *catabolism.* Figure 2.10 shows their relationship.

In **photosynthesis,** the plant uses light energy to make organic acids from carbon dioxide and water. The organic acids are later converted to sugars or amino acids. Photosynthesis is the primary source of energy for the construction of plants. It is also the primary way in which carbon is brought into the living world. Photosynthesis in plants is especially important to humans because our own tissues cannot use light energy. Therefore, we must use other organisms, including plants, for food. Chapter 8 describes photosynthesis in detail.

Anabolic pathways build complex molecules from sim-

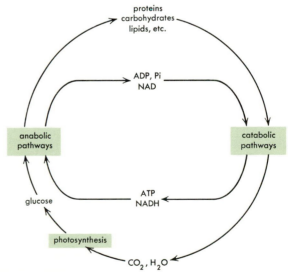

Figure 2.10. Phases of metabolism. Anabolism produces complex, high-energy products from simpler raw materials. Catabolism breaks down fuels to form compounds such as ATP and NADH that are needed for anabolism. Photosynthesis retrieves carbon from CO_2 and captures light energy.

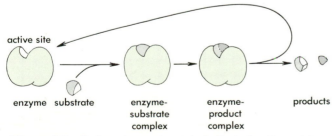

Figure 2.9. Outline of enzyme catalysis, illustrating the match between the substrate and the active site. The process can be reversed.

pler ones. (Photosynthesis could have been put in this category, but its unique use of light energy calls for separate treatment.) Anabolism is the basis of growth, development, and reproduction. An example is the construction of proteins from amino acids, which is discussed on page 13.

Catabolic pathways have two functions. First, they extract energy from storage compounds and put it into forms that can power anabolism and movement of materials around the plant body. Secondly, catabolic pathways recycle molecules and make the components available for building new molecules. Catabolism is discussed below.

CONNECTIONS BETWEEN PHASES: ATP

The anabolic and catabolic paths are linked by several kinds of molecules that carry energy and building materials. Catabolic paths load the carriers, and anabolic paths unload them. Two of the most prominent of the loaded carriers are *ATP,* which carries phosphate, and *NADH,* which carries hydrogen. Both molecules also carry energy. NADH will be discussed in the next section; here, let us examine ATP.

ATP, or **adenosine triphosphate,** is the main energy-carrying compound in cells. Almost everywhere that biological work is being done, ATP provides the energy. In the process, ATP is broken into two smaller parts.

Figure 2.11 shows a molecule of ATP. The portion labeled *adenosine* is simply a handle that enzymes can recognize. ATP stores energy because it has three phosphoryl groups in a chain. This arrangement is unstable because the 10 oxygen atoms in the triphosphate chain compete with one another for electrons, and the oxygen atoms are left with a substantial ability to attract electrons from other molecules. Thus when the opportunity arises, ATP will trade one or two phosphoryl groups to other molecules:

$$ADP—OP + R—OH \longrightarrow ADP—OH + R—OP$$
$$(\text{``ATP''}) \qquad\qquad (\text{``ADP''})$$

We have symbolized the "other molecule" as R—OH; R can be many different compounds. The "—OP" represents a phosphoryl group, as shown on page 5. The substitution of —OH for —OP releases energy because it relieves the competition between the oxygen atoms. The part of the ATP molecule that remains after the transfer is called **ADP,** or **adenosine diphosphate.** Later, the catabolic system attaches an inorganic phosphate to ADP, regenerating ATP:

$$ADP—OH + P—OH \longrightarrow$$
$$ADP—OP \text{ (ATP)} + H—OH \text{ (water)}$$

HOW CELLS PRODUCE ATP

The main paths of catabolism are summarized in Fig. 2.12. Their principal result is to join ADP and inorganic phosphate (symbolized as **Pi**) into ATP, using energy that is released by the controlled oxidation of fuels. In most organisms, over 95% of the ATP is made in a process called **cellular respiration.** The paths of cellular respiration use **organic acids**

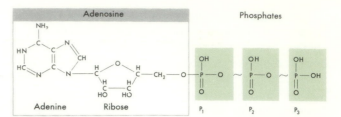

Figure 2.11. The structure of ATP (adenosine triphosphate). ADP has only two phosphate groups and AMP has only one. AMP is an example of *nucleotide.*

as fuels—compounds such as malic and citric acids (the cause of the sour taste in green apples and lemons, respectively). Cellular respiration oxidizes the organic acids and converts them to CO_2 and H_2O.

Organic acids are usually made from sugars, fats, and amino acids. Many pathways generate fuel acids, but the most vigorous and important path is **glycolysis,** the pathway that converts glucose into pyruvic acid.

GLYCOLYSIS

Glycolysis consists of the steps shown in Fig. 2.13. First, ATP donates a phosphate group to the glucose molecule. A rearrangement follows, a second phosphate is added, and the sugar-phosphate is cleaved into two 3-carbon compounds. Another reaction makes the two products identical. Then several enzymes work in turn to change the 3-carbon units into **pyruvic acid,** or **pyruvate,** the main product of glycolysis. During these steps two molecules of NAD^+ pick up hydrogen and electrons, and four molecules of ATP are made by passing phosphate from 3-carbon compounds to ADP.

CELLULAR RESPIRATION

Cellular respiration consists of many steps, arranged in three distinct phases that occur in different parts of the cell. Glycolysis, which we have just discussed, occurs in the fluid between the organelles. By contrast, cellular respiration occurs in organelles called **mitochondria,** which are present in every plant cell. Mitochondria vary in form, but often they are sausage-shaped. Viewed more closely (Fig. 2.14), a mitochondrion has two surrounding membranes that enclose a fluid called the **matrix.** Both the matrix and the inner membrane take part in cellular respiration.

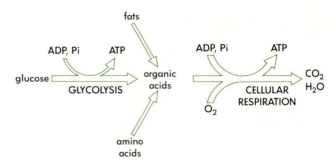

Figure 2.12. Overview of the main energy-yielding pathways of catabolism.

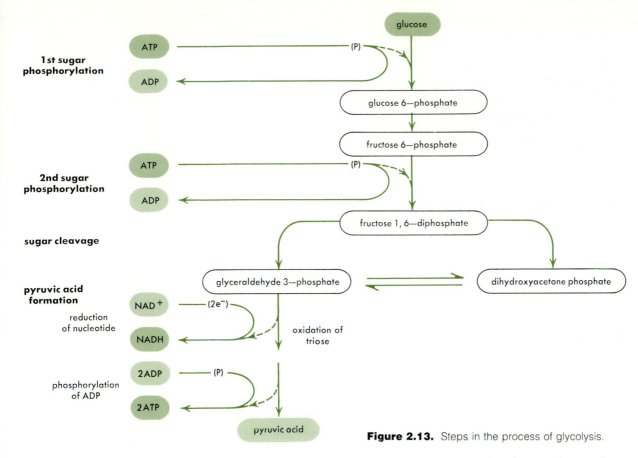

1st sugar phosphorylation

2nd sugar phosphorylation

sugar cleavage

pyruvic acid formation

reduction of nucleotide

phosphorylation of ADP

glucose

ATP

ADP

(P)

glucose 6—phosphate

fructose 6—phosphate

ATP

ADP

(P)

fructose 1, 6—diphosphate

glyceraldehyde 3—phosphate

dihydroxyacetone phosphate

NAD+

(2e⁻)

NADH

oxidation of triose

2ADP

(P)

2ATP

pyruvic acid

Figure 2.13. Steps in the process of glycolysis.

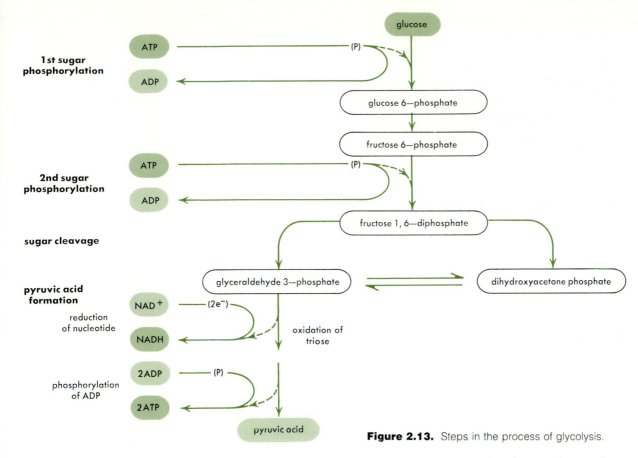

Figure 2.14. Sketch of a mitochondrion, simplified and cut open to show where the phases of cellular respiration occur.

inner membrane (performs electron transport, creates proton gradient, and makes ATP)

outer membrane

fluid matrix (citric acid cycle occurs here)

Reactions in the Matrix

Cellular respiration begins when organic acids pass through the mitochondrial membranes and enter the matrix. There, enzymes oxidize the organic acids in a stepwise manner, and some of the intermediate compounds can be used to build amino acids and lipids. Thus the matrix is an important crossroads in metabolism, where paths of breakdown and synthesis come together.

Figure 2.15 outlines the steps of the matrix reactions. First, enzymes remove CO_2 from pyruvic acid and bind the remaining **acetyl group** to a carrier known as **coenzyme A.**

Next, an enzyme transfers the acetyl group from coenzyme A to an acid with four carbon atoms called **oxaloacetate.** The product, **citric acid,** has six carbon atoms. This event is the first step in a cycle of reactions called the **citric acid cycle** (sometimes also called the **Krebs cycle,** in honor of its discoverer). In the citric acid cycle, the molecule that began as citric acid is gradually trimmed and reshaped until it becomes another molecule of oxaloacetate. This is the same kind of molecule that started the cycle. Overall, each round of the citric acid cycle consumes one acetyl group and three water molecules. It makes two molecules of CO_2, three of NADH, and one molecule of $FADH_2$. The cycle also makes one molecule of GTP per turn. GTP is similar to ATP and carries the same amount of energy.

The "loaded" carriers, NADH and $FADH_2$, are the main products of the matrix. They serve as fuels for the final steps of respiration that are discussed in the next section.

Electron Transport and the Proton Gradient

The last step in extracting energy from fuel molecules occurs at the inner membrane of the mitochondrion. This membrane passes electrons from NADH and $FADH_2$ to oxygen, forming water. The transfer releases a great deal of energy because it satisfies the very strong tendency of oxygen to take up electrons. Much of the energy is converted to heat, which is a waste product that organisms cannot use. However, one-third to one-half of the energy is conserved in the form of ATP.

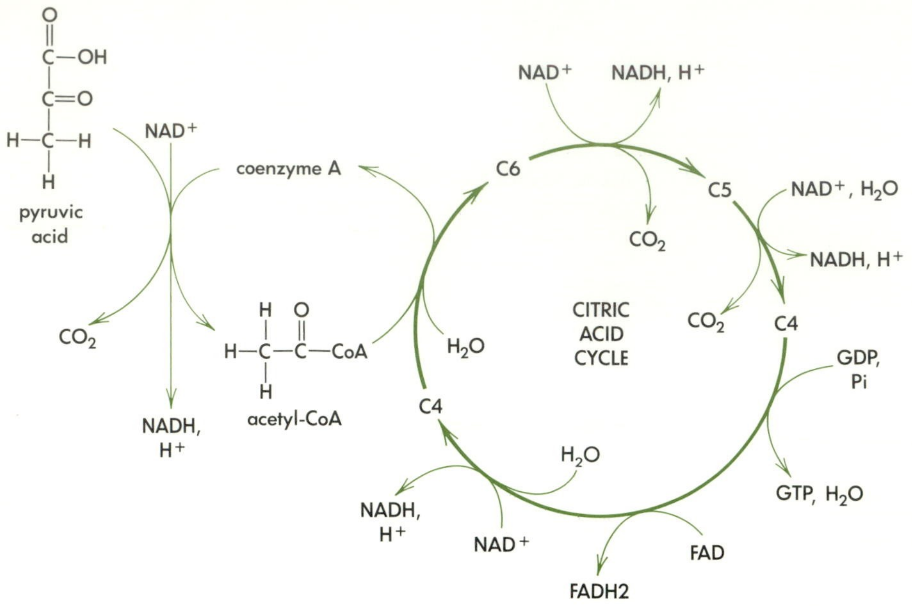

Figure 2.15. Reactions of the mitochondrial matrix. Organic acids that have 4, 5, and 6 carbon atoms are represented, respectively, by C4, C5, and C6. These symbols are used instead of chemical formulas to emphasize the change in number of carbon atoms as molecules move around the citric acid cycle. All the carbon atoms of pyruvate are released as CO_2, while electrons and protons (H^+) from pyruvate and water are transferred to NAD^+ and FAD. GTP is equivalent to ATP.

Figure 2.16 illustrates electron transport and how it leads to ATP synthesis. Many electron-carrying compounds are embedded in the inner mitochondrial membrane. The carriers work in chains, and thousands of chains are present in the membrane (for clarity, Fig. 2.16 shows just one chain of carriers). The carrier at one end of a chain takes electrons from NADH, and the carrier at the other end passes them to O_2. Intermediate carriers act as a bucket brigade, each one passing electrons to its neighbor. Energy is released at each step along the chain.

Until recently it was not clear how the inner membrane uses electron transport to make ATP. However, we are reasonably sure now that the electron transport chains act as **proton pumps:** some of the released energy is used to drive protons (H^+ ions) through the membrane and out of the matrix. The result is an energy-storing **proton gradient,** a difference in proton concentration across the membrane. Special enzymes called **ATPases** draw energy from the gradient to make ATP.

How can a proton gradient store energy? There are two aspects: one of them relates to *electrical charge,* the other to *diffusion.*

The proton gradient stores *electrical energy* because each proton has a positive charge. A resting mitochondrion has the same number of negative and positive charges. When protons are pumped out, the matrix is left with a negative charge. The negative matrix strongly attracts the positive protons, but the membrane prevents the protons from moving in.

The proton gradient stores *diffusional energy* because all particles (not just charged ones) tend to distribute themselves uniformly in space through diffusion. If the membrane were not in the way, protons would rush into the matrix where their concentration is low.

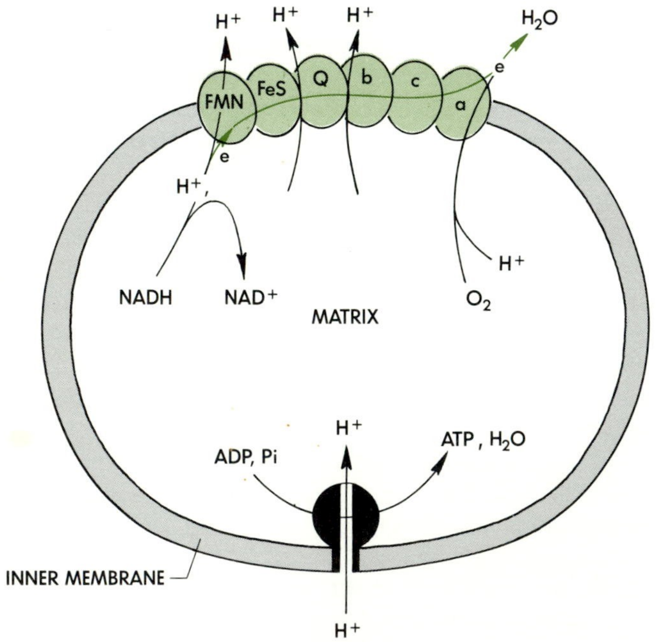

Making ATP

ATP synthesis, or **oxidative phosphorylation,** is the last step in cellular respiration. As mentioned, protons would rush into the matrix if the inner membrane did not block their path. (You might compare the membrane to a dam that prevents water from rushing downhill.) To harvest the energy of the proton gradient, many ATPase units are bound to the

Figure 2.16. Electron transport and ATP synthesis in the mitochondrion (schematic). Many ATPase units and chains of electron carriers are embedded in the inner membrane; only one chain of electron carriers (green circles) and one ATPase (black circle) are shown. Electrons move from NADH and $FADH_2$ to the first electron carrier and then through the chain to O_2. Energy is released in the process, and some of it is used to drive H^+ across the membrane. The resulting proton gradient provides energy for ATP synthesis.

inner membrane (Fig. 2.16). Each ATPase has several enzymes that cooperate to make ATP. We do not know precisely how the ATPases work. In principle, they allow protons to pass through the membrane, discharging the gradient. The transfer is done in such a way as to combine ADP and Pi into ATP.

Using these mechanisms, the inner membrane makes about 32 molecules of ATP for every molecule of glucose that is consumed—an impressive yield, compared to the 2 ATPs that arise in glycolysis and the 2 GTPs that are made in the citric acid cycle.

ALCOHOLIC FERMENTATION

When deprived of molecular oxygen (O_2), most higher plants begin to die in a few hours because ATP is no longer produced in sufficient amounts. However, yeasts and a few other microorganisms can thrive without O_2, and even some tissues and organs of plants (apples, for example) can survive for a while with little or no O_2. They are making ATP by way of **alcoholic fermentation,** in which sugar is converted to ethanol and CO_2:

$$C_6H_{12}O_6 + 2ADP + 2Pi \longrightarrow$$
$$2CO_2 + 2C_2H_5OH + 2ATP + 2H_2O$$

Alcoholic fermentation is much simpler than cellular respiration. It uses glycolysis with the two added steps shown in Fig. 2.17. Glycolysis does not require O_2, and it makes two molecules of ATP for every glucose molecule it consumes. However, glycolysis requires NAD^+, and it soon stops unless there is a way to "unload" the NADH and recycle the NAD^+. This is where alcohol production becomes important. When O_2 is scarce, an enzyme attacks pyruvate and splits it into CO_2 plus a molecule of **acetaldehyde.** Then a second enzyme passes electrons and H^+ from NADH to acetaldehyde. Alcohol and NAD^+ are the products.

In short, alcoholic fermentation recycles NAD^+ so that glycolysis can continue with its small yield of ATP. But fermentation is inefficient: it yields only 2 ATP per glucose, whereas respiration yields 36 ATP. Also, alcohol is toxic and

will kill cells if it accumulates in large amounts. Even yeast, which is used to produce alcoholic beverages, becomes inactive as alcohol accumulates (which is the reason natural wines have only about 12% alcohol).

NUCLEIC ACIDS AND HEREDITARY INFORMATION

The plant body is composed of water and metabolic products that accumulate during its development. To produce an adult body that is typical of the species, the metabolic system requires guidance from inherited information.

Hereditary information is stored in molecules of **DNA (deoxyribonucleic acid).** DNA indirectly controls metabolism by guiding the production of enzymes and other proteins. The following pages will show the nature of DNA, how it stores information, and how it is duplicated so that every cell has a copy. The next section will show how DNA guides protein synthesis.

HOW DNA STORES INFORMATION

DNA belongs to a class of polymers called **nucleic acids,** that is, polymers made of subunits called **nucleotides.** Figure 2.18 shows two nucleotides linked as in a nucleic acid polymer. Each nucleotide contains a sugar, a phosphate group, and a unit called a base. The nucleotides join to give DNA a sugar-phosphate backbone, and the bases jut out to the side. DNA has four kinds of bases, which can occur in any order along the polymer. The bases are adenine (A),

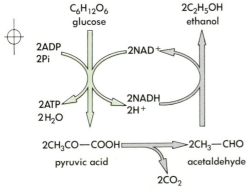

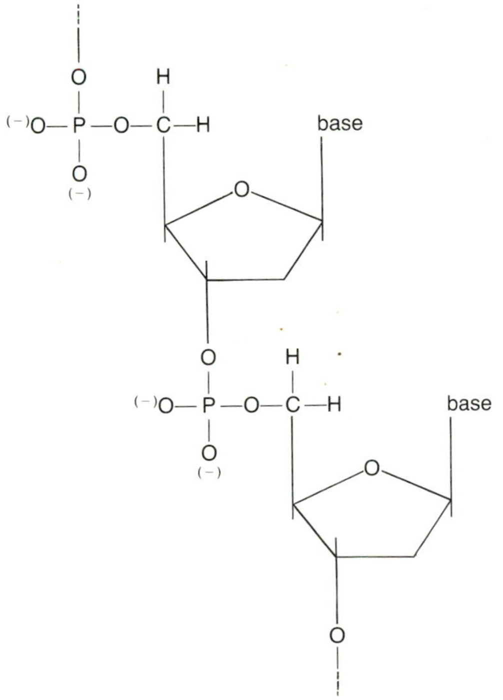

Figure 2.17. Main events of alcoholic fermentation. Glycolysis (green arrow) makes ATP while converting glucose to pyruvate. Two added steps convert pyruvate to CO_2 and alcohol as a way of regenerating the NAD^+ needed for glycolysis.

Figure 2.18. Two nucleotides linked as in a nucleic acid polymer. Each nucleotide has a base, a sugar (the ring), and a phosphoryl group.

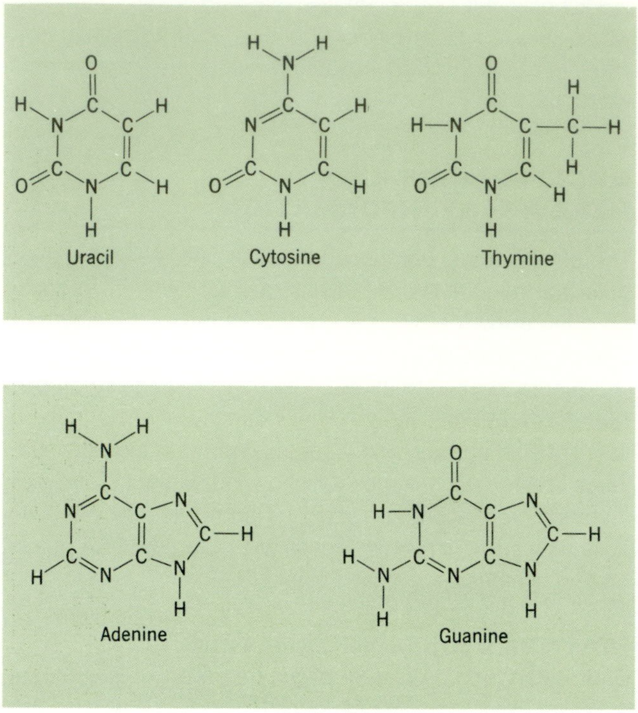

Uracil

Cytosine

Thymine

Adenine

Guanine

Figure 2.19. The bases in DNA and RNA.

cytosine (C), guanine (G), and thymine (T). Their structures are shown in Fig. 2.19.

DNA can store information because its four bases (A, C, G, and T) can be arranged along the polymer in varied sequences, almost like a message written with an alphabet of four letters. Most DNA molecules contain millions of nucleotides and have room to store many messages concerning protein synthesis. These hereditary messages, stored in DNA, are called **genes.** The best-understood genes describe proteins; they specify sequences of amino acids. It takes 3 nucleotides to code for 1 amino acid. Thus, in principle, a protein with 150 amino acids could be spelled out by about 450 nucleotides along a DNA molecule. At this rate, a million-base DNA molecule could describe about 2000 proteins, each with 150 amino acids. However, only about 10% of the DNA actually codes for proteins. The rest of the DNA is highly repetitive, and its function is still not clear.

How does a cell read its DNA? Biochemists began to see the answer when they found that *DNA molecules occur in pairs.* As you can see in Fig. 2.20, the paired molecules (called **strands**) wrap around each other in a **double helix.** The bases are turned inward toward the axis of the helix. Most importantly, hydrogen bonds join each base of one polymer to a base of the other polymer. The **base pairs,** visible in Fig. 2.20, are the key to the reading system. Only two kinds of base pairs occur in DNA: G always pairs with C, and A always pairs with T. This fact is called the **base-pairing rule.** The base sequence of one DNA polymer can be predicted if you know the sequence of its partner in the helix. For example, when one strand has ATC, the other must have TAG. The two strands are said to have **complemen-**

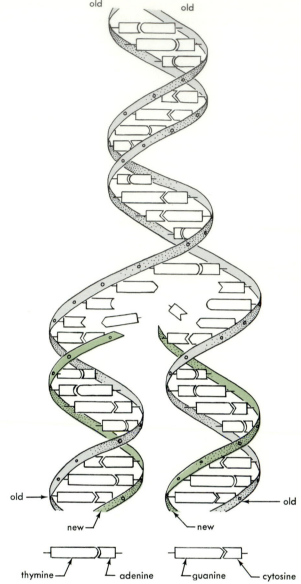

old old

old old

new new

thymine adenine guanine cytosine

Figure 2.20. Schematic diagram showing how one "old" double-stranded molecule of DNA can be replicated into two new double-stranded molecules of DNA. Each new molecule has one old and one newly synthesized strand.

tary sequences and are related like a photographic negative and a matching print.

COPYING THE DNA

Figure 2.20 shows how the base-pairing rule makes it possible to copy DNA. First the two polymers unwind, exposing the bases. Enzymes line up free subunits along each strand, obeying the base-pairing rules. One by one, subunits are fitted, tested, and joined to make a new polymer. Two double helices are formed, each exactly like the original. Notice that the pairing rule allows each strand to serve as a model, or **template,** for lining up subunits in the correct sequence. This is how the information stored in DNA—the base sequence—can be duplicated.

Actually, DNA cannot duplicate itself. Several proteins

are needed, and their combined action is so complex that we can only sketch it briefly. To start the duplication, *initiation proteins* bind to the DNA at special initiation points. *Unwinding proteins* cause part of the helix to unwind, exposing the bases. The unwinding causes kinks (called supercoils) to occur elsewhere, and an enzyme called *gyrase* straightens the kinks by temporarily breaking the DNA so that the DNA can unwind. *Polymerase* enzymes make short segments of new DNA along the old strands. *Ligase* enzymes link the new segments together. These events occur repeatedly as duplication progresses along the DNA. Setting aside the details, let us emphasize the role of base-pairing: accurate duplication is possible only because the enzymes obey the base-pairing rule when they attach subunits to a growing DNA polymer.

HOW PROTEINS ARE MADE

In the anabolic systems that build proteins we find the central processes of biology. This is true not only because life depends on proteins, but also because heredity (a prime characteristic of life) is based on the control of protein synthesis. The following pages will show how the information stored in DNA guides protein synthesis.

The path from DNA to protein has two phases: *transcription* and *translation* (Fig. 2.21). In transcription, an enzyme makes a molecule of the substance RNA to match a selected DNA segment. (We shall describe RNA in a moment.) In translation, subcellular units called **ribosomes** read the RNA and make proteins with the specified amino acid sequences.

READING DNA: RNA SYNTHESIS

Enzymes read DNA in the process called **transcription.** The resulting RNA molecule carries the same information as DNA. Thus RNA might be compared to a photocopy of a book.

RNA, or **ribonucleic acid,** resembles DNA except for three differences: (1) the nucleotides of RNA contain *ribose*

whereas DNA contains deoxyribose; (2) RNA contains the base *uracil* (Fig. 2.19) whereas DNA has thymine; and (3) RNA occurs as a *single strand* rather than two complementary strands.

Transcription is shown in Fig. 2.22. It works on the same principle as DNA replication, except that only one strand of the double helix serves as a template. First the helix uncoils near the start of a gene. An enzyme called **RNA polymerase** attaches to one of the two DNA strands. The enzyme moves along the DNA, lining up free subunits of RNA according to the base-pairing rules: where DNA has the base A, an RNA subunit with base U will attach; where DNA has C, RNA will have G; and so on. The resulting RNA polymer has a base sequence complementary to DNA, with U replacing T (for instance, DNA with the sequence ACGT will result in RNA with the sequence UGCA). When the enzyme reaches the

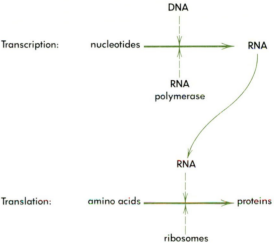

Figure 2.21. The two phases of gene-reading: *transcription* and *translation*. Each dotted arrow points from a catalyst or controlling agent to its target of action. *Transcription:* RNA polymerase links nucleotides into RNA molecules with base sequences complementary to DNA. *Translation:* ribosomes use the information stored in RNA to make proteins with specific amino acid sequences.

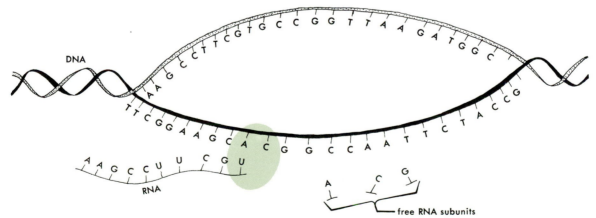

Figure 2.22. Transcription. A segment of DNA about to be read has uncoiled from the helix. The segment contains genes for one or several proteins. The RNA polymerase enzyme, shown as an oval, has built part of an RNA molecule and is waiting for the next free subunit (which should carry the base guanine) to arrive.

end of the DNA segment that was to be read, the enzyme detaches itself from the DNA and releases the RNA product.

Depending on the gene that is transcribed, three kinds of RNA may be formed. They have the same general structure but differ in chain length and function. When the recipe for a protein is transcribed, the resulting RNA is called **messenger RNA,** or **mRNA,** because it carries information from DNA to the ribosomes. The other two kinds of RNA (rRNA and tRNA) are not messages for making specific proteins but serve more general functions in protein synthesis, as described below.

READING RNA: PROTEIN SYNTHESIS
In **translation,** ribosomes make proteins according to the information carried in mRNA. A living cell contains thousands of ribosomes, each one a complete protein-builder. A ribosome is a complex unit made of **ribosomal RNA,** or **rRNA,** plus about 70 protein molecules.

The Genetic Code
To understand how a ribosome reads the mRNA, we must start by learning the language in which mRNA messages are "written." An English message describing a protein might be written as "met-ala-gly-leu . . ." and so on, where "met . . ." are names of amino acids. This is how mRNA describes

a protein. The "words" in RNA language are groups of bases and are called **codons.** Every codon has three bases; the language of RNA is written in three-letter words.

Sixty-four different codons can be formed by taking the four kinds of bases in all possible groups of three (AAA, AAG, and so on). Each of the 64 codons has a meaning, and the table of meanings is called the **genetic code.** Most codons name amino acids: for instance, the codon *AAA* calls for the amino acid *lysine.* However, 4 codons serve as punctuation marks: a "start" codon (AUG) tells the ribosome where to begin reading along the mRNA molecule,[3] and any of 3 alternative "stop" codons can tell the ribosome where the description of a protein ends. An mRNA message includes a start codon, followed by many codons that name amino acids in the desired sequence, and finally, a stop codon.

Making Amino Acids Recognizable: tRNA
How does a ribosome read the genetic code? The surprising fact is that ribosomes alone cannot decipher the mRNA message. They cannot distinguish between the codons, and they cannot recognize amino acids. Something extra is needed to match up amino acids with codons. The "extra"

[3]AUG also names the amino acid *methionine.*

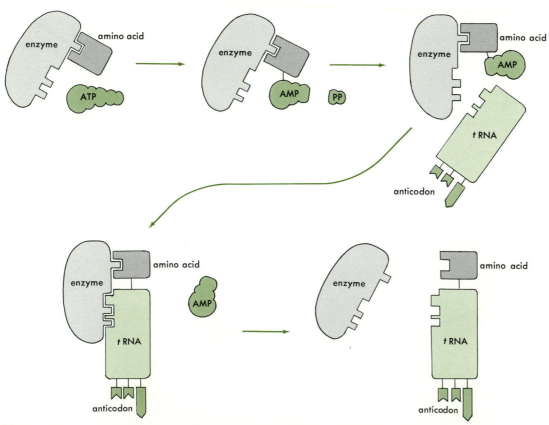

Figure 2.23. How amino acids and tRNA are joined in the correct combinations. The key is a set of enzymes called *amino-acyl-tRNA synthetases.* There is one such enzyme to fit each amino acid (20 in all). The enzyme's active site recognizes both the amino acid and the correct tRNA. Energy from ATP is spent in joining the amino acid to the tRNA.

is a set of small **transfer RNA,** or **tRNA,** molecules and a matching set of enzymes (Fig. 2.23). The tRNAs can recognize codons; the enzymes can recognize tRNAs and bind them to the correct amino acids.

To recognize the correct codon, each tRNA has a unique set of three bases called the **anticodon.** The anticodon obeys the base-pairing rule when it binds to mRNA; for instance, a tRNA molecule with the anticodon AGA will bind to the codon UCU. However, this precise pairing would be useless if the tRNA were to carry the wrong amino acid. Thus the enzymes that attach amino acids to tRNA are immensely important in life; they actually put the genetic code into effect. Their action is illustrated in Fig. 2.23.

The Action of Ribosomes

We have seen the preparations for protein synthesis; now let us turn to the ribosome. In brief, the ribosome moves along the mRNA one codon at a time and builds a protein by linking one amino acid after another into a chain.

The ribosome has three binding sites (Fig. 2.24). One site holds the mRNA molecule; the second site holds a tRNA with the protein that is being built; and the third site holds a tRNA with an amino acid that is to be added to the protein. The construction of a new protein molecule begins when existing proteins called **initiation factors** couple the ribosome to an mRNA and put the first two tRNA-amino acids into the binding sites. Then the ribosome transfers the first amino acid from tRNA to the tail of the second amino acid. Next, the ribosome moves along the mRNA by one codon. This ejects the first (now empty) tRNA and also opens up a site to accept another tRNA-amino acid. By repeating these events, the ribosome adds one amino acid after another to the protein. Finally, the ribosome reaches a "stop" codon. Then a protein called a **release factor** causes the ribosome to release its finished protein and to detach from the mRNA. The ribosome, mRNA, tRNA, and translation factors may be reused many times to make many protein molecules.

THE CONTROL OF METABOLISM

Within each cell, metabolism builds and destroys thousands of compounds at once. One might expect all this activity to produce chaos—yet, step by step, the metabolic system in a seedling builds the organized body of an adult plant. Clearly, elaborate controls must coordinate all the pathways of metabolism.

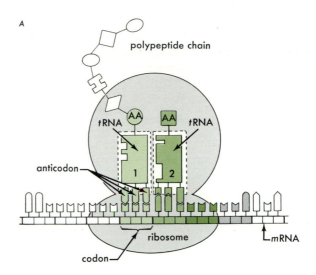

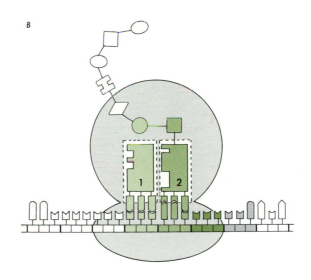

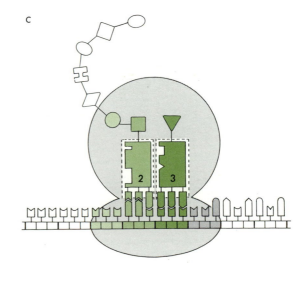

Figure 2.24. Steps in translation. In *A*, the ribosome binds an mRNA molecule with two codons in working positions. A partially completed protein is bound to tRNA 1, and another tRNA with its amino acid has just arrived at the second working site. In *B*, as the next step, the ribosome shifts the protein from tRNA 1 to the newly arrived amino acid; tRNA 1 is now free to depart. In *C*, the ribosome has moved along the mRNA molecule a distance of one codon, so that tRNA 2 is now in the left-hand working site. A third tRNA (3) with its amino acid has bound to the next codon. Step *B* can now take place again. This cycle of events occurs each time a new amino acid is added to the protein.

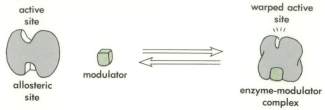

Figure 2.25. Control of an allosteric enzyme. On binding, the modulator changes the shape of the enzyme. Depending on the enzyme and modulator, the newly shaped active site may work either faster or slower than before the binding occurred. The enzyme returns to its original shape when the modulator leaves the allosteric site.

Metabolic control occurs at two levels: (1) existing enzyme molecules may be speeded or slowed to give moment-to-moment adjustments; and (2) hereditary instructions lead the system to build new kinds of enzymes and eliminate others.

CONTROL OF EXISTING ENZYMES

Several mechanisms can control existing enzyme molecules. Enzymes may be isolated within organelles, or chemically modified, or blocked by molecules that fit crudely into the active site. But most importantly, key enzymes may be turned on or off by reversibly binding special signal molecules known as **modulators** (Fig. 2.25). The modulators usually do *not* bind at the active site; rather, the enzyme has a special **allosteric site** that binds the modulator. The effect is much like turning a machine on or off with a key. The modulator changes the shape of the enzyme, altering its ability to bind or react with the substrate. Depending on the enzyme and modulator, the attachment may either speed or slow the enzyme's action.

Allosteric enzymes occur at the start of many pathways, as in Fig. 2.26. They are usually inhibited by the products of the pathway. This leads to a supply-and-demand control, for as products accumulate, they bind and inhibit the allosteric enzymes.

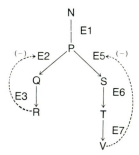

Figure 2.26. The principle of feedback inhibition. Enzymes at the heads of pathways (E2 and E5) are allosteric, and are inhibited by the final products of the pathways (R and V). E's represent enzymes; N through V are substrates and products. Dotted arrows with attached (-) indicated inhibitory effects. Production along a pathway will stop if products accumulate.

CONTROL OF PROTEIN SYNTHESIS

Metabolism can also be controlled by making and breaking down enzyme molecules. The breakdown is accomplished by enzymes called **proteases,** which are nearly always present in cells. If a given enzymatic activity is to continue, new copies of the enzyme must be made to replace those that are broken down. The course of metabolism can be changed by building new kinds of enzymes and allowing others to disappear. This happens as a plant grows and is responsible for developing the shape of the plant body.

We know that the control of enzyme synthesis is based on reading the DNA. Unfortunately, we still know little about the systems that tell a plant when a given gene should be read. For a model of what might happen, we must turn to the bacteria, in which the control systems are better known.

Bacteria control protein synthesis in several ways, of which the best known is the **operon system** (Fig. 2.27). An operon is a segment of DNA, perhaps containing several genes, that is controlled as a unit. The operon is headed by a **promoter site,** a segment of DNA to which RNA polymerase can bind to start transcribing. Beside the promoter site is an **operator site,** which is the region that controls transcription. Next come one or more **structural genes,** which are templates for proteins. And finally, a **termination sequence** ends the operon.

The control of the operon depends on a special protein called a **repressor.** The repressor is shaped to bind at the operator site. When it is so bound, it blocks the path of RNA polymerase and will not allow transcription. One might compare the repressor to a lock on a file cabinet. And like a lock, a repressor can be attached or removed by means of a specific key. The "keys" are small molecules that are formed in the cell or that enter from the environment.

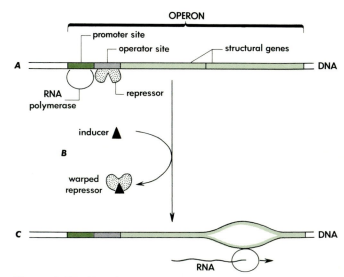

Figure 2.27. How the operon system controls protein synthesis. *A,* with a repressor attached to the operator site, the polymerase enzyme cannot reach the structural genes that code for proteins. *B,* an inducer binds and changes the shape of the repressor so that it detaches from the operator site, allowing *C,* the enzyme to transcribe the genes.

Operons typically respond to the environment so as to read the genes only when there is a need for their products. The classic example is the *lac* operon, which occurs in the bacterium *Escherichia coli.* The *lac* operon governs the production of several proteins, most notably an enzyme that breaks down lactose (milk sugar). Many *E. coli* cells live in the human gut and subsist on the foods that we ingest. Within half an hour after the host consumes milk, *E. coli* produces enzymes to attack lactose. A slightly modified form of lactose acts as an **inducer** to turn on the *lac* operon. When this compound is present, it binds to an allosteric site on the repressor. This warps the repressor and causes it to fall off the DNA. Then transcription occurs, mRNA is formed, and ribosomes make the enzyme to break down lactose. After the milk meal has been digested, there is no more lactose to bind with the repressor. The repressor returns to its normal shape, binds to the operator site, and transcription stops. In this way, the *lac* operon is read only when there is a need for the enzyme to break down lactose.

We know much less about gene control in plants and other advanced organisms. However, some of the controls remind us of bacterial operons: there are cases in which signal molecules bind to special proteins, which in turn cause the DNA to unwind for transcription. Our knowledge is growing rapidly in this area, and we can expect some great insights in the next few years.

SUMMARY

1. The plant body is produced and controlled by a complex system of reactions called metabolism.

2. The raw materials used in metabolism are carbon dioxide, water, and minerals.

3. Carbohydrates, proteins, and nucleic acids are important biological polymers and are made, respectively, of sugars, amino acids, and nucleotides. Other important products include lipids and organic acids.

4. Nearly all metabolic reactions require catalysis by enzymes. Complex syntheses involve metabolic pathways, which are sequences of reactions.

5. Three major phases of metabolism are anabolism, catabolism, and photosynthesis. Anabolic pathways synthesize complex molecules, while catabolic pathways break down molecules to provide energy and materials for anabolism and transport processes. Photosynthesis uses light energy to build fuel molecules. Energy and building materials are carried from catabolic to anabolic reactions by recyclable carrier molecules such as ATP and NADH.

6. Glycolysis (in the cytoplasm of cells) and respiration (in the mitochondria) are the principal pathways of catabolism. Glycolysis converts glucose to pyruvic acid, forming a small amount of ATP and NADH. Respiration consists of the citric acid cycle, an electron transport system, and enzymes that form ATP. The citric acid cycle passes electrons from organic acids to carriers that in turn feed electrons to the transport system. Electron transport leads to a proton gradient across the mitochondrial membrane. Energy for ATP synthesis is obtained by discharging the proton gradient.

7. Without O_2, respiration cannot occur, but alcoholic fermentation may produce ATP by converting glucose to ethanol and carbon dioxide. This involves glycolysis plus two extra steps.

8. Protein (enzyme) synthesis is guided by hereditary information that is stored in the base sequences of DNA. Information is transferred by mechanisms that depend on a specific pairing between bases.

9. DNA occurs as a double helix of two molecules that have complementary base sequences. To copy DNA, enzymes separate the two DNA molecules; free subunits line up along each molecule according to a base-pairing rule; and enzymes join the subunits to form new DNA molecules.

10. Protein synthesis begins with *transcription,* in which DNA serves as a template for producing mRNA with base sequences complementary to the DNA. In *translation,* ribosomes then assemble amino acids into proteins, using mRNA molecules as templates.

11. The rates of metabolic reactions are controlled by regulatory enzymes that respond to molecular signals, by having enzymes localized within organelles, and by the production and breakdown of enzyme molecules. Feedback signals help to pace reactions.

12. The control of protein synthesis by the operon system, found mainly in bacteria, is based on blocking transcription through the reversible attachment of proteins to DNA. The systems that control protein synthesis in more complex organisms are still unknown.

3 THE PLANT CELL

Galen, the last of the great Greek doctors, who lived in Asia Minor during the second century A.D., thought that all organs—such as the spleen, brain, and kidney of animals, and the leaves, stems, and roots of plants—were a "sensible element, of similar parts all through, simple and uncompounded." Others before him had thought that animal tissues were simple coagulated "juices" seeping through the walls of the intestine. No one dreamed that these tissues and their plant counterparts had an astonishing and complicated structure.

In 1590, Zacharias Jansen invented the microscope, and in the mid-1600s, Robert Hooke improved this instrument and used it to examine all sorts of natural objects. Among the objects he examined were thin slices of cork (the dead outer bark of an oak). Figure 3.1 shows cork tissue as Hooke saw it under his microscope. This illustration was published in 1664 in an article entitled "Micrographia; or Some Physiological Descriptions of Minute Bodies Made by Magnifying Glasses." The term *cell* was first used by Hooke to denote in cork the "little boxes or cells distinct from one another . . . that perfectly enclosed air." He estimated that a cubic inch of cork would contain about 1259 million such cells.

Because of the prominence of cell walls in plant tissue, the cell was soon considered the unit of structure and of life in plants. However, during these early years, zoologists con-sidered the tissues as the true centers of life in the animal body. There were supposed to be 21 different tissues, able to change within themselves, depending upon the organ in which they were located; and life was thought to reside in these tissues. Numerous observations of protozoa and animal tissues led to a gradual accumulation of evidence that cells also existed in animals.

In 1838, the German zoologist Theodor Schwann and the botanist Matthias Jacob Schleiden collaborated in a paper entitled, "Microscope Investigations on the Similarity of the Structure and Growth in Animals and Plants." This paper established on a firm basis the theory that the cell is a basic unit of structure in both plants and animals. Another 30 years of research were necessary for the general acceptance of this idea: that all organisms are composed of cells, and that cells are indeed the basic units or building blocks of life.

This chapter is divided into two parts. The first examines plant cell structure, and the second summarizes cell division and the cell cycle.

CELL STRUCTURE

As stated in Chapter 1, organisms can be classed as eukaryotes or prokaryotes, depending on their cell structure. Prokaryotes include all the bacteria and the blue-green algae (cyanobacteria). Their cells are small and relatively simple in structure, with little division into internal compartments. A prokaryotic cell is pictured in Chapter 16, page 246.

All other forms of life, including plants, fungi, and animals, are eukaryotes. Their cells are much larger and more complex than prokaryotic cells, with many distinct internal parts called **organelles.** The organelles provide a division of labor within the cell; each organelle has a distinctive function (Table 3.1).

Figure 3.2 shows a leaf cell of the aquatic plant *Elodea*, as it would appear in a classroom microscope. From such a view, an experienced microscopist might produce a drawing such as Fig. 3.3. Outside the cell is a wall, which was secreted by the cell and its neighbors. Just inside the wall, not visible with this kind of microscope, is a thin membrane called the **plasmalemma.** The plasmalemma is the true surface of the cell and is responsible for controlling the movement of materials in and out of the cell. Biologists have learned how to remove cells from their walls for culture and observation. Such naked cells, or **protoplasts,** are sometimes used in genetic engineering experiments when foreign DNA is to be injected into the cell (see Chapter 11).

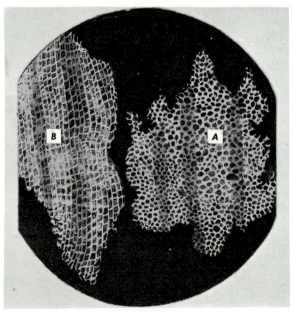

Figure 3.1 Cork tissue as Robert Hooke observed it under his microscope.

Table 3.1. The Parts of the Cell[a]

Cell Part	Dimension	Function
Plant cell, **1880**		
Primary cell wall, **1665**	2–5 μm	Protection, strength
Cellulose microfibrils, 1948	10–25 nm (indefinite length)	Mechanical strength
Cellulose molecules, 1922 (X ray), 1949	0.834 × 0.8 nm	Strength
Amorphous matrix		
Middle lamella	2 μm	Adhesion between cell walls
Protoplast, **1846**	0.025–2 mm	
Protoplasm, **1840**		Site of metabolic reactions under direction of nucleus
Cytoplasm, **1882,** 1957		
Chloroplast, **1702,** 1953	2–20 μm	Photosynthesis
Chromoplast, **1900,** 1958	2–10 μm	Accumulation of carotenes and similar pigments
Amyloplast, **1884,** 1955	2–25 μm	Starch storage
Leucoplast, **1883,** 1957	2–10 μm	
Mitochondrion, **1897,** 1947	0.5–2 × 2–10 μm	Respiration
Dictyosomes, **1927,** 1956	3 μm	Enzyme synthesis
Endoplasmic reticulum, 1957	17 nm (indefinite length)	Protein synthesis and movement
Ribosomes, 1955	20 nm	Protein synthesis
Spherosomes, **1919,** 1967	2 μm	Lipid synthesis and storage
Microbodies, 1965	0.1–2.0 μm	Compartmentalization of various enzymes
Microtubules, 1960	18–25 nm (variable length)	Controls cell wall shape, plane of new cell division, and chromosome movement
Microfilaments	4–7 nm (diameter)	Involved in cytoplasmic movements
Tonoplast, **1877,** 1960	8 nm	Regulation of exchange between vacuole and cytoplasm
Plasmalemma, **1880**(?), 1954	8 nm	Regulation of exchange between cytoplasm and external solution
Crystals, 1963	10 μm	Unclear, possibly storage
Plasmodesmata, **1879,** 1957	2 μm	Protoplasmic bridge between cells
Nucleus, **1831**	5–30 μm	Contains genetic information necessary for normal cell development and activity
Nuclear envelope, **1907,** 1955	25 nm	Separates nucleoplasm from cytoplasm
Nucleoplasm, **1879**		
Nucleoproteins, **1869**		
Nucleic acids, **1889**		
DNA, **1924**		
DNA helix, 1953	1.8 nm (indefinite length)	Carries genetic code
Unit fibers, 1963	12.5 nm (indefinite length)	Encompasses DNA helix and nucleoproteins
Nucleolus, **1882,** 1958 (animal, 1952)	1–5 μm	RNA synthesis
RNA		Transferal of information from DNA to cytoplasm
Chromosome, **1888,** 1955	2–200 μm	Vehicle carrying DNA helix in replication and distribution
Kinetochore		Region of chromosome to which spindle fibers are attached
Centromere, **1925**		
Chromatid, **1900**	1–10 μm	One-half of a chromosome
Spindle fibers, **1881,** 1960	Various indefinite lengths	Cytoplasmic structure involved in moving chromosomes during mitosis
Vacuole, **1835,** 1957		Various functions important in water economy of cell

[a]Date in **boldface type** refers to the year first reported with the light microscope; date in lightface type indicates the year first seen with the electron microscope.

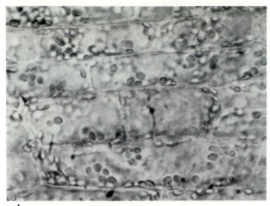

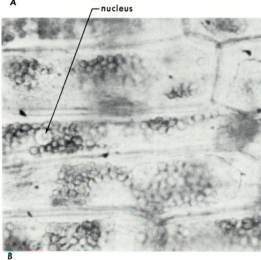

A

nucleus

B

Figure 3.2 Leaf cells from the aquatic plant *Elodea. A,* living untreated cells; note the distribution of chloroplasts around sides, top, and bottom of the cell, with a vacuole occupying the rest of the cell. *B,* in cells recently killed with formaldehyde, the protoplast may withdraw slightly from the cell wall, chloroplasts may form irregular clumps and the nucleus frequently becomes visible.

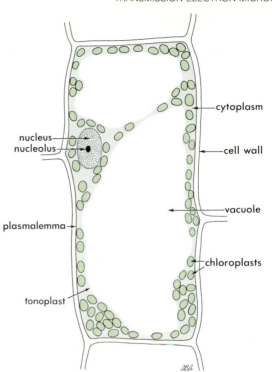

Figure 3.3 Diagram of a living cell from an *Elodea* leaf. When viewed on the light microscope, organelles in living cells, other than chloroplasts, lack contrast. The nucleus and nucleolus are generally visible. Mitochondria are seen only with perfect lighting conditions. × 700.

The largest organelle in the leaf cell is the **vacuole,** which is a mass of fluid surrounded by a thin membrane, the **tonoplast.** The vacuole stores food and wastes, and adds bulk to the cell. The rest of the material in the cell (the green and gray areas in Fig. 3.3) is called **protoplasm.** It may form strands across the vacuole, and it often moves about in a circulating motion called **protoplasmic streaming.**

Numerous rounded, green organelles are embedded in the protoplasm of the leaf cell. These **chloroplasts** capture sunlight and use its energy to make foods by means of photosynthesis.

Also visible in the leaf cell is the **nucleus,** a rounded, transparent organelle. It contains about 99% of the cell's DNA and is the principal center of control in the cell. When cell biologists realized the importance of the nucleus and began studying it intensively, they began to need a word to indicate the rest of the protoplasm that is outside the nucleus; the term used is **cytoplasm.**

The organelles are suspended in a fluid medium called the **ground cytoplasm,** which not only supports the organelles but also has metabolic functions of its own.

This brief account has covered only the organelles that are easily visible with the ordinary microscope. To see the rest of the cell contents and to delve more deeply into the structures already mentioned, we must take up the subject of electron microscopy.

TRANSMISSION ELECTRON MICROSCOPY

The light microscope with glass lenses is not capable of resolving the details of cellular organelles. This limitation was overcome in the 1940s and 1950s, when the **transmission electron microscope,** or **TEM,** was developed. Using an electron beam instead of light and magnets instead of glass lenses, the TEM can show details about 1,000 times smaller than what can be seen with a light microscope.

For viewing with the TEM, tissues are prepared by killing

them chemically to fix their structural details; the tissues are then dehydrated and embedded in a plastic resin. Very thin sections of the plastic-embedded tissue are cut with a diamond knife on a special instrument called a microtome. These sections, usually no more than 70 nm thick, are stained and then examined. The resulting images are not true to life because the specimens are chemically modified during preparation. But with careful interpretation, these images have revealed details of cell fine structure that were never suspected before the invention of the electron microscope. Figures 3.5 through 3.14 illustrate the products of electron microscopy.

CELL ORGANELLES AND OTHER STRUCTURES

MEMBRANES

Perhaps the most notable characteristic of cells at high magnification is their compartmentation into membrane-bound organelles. The protoplasm itself is bounded externally by the plasmalemma and internally by the tonoplast (Fig. 3.3). The plasmalemma separates the protoplasm from the external cell wall and acts as a regulating barrier to control the movement of materials into and out of the living cell. The tonoplast separates the protoplasm from the vacuole. In an organized living system like a cell, membranes are necessary, for they isolate high-energy regions from low-energy regions.

What is the nature of the biological membrane? Because of the chemical damage that membranes suffer in preparation for electron microscopy, we cannot be sure of their exact structure. However, numerous experiments support the **lipid-globular mosaic** model of membrane structure that is shown in Fig. 3.4. To interpret the image, bear in mind that water bathes both faces of the membrane. Phospholipid molecules make up about half the mass of the membrane; these molecules are oriented in two layers, forming a **lipid bilayer.** Each lipid molecule has its hydrophilic head toward the adjacent water, and the hydrophobic tails point away from the water, toward the opposite lipid layer. Nestled between the hydrophobic tails, there may be sterol molecules, such as cholesterol. To complete the membrane, globular (rounded) protein molecules are attached to the lipid bilayer. Some proteins are loosely bound at the surface, while others extend into the lipid bilayer and may actually project into the water on both sides. Some of the bound proteins are enzymes, while others may control the passage of solutes through the membrane.

MITOCHONDRIA

Among the most abundant organelles are **mitochondria,** of which there may be hundreds per cell. As discussed in Chapter 2, mitochondria perform aerobic respiration; they oxidize fuel molecules to form ATP. Mitochondria vary in shape but are most often rodlike. They have two surrounding membranes. The inner membrane folds inward to form **cristae** (Fig. 3.5; Fig. 2.14, p. 9). The fluid enclosed by the inner membrane is called the **mitochondrial matrix.** Both the cristae and the matrix contain enzymes that take part in respiration.

PLASTIDS

Chloroplasts, the green structures of the plant cell, belong to a class of organelles called **plastids** (Figs. 3.6, 3.7, and 3.8). Like mitochondria, all plastids are bounded by an envelope of two membranes (Figs. 3.7 and 3.8). The envelope encloses a material called the **stroma,** which looks granular in the electron micrograph. Some of the granules are ribosomes, units that manufacture proteins (see Chapter 2). DNA also occurs in the stroma; sometimes it appears as a mass of slender fibrils, and sometimes its location merely appears as a clear area.

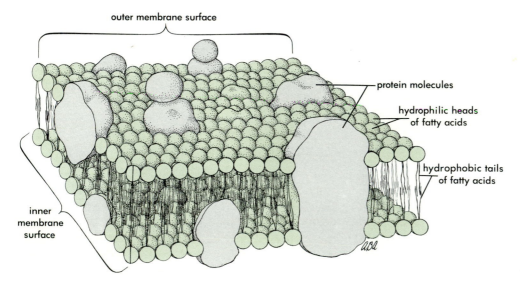

outer membrane surface

protein molecules

hydrophilic heads of fatty acids

hydrophobic tails of fatty acids

inner membrane surface

Figure 3.4 The lipid-globular mosiac model of the membrane.

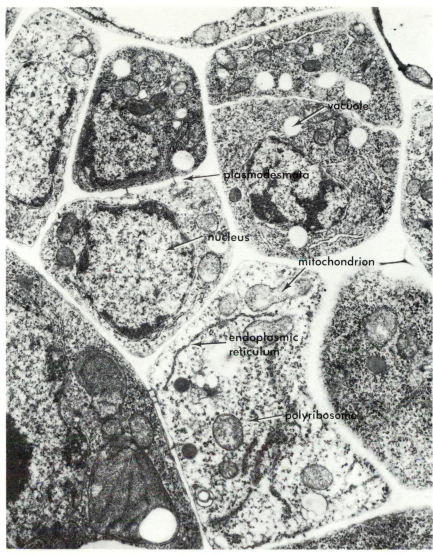

Figure 3.5 Electron microscope photograph of an immature wheat (*Triticum aestivum*) leaf. Numerous organelles are present. Note especially the coiled polyribosomes (arrows), ×9000.

The chloroplast differs from other kinds of plastids in that it contains many flat membranous sacs called **thylakoids.** The thylakoids capture light and store its energy by making ATP and other energy-rich molecules (see Chapter 8). Often, thylakoids form cylindrical stacks called **grana** (Fig. 3.7). Thylakoids may extend from one granum to another, and parts that are located between grana are sometimes called **frets.**

Other Kinds of Plastids

Plastids can change their structure, contents, and function in response to environmental conditions (Fig. 3.8). For example, dark-grown seedlings contain plastids with an elaborate crystal-like structure called a prolamellar body. Embryonic tissues contain small plastids with few internal membranes, which can develop into chloroplasts if they are exposed to light; these are called **proplastids.** Mature epidermal cells and white storage tissues, such as the flesh of an onion or apple, often contain colorless **leucoplasts,** which store food. Some of these plastids contain large amounts of starch and are called **amyloplasts.**

As fruits ripen on trees and as leaves prepare to fall at the end of summer, they change from green to red, orange, or yellow. This results from the destruction of chlorophyll, accompanied by an accumulation of yellow or red pigments known as **carotenoids.** The plastids with a dominance of red and yellow pigments are **chromoplasts.** A chromoplast from a ripe tomato is shown in Fig. 3.8 (right).

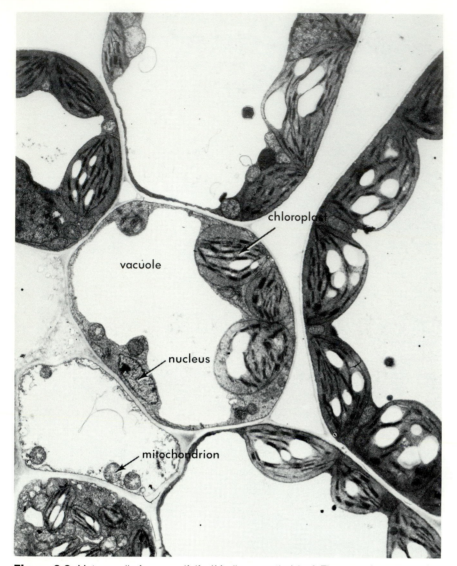

Figure 3.6 Mature cells from an alfalfa (*Medicago sativa*) leaf. The vacuoles are much larger in these old cells. Note also that the nucleus is proportionally smaller than in immature cells. Chloroplasts and mitochondria are abundant, ×6000.

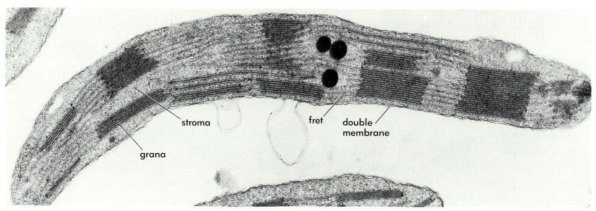

Figure 3.7 Chloroplast from leaf of *Senecio vulgaris*, ×35,400.

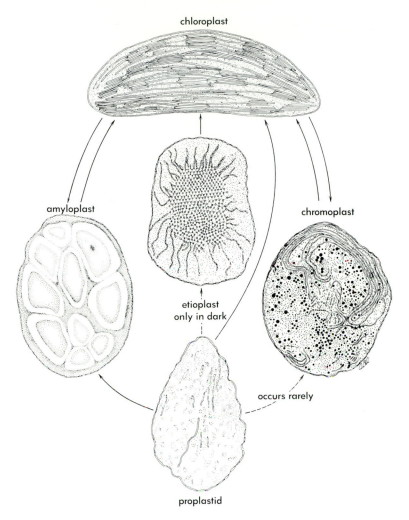

chloroplast

amyloplast

chromoplast

etioplast
only in dark

occurs rarely

proplastid

Figure 3.8 Diagram to show the interconversions of plastids. Bottom center, proplastid; center, etioplast; top center, chloroplast; left, amyloplast; and right, chromoplast.

THE NUCLEUS

Like the preceding organelles, the nucleus also has an envelope of two membranes. However, the nuclear envelope is unique in being punctured by many large openings called nuclear pores. They are especially evident in preparations in which cells are frozen and then fractured so as to expose the surface of the nucleus (Fig. 3.9a and 3.9b). Sections through the nucleus (Fig. 3.9b, top) show that the inner and outer membranes of the envelope are fused together around the rim of the pore. In some preparations the nuclear pores appear to be plugged by a structure called an annulus. Opinions differ as to whether the pores actually serve as passageways for molecules to enter and leave the nucleus.

Within the nucleus are several chromosomes, which store hereditary information. Each chromosome is made of **chromatin,** which consists of a long double-helix of DNA to which many protein molecules are attached. When the nucleus is dividing, the chromatin is coiled tightly so that each chromosome becomes visible (when stained) with the light microscope. Between divisions, the chromatin uncoils and appears as a mass of slender fibrils that are too thin to be seen except with the electron microscope. These changes are described in detail later in the chapter.

Also present between nuclear divisions is a rounded body called a **nucleolus** (Fig. 3.10). This is a region in which parts of ribosomes are produced, following instructions that are stored in the DNA. Most of the nucleolus consists of incomplete ribosomes.

RIBOSOMES

The cytoplasm, after chemical fixation, is moderately electron transparent and has a soft gray, granular appearance. In it always appears at least one type of densely stained granule, roughly angular and from 17 to 20 nm in diameter (Fig. 3.5). Treating tissue with the enzyme RNAase, which specifically degrades RNA, results in the disappearance of these granules, with no change taking place in the other organelles. This is good evidence that RNA is present in these particles, the *ribosomes* (see Chapter 2). Ribosomes are present in plastids and mitochondria, as well as in the cytoplasm. They appear to be lacking in nuclei, but some micrographs show them attached to the cytoplasmic side of the nuclear envelope. If they do occur within the nucleus, they are probably confined to the nucleolus. Ribosomes are frequently associated with the cytoplasmic-membrane system, the endoplasmic reticulum (Fig. 3.5). Under certain cir-

Figure labels: nuclear pore, nucleus, mitochondrion, cell wall, vacuole, endo-plasmic reticulum, dictyosomes, nuclear pores, nucleus

Figure 3.9 Freeze-etch preparation of an onion root tip cell, showing the organelles as they appear when the cells are examined after freezing, rather than after chemical fixation. Bacteria and other cells, after a similar cold treatment, are viable. The similarity between the chemical fixation image and the freezing (living) image is striking. A, freeze-etch preparation × 10,000. B, diagram of A with labels (ER represents endoplasmic reticulum), × 7400. (A courtesy of D. Branton.)

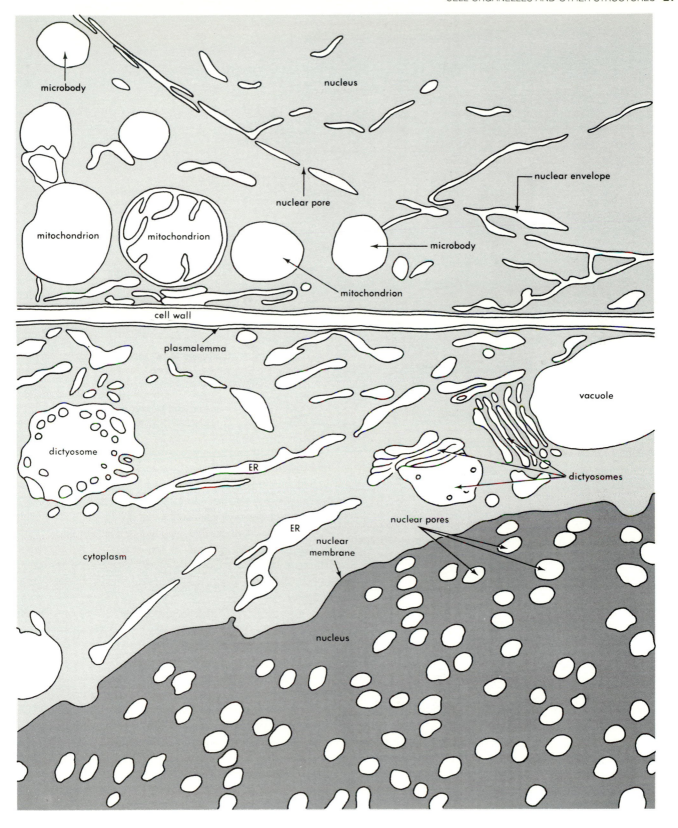

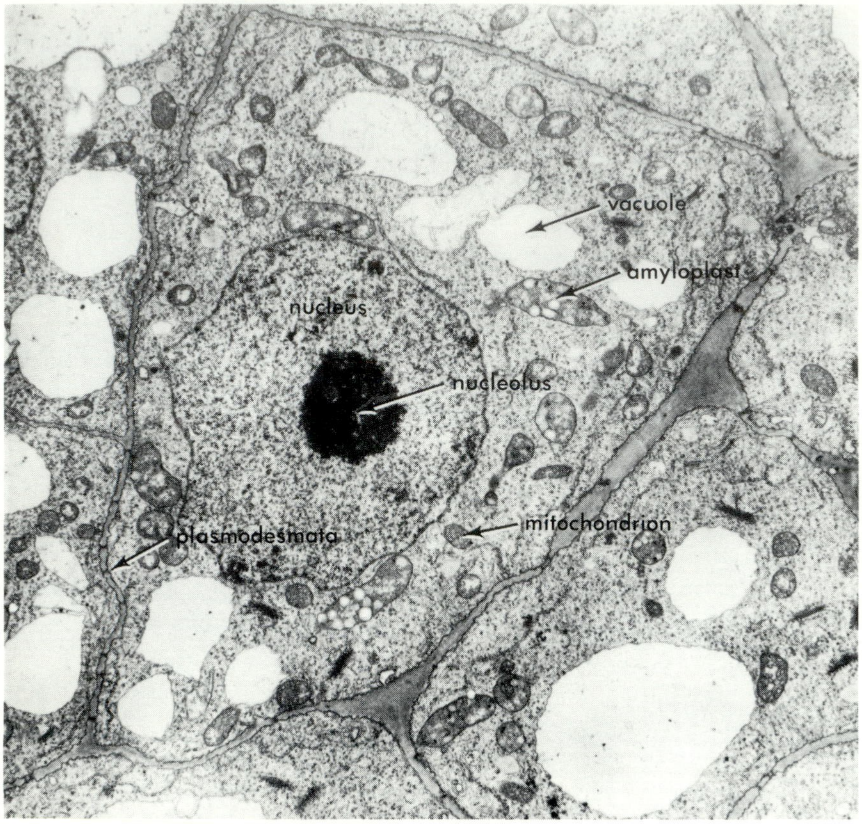

Figure 3.10 Immature cells from young root tip of *Gladiolus* sp. Notice the similarities to the immature leaf cells in Figure 3.14; small vacuoles, amyloplasts, mitochondria, and dictyosomes are apparent, ×5000.

cumstances they appear in groups, frequently as helical aggregations, which are called **polyribosomes,** or simply **polysomes** (Fig. 3.5).

THE ENDOMEMBRANE SYSTEM

An extensive set of membranes called the **endomembrane system** occurs in the ground cytoplasm between the organelles discussed in the previous sections. This system includes three kinds of components: the *endoplasmic reticulum, dictyosomes,* and *vesicles.* Together, they have a variety of functions concerned with manufacturing products and exporting materials from the cell.

THE ENDOPLASMIC RETICULUM

The endoplasmic reticulum consists of units called **cisternae.** A cisterna is a flat sac, lined by a membrane (Figs. 3.5, 3.9, and 3.11). In cross section, a cisterna appears as two parallel membranes spaced about 4 nm apart, with a light space between them. Note that these membranes form a closed unit; their ends are never open to the ground cytoplasm. Many cisternae may be connected to one another at their edges to form a large system.

Most cisternae have ribosomes attached to the surface, as in Figs. 3.5 and 3.11. When ribosomes are attached, the cisternae are said to make up a **rough endoplasmic reticulum,** because the ribosomes give the cisternae a rough appearance. Cisternae without ribosomes make up the **smooth endoplasmic reticulum,** which is less abundant.

The rough endoplasmic reticulum is concerned with manufacturing membranes and manufacturing proteins for export from the cell. The attached ribosomes produce proteins, which enter the membrane of the cisterna and either remain and become part of the membrane surface or pass into the interior, where they are stored for later export. The enclosed proteins migrate to the edges of the cisternae. There, small portions of the cisterna pinch off to form units called **vesicles.** Each vesicle carries some of the stored proteins.

The function of smooth endoplasmic reticulum is less well understood in plant cells.

Dictyosomes are units made of from 5 to 15 flat, circular cisternae, aligned in a stack (Fig. 3.12). In cross sections, many small vesicles appear at the margins of the cisternae. However, face views reveal that the margins of each cisterna form a coarse net with true vesicles only at the extreme outer regions (Fig. 3.12). The peripheral vesicles often contain granules of material that have been produced and collected by the dictyosome (Fig. 3.11). Vesicles from the dictyosomes are thought to migrate to the cell surface, where they fuse with the plasmalemma and deposit their contents into the cell wall. Thus dictyosomes are concerned with producing and carrying materials for cell wall synthesis.

Dictyosomes appear to be polarized; that is, there is a forming face and a concave disappearing face that is somehow used up in the formation of the vesicles. Dictyosomes are thought to take in new material at the forming face by joining with vesicles from the endoplasmic reticulum.

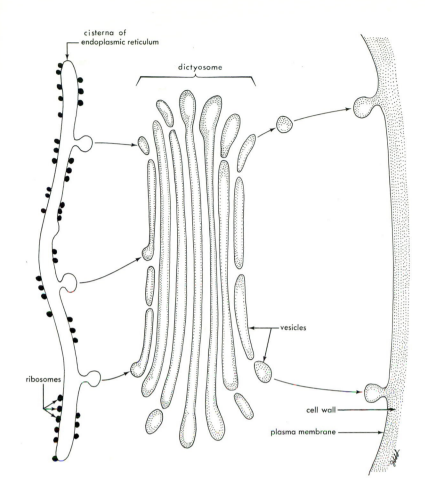

Figure 3.11 Diagram to show membrane flow from ER through a dictyosome to the plasmalemma.

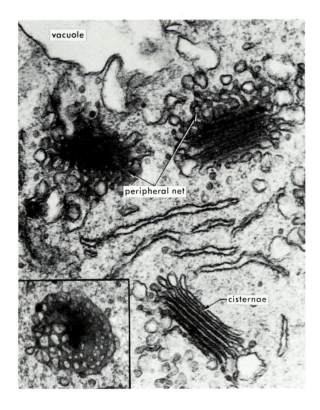

Figure 3.12 Four dictyosomes in a cell from the root tip of a water plant (*Hydrocharis*). One dictyosome was sectioned at right angles to its cisternae, and another one was sectioned almost parallel to its cisternae. Note the peripheral net around this second dictyosome. The two other dictyosomes were sectioned obliquely to the cisternae, and some material appears to be passing from them to the vacuole, × 15,000.

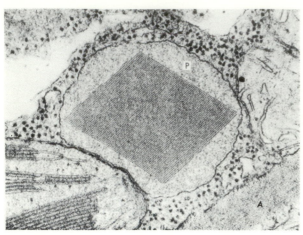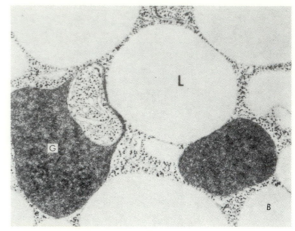

Figure 3.13 A, electronmicrograph of a type of microbody, the peroxisome (P), which is found in leaf cells, ×54,500. B, a second type of microbody, the glyoxysome (G), is found in seeds and is associated with lipid (L) digestion during seed germination, ×20,714.

To summarize, the endomembrane system appears to be concerned largely with manufacturing materials for export from the cell, much of it to be used in wall formation (Fig. 3.11). The rough endoplasmic reticulum manufactures proteins and releases vesicles that join the dictyosomes. The dictyosomes add cell wall materials to the contained proteins and release vesicles that will migrate to the wall.

MICROBODIES

All eukaryotic cells contain a variety of spherical organelles called by the general term *vesicles*. Their common feature is that each vesicle has a single surrounding membrane. We have already discussed the formation of small vesicles by the endoplasmic reticulum and the dictyosomes. These vesicles are concerned with storing and transporting materials. Now let us consider the largest of the vesicles— **microbodies.**

Microbodies are about the size of mitochondria, but they differ in that they have only a single membrane and they lack cristae. The interior often appears rather dense and may contain a variety of crystals (Fig. 3.13a).

Unlike the secretory vesicles mentioned before, microbodies remain in the cell. In general, they perform enzymatic functions, but the exact function varies from one case to another. For example, microbodies isolated from leaves are called peroxysomes because they have enzymes that break down peroxides (Fig. 3.13a). These microbodies may serve to destroy toxic peroxide compounds that are produced during photosynthesis. By contrast, microbodies isolated from oily seeds, such as castor beans, take part in breaking down fats and are called glyoxysomes (Fig. 3.13b). Still other kinds of microbodies probably remain to be discovered.

VACUOLES

In many young cells the cytoplasm occupies much of the space within the cell, but small vacuoles are present (Figs.

3.5 and 3.14). As the cell expands, the vacuoles increase in size and coalesce (Fig. 3.15a, 3.15b, and 3.15c). Finally, when the cell has attained its mature size, only a few large vacuoles, or even only one, may remain (Figs. 3.6 and 3.15d). The mature vacuoles are often so large that the rest of the

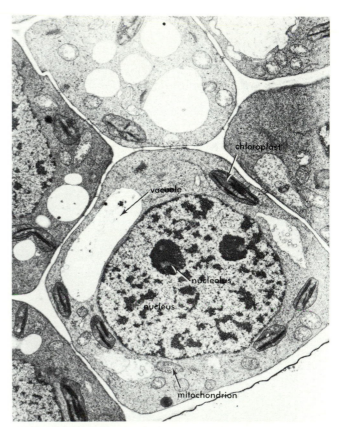

Figure 3.14 Immature cells from the apex of a sunflower (*Helianthus annus*). Note the small vacuoles (v), large nucleus (n), and other organelles—mitochondria (m), chloroplasts (c) and dictyosomes (d), ×6416.

cytoplasm, pressed against the cell wall, may be overlooked by a casual observer.

Vacuoles contain not pure water but a solution of many substances called **cell sap.** At times the solute concentration may become quite high: in wine grapes and sugar beets, for example, the vacuoles may contain 20% sugar. More often, however, the cell sap is dilute. Among the dissolved substances are (a) atmospheric gases, including nitrogen, oxygen, and carbon dioxide; (b) inorganic ions such as nitrate, sulfate, phosphate, chloride, sodium, potassium, calcium, iron, and magnesium; (c) organic acids such as malic, oxalic, tartaric, citric; (d) insoluble salts of organic acids, such as calcium oxalate; (e) sugars, such as glucose and sucrose; (f) proteins, alkaloids, and pigments. The solute content can vary in time and between tissues.

The most common pigments of the vacuolar sap are the **anthocyanins.** These pigments are responsible for the red, purple, or blue of many flower petals and other plant parts. The red color of beets, for instance, is due to a pigment called betacyanin.

Cells with crystals are found in almost all plants and in many different plant tissues. Crystals form within vacuoles and vary in chemical composition and in form (Fig. 3.16). The most common crystals are of calcium oxalate; it is gen-

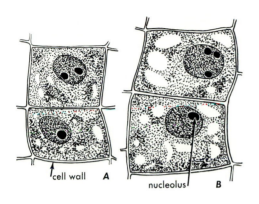

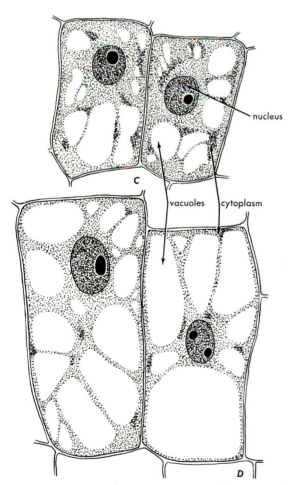

Figure 3.15 Stages in the growth of a cell. Progressively older cells shown from A through D, ×2000.

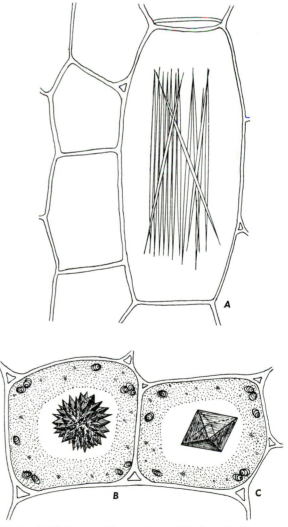

Figure 3.16 Types of inorganic crystals found in the vacuoles of living cells and in the cell walls of older nonliving cells. A, raphides; B, a cluster of crystals; C, a single crystal, ×2000.

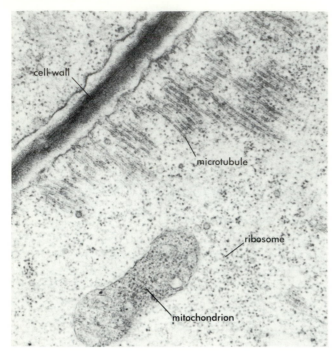

Figure 3.17 Microtubules in root cells of the water fern, *Azolla pinnata,* ×40,000.

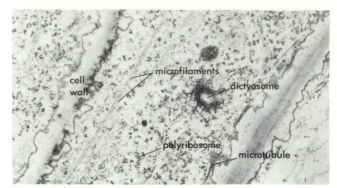

Figure 3.18 Radish (*Raphanus sativus*) root cell showing microfilaments, ×7800. Used by permission by Dr. Robert Seagull.

erally believed that they are an excretory product of the protoplast formed by the union of calcium and oxalic acid. This acid is soluble in cell sap and is toxic to the protoplasm if it attains a high concentration in the cell. By its union with calcium, the soluble oxalic acid is converted into the highly insoluble calcium oxalate, which will not injure the protoplasm. In addition to calcium oxalate, crystals of calcium sulfate, calcium carbonate, or protein are sometimes found.

MICROTUBULES AND MICROFILAMENTS

The cytoplasm of almost all kinds of cells contains three types of submicroscopic fibrils: **microtubules,** which are between 18 and 25 nm in diameter; **microfilaments,** 4 to 7 nm in diameter; and other filaments, 8 to 10 nm in diameter. Some of the physical properties of cytoplasm, such as elasticity, and many of its structural characteristics appear to be due to the presence of these structures. Microtubules are tubular fibrillar structures of indefinite length, with a cylindrical shape of highly organized globular protein (tubulin) subunits surrounding an electron-transparent core about 10 nm in diameter (Fig. 3.17). Various cellular functions frequently associated with some aspect of motility are now generally accepted as being dependent on microtubule action. Microtubules occur in the cilia or flagella of motile cells of all eukaryotes. They are also involved in chromosomal movement during cell division (see next section). Microtubules and microfilaments (Fig. 3.18) play roles in the movement of vesicles through the cytoplasm. Their close association with cell walls, just internal to the plasmalemma, and their orientation parallel with each other and at right angles to the long dimension of the cell wall of many elongating cells have led to the suggestion that they may be associated

in some way with the direction of deposition of cellulose microfibrils during wall synthesis.

THE CELL WALL

The protoplast is surrounded by a plasmalemma or cytoplasmic membrane. Outside this membrane, and surrounding the entire protoplast, is a rigid wall synthesized by the protoplast that it encloses. Walls of adjacent cells are cemented together by an intercellular substance, the **middle lamella** (Fig. 3.19a), which is composed of pectin (polymers of various sugars, especially rich in partly oxidized galactose) and certain other substances. The first wall formed by the protoplast is the **primary wall,** which is made of cellulose and other carbohydrates. When the cell ages and ceases elongation, the protoplast may deposit more wall material on the primary wall. Thus a **secondary wall** (Fig. 3.19d and 3.19e) is formed, and the completely mature cell wall may have a thickness many times greater than the primary wall. In some tissues, the secondary wall is stratified and composed of several layers. In others, the cells do not lay down any secondary wall material. The secondary wall may be of cellulose or of cellulose impregnated with other substances. Some of these substances, notably **lignin,** strengthen the wall; others, like **cutin** and **suberin** (Fig. 3.19b and 3.19c), are waxy and protect leaves and stems against water loss. In addition, certain other materials may enter into the composition of the cell wall—gums, tannins, minerals, pigments, proteins, fats, and oils. It should be emphasized that in tissues such as wood, lignin may be deposited not only in the secondary wall but also in the primary wall and middle lamella.

Although the walls of cells vary considerably in composition in different species, and from one part to another in the same individual plant, **cellulose** constitutes the greatest percentage of the material of which most cell walls are made. It is synthesized by the protoplast and deposited across the plasmalemma by some mechanism.

Living cells are interconnected with each other through cytoplasmic channels called **plasmodesmata** (Fig. 3.5). The plasmalemma lines these small channels, and endoplasmic reticulum can often be observed within them. Pits are channels through secondary cell walls; they will be discussed in Chapter 4.

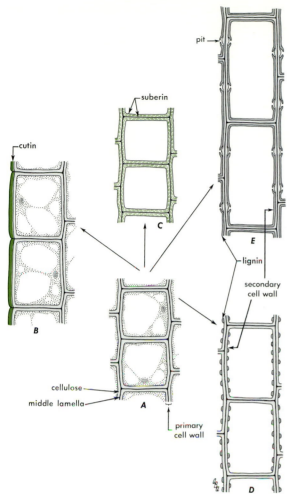

labels: pit, suberin, cutin, lignin, secondary cell wall, cellulose, middle lamella, primary cell wall, A, B, C, D, E

Figure 3.19 Development of cell walls during primary growth while stem is elongating. *A,* cell with cellulose walls. *B,* cutin may be laid down on the outside of an epidermal cell. *C,* in another location, suberin may be deposited within the cellulose primary cell wall to form a cork cell. Lignin is deposited within the primary wall and between the primary cell wall and the protoplast. *D,* during rapid elongation, lignin is laid down as rings or as a spiral band. *E,* when elongation stops (except for pits), the lignin forms a complete secondary wall around the cell.

SUMMARY

1. A cell may be divided into protoplast and cell wall.

2. Protoplasm is divided into an array of particles, membranes, and organelles.

3. The plasmalemma is the membrane separating the protoplast from the cell wall, and the tonoplast is the membrane separating the vacuole from the cytoplasm.

4. The lipid-globular mosaic model proposes that membranes are composed of globular proteins embedded in a lipid bilayer.

5. Ribosomes are small particles that have a high RNA content. Polyribosomes are aggregates of ribosomes that are involved in protein synthesis.

6. The endoplasmic reticulum consists of an extensive ar-

ray of flattened vesicles. It may or may not have ribosomes associated with the outer surface of the vesicles and appears to be involved in protein synthesis.

7. Mitochondria are small organelles as small as 0.5 μm in width and from 1 to 3 μm in length. They are bounded by a double membrane envelope. The inner component invaginates to form cristae, which are surrounded by a homogeneous matrix. Respiration is localized in the mitochondria.

8. Chloroplasts are approximately 5 $\times$ 10 μm in size. They are bounded by a double membrane envelope. The internal chloroplast lamellae aggregate to form cylindrical grana, which are connected by intergranal lamellae (frets). The photochemical reactions of photosynthesis take place on the membranes; the enzymatic reactions of photosynthesis are located in the stroma.

9. In addition to chloroplasts, cells may contain leucoplasts, amyloplasts, proplastids, and chromoplasts.

10. Dictyosomes are stacks of from 3 to 10 flattened cisternae. Each cisterna is surrounded by a peripheral net. Dictyosomes appear to be involved in the synthesis of various cellular products.

11. Microtubules are of indefinite length; they are about 28 nm in diameter and have an internal core about 8 nm in diameter. They are known to be involved in chromosome movements, to possibly regulate the pattern of cell wall fibril deposition, and to somehow control cell motility. Microfilaments are 4 to 7 nm in diameter, variable in length, and are involved in the mechanisms of cytoplasmic movement.

12. Microbodies are all bounded by a single membrane. They are variable in size and in morphology. They are closely associated with various types of intracellular enzyme activities.

13. The nucleus is bounded by a double membrane envelope provided with pores. The nucleoplasm appears to be characterized by a closely packed array of unit fibers about 22.5 nm in diameter and of indefinite length. Cells may live and even differentiate for a short time without a nucleus; however, a nucleus is required for the continued life of a cell and for cell division.

14. Vacuoles contain aqueous solutions within the protoplast and are separated from the cytoplasm by the tonoplast. Inorganic crystals, when present, are located in vacuoles.

15. The cell wall, which bounds the protoplast, is formed of cellulose fibrils embedded in an amorphous matrix. Other compounds (suberin, pectin, cutin, and lignin) may be present. Plasmodesmata are cytoplasmic connections between cells through the primary cell wall.

16. The type of cell just summarized is highly compartmentalized; it is known as a eukaryotic cell. Prokaryotic cells, which are more primitive, are not compartmentalized; DNA, photosynthetic processes, and respiratory activity all share a common cytoplasm.

THE CELL CYCLE

DESCRIPTION

Organs of the plant body are composed of cells and aggregates of cells called tissues (see Chapter 4). **Meristems,** which are regions that initiate these tissues, consist of cells that are actively dividing and beginning to enlarge and differentiate. Cells that are dividing or are preparing to divide are said to be progressing through the **cell cycle.** One cell cycle equals the time from the formation of an initiating cell to its subsequent division to form two new derivative cells (Fig. 3.20). The four stages of the cell cycle were discovered and named in 1953 by two scientists, A. Howard and C. Pelc. Their terminology for the stages of the cell cycle **(G$_1$-S-G$_2$-M)** have been universally adopted for both animal and plant cells. Traditionally, the first three phases of the cycle are referred to collectively as **interphase.** It is appropriate to consider interphase as the preparation period for division. **Cell division** is composed of two stages—**mitosis (nuclear division)** and **cytokinesis (cytoplasmic division).** We shall now consider the entire cell cycle in detail.

G$_1$-S-G$_2$ (INTERPHASE)

The stages of interphase are as follows (Fig. 3.20): The **G$_1$ phase,** or **pre-DNA synthetic phase,** represents the time during the cell cycle in which the cell prepares itself metabolically to go through DNA synthesis. This preparation involves such things as the accumulation and synthesis of specific enzymes needed to control DNA synthesis and the production of the DNA base units so that a supply or pool is on hand when synthesis begins.

The second phase of the cycle is the **S phase,** or **DNA synthesis phase.** It is during this portion of the cycle that the DNA molecules are actually duplicated. If the preparatory events in G$_1$ are not completed, the S phase cannot occur. Specific events of the S phase were discussed in Chapter 2.

The **G$_2$ phase,** or **premitosis phase,** follows DNA synthesis and is the time when the cells are involved in metabolic preparations for mitosis. For example, energy storage for chromosome movement, the production of mitosis-specific proteins and RNA, and the formation of microtubule subunits needed to make spindle fibers must be completed before mitosis can begin.

CELL CYCLE DURATION

Each plant species has a characteristic average amount of DNA. Pea *(Pisum sativum)* plants, for example, have 7.9 pg (1 picogram = 10^{-9}g) DNA per nucleus; longpod bean *(Vicia faba)* plants have 24.3 pg per nucleus. Cytologists (scientists who study cells) have established the generalization that the more DNA a cell has in its nucleus, the longer it will take for that cell to progress through the cell cycle. As an example, pea cells take 14 hours to progress once through the cell cycle, and longpod beans take approximately 18 hours. Each stage of the cell cycle generally takes a characteristic proportion of the total cell cycle time. G$_1$ usually is the most variable in length; mitosis is usually the shortest period. Table 3.2 gives the durations of different stages of the cell cycle and the total cycle time (CT) in three plants.

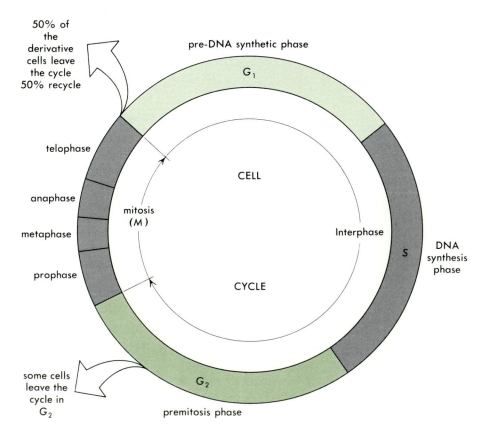

Figure 3.20 Diagram of the plant cell cycle.

Table 3.2. Cell Cycle Durations

	DNA Amount (pg)	CT (hours)	G₁	S	G₂	M
Helianthus annuus (sunflower)	6.3	7.8	1.2	4.5	1.5	0.6
Pisum sativum (pea)	7.9	14	5	4.5	3	1.2
Vicia faba (longpod bean)	24.3	18	4	9	3.5	1.9

Source: Jack Van't Hof, "The Duration of Chromosomal DNA Synthesis, the Mitotic Cycle, and Meiosis in Higher Plants," in *Handbook of Genetics*, vol. 2, ed. R.C. King (New York: Plenum, 1974).

MITOSIS

Cell division is apparently required because a single nucleus with a full complement of genes can control only a small amount of cytoplasm. Increase in size and complexity of the plant body requires cell division. The resulting increase in the number of cells makes possible the specialization of tissues for the different structural and physiological activities including absorption, conduction, reproduction, photosynthesis, and support.

The formation of new tissue cannot take place without cell division. During the period of division, both nucleus and cytoplasm divide. It is customary to designate the period of nuclear division as **mitosis** and to divide it into four phases: (1) **prophase,** (2) **metaphase,** (3) **anaphase,** and (4) **telophase** (Fig. 3.21). The division of the cytoplasm is known as **cytokinesis.** Mitosis gives rise to derivative nuclei having identical gene complements. Cytokinesis gives rise to two new parcels of cytoplasm that are similar but probably never identical.

Prophase

The DNA strands in the interphase nucleus are long, slender, and seem tangled. The onset of mitosis is heralded by the presence of definite chromatin threads (Figs. 3.21*a* and 3.22*a*). These threads gradually shorten and thicken and become easier to see (Fig. 3.21*b* and 3.21*c*), and they stain more heavily with certain dyes. For this reason they are called colored *(chromo)* bodies *(soma),* or **chromosomes.** Each chromosome consists of two strands of chromatin. The nucleolus slowly decreases in size and finally disappears during this stage (Fig. 3.21*a*, 3.21*b*, 3.21*c*, and 3.21*d*).

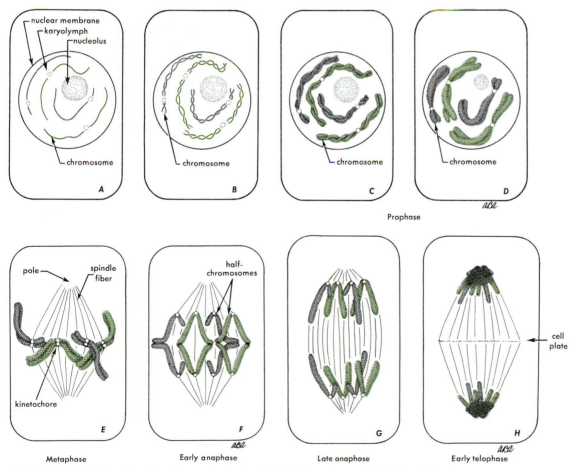

Figure 3.21 Mitosis, diagrammatic representation A to D, stages in prophase. Two pairs of chromosomes are represented; the "green" pair have a median kinetochore, while in the "gray" pair the kinetochore is close to one end. The chromosomes shorten and thicken, and chromatids become apparent. E, metaphase; F, early anaphase; G, late anaphase; H, early telophase.

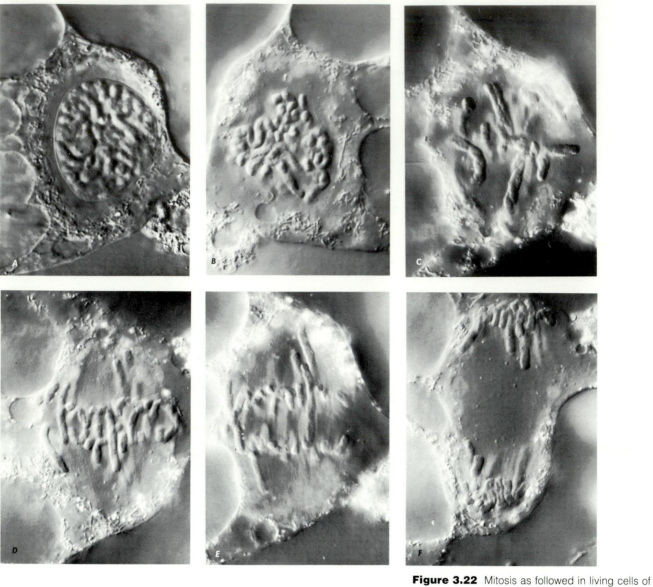

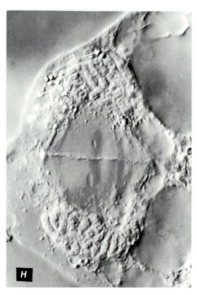

Figure 3.22 Mitosis as followed in living cells of the endosperm of seeds of the blood lily (*Haemanthus katherinae*). (Photographs taken with Nomarski optics.) *A,* early prophase; the nuclear membrane is still present. Note the clear zone of cytoplasm surrounding the nucleus. *B,* the nuclear envelope has disappeared, and the clear zone still surrounds the chromosomes. A few short spindle fibers are already present. *C,* full metaphase; the coiled chromatids are distinct. The two chromatids moving to the upper pole have separated and a few spindle fibers can be seen. *D,* early anaphase; the chromatids are moving to opposite poles of the cell. *E,* mid-anaphase; spindle fibers are in evidence between the chromosomes and the poles of the cell. *F,* telophase; the chromatids have aggregated at the opposite poles of the cell. Nuclear envelopes have not yet formed. Time intervals after *A; B,* 14 min; *C,* 64 min; *D,* 74 min; *E,* 93 min; *F,* 107 min, ×2000. *G,* telophase; the chromosomes are tightly clustered at opposite poles of the cell, and the cell plate has started to form; ×2000. *H,* late telophase; the cell plate is almost continuous, and fibrils can be seen extending poleward a short distance from the plate; ×2000.

It becomes apparent that at late prophase each chromosome is composed not of one but of two threads coiled about each other (Fig. 3.21c). The nuclear membrane disappears toward the end of prophase, and the spindle apparatus develops.

Metaphase

Forces active within the cell now arrange the chromosomes, or at least a specialized portion of each chromosome (the **kinetochore,** or **centromere**), in the equatorial plane of the cell (Figs. 3.21e and 3.22d). **Spindle fibers** composed of bundles of microtubules (Fig. 3.23) are attached to the kinetochores. As the nucleoplasm elongates, the chromosomes, or at least the kinetochores, are moved to the equatorial plane of the cell. Spindle fibers extend from the chromosomes to the opposite poles of the cells. Other fibers apparently reach to the poles but are not attached to the chromosomes.

The chromosomes are now visibly composed of two closely associated halves, each half being known as a **chromatid** (Fig. 3.21f). In plants such as corn, which has been intensively studied, each chromosome can be recognized and numbered. There are 20 chromosomes in corn but only 10 different types that can be distinguished by their size and form. There are thus two chromosomes of each type. The 20 chromosomes of corn may be arranged in 10 pairs. Maps showing the relative positions of the genes along corn chromosomes have been prepared. Since the chromosomes split longitudinally, each gene is replicated and each chromatid contains a full set of genes.

Anaphase

The chromosomes do not remain long in the equatorial plane. The chromatids soon separate from each other and move to opposite poles of the cell by a physical-chemical mechanism involving an interaction of the spindle microtubules (Fig. 3.21g). This phase of mitosis is called anaphase (Figs. 3.21f, 3.21g, 3.22d, and 3.22e).

Telophase

When the divided chromosomes have reached the opposite poles of the cell, they group together and the nuclear membrane and the nucleolus again become apparent. This period is known as telophase (Figs. 3.21h, 3.22f, 3.22g, and 3.22h). Cells in telophase begin to recycle through the cell cycle, starting again in G_1 (Fig. 3.20).

CYTOKINESIS

In the great majority of cases, the division of the nucleus is followed by the division of the cytoplasm. Light microscopy has demonstrated that a cell plate forms at the equatorial plane of the cell at right angles to the spindle fibers (Figs. 3.21h and 3.24). From electron microscopy we know that the spindle fibers are groups of microtubules. At the equatorial region, the spindle fibers are surrounded by an amorphous material. Vesicles appear in this region and eventually fuse to form the first barrier dividing the derivative protoplasts (Fig. 3.24a and 3.24b). This structure, the **cell plate,** is formed of pectin and will become the **middle lamella.** The new cell wall is formed by the deposition of cellulose by each protoplast on its side of the middle lamella. The resulting structure is the **primary cell wall.**

The formation of the primary wall and, subsequently, the **secondary wall** poses an interesting problem of genetic control. The cellulose is laid down outside the protoplast, apparently resulting from the polymerization of many sugar molecules to form long, unbranched molecules of cellulose. The molecules are organized into a crystalline array by hy-

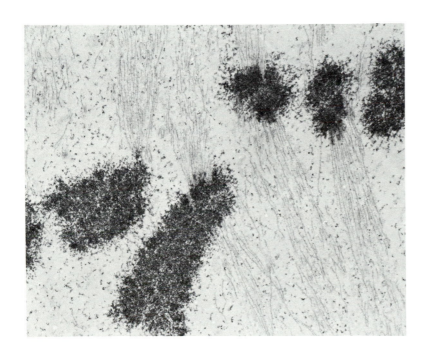

Figure 3.23 Electron micrograph of a thin section of metaphase chromosomes in the endosperm of *Haemanthus katherinae,* showing the aligned chromosomes with microtubules (spindle fibers) attached to the kinetochore. Magnification ×40,000.

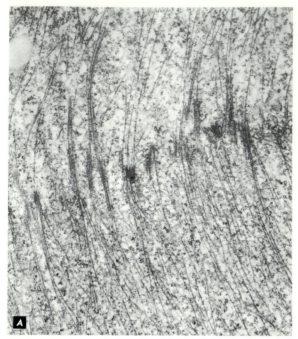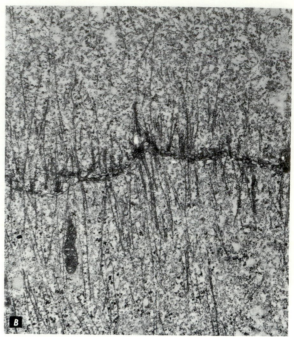

Figure 3.24 The formation of the cell plate in the endosperm of *Haemanthus katherinae. A,* early stage in plate formation; the microtubules are clustered into groups that are associated with a denser material, apparently in the plane of the future cell plate, ×26,000. *B,* Late stage; the microtubules are fewer in numbers and are not clustered; vesicles are present and apparently fusing to form a continuous separation phase between the sister cells, ×26,000.

drogen bonds to form long unbranched **fibrils** that are visible in the electron microscope.

After removal of the amorphous materials from the wall, the array of fibrils can be studied with the electron microscope. When first deposited by the protoplast, the fibrils are in parallel array and form a band around the protoplast (Fig. 3.25a). The mechanism bringing about this precise arrangement is not understood. However, the microtubules in the cytoplasm are also present in a similar parallel array (Fig. 3.26). It is logical to postulate that the cytoplasmic microtubules may be involved in the orientation of the cellulose fibrils.

As the cell lengthens, the deposition of fibrils by the protoplast continues, with no change in the orientation of the newly deposited fibrils. As elongation continues, the fibrils, now further removed from the protoplast, assume a position more in line with the axis of the elongating cell.

A section through a primary cell wall after completion of elongation shows the oldest cellulose fibrils to be parallel to the long axis of the cell; the most recently deposited cellulose fibrils encircle the cell at right angles to the long axis (Fig. 3.25a, 3.25b, and 3.25c). Intermediate fibrils have intermediate positions.

POLARITY OF CELL DIVISION AND THE MICROTUBULE CYCLE

If the spindles in a meristematic tissue were oriented at random, an irregular mass of tissue would result. This occurs when a single cell or small group of cells from a carrot root

in tissue culture produces an irregular mass of cells called a **callus** (Fig. 3.27a). For orderly growth there must be a precise orientation of the mitotic spindles. A polarity is established so that, in general, the axes of spindles are parallel with each other and with the axis of the shoot or root (Fig. 3.27b). When the spindle axes are oriented parallel to the root-shoot axis, the divisions are called **anticlinal.** When the axes are perpendicular to the root-shoot axis, the divisions are called **periclinal** (Fig. 3.27b).

Polarity seems to be inherent in cells and in tissues of which they are a part. Changes in polarity may be induced by hormones. Occasionally, the orientation to be assumed by the spindle can be detected in plants at the cellular level before metaphase. In some instances, the cellular organelles pass largely to one end of the interphase cell. According to some observations, microtubules form a preprophase band at the equatorial plane of the cell during interphase, just before the onset of mitosis (Fig. 3.28). Positioning of the preprophase band occurs at some time in interphase well before mitosis actually occurs. The orientation of the preprophase band is a critical factor in development because it marks the precise position of the cell plate. This observation means that some type of control mechanism exists in tissues that determines the orientation of new cells many hours before they are formed by mitosis.

The study of the roles of microtubules in plant development and the factors influencing their position and synthesis is rapidly expanding. One technique that shows tremendous promise involves the formation of antibodies in

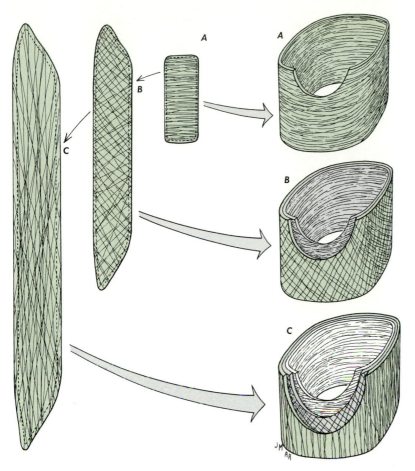

Figure 3.25 Diagram to show deposition and change in orientation of the cellulose fibrils in an elongating primary cell wall. *A*, *B*, and *C* represent increasing age and length of the same cell. The cellulose fibrils are first deposited parallel to the circumference of the protoplast. They are then pulled out of this orientation as the cell wall elongates. (In the actual wall, the different stages of fibril orientation would grade into each other.) Green represents the earliest fibrils deposited, and light gray represents those most recently deposited.

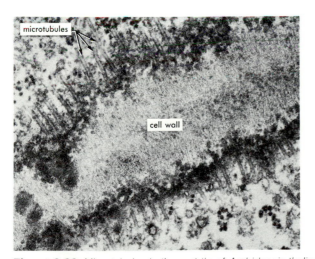

Figure 3.26 Microtubules in the root tip of *Arabidopsis thaliana:* The microtubules are oriented parallel to the circumference of the cell wall and are thus parallel to the cellulose fibrils, ×50,000.

mammals (such as rabbits) against antigens, like microtubule protein, isolated from plants and other organisms. Antibodies produced in this way can be isolated and attached to (tagged by) fluorescent stains. These tagged antibodies can then be added to fixed cells containing the original antigen. When these cells are viewed with a fluorescence microscope, the exact position of the antigen (microtubules in this case) can be observed.

Figure 3.29 shows an application of this technique using onion (*Allium cepa*) root cells reacted against microtubule protein antibodies that are fluorescently tagged. The cycle of microtubules can be followed in this way throughout the entire cell cycle. Figure 3.29*a*, 3.29*b*, and 3.29*c*, show the position of the preprophase band of microtubules in interphase. In prophase (Fig. 3.29*d* and 3.29*e*), the microtubules are repositioned into the mitotic spindle from a source at the spindle poles. The spindle is again repositioned during metaphase (Fig. 3.29*f*). At anaphase the chromosomes migrate back to the poles, and the spindle microtubules are dispersed (Fig. 3.29*g*). As discussed, in telophase, the microtubules play a role in the deposition of the cell plate in the location and orientation previously indicated by the preprophase band (Fig. 3.29*h* and 3.29*i*). Understanding the forces involved and the mechanisms needed to control microtubules in cells is a major scientific challenge for the next decade.

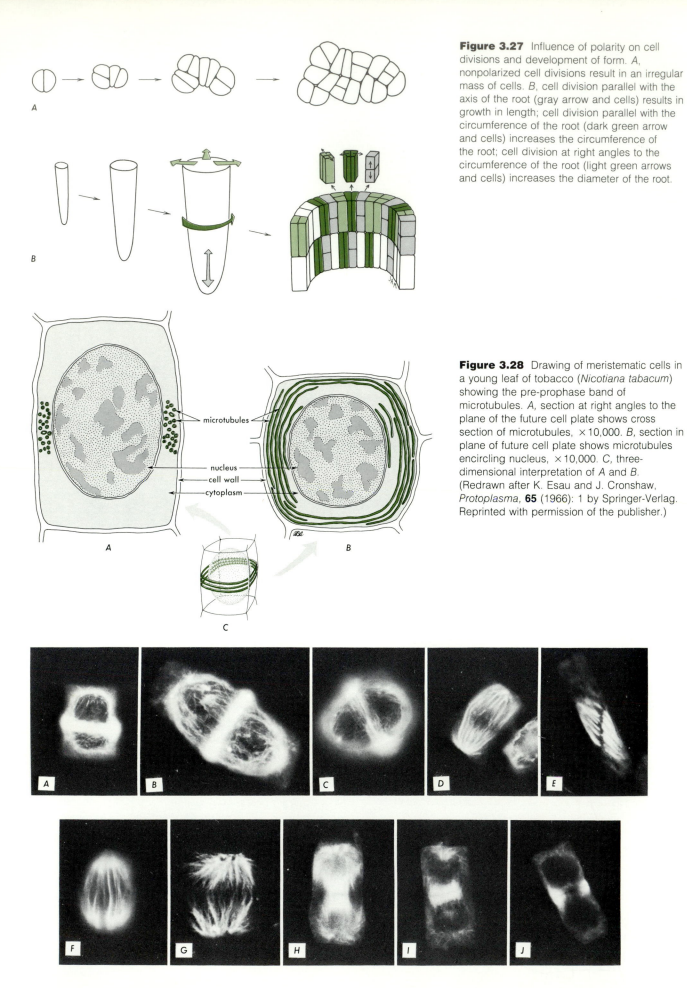

Figure 3.27 Influence of polarity on cell divisions and development of form. *A,* nonpolarized cell divisions result in an irregular mass of cells. *B,* cell division parallel with the axis of the root (gray arrow and cells) results in growth in length; cell division parallel with the circumference of the root (dark green arrow and cells) increases the circumference of the root; cell division at right angles to the circumference of the root (light green arrows and cells) increases the diameter of the root.

Figure 3.28 Drawing of meristematic cells in a young leaf of tobacco (*Nicotiana tabacum*) showing the pre-prophase band of microtubules. *A,* section at right angles to the plane of the future cell plate shows cross section of microtubules, ×10,000. *B,* section in plane of future cell plate shows microtubules encircling nucleus, ×10,000. *C,* three-dimensional interpretation of *A* and *B.* (Redrawn after K. Esau and J. Cronshaw, *Protoplasma,* **65** (1966): 1 by Springer-Verlag. Reprinted with permission of the publisher.)

microtubules

nucleus

cell wall

cytoplasm

SUMMARY OF THE CELL CYCLE

1. The cell cycle consists of G_1-S-G_2-M. G_1 and G_2 are important because it is during this time that cells prepare for DNA synthesis (S) and mitosis (M).

2. Cell division increases the number of cells.

3. All cells have the potentiality to divide but are normally blocked from doing so in mature tissues.

4. Nuclear division is called mitosis; cytoplasmic division is called cytokinesis.

5. The derivative nuclei are genetically identical.

6. In general, mitotic spindles are oriented parallel to each other and to the axis of the shoot. A definite, predetermined polarity produces this orientation.

7. The preprophase band of microtubules forms during interphase to predetermine the orientation of the new cell plate.

8. The stages of mitosis are prophase, metaphase, anaphase, and telophase.

9. The spindle, which somehow directs chromosome movement, is constructed from many microtubules.

10. The middle lamella is the first layer to separate the two derivative protoplasts.

11. Vesicles collect around the microtubules in the equatorial plane of the cell. They fuse to form the middle lamella. Cellulose formed by the two daughter protoplasts and deposited on the middle lamella forms the primary wall.

12. Microtubules lying parallel to the cell wall orient the position of new cellulose microfibrils.

Figure 3.29 Stages of mitosis in onion (*Allium cepa*) root tips, photographed by fluorescence microscopy. Root tip cells were treated with fluorescently tagged spindle protein antibodies so that the photographs show only the microtubules. *A, B,* and *C,* preprophase—shows the preprophase band of microtubules plus the microtubules around the cortex of the cell necessary for cell wall formation: $A = \times 620$, $B = \times 850$, $C = \times 820$, *D* prophase—preprophase band now dispersed and the spindle apparatus is in position; $\times 1150$. *E,* prometaphase—chromosomes are now fully coiled, the spindle apparatus is brightly fluorescent; $\times 770$. *F,* metaphase—chromosomes are located in the center of the cell; $\times 850$. *G,* anaphase—chromosomes migrate back to the poles and the spindle apparatus becomes dispersed; $\times 770$. *H* and *I* telophase—note the dense accumulation of fluorescent microtubules and the dark line in the center of the cell showing the new cell plate; $H = \times 880$, $I = \times 770$. *J,* late telophase—shows the spindle microtubules only at the leading edge of the cell plate; $\times 800$. (Used by permission of Dr. S. Wick, Jour. Cell Biol. 89:685–690 (1981), © The Rockefeller Univ. Press, N.Y.)

4 THE STRUCTURE AND DEVELOPMENT OF THE PLANT BODY, AND THE ANATOMY OF STEMS

The flowering plant body consists of several structures shown in Figure 4.1. The flower and fruit are reproductive structures specialized for producing and disseminating seeds. They will be described in Chapters 12 and 13. Stems, leaves, and roots are called vegetative organs because they are concerned chiefly with nutrition and growth rather than reproduction. In particular plants, any of the vegetative organs can produce and store food, provide mechanical support for other plant parts, or even serve a reproductive role in special instances. However, in most cases each organ has a characteristic set of main functions. Roots typically anchor the plant, hold it erect, and absorb water and minerals from the soil. Leaves provide energy and new organic building materials (chiefly sugars) by means of photosynthesis. Stems give rise to new leaves and flowers, and they transport food and water between organs. At the tip of the stem axis is the shoot apex (Fig. 4.1D), which is the site for the formation of new leaves and new cells needed for growth of the plant body. The root apex, at the opposite end of the plant axis, produces new cells for the root (Fig. 4.1E).

Since we are surrounded by plants, everyone can recognize the most common forms of plant organs. But plants vary widely, and a given organ may become so highly modified that its nature is hard to determine without close study. Fortunately, in most cases there is a relatively easy criterion for distinguishing between the three vegetative organs. As shown in Fig. 4.1, leaves occur at intervals along the stem. The point on the stem where a leaf attaches is called a node, while the segment of stem between leaves is an internode. Usually, a small dormant shoot tip (a bud) occurs just above each leaf, in a position called the axil of the leaf: these buds are called axillary buds. In brief, then, stems have internodes, and they have nodes where leaves and buds are borne. These features are not usually present in leaves and roots, though roots may sometimes form buds upon injury.

In the next several chapters we will examine the external form (morphology) and internal structure (anatomy) of the vegetative organs. We will find that plants form several groups with quite different features, so generalizations are not always safe. For convenience, let us establish names for these

groups before discussing anatomy. The angiosperms are plants that produce flowers and seeds. They include plants with showy flowers as well as many species (such as grasses and shade trees) whose flowers are green and inconspicuous. By contrast, gymnosperms form seeds but not flowers. They include the conifers, such as pines and firs, as well as many less familiar forms. The ancestors of angiosperms were probably gymnosperms.

The angiosperms can be further divided into two subdivisions: the dicotyledonae (dicot plants) and the monocotyledonae (monocot plants). Dicots include most broadleafed trees and shrubs as well as numerous herbs. Monocots include the grasses, palms, orchids, and many others. Dicots and monocots differ in many ways, but the defining difference is in the very young plant (the embryo) within the seed. A monocot embryo has a single food-absorbing leaf, or cotyledon; hence the name (mono = "one"). A dicot embryo has two cotyledons. These groups of plants differ so much that we are forced to discuss them separately when we introduce the anatomy of the plant. We will begin with the anatomy of the dicots.

A serious problem in presenting this material is that photographs and drawings are two-dimensional, and words are often inadequate to explain the pattern and symmetry of plant parts. So, while reading these sections, visualize the plant's internal structure in three dimensions. Try to use your "mind's eye" as a kind of X-ray vision to see the length, width, and breadth of cells and tissues and their interconnections. As you study plant structure, it will become apparent that all organs are integrated functionally and structurally. The entire plant acts as one unit and is not simply an aggregate of isolated parts.

DEVELOPMENT OF TISSUES OF THE PRIMARY PLANT BODY

THE PRIMARY MERISTEMS

Plant organs (leaves, stems, roots, and flower parts) are obviously different from each other morphologically (i.e.,

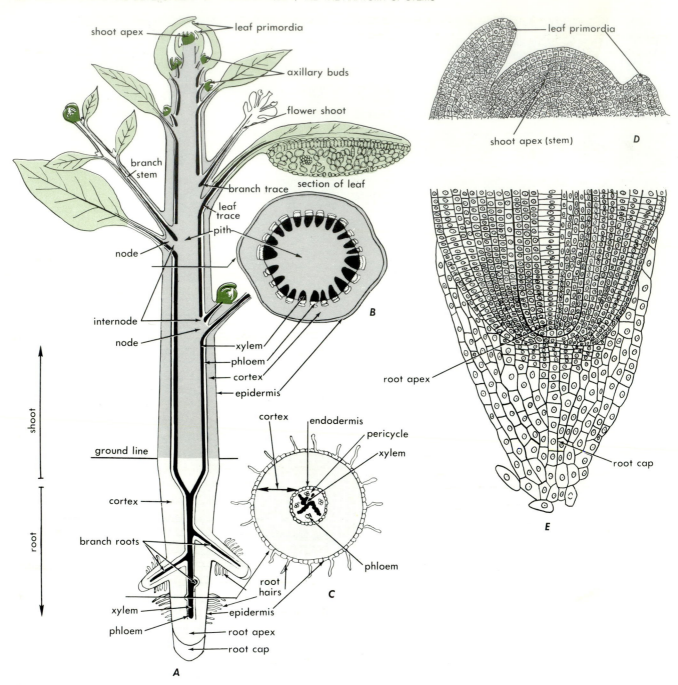

Figure 4.1 *A,* the principal organs and tissues of the body of a seed plant; *B,* cross section of stem; *C* cross section of root; *A,* redrawn from R. M. Holman and W. W. Robbins, *A Textbook of General Botany. John Wiley & Sons,* 1924. New York: *D,* enlarged view of the shoot apex showing the cells of the meristem and two leaf primordia; *E,* enlargement of the root apex and the root cap.

based on their external form). But, if we examine their internal anatomy, it is apparent that all organs are composed of similar structural units—cells and tissues—that are dissimilarly arranged. Each **cell type** is modified to make it ideally suited to perform one or more specific functions. **Tissues are organizations of cells of one or more types that have a common origin and a common collective function.**

Cell types and tissues develop by a process called **differentiation.** A differentiating cell progresses through a series of steps that result in the cell becoming mature and functional. Each cell type "experiences" slightly different differentiation steps. The formation of new cells and the initiation of differentiation in plants takes place in specific regions called **meristems.** The **apical meristems** occur in the **shoot** and **root tips.** Apical meristems are the source of all other meristems. They form the three primary meristematic tissues, or **primary meristems: protoderm, ground meristem,** and **procambium.** These three primary meristems differentiate into the three **primary tissues: epidermis, ground tissues (pith** and **cortex)**, and **vascular tissues (phloem** and **xylem)**. We will now consider in detail each primary meristem and the tissues that arise from it.

Protoderm

The outermost layer of cells at the shoot apex is called the protoderm (Figs. 4.1A, 4.2, and **4.3, Color Plate 3**). It develops into **epidermis**—the special primary tissue that covers and protects all underlying primary tissues. The epidermis prevents excessive water loss and yet allows for the exchange of gases necessary for respiration and photosynthesis.

Ground Meristem

The ground meristem comprises the greater portion of meristematic tissue of the shoot tip (Figs. 4.2A, 4.2D, and **4.3,**

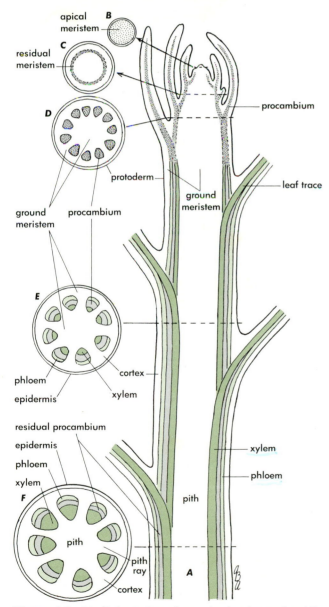

Figure 4.2 A to F, the pattern of vascular development and the positions of the three primary meristems.

Color Plate 3). Its cells are relatively large, thin-walled, and isodiametric. The regions forming from the ground meristem are (a) **pith,** in the very center of the stem, and (b) **cortex,** in a cylinder just beneath the epidermis and surrounding the vascular tissue. A note of clarification must be added here. Traditionally speaking, the pith and cortex are not tissues but instead are stem and root regions that are composed of the ground tissues called parenchyma, sclerenchyma, and collenchyma. These will be discussed in the following section.

Procambium

Procambium cells usually appear first as strands among ground meristem cells (Fig. 4.2). In cross section (Fig. 4.2D) the strands appear as isolated groups of cells arranged in a circle. Sometimes a continuous **procambium cylinder** is formed. Procambium cells are smaller than those of the surrounding ground meristem; in longitudinal section they are much longer, and some of them may be pointed at the ends.

The procambium differentiates into two primary vascular tissues: **phloem** and **xylem** (Figs. 4.2E, 4.2F, and **4.4, Color Plate 3**). Vascular tissues are primarily specialized for conducting materials, although they also provide mechanical strength to organs. The phloem conducts foods that are synthesized in photosynthesis, while the xylem conducts water and mineral salts from the soil to the aerial parts of the plant.

PRIMARY TISSUES

Primary tissues of the stem are differentiated from the three primary meristematic tissues—protoderm, ground meristem, and procambium. In woody plants, primary tissues of the stem are located only a very short distance behind the stem tip. Even before the end of the first season's growth, these primary tissues complete their differentiation, and **secondary tissues** may be formed in abundance. Secondary tissues are the result of production of new cells by the **vascular cambium** and by the **cork cambium.** These two types of cambia are discussed later.

The Epidermis

The epidermis is usually a single superficial layer of cells that covers all other primary tissues, protecting them from drying out and, to some extent, from mechanical injury. It is the limiting layer of cells between the plant and its environment. In surface views, epidermal cells are elongated in the direction of the organ's length; in transverse section, they are usually isodiametric (Fig. 4.5).

The outer tangential wall of cells exposed to air is often thicker than the other walls, and its surface layer is usually coated with a waxy substance called **cutin.** This superficial layer, the **cuticle,** is quite impermeable to water and gases. However, there may be cracks or other imperfections in the cuticle, and water vapor may pass out of the plant at these points. The inner walls, parallel to the stem surface, are thinnest; and radial walls, at right angles to the surface, often taper in thickness toward the inner wall.

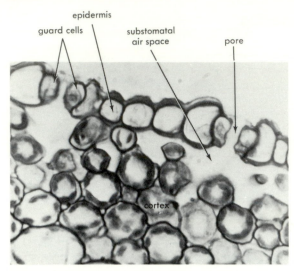

epidermis

guard cells

substomatal
air space

pore

cortex

Figure 4.5 Epidermis; cross section of alfalfa (*Medicago sativa*) stem showing epidermal cells, guard cells, and cortex with a substomatal air space, ×400.

Young stems usually have specialized epidermal cells called **guard cells.** Between each pair of guard cells is a small opening, or **pore,** through which gases enter and leave the underlying stem tissues. Two guard cells plus the pore comprise one **stoma** (plural, **stomata**) (Fig. 4.5). Guard cells differ from other epidermal cells by their crescent shape and the fact that they contain chloroplasts. Stomata are common in the epidermis of leaves, floral structures, and fruits as well as stems. Stomata are discussed more thoroughly in Chapter 5.

Epidermal appendages, such as hairs, may occur on young stems. These structures will be discussed in the chapter on leaves.

The Cortex

The cortex is a complex region that is derived from ground meristem. It forms beneath the epidermis as a cylindrical zone that extends inward to the primary phloem (Figs. 4.2E and **4.4, Color Plate 3**). The following tissues or cell types may be found within the cortex: **parenchyma, collenchyma, sclerenchyma,** and **secretory tissue.**

Parenchyma. The principal tissue of the cortex is parenchyma (Fig. 4.6G). It consists of isodiametric cells with thin walls—made mostly of cellulose—and protoplasts that remain alive for a long time. We can speak either of a parenchyma tissue, where many parenchyma cells are found together, or of individual parenchyma cells.

Parenchyma tissue is characterized by the presence of intercellular air spaces, which vary greatly in size; in some parenchyma tissues they are difficult to find, while in others they are very apparent. Because parenchyma cells retain active protoplasts, they function in the storage of water and food, in photosynthesis, and sometimes in secretion. When the parenchyma cells contain chloroplasts, they are collectively referred to as **chlorenchyma.** Parenchyma cells can be reprogramed to differentiate into different cell types.

Parenchyma cells are not confined only to the cortex and pith of the stem but also occur in practically all other types of tissue and in other organs of the plant.

Collenchyma. The outermost cells of the cortex of young stems, lying just beneath the epidermis, often constitute a tissue known as collenchyma. This tissue may form a complete cylinder, or it may occur in separate strands. Collenchyma cells are elongated, often contain chloroplasts, and are living at maturity. The walls of collenchyma cells are composed of alternating layers of pectin and cellulose. In the most common type of collenchyma, the cell walls are thickened at the corners (Figs. 4.6D and **4.7A, Color Plate 4**). These thickenings are quite flexible and will stretch, in much the same way as heated plastic. Collenchyma is, therefore, an ideal strengthening tissue because it strengthens and at the same time allows normal tissue growth. Collenchyma serves as a strengthening tissue in young expanding stems and also in the petioles of leaves.

Sclerenchyma. The main functions of sclerenchyma cells are support and, in many cases, protection. Their shape and the thickness and toughness of their walls contribute to the ability of sclerenchyma cells to support and protect the stem. Thickness and toughness of walls are increased by deposition, within the original cellulose wall, of a substance known as **lignin.** Lignin is made by linking together small molecules that have been secreted into the wall by the protoplast. Thus lignin is a tough, highly branched polymer that fills most of the spaces between the cellulose microfibrils in the completed wall. When this secondary wall is completely formed, the protoplast usually dies.

There are two types of sclerenchyma cells: **sclereids** and **fibers** (Figs. 4.6A, 4.6B, 4.6C, 4.6F, 4.6H; and **4.7C, Color Plate 4**).

Sclereids occur not only in the cortex of stems but also in the hard shells of fruits, seed coats, and bark, in pith of some stems, and in certain leaves.

Of various types of sclereids, the most common are stone cells **(Fig. 4.7B, Color Plate 4).** Other types of sclereids are branched, resembling very irregular stars (Figs. 4.6F and **4.7C, Color Plate 4**). Some sclereids are derived from parenchyma cells by a pronounced thickening of cell walls; others arise from separate meristematic cells.

Fibers are elongated, thick-walled, and usually pointed at the ends (Fig. 4.6A and 4.6B). Their walls may or may not be lignified. The walls may become so thick that the cell cavity, the **lumen,** almost disappears. Simple pits form in their thick walls (Fig. 4.6A). Fibers are sometimes very elastic and can be stretched to a great degree without losing their ability to return to their original length. Protoplasts of fibers often die as the cells attain maturity.

Secretory Cells. Secretory cells are parenchymalike and contain dense protoplasm. They secrete various substances, such as resinous materials (see Fig. 4.28) and nectar in many flowers. Many epidermal hairs are secretory cells.

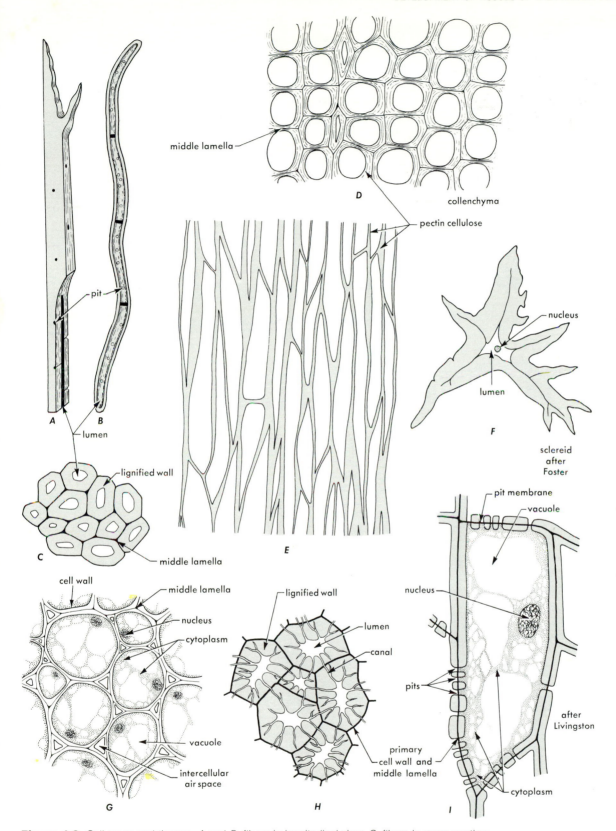

Figure 4.6 Cell types and tissues. *A* and *B*, fibers in longitudinal view; *C*, fibers in cross section; *D*, collenchyma in cross section; *E*, collenchyma in longitudinal view; *F*, sclereid; *G*, parenchyma; *H*, stone cells; *I*, woody parenchyma.

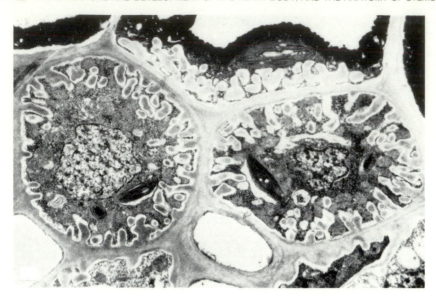

Figure 4.8 Transfer cells around vascular bundle in leaf of *Armeria corsica*, × 2900.

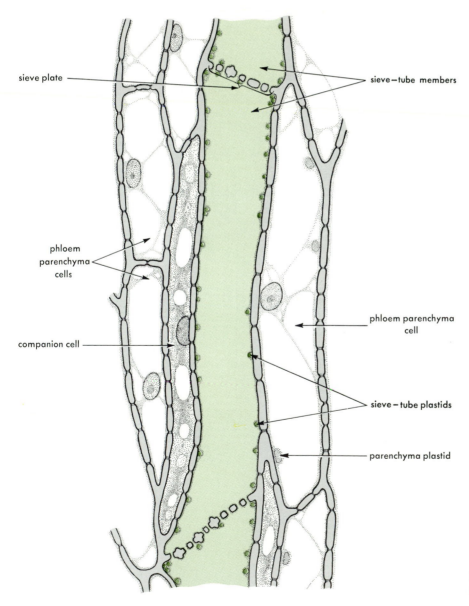

sieve plate

sieve—tube members

phloem parenchyma cells

phloem parenchyma cell

companion cell

sieve—tube plastids

parenchyma plastid

Figure 4.9 Phloem tissue from the stem of tobacco (**Nicotiana Tabacum**).

Transfer Cells. Frequently, cells located in positions of active solute transfer will show an irregular extension of the cell wall into the protoplast (Fig. 4.8). Since the plasmalemma follows the contour of the wall, its surface is greatly extended. Mitochondria appear to aggregate adjacent to these areas of increased membrane surface. This suggests an adaptation that facilitates transport from one cell to another, or from the interior to the exterior of the plant. These cells are called transfer cells.

The Pith

The pith makes up the central core of many stems. It is composed mostly of parenchyma cells that store food products such as starch. Sclerenchyma cells (especially sclereids) can also be found in the pith. The cellular region between vascular bundles is sometimes called a pith ray (Fig. 4.2F).

THE PRIMARY VASCULAR TISSUES

In vascular plants, water and various water-soluble inorganic salts from the soil, as well as food substances, are conducted throughout the plant in well-defined vascular tissues.

In a young stem, very near the tip (Figs. 4.2 and **4.3, Color Plate 3**), vascular tissues occur as separate bundles—**primary vascular bundles.** Each primary vascular bundle is differentiated from a procambium strand and consists of primary xylem and primary phloem.

The Primary Phloem

In flowering plants, phloem may possess several types of cells: **sieve-tube members, companion cells, fibers, sclereids,** and **parenchyma.** A **sieve tube** is a vertical row of elongated cells; each cell is known as a **sieve-tube member.** These are the conducting elements of phloem (Fig. 4.9) that translocate sugars, produced in the leaves by photosynthesis, to other plant parts (see Chapter 7).

Among angiosperms, a sieve-tube member and a companion cell originate by division from the same procambial cell. Young sieve elements have the usual complement of organelles: nucleus, plastids, mitochondria, and dictyosomes. As the element matures, its protoplast becomes greatly modified. Its nucleus is thought to disintegrate. Plastids lose most of their internal membranes but usually retain starch. The mitochondria become small. The cytoplasm becomes reduced to a thin peripheral layer. The central part of the cell is occupied by a mass of strands or tubules. This mass, which has been called **slime,** can be seen with the light microscope; it is more correctly referred to as **P-protein** (Fig. 4.10).

At maturity, one or more companion cells lie adjacent to each sieve-tube member (Fig. 4.9). Since companion cells have a normal protoplast with a full complement of organelles (Fig. 4.10A), these cells possibly regulate the metabolic activity of the sieve-tube members since they have no nucleus. Plasmodesmata connect the protoplasts of companion cells and sieve-tube members.

A characteristic structural feature of mature sieve-tube members is the **sieve plate.** It may occur in the end or side walls (Figs. 4.9 and 4.10). The end wall between two adjacent sieve-tube members is thickened, and strands of cytoplasm pass through pores adjoining them. Sieve-tube members live and function about 1 to 3 years, except in a few trees such as palms, where they live longer.

In many studies on the structure of mature sieve-tube members, a carbohydrate known as **callose** is seen around the margins of pores in the sieve plate (Fig. 4.10B and 4.10C). In some instances, protein may also collect at the sieve plate. Obviously, this would block the pores of the sieve plate and obstruct movement of food materials. It has been demonstrated in other tissues that callose forms very rapidly in response to wounding and other stresses (Fig. 4.10C). It may help to limit the loss of sap from injured sieve tubes.

Gymnosperms have **sieve cells** rather than sieve-tube members. Sieve cells have tapered end walls without sieve plates and do not connect to form sieve tubes.

The Primary Xylem

The conducting cells that occur in primary xylem of vascular plants are **tracheids** and **vessel members.** These cells conduct water and mineral salts. Associated with them may be xylem fibers and xylem parenchyma.

A tracheid is a single elongated cell more or less pointed at its ends (Fig. 4.11A). Functioning tracheids are not alive. The entire tracheid wall is lignified and thickened except for numerous small, circular or oval areas called **pits** (Fig. 4.11E). There are two types of pits in xylem cells: **simple pits** and **bordered pits.** Simple pits, as shown in Fig. 4.6I, occur in fibers, sclereids, and in parenchyma cells when they have secondary walls. Pits form opposite each other in the secondary walls of adjacent cells. A pit-pair is not a hole in the wall, since the primary wall and the middle lamella of the two communicating cells remain intact. These primary layers, however, are penetrated by plasmodesmata while the cells are living. The type of pit known as a bordered pit (Fig. 4.11A and 4.11E) occurs in tracheids, vessel members, and some xylem fibers. This type of pit is more structurally complex. It consists of an expanded border of the secondary cell wall that extends over a small pit chamber. Within the chamber is a diaphragmlike primary cell wall. In gymnosperms, the center portion—the torus—is thickened and impregnated with a waxy material.

A **vessel member** is a single cell with oblique, pointed, or transverse ends. A **vessel** is a series of vessel members differentiating end to end, with perforated end walls. Vessels are often several centimeters long, and in some vines and trees they may be many meters in length. Before the protoplasts disappear, vessel walls become thickened, forming a secondary wall (Fig. 4.12); the thickening material is laid down on primary walls in various patterns so that in some places the secondary walls are thick and in others they are thin. The material deposited is cellulose; later, the layers of cellulose become lignified. The end walls of the vessel members dissolve before the protoplasts disappear. Thus, the

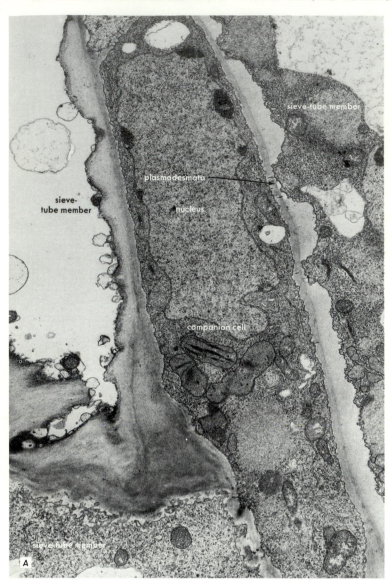

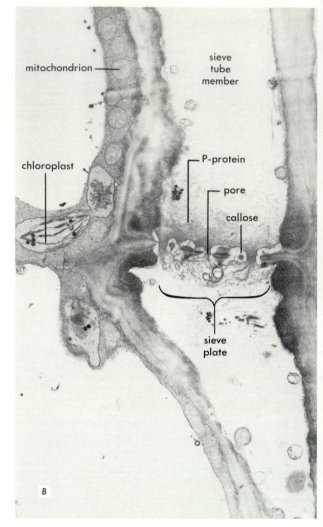

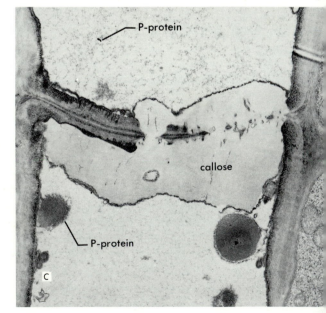

Figure 4.10 *A*, longitudinal section of three sieve-tube members and one companion cell. The nucleus and various organelles are present in the central companion cell. Most organelles are missing in the left sieve-tube member but are present in the one on the right. *Cucurbita maxima,* × 8300. *B, Beta vulgaris* petiole phloem to show P-protein and callose around pores of the sieve plate, × 10,600. *C, Beta vulgaris* petiole 1 hour after cold treatment stress, (note the massive amount of callose plugging the sieve plate), × 10,200 (*B* and *C* courtesy of V. Franceschi.)

Figure 4.3. Longitudinal section of a shoot apex, ×90.

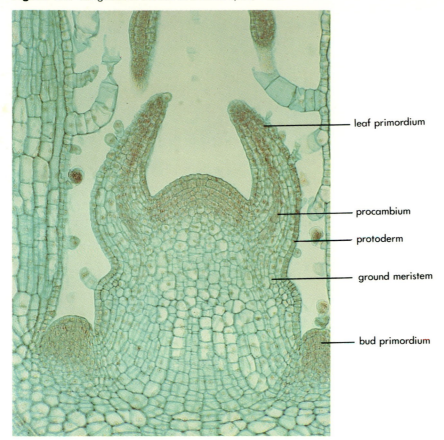

leaf primordium

procambium

protoderm

ground meristem

bud primordium

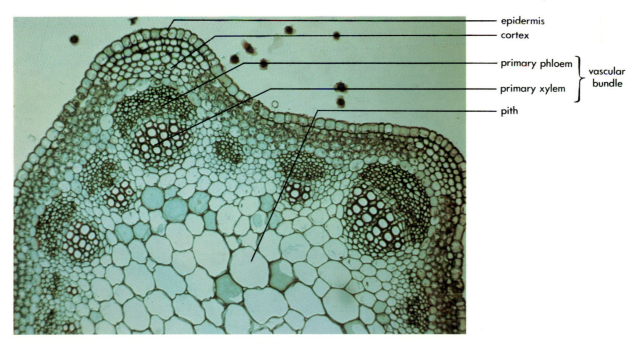

epidermis

cortex

primary phloem ⎤
             ⎬ vascular bundle
primary xylem ⎦

pith

Figure 4.4. Cross section of primary stem of alfalfa *(Medicago sativa)*. Note that the lignified walls are stained red and cellulose is stained blue-green. When the section includes an entire transverse wall of a cell, the entire cell appears to be blue-green.

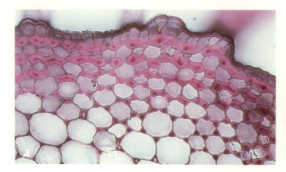

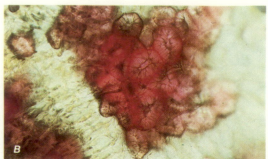

Figure 4.7. *A,* section of Marigold *(Calendula* sp.) stem showing collenchyma cells with thick pink-stained walls. *B,* red-stained sclereids (stone cells) from the fruit of pear *(Pyrus* sp.). *C,* sclereid (star-shaped) from phloem of Douglas fir stem *(Pseudotsuga menziesii).*

Figure 4.13. Stems vary in structure and function. Some are annuals, others are perennials of great age. *A,* redwood *(Sequoia sempervirens);* some are vines. *B,* fig *(Ficus pumila);* some are prostrate. *C,* dandelion *(Taraxacum vulgare).*

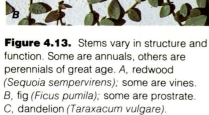

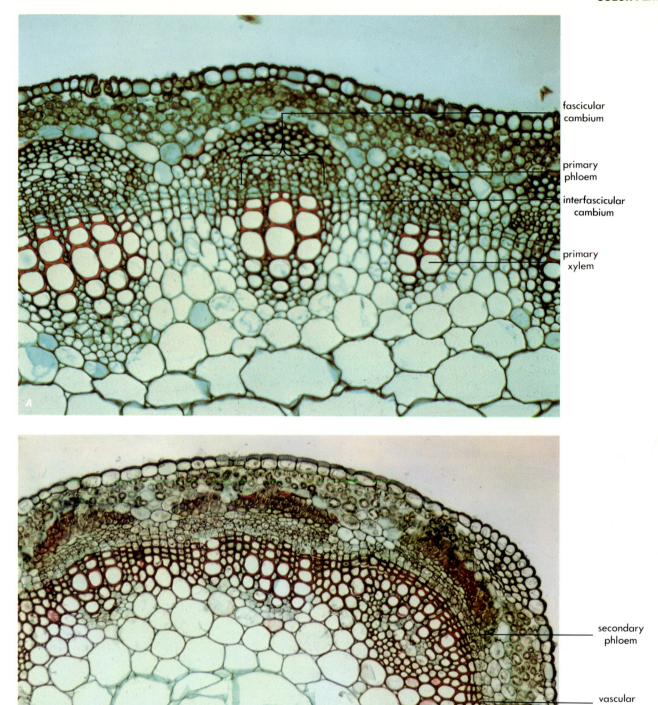

fascicular
cambium

primary
phloem

interfascicular
cambium

primary
xylem

secondary
phloem

vascular
cambium

secondary
xylem

Figure 4.18. Cross sections of a stem of alfalfa *(Medicago sativa)*. *A*, showing the fascicular cambium within the bundle and interfascicular cambium between the bundles. *B*, some secondary vascular tissue has been produced, ×300.

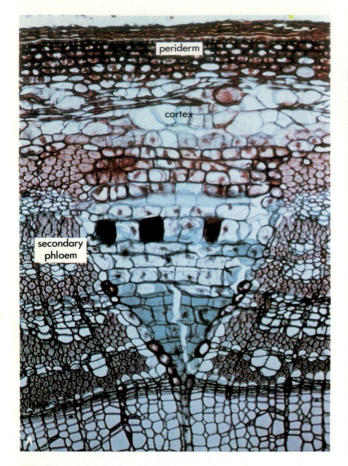

periderm

cortex

secondary phloem

secondary phloem

secondary xylem

vascular cambium

annual ring

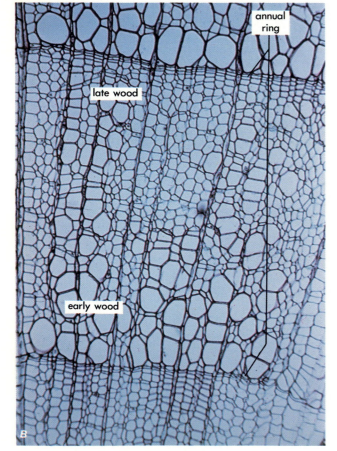

annual ring

late wood

early wood

Figure 4.22. *A,* cross section of basswood stem *(Tilia americana)* through the secondary phloem. From the outside are layers of cork cells, followed by cortical parenchyma. *A,* wedge of parenchyma cells (a dilated ray) lies between two regions of active phloem cells. These cells consist of alternating layers of red thick-walled fibers, thin-walled sieve-tube members, and companion cells. *B,* cross section of basswood through secondary xylem showing an annual ring and portions of two others. *C,* lower magnification of a three-year-old basswood stem to show the annual rings and ring porous wood.

Figure 4.23. Maple *(Acer saccharum)* cross section of diffuse porous wood.

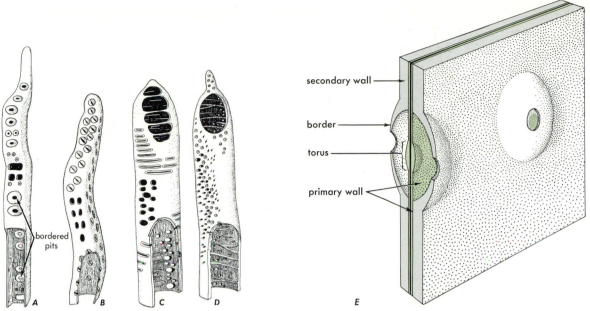

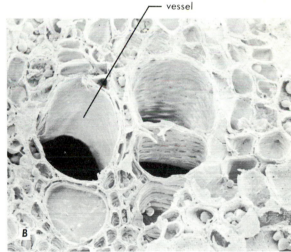

Figure 4.11 Tracheids and vessel members from secondary wood. *A*, tracheid from spring wood of white pine (*Pinus*); *B*, tip of tracheid from wood of oak (*Quercus*); *C*, tip of vessel member from wood of Magnolia; *D* tip of vessel member from wood of basswood (*Tilia*); *E*, diagram of bordered pit.

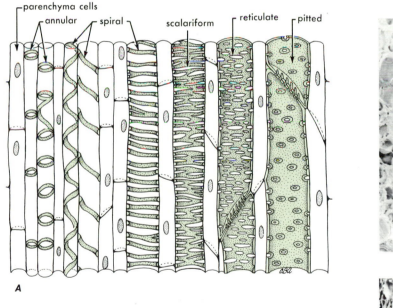

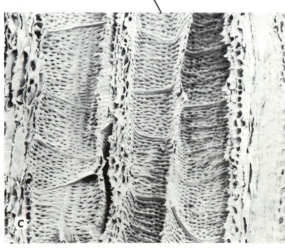

Figure 4.12 Types of vessel members in primary xylem of an elongating stem. *A*, Longitudinal view. Annular vessel members were formed first and therefore the oldest and most stretched. When elongation has stopped, the pitted vessel members, formed last, will not be stretched. *B*, cross section; *C*, longitudinal section. (Note that the rims of old cross walls are now no longer present.) (B and C Courtesy of D. Hess.)

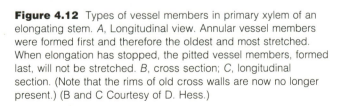

deposition of thickening material forming the secondary walls and the dissolution of end walls are functions of living cells. After these events have taken place, the protoplast dies.

The secondary walls of vessels in angiosperm stems are deposited in several different patterns (Fig. 4.12A): **annular, scalariform, reticulate,** and **pitted.** The ends of the vessel members (Fig. 4.11C and 4.11D), are generally on a slant and, although open, they may have bars of wall material across them. The shape of vessel members may indicate evolutionary relationships among plants. Vessel members do not occur in small veins in leaves and are lacking in most gymnosperms and in the lower vascular plants. In these forms, tracheids occur in elongating regions, and they may have annular and spiral types of secondary walls.

Xylem parenchyma cells store water and food, and they may also conduct materials for short distances. Xylem fibers are similar to phloem fibers.

SUMMARY OF PRIMARY TISSUES

1. Primary growth lengthens shoots and roots, and establishes the basic pattern of cells and primary tissues characteristic of the plant body.

2. Primary meristematic tissues are apical meristem, protoderm, ground meristem, and procambium.

3. The primary plant body is composed of (a) epidermis, which may be differentiated into three cell types: epidermal, guard cells, and epidermal hairs; (b) cortex, which is composed of collenchyma, sclerenchyma (fibers and sclereids), and parenchyma; (c) vascular tissue, composed of xylem (fibers, tracheids, vessel members, and parenchyma) and phloem (fibers, sieve-tube members, companion cells, and parenchyma); (d) pith, composed largely of parenchyma and sometimes accompanied by sclereids; and (e) pith rays, composed of parenchyma cells.

4. Parenchyma stores water and food, and conducts materials for short distances. Collenchyma, sclereids, and fibers are strengthening or mechanical tissue elements. Tracheids and vessels (series of vessel members) conduct water and mineral salts. Sieve tubes (series of sieve-tube members) conduct foods.

5. Collenchyma and sclerenchyma may occur in patches or completely surround the stem just underneath the epidermis. Sclerenchyma is frequently associated with vascular bundles.

STEMS

Stems provide *mechanical support* for leaves in erect plants and are an axis for attached leaves in horizontal plants. Flowers and fruits are also produced in positions on stems that allow for pollination and seed dispersal.

Stems provide a pathway for the *conduction* of water and mineral nutrients from roots to leaves and for transfer of foods, hormones, and other metabolites from one part of the plant to another.

The usual life span of plant cells is from 1 to 3 years. Water and mineral salts in dilute solution move in dead cells, but this movement depends upon the activity of living cells in leaves and roots that are generally less than 3 years old. Stems in herbaceous perennials **(Fig. 4.13, Color Plate 4)** and in 2000-year-old redwoods annually *provide new living tissue* for normal metabolism of the plant. Other stems are modified for the *storage* of plant products.

Stems have four major functions: (1) support; (2) conduction; (3) production of new living tissue; and (4) storage.

THE DICOTYLEDONOUS STEM

Primary growth

The tip of the stem consists of small immature leaves enclosing a dome-shaped apical meristem. The apical meristem is composed of dividing cells, arranged in various ways, that give rise to the leaves, buds, and the primary meristematic tissues (Fig. 4.14, and **Fig. 4.3, Color Plate 3**). The apical meristem is said to be indeterminate; that is, if conditions were ideal, the apex could grow continuously. We know, however, that this doesn't happen in nature either because environmental factors are limiting or because the onset of flowering usually causes vegetative growth to stop. In addition, each plant species is genetically programed to develop within a certain size/age range.

Beneath the apical meristem are the three primary meristematic tissues—protoderm, ground meristem, and procambium **(Fig. 4.3, Color Plate 3)**. The ground meristem starts to differentiate first. These tissues form a cylinder near the outside of the stem and a core in the inside (Fig. 4.2D), which will continue to differentiate into the cortex and pith, respectively.

Between these two tissues, near the apex, is a ring of residual meristem cells that retain the cellular characteristics of the apical meristem (Fig. 4.2C). This ring will become a cylinder of discrete procambium strands that later differentiate into primary xylem and phloem (vascular bun-

Figure 4.14 Stereoscan view of a shoot tip of *Adonis aestivales.* (Courtesy of Dr. J. Lin.)

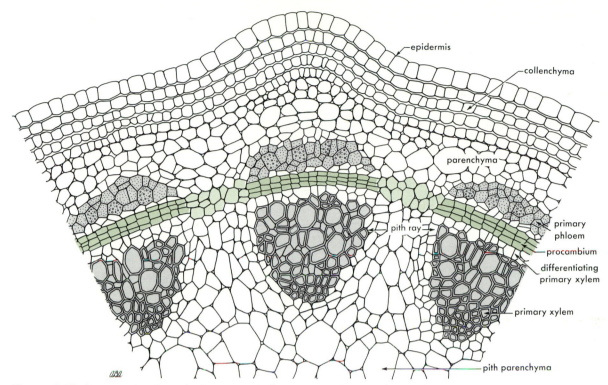

Figure 4.15 A stem at the close of primary growth. Some procambium is still present, and the last primary xylem and primary phloem cells are still undergoing differentiation. Ray parenchyma separates vascular bundles. Cortex consists of collenchyma and parenchyma. Epidermis is intact.

dles). Each bundle is separated from others by regions of parenchyma cells which are sometimes called pith rays (Figs. 4.2D, 4.2F, and 4.15). The formation of vascular bundles from the residual meristem is apparently in response to leaf development. The mechanism is not exactly understood, but it may involve the production of some substance by the leaf primordia (immature leaves on the shoot tip) and its transport downward to the residual meristem ring. When the substance reaches a certain critical concentration, it apparently induces the residual meristem cells to form bundles of procambium and then xylem and phloem. Each bundle is called a **leaf trace** and, as it matures, will lead from the stem into a leaf, connecting it to the axis of the stem (Fig. 4.1A). The vascular system of the stem actually consists, for the most part, of interconnected leaf traces (Fig. 4.16). Other traces, however, connect to buds (bud traces), and some traces end at the apical meristem without connecting to either a leaf or a bud.

The role that leaves play in vascular differentiation and in the pattern of vascular bundles has been determined by elegant experiments. In one of these experiments, the leaves and leaf primordia were removed around the apex. As the stem elongated the new leaf primordia that formed were destroyed. Anatomical examination of the stem that differentiated during the weeks of the experiment revealed that the "vascular tissue" remained as an unbroken cylinder. The cells making up the cylinder did not differentiate completely,

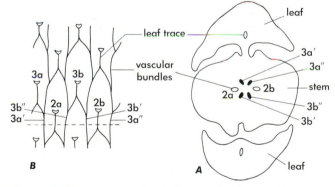

Figure 4.16 A, cross-sectional diagram of a stem and two leaves taken near the shoot apex. The vascular bundles in the stem are labeled so that their position can be mapped in B. B, longitudinal diagram showing the main vascular bundles and interconnected leaf traces. The diagram was made by splitting the stem between bundles 3a′ and 3a″ and laying the stem open. Only the bundles shown in A are labeled. Section A was made at the level marked by the dotted line. (Redrawn from Namboodiri and Beck, 1968, Amer. J. Bot. 55:460, used with permission.)

and vascular bundles did not form. If leaves were later permitted to develop, the newly formed vascular tissue did have bundles and did differentiate into xylem and phloem. This experiment demonstrated that leaves are needed for vascular differentiation and for the formation of procambial strands and vascular bundles.

Secondary Growth

The Source of Secondary Vascular Tissue. During primary growth, stems increase in length. Perennial stems and some annual stems, such as those of tomato *(Lycopersicon)*, sunflower *(Helianthus)*, and alfalfa *(Medicago)*, also increase in diameter. This lateral thickening involves the activation of a secondary meristem, the **vascular cambium** (Fig. 4.17). The vascular cambium is composed of two parts—the **fascicular cambium,** which forms from within the vascular bundles, and the **interfascicular cambium,** which originates from parenchyma cells that lie between vascular bundles (Figs. 4.17*A*, 4.17*D*, and **4.18*A*, Color Plate 5**).

Recall that the first stage of development of vascular tissues in stems was the formation of a cylinder of residual meristem. Next, the newly formed leaves acted on this cylinder to induce portions of it to become procambium strands and then vascular bundles of xylem and phloem. After development of the primary xylem and phloem, in some plants a portion of the procambium remains undifferentiated. At some time during the plant's life cycle, very early in the case of woody plants and later in the case of herbaceous plants that develop secondary growth, this residual procambium reinitiates divisions to become the fascicular cambium (Fig. 4.17*A* and 4.17*B*). Simultaneously, parenchyma cells adjacent to the fascicular cambium, but between vascular bundles, are also stimulated to divide, and they become the interfascicular cambium. Together, both components form the vascular cambium that will make secondary phloem to the outside and secondary xylem to the inside of the stem **(Fig. 4.18*B*, Color Plate 5)**.

The Axial and Ray Systems. Close examination of a woody stem reveals that the cells that make up wood are actually oriented in two ways (Figs. 4.19, 4.20*A*, and 4.20*B*). Some cells are elongated parallel with the axis of the stem, making the **axial system.** The axial system is composed of vessel

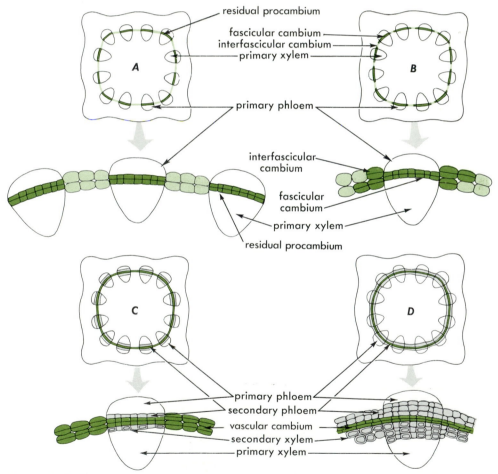

Figure 4.17 Formation of a complete vascular cambium. *A*, at completion of primary growth, some meristematic cells remain between primary xylem and primary phloem. This is the residual or "leftover" procambium shown in dark green. Parenchyma cells appear in pith rays between vascular bundles (light green). *B*, the residual procambium becomes reactivated to form the fascicular cambium; some of the parenchyma cells of the pith ray become meristematic to form the interfascicular cambium (dark green). *C*, the fascicular and interfascicular cambia have joined to form a complete cylinder of vascular cambium (dark green). *D*, cylinders of secondary xylem and secondary phloem (light gray) have been formed by the vascular cambium. Parenchyma cells are shown in light green, vascular cambium in dark green, and secondary tissues in light gray.

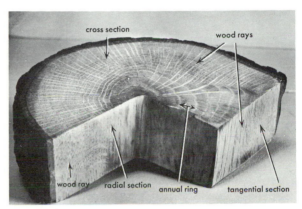

bark — wood
vascular cambium

wood ray

phloem ray

fusiform initial
ray initials } vascular cambium

Figure 4.19 Diagram showing the relationship of the cambium to the stem. The cells of the axial system are elongated vertically; those of the ray system are elongated horizontally relative to the stem axis.

members, tracheids, fibers, and parenchyma. In the phloem, sieve-tube members, fibers, companion cells, and parenchyma make up the axial system. The source of these cells in the vascular cambium are **fusiform initials.** A second system, the **ray system,** is composed of cells oriented at right angles to the axis of the stem (Figs. 4.19, 4.20A, and 4.20B). The **rays** that make up this system develop from **ray**

initials in the vascular cambium. Rays are living channels through which nutrients and water move laterally in stems with secondary growth. Xylem rays are made up of ray tracheids and ray parenchyma, and phloem rays are made of phloem parenchyma. Ray cells may remain alive for several years, perhaps 10 or more.

Anyone who has examined the stump of a cut tree has observed **annual rings** (Figs. 4.21 and **4.22C, Color Plate 6**). One annual ring represents the amount of secondary xylem growth for one season. Thick rings mean maximum growth has occurred during a season, and thin rings mean minimal growth. The density (width) of these rings indicates past climate conditions.

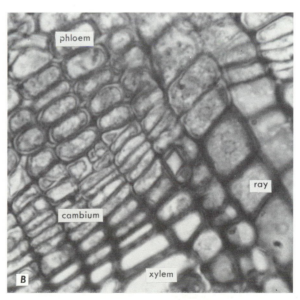

cross section — wood rays

wood ray — radial section — annual ring — tangential section

Figure 4.21 Portion of stem of oak (*Quercus*) showing cross, radial, and tangential sections and their gross characteristics, ×½.

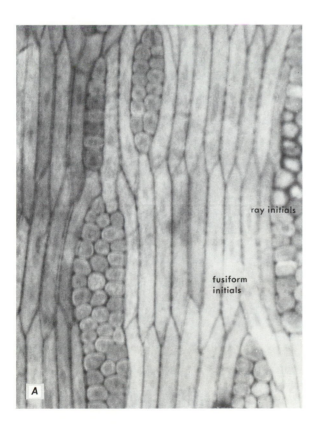

ray initials

fusiform initials

Figure 4.20 Vascular cambium. *A*, tangential section showing both fusiform initials and ray initials. The fusiform initials produce cells of the axial system in both secondary xylem and phloem. *B*, cross section of the cambium in locust (Robinia); ×300.

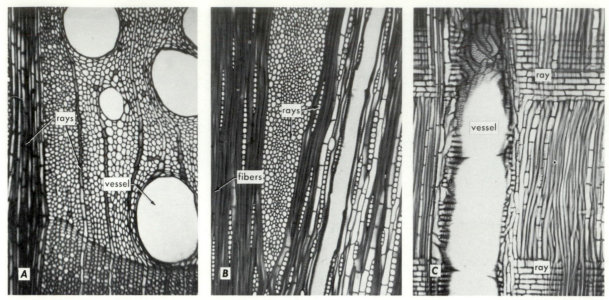

Figure 4.24 Sections of wood of oak (*Quercus borealis*). *A*, cross section; *B*, tangential section; *C*, radial section, ×40.

Microscopic examination of an individual growth ring shows that during the early part of the growing season, cells are large and have relatively thin walls. This part of the growth ring is called the **early wood** (spring wood). Later in the season the cells become smaller in diameter and have thicker walls; this is the **late wood** (summer wood) **(Fig. 4.22B, Color Plate 6).** In certain trees, large vessel members form only in the early wood and only small vessel members occur in the late wood. This organization is called **ring porous (Fig. 4.22B).** In other trees the occurrence of vessel members is uniform throughout the growing season; this is **diffuse porous** wood **(Fig. 4.23, Color Plate 6).** Many trees in tropical areas, where growing conditions are uniform throughout the year, show no annual rings in their wood.

A woody stem can be sectioned into three planes of view (Fig. 4.21): radial, tangential, and cross. Figures 4.24 and 4.27 show these views microscopically.

After a tree or branch is several years old, the inner part of the stem usually becomes inactive and filled with resins. This dense, central core is called the **heartwood.** Good barrels and wooden tubs are made from heartwood because the resin makes this wood impermeable to water. Outside the heartwood is a light-colored, less dense region of active wood called **sapwood** (Fig. 4.25).

Two things contribute to the formation of heartwood. One is the curious formation of structures called **tyloses** (Fig. 4.26). Tyloses are ingrowths of the primary wall of parenchyma cells that grow through the pits of vessel members. The walls of tyloses eventually become enlarged and may form secondary walls that completely plug up the xylem. Species of trees that have abundant parenchyma cells adjacent to vessel members readily form tyloses; however, other species with more scattered parenchyma do not. The second contributing factor to forming heartwood is the activity of rays. Rays apparently play an important role in trans-

Figure 4.25 Cross section of a branch of mulberry (*Morus*) showing heartwood and sapwood, ×½.

porting resins from the active sapwood into the central heartwood. Heartwood is a repository for metabolic by-products, perhaps serving a function parallel to the excretory system of animals.

Gymnosperm wood is anatomically simpler than angiosperm wood (Fig. 4.27). It is composed almost entirely of tracheids in the spring wood and of fiber-tracheids (cells that are intermediate between fibers and tracheids) in the summer wood. The rays in gymnosperms are usually only one cell-layer thick, and axial parenchyma is rare. Vessels are absent, but there are many **resin ducts.** These secretory ducts are long hollow tubes containing an internal layer of cells that produce resin (Fig. 4.28).

Formation of Cork (Periderm). After the vascular cambium has begun to form secondary xylem and phloem, **cork cambia** often develop within the cortex or phloem. A cork cambium produces tissues that can protect the plant from injury.

xylem vessel member   fiber

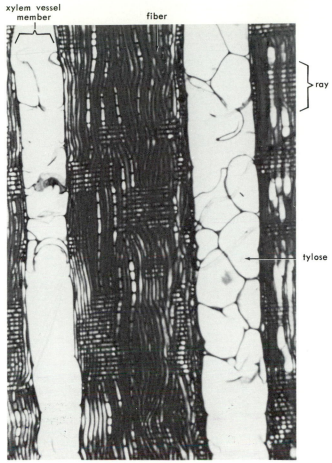

ray

tylose

Figure 4.26 Radial section of wood from white oak (*Quercus alba*) that shows tyloses. Tyloses are ingrowths of parenchyma cells that form secondary cell walls and may plug up xylem vessel members, ×35.

Cork Cambium. Figure 4.29 shows the origin of cork cambium and the development of secondary tissues from it. Cork cambium in most plants arises from outer cortical cells. The outer derivative cells from the cork cambium generally differentiate into cork cells, thus forming a layer of **cork** beneath the epidermis. The inner cells are known as **phelloderm,** and they are parenchymalike cells.

Cork tissue (Figs. 4.29 and 4.30) is composed of flattened, thin-walled cells with no, or small, intercellular spaces. A waxy substance called **suberin** is deposited in the walls, rendering the cells almost impermeable to water and gases. Hence, this tissue provides protection for the stem against excessive loss of water and also against mechanical injury. The protoplasts of cork cells are short-lived.

Bark. In a young stem, the bark is made up of the following tissues, in order, from the *outside* to the *inside*: cork, cork cambium, phelloderm (if present), cortex, and phloem (Figs. 4.29 and 4.30). Microscopic examination may reveal the presence of epidermal cells still clinging to the cork. In old stems, the epidermis, cortex, and primary phloem become separated from the adjacent inner tissues by successively deeper layers of cork formation. As this happens,

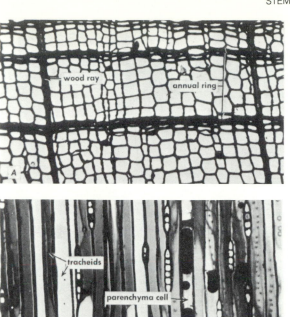

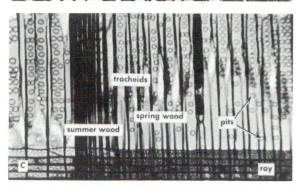

Figure 4.27 Sections of redwood stem (*Sequoia sempervirens*). *A,* cross section; *B,* tangential section; *C,* radial section, ×100.

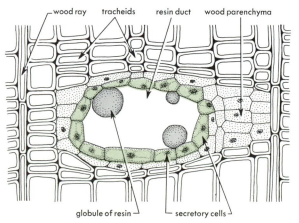

wood ray   tracheids   resin duct   wood parenchyma

globule of resin   secretory cells

Figure 4.28 Resin duct in pine (*Pinus*) wood as seen in cross section.

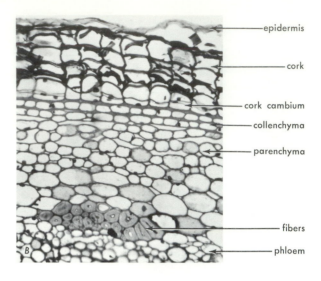

cuticle
epidermis
first cork cell
cork cambium
phelloderm

cortex

cuticle
epidermis

cork

cork cambium
phelloderm cell

cortex

A

epidermis

cork

cork cambium
collenchyma

parenchyma

fibers

phloem

B

Figure 4.29 *A,* cross-sectional diagrams showing origin of the first cork cambium and the first layers of cork. Above, early stage; below, later stage. New cork cambia and new layers of cork form in a similar manner each spring. *B,* light micrograph showing cork cambium and newly formed cork in a cross section of elderberry (*Sambucus*) stem; ×200.

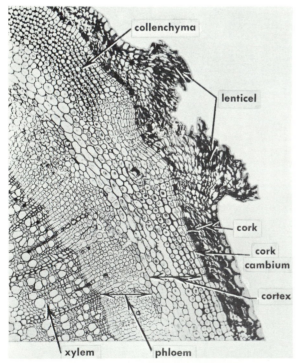

collenchyma

lenticel

cork

cork
cambium

cortex

xylem   phloem

Figure 4.30 Cross section of elderberry (*Sambucus*) stem showing secondary tissues. From the outside to the inside the following tissues occur: cork, cork cambium, cortex, secondary phloem, and xylem. The vascular cambium is at the boundary between the phloem and xylem. Note the lenticel in the cork. The bark is composed of everything outside the vascular cambium, ×100.

tissues outside newly formed cork die for lack of water and nutrients. They dry up and eventually wither away.

Lenticels. An impervious layer of cork would cut off the oxygen supply of the living tissues beneath it if groups of parenchyma cells called **lenticels** did not develop in various places (Fig. 4.30). They frequently originate beneath the epidermal stomata of young stems. When a cork cambium is formed, it produces ordinary parenchyma cells below these stomata in an outward direction. The resulting loose aggregation of parenchyma tissue bursts through the epidermis to form the lenticel. The air spaces between the parenchyma cells permit gas interchange.

STEM MORPHOLOGY

Arrangement of Leaves and Buds

A 3-year-old twig of walnut in winter condition is shown in Fig. 4.31. The tip of the twig generally bears a large **terminal leaf bud.** At regular intervals along the stem, other buds may be seen; they are called **lateral buds.** Note that below the base of each lateral bud there is a scar that was made when a leaf fell from the twig; this is a **leaf scar** (Fig. 4.33). Vascular bundle scars may be seen within each leaf scar; strands of food-conducting and water-conducting tissues passing from the stem into the leaf were broken when the leaf fell, leaving these scars. Buds and leaves are usually borne in this relationship to each other; buds form in the angle made by the stem and the leaf stalk. This angle is termed the **leaf axil,** and consequently, these buds may also be called axillary buds.

Protecting the young immature leaves and cells within the bud is a series of overlapping scales—**bud scales.** They are usually shed when the bud develops into a new shoot, and they also leave scars—**bud-scale scars.** The part of a stem or twig between sets of terminal-bud-scale scars is generally formed during one growing season. For instance, growth made by the twig this year is set off from growth made last year by means of a ring or girdle or terminal-bud-scale scars (Fig. 4.31). When scales of a terminal bud fall off in spring, they leave a number of closely crowded scars that form a distinct ring. Examination of twigs several years old shows that growth in length may vary from year to year, as revealed by the different spacings between the terminal-bud-scale scars. There is no increase or decrease in the length of any portion of a stem after that portion is one year old.

Position of Buds on a Woody Twig

In the walnut twig there is just one leaf bud and one leaf at each node. This arrangement of buds and leaves on the stem is spoken of as **alternate** (Figs. 4.32A and 4.33A). It is the most common type of bud and leaf arrangement. Ash, maple, lilac, and many other plants have two leaves opposite each other at each node and a bud in the axil of each leaf (Figs. 4.32B and 4.33F). This arrangement of leaves and buds is spoken of as **opposite.** When three or more leaves and buds occur at each node, as in *Catalpa,* leaf arrangement and bud arrangement are said to be **whorled** (Fig. 4.32C).

Some plants have several buds in or near the leaf axil. For example, apricot (*Prunus* sp.) often has a group of three buds in the leaf axil: a central bud, which develops into a side branch, and two lateral ones, which are called **accessory buds.** Walnut (*Juglans* sp.) may also have more than one bud in the leaf axil (Fig. 4.32A).

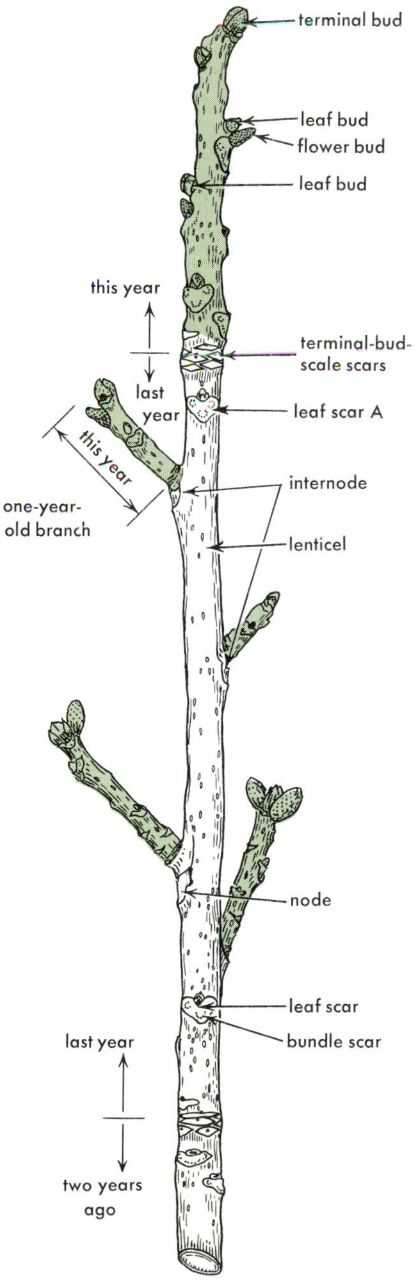

Figure 4.31 Three-year-old twig of walnut (*Juglans regia*), ×1.

Labels on Figure 4.31: terminal bud; leaf bud; flower bud; leaf bud; this year; terminal-bud-scale scars; last year; this year; leaf scar A; internode; one-year-old branch; lenticel; node; leaf scar; bundle scar; last year; two years ago

Figure 4.32 Twigs showing three types of bud and leaf arrangement. The position of leaves is shown by the leaf bases and scars. A, alternate, walnut (*Juglans regia*); B, opposite, lilac (*Syringa vulgaris*); C, whorled, *Catalpa,* ×½.

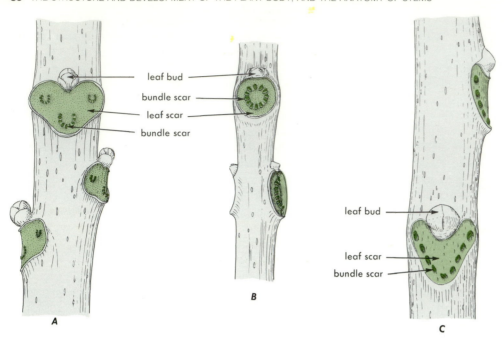

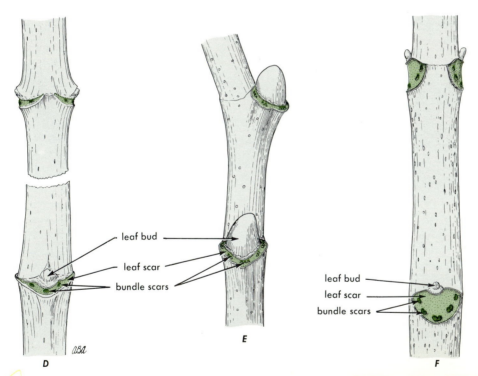

Figure 4.33 Leaf scars and bud arrangement of different species of woody plants. *A*, walnut (*Juglans regia*), ×2; *B*, catalpa (*Catalpa bignonioides*), ×2½; *C*, tree of heaven (*Ailanthus altissima*), ×2; *D*, box elder (*Acer negundo*), ×2; *E*, European plane (*Platanus acerifolia*), ×3; *F*, buckeye (*Aesculus californica*), ×2.

Buds may arise at places other than leaf axils: they may appear on stems, roots, or even leaves, and give rise to new shoots. Such buds are called **adventitious buds.** Their formation may be stimulated by injury, such as occurs in pruning.

Dormant, or **latent, buds** arise in a regular fashion in the leaf axil, but their development is usually inhibited by the dominance of the terminal bud. This mechanism, **apical dominance,** is fully discussed in Chapter 9.

From the above discussion it is seen that buds can be classified by their arrangement on the stem, which may be (a) alternate, (b) opposite, or (c) whorled; by their position on the stem, which may be (a) terminal, (b) lateral (axillary), (c) accessory, or (d) adventitious; and by the nature of the organs into which they develop, which may be (a) leaf, (b) flower, or (c) mixed.

As a rule, the terminal bud of a stem is the most active and grows more vigorously than any of the axillary buds. Usually, the lowest lateral buds on a year's growth of the shoot remain dormant and do not develop into branches. If the terminal bud is removed, however, as may be done in pruning, lateral buds, otherwise dormant, may become active.

THE MONOCOTYLEDONOUS STEM

Primary Stem Anatomy

Monocot stems differ in several ways from the picture we have formed of the dicot stem. One prominent difference is that the vascular bundles often have a characteristic anatomy and are usually scattered more or less randomly across the monocot stem, rather than being arranged in a ring **(Fig. 4.34A and 4.34C, Color Plate 7).** With this arrangement, the terms *pith* and *cortex* are often not applicable to monocot stems. Instead, the more general term **ground tissue** is applied to the region between the bundles. However, there are exceptions: some monocot stems have a hollow core, so that the bundles are necessarily confined to a ring **(Fig. 4.34B).** At the nodes, the monocot stem has a complex network of branching bundles called a **nodal plate (Fig. 4.34D).** This happens because many vascular traces enter each monocot leaf. By contrast, in dicots only a few trace bundles enter a leaf.

Some monocots achieve a large stem diameter by means of a unique form of primary growth: they form a **primary thickening meristem** (Fig. 4.35). The shoot apex is a small dome in a broad depression; young leaves are produced from the apical meristem, grow out of the depression, and arch over the apex. Beneath this leaf-forming zone is the primary thickening meristem, which is shaped like an inverted saucer. It may extend down the stem a short way, so that it forms new cells upward to allow for an increase in stem length and outward to allow for an increase in girth. Leaf traces pass through the primary thickening meristem and connect with already existing vascular bundles. In this way, monocots such as palms can achieve a diameter of many centimeters while the stem is still very short.

Secondary Growth

Most monocots lack a vascular cambium and cannot form secondary xylem or phloem. Such monocots are unable to support large and highly branched vertical shoot systems. Rhizomes are especially common among the herbaceous monocots (Fig. 4.38A), and they provide a way for the plant body to become large without encountering problems of conduction and mechanical strength. Small vertical shoots

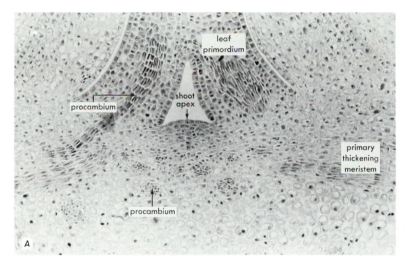

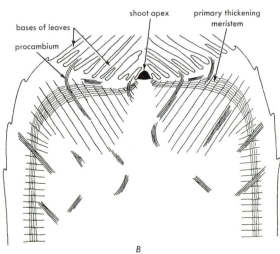

Figure 4.35 *A*, longitudinal section of *Iris* shoot apex to show the primary thickening meristem, ×88. *B*, longitudinal diagram of the shoot apex of corn (*Zea mays*).

Figure 4.36 Arborescent (treelike) mono-cotyledonous plants. *A*, California fan palm (*Washingtonia filifera*); palms are un-branched and lack true secondary growth. *B*, *Pandanus* sp. is branched and lacking in true secondary growth. *C*, the Joshua tree (*Yucca brevifolia*) has a branched stem and true secondary growth.

and adventitious roots can form at any node, so that the few vascular bundles of the primary stem (rhizome) serve and support only a small part of the plant body. By this means, a plant without secondary growth can attain enormous size.

Treelike (arborescent) monocots do occur, especially in deserts and swampy tropical areas. They are not found in cold climates, perhaps because they cannot form the dormant terminal buds with protective bud scales that are found in dicots of cold areas. The monocots of temperate climates are usually herbs in which bulbs or corms or rhizomes are sheltered underground through the winter.

There are three types of arborescent monocot plants: (1) palms (Fig. 4.36A), which are unbranched, do not have secondary growth, and lack a cambium; (2) pandans (Fig. 4.36B), which have branched stems but also lack a true cambium; and (3) tree lilies, *Yucca,* and others (Fig. 4.36C), which have branches and form a true cambium.

Palms rely throughout life on the primary vascular bundles that are formed at the shoot apex. Since the number of bundles is fixed, the stem cannot add branches, and old leaves must be shed as new ones are added. This is why palms have bare flanks and a tuft of leaves at the top. The trunk has a nearly uniform diameter because there is no cambial activity. Some thickening may occur through a continued growth and division of parenchyma cells, a process called **diffuse secondary growth.** However, palms lack the strikingly tapered form of dicot trees. (Some palms become enlarged at the base because they form numerous adventitious roots.)

Pandans (*Pandanus* sp.) also show diffuse secondary growth, but the process is complicated by the formation of branches near the apical meristem.

Other monocot plants such as *Agave, Cordyline, Sansevieria,* and *Dracaena* have true secondary growth that involves a vascular cambium, or **secondary thickening meristem.** The cambium arises in the periphery of the stem, in connection with the primary thickening meristem, and extends to the base of the plant. It is not initiated from fascicular and interfascicular cambium as in the dicots. The cambium forms only parenchyma cells to the outside **(secondary cortex).** To the inside, it forms distinct vascular bundles that are embedded in parenchyma (Fig. 4.37). The secondary vascular bundles consist of a ring of xylem surrounded by phloem, whereas the primary vascular bundles consist of xylem and phloem side-by-side.

Some monocot roots and stems form a barklike layer. This tissue, called "storied cork," is derived from cortical parenchyma cells that divide to form files of suberized parenchyma cells.

Only one monocot—*Dracaena Draco,* the dragon's blood tree—is known to have secondary growth in the roots.

STEM MODIFICATIONS

Stems may become adapted for functions other than support, conduction, and production of new growth. They may, for instance, become attachment organs for vines; they may carry on photosynthesis, store food or water, and develop protecting devices.

Rhizomes

A rhizome is a horizontal, underground stem. In most *Iris* species, leaves and flowering stalks are produced at the growing rhizome tip. The leaves die a relatively short distance back from the growing tip. Roots are also formed at nodes, and they may remain for the life of the rhizome. Other rhizomes, as in *Canna* (Fig. 4.38A), produce upright leafy stems with terminal flowers at every third node. The intervening nodes are marked by only small sheath leaves.

Corms

A corm is a short, vertical, thickened underground stem (Fig. 4.38B). In *Gladiolus,* it consists of a short stem filled with stored food. Nodes are, as usual, indicated by leaves; some bases are shown in Fig. 4.38B. Small buds occur in axils of some of these leaves. In a median section of a corm (Fig. 4.38C) one can distinguish between stored food and the central portion containing a single bud that will produce a single leafy, flowering shoot. Food stored in a corm is used in the production of the leafy shoot. New corms will develop from axillary buds. In addition, short underground stems may form, each giving rise, at its tip, to a single small corm.

Bulbs

A bulb differs from a corm in that food is stored in leafy scales. The stem portion is small and has at least one central terminal bud that will produce a single upright leafy stem. In addition, there is at least one axillary bud that will produce a bulb for the subsequent year. In the longitudinal section of the sprouting daffodil bulb shown in Fig. 4.38D, the stem is producing three leafy stalks, one of which is forming a new bulb.

Food stored in the leafy scales of a bulb is used up by the initial growth of a leafy shoot. Food to be stored for a new bulb is supplied from a leafy shoot. The table onion is a good example of a commercially valuable bulb.

Tubers

Tubers are enlarged terminal portions of slender rhizomes (Figs. 4.38E and 4.39). The potato (*Solanum tuberosum*) is a good example. The potato plant possesses three types of

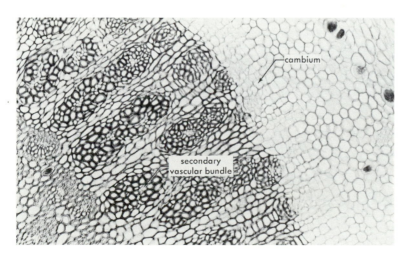

Figure 4.37 Cross section of a portion of monocotyledonous stem, *Dracaena* sp., that has true secondary growth. Note that the cambium forms discrete secondary vascular bundles and parenchyma cells to the inside of the stem, and parenchyma cells only toward the outside, ×13.

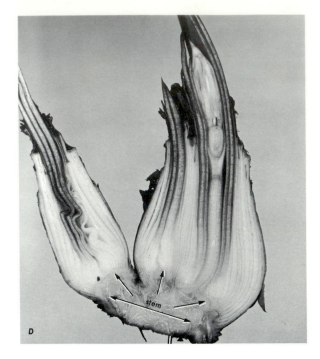

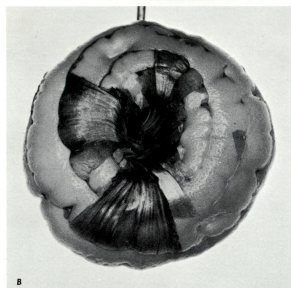

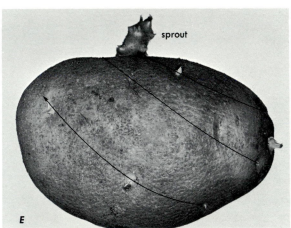

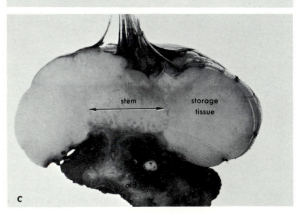

Figure 4.38 Plants illustrating continued development with only primary growth. *A*, roots and shoots are produced at every third node by the rhizome of *Canna*, ×½; *B* corm of *Gladiolus*, ×1; *C*, corm sectioned to show short stem with storage tissue of current corm and disintegration of corm of preceding year, ×1; *D*, longitudinal section of young daffodil (*Narcissus*) plant, showing two bulbs, one with two shoots, united by the short stem, ×1; *E*, potato tuber (*Solanum*), note spiral arrangement of the eyes and the short sprout, ×1.

stems: (1) ordinary aerial stems, (2) slender underground rhizomes, and (3) tubers, the enlarged tips of rhizomes. In the mature potato, the scar left where the tuber was broken from the rhizome is clearly visible. The potato tuber has nodes and internodes, lateral buds, and a terminal bud. Buds develop into stems (Fig. 4.38*E*). The "eyes" of the tuber are groups of buds; each group along the sides represents a lateral branch with undeveloped internodes. At the unattached "seed end" of the tuber, the "eye" is in reality a terminal branch on which only one bud is strictly terminal.

Stem Tendrils

Tendrils are slender, coiling structures that are sensitive to contact stimuli and attach the plant to a support. Tendrils are of two morphological types: leaf and stem. In the trumpet flower (*Bignonia*), for example, several uppermost pairs of

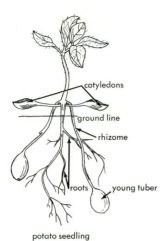

Figure 4.39 A potato seedling showing development of young tubers at the end of slender rhizomes.

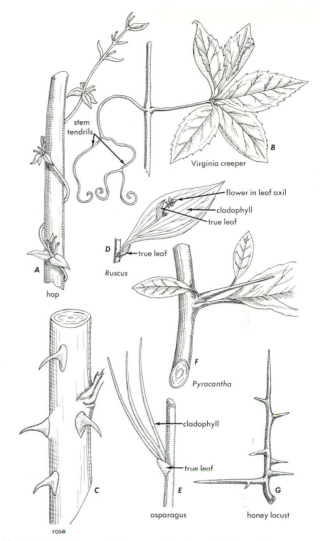

Figure 4.40 Types of stem modification. *A*, twining vine; *B*, stem tendril; *C*, rose stem prickles; *D* and *E* leaf like stems called cladodes (*cladophylls*); *F* and *G* thorns.

leaflets have no blades but instead form very slender leaf tendrils. In grapes *(Vitis)* and Virginia creepers *(Parthenocissus quinquefolia)*, tendrils are modified stems (Fig. 4.40*B*), as evidenced by their presence at nodes in leaf axils.

In the Virginia creeper, each tendril ends in a knob that flattens out when it comes in contact with a surface to which it adheres.

Cladodes

These are stems that are leaflike in form, are green, and perform the functions of leaves. They may bear flowers, fruit, and temporary leaves. Examples of plants with cladodes are *Ruscus* (Fig. 4.40*D*), *Asparagus* (Fig. 4.40*E*), *Smilax,* various species of cacti, and some orchids (e.g., *Epidendrum*).

Spines and Thorns

Many plants have thorny outgrowths, which may derive from either stems or leaves. When derived from leaves, as in barberry *(Berberis)* and black locust *(Robinia pseudoacacia)*, they are usually called **spines.** By contrast, true thorns are modified stems. Examples are found in firethorn *(Pyracantha)* (Fig. 4.40*F*) and honey locust *(Gleditsia)* (Fig. 4.40*G*). Thorns, like ordinary branches, are borne in the axils of leaves. Some thorns have leaves, which is further evidence that they are stems. Prickles, such as those on rose stems, are merely epidermal outgrowths somewhat analogous to hairs (Fig. 4.40*C*).

Stolons

Bermuda grass *(Cynodon dactylon)* has aboveground horizontal stems called stolons. These stems creep along the ground, and at each node, shoots and roots arise. In strawberry *(Fragaria)* (Fig. 4.41), stolons, roots, and leaves arise at every other node.

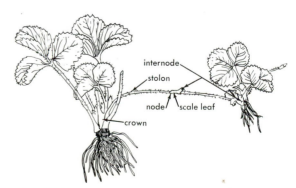

Figure 4.41 Runner of strawberry *(Fragaria)*, $\times \frac{1}{8}$; roots and shoots are produced at every other node.

SUMMARY OF STEMS

1. Buds are characteristic of woody stems. Bud scales enclose and protect rudimentary leaves surrounding an apex. The apex may be either a vegetative shoot or a floral apex. A floral apex terminates growth. Woody plants each year require new vegetative growth, which is normally accompanied by flowers.

2. The predominantly short life span of 1 to 3 years for cells, in general, limits the size and longevity of individual plants having only primary growth. The mechanism of secondary growth overcomes this limitation.

3. If all meristematic cells in a procambium strand become differentiated into primary vascular tissue, the vascular strand is closed to further growth.

4. If an active meristematic region remains between the primary xylem and primary phloem, continued growth is possible. These meristematic cells become the fascicular cambium.

5. An interfascicular cambium forms from ray parenchyma cells between vascular strands. Union of fascicular and interfascicular cambium produces a complete cylinder of vascular cambium. Divisions in cambium are longitudinal, so that the stem now increases only in girth.

6. Cork cambium is short-lived; new cork cambia may arise each year, producing new layers of cork. Cork cambium may originate in successively deeper tissues from epidermis, cortex, and phloem.

7. The production of secondary vascular tissue thus makes possible the attainment of great size and great age by individual trees, even though the functioning cells may be no older than those in a perennial herbaceous plant such as an *Iris*.

8. Most monocot stems have only primary growth that arises from a special primary thickening meristem. This meristem causes increase in both height and girth.

9. Palms supplement their lateral growth by the division of parenchyma cells throughout the stem.

10. Other monocots, such as *Agave* and *Sansevieria*, have true secondary growth, with a type of cambium that produces secondary vascular bundles and parenchyma cells.

11. In herbaceous dicot stems, vascular tissues are arranged in bundles, generally forming a circle. In monocotyledonous plants, bundles are irregularly distributed throughout the stem.

12. Rhizomes, bulbs, corms, stolons, and tubers are all modified stems.

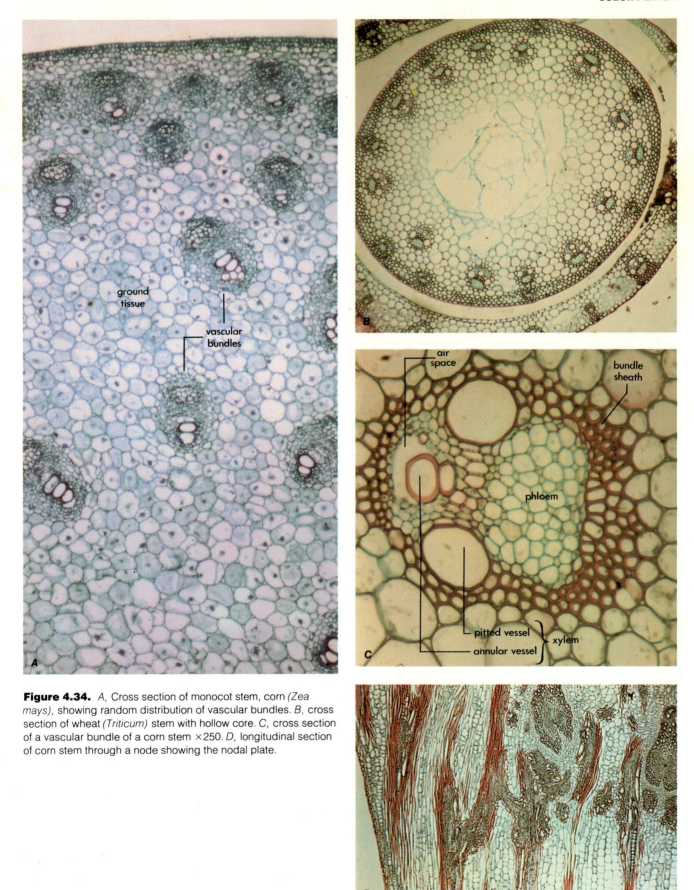

Figure 4.34. *A*, Cross section of monocot stem, corn *(Zea mays)*, showing random distribution of vascular bundles. *B*, cross section of wheat *(Triticum)* stem with hollow core. *C*, cross section of a vascular bundle of a corn stem ×250. *D*, longitudinal section of corn stem through a node showing the nodal plate.

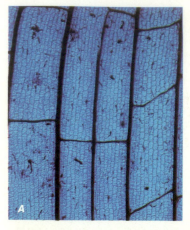

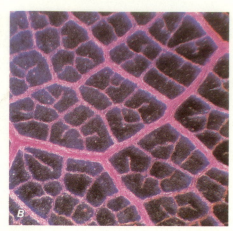

Figure 5.3. Venation patterns in leaves. *A,* parallel venation in the monocotyledonous plant, *Orthoclada* sp. *B,* netted venation, *Acer* sp. *C,* netted venation *(Oxalis, sp.)* to show the vein ending.

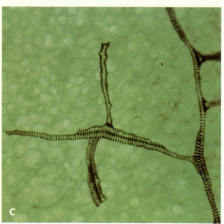

Figure 5.8. Glandular hairs on the stem surface of *Phaseolus* (also see Fig. 10.13*C*).

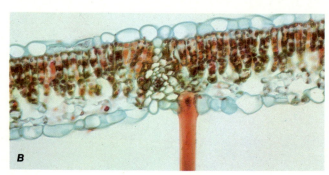

Figure 5.11. Transverse section of a corn leaf showing large bundle sheath parenchyma cells surrounding the small veins.

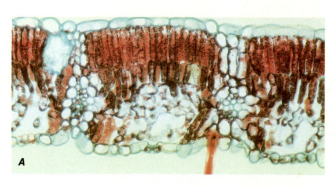

Figure 5.16. *A,* transverse section of sun leaf (*Acer* sp.). *B,* a shade leaf.

Figure 5.19. Insectivorous plant leaves. *A*, Venus fly trap *(Dionaea muscipula). B*, the trap, a modified leaf, will close on an insect that trips the trigger hairs on the leaf surface. Hairs secrete enzymes that digest the insect. *C, D*, sundews. *C, Drosera capillaris. D, D. rotundifolia* growing in a sphagnum bog. Glandular hairs on leaf surfaces secrete a sticky substance attractive to insects. Insects become stuck; secretory enzymes produced by hairs digest insects.

Figure 5.20. Insects fall into the attractive pitchers that contain water and digestive juices. *A, Sarracenia* sp. *B, Sarracenia purpurea. C, D. Darlingtonia californica. D*, spiny hairs inside pitcher on walls inhibit insects from escaping.

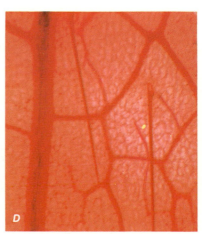

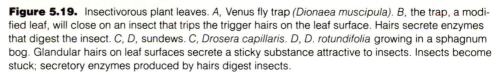

Figure 6.1. Specialized roots. *A*, banyan *(Ficus bengalensis)* tree showing an extensive adventitious root system. (The tree is growing in Honolulu). *B*, extensive adventitious root system of mangrove *(Rhizophora mangle)*. Note the many air roots sticking up from the mud. (The tree is growing in the tidal zone of Australian tropical coast.) *C*, adventitious roots of *Ficus pumila* forming adhesive pad.

5 THE STRUCTURE OF LEAVES

Green plants and a few species of bacteria are the only organisms that use sunlight as an energy source; all other inhabitants of the earth consume what the green plants produce. In seed plants, leaves are the principal organs of production. **Chloroplasts** within leaf cells trap light energy; they convert light into chemical energy by means of **photosynthesis.**

Leaves provide large surfaces for the absorption of light and carbon dioxide, both of which are required for photosynthesis. In general, leaves are thin and therefore no cells lie far from the surface. This leaf form facilitates light absorption, but at the same time, it promotes the loss of water because any opening that will pass carbon dioxide will also allow water to evaporate. Thus leaves are the primary source of water loss from the plant. Evaporation helps to cool the leaf, just as perspiration cools the human body. However, excessive transpiration places the plant in serious danger of dehydration. To a great extent, leaf form and anatomy are a compromise between capturing light and carbon dioxide, and conserving water. Water-saving features become particularly marked in plants that occupy hot, dry regions, such as rocky cliffs and deserts where water loss is a great hazard. The following pages will show how the form and anatomy of leaves are related to these functional problems.

LEAF COMPONENTS

The leaves of dicots are generally different from those of monocots. A typical foliage leaf of a dicot is composed of two parts: **blade** and **petiole** (Fig. 5.1A). The petiole is usually slender, while the blade is usually thin and flat. **Veins** composed of xylem and phloem traverse the petiole and form a network throughout the blade. In addition to supporting the softer tissues of the blade, the veins carry water, mineral salts, and food to and from the leaf.

THE PETIOLE

The petiole improves the photosynthetic efficiency of the leaf in several ways. It extends the blade away from the stem, reducing the extent to which leaves shade one another. It also allows the blade to move in response to air currents. This is important because a layer of stagnant air—the **boundary layer**—forms at the leaf surface. During photosynthesis the boundary layer becomes depleted of carbon dioxide. Leaf movements help to bring in fresh air with higher amounts of carbon dioxide and may also cool the leaf.

Petioles vary widely among plants. The petiole may be long or short, cylindrical or flat. It is usually attached to the base of the leaf blade, but in plants such as castor bean (*Ricinus*), it attaches to the middle of the blade on the underside (Fig. 5.1B); such a leaf is said to be **peltate.** Other leaves lack petioles, and the blade directly joins the stem (Fig. 5.15B); in this case the leaf is said to be **sessile.**

Some leaves have accessory structures called **stipules** at the base of the petiole (Fig. 5.1D and 5.1F). The stipules sometimes have a bladelike form and are photosynthetic. This is true of the garden pea (*Pisum*).

Monocotyledonous leaves such as grasses do not have petioles. Instead, the leaf is divided into two parts: **sheath** and **blade.** The blade is the typical thin, expanded portion. The sheath is green, perhaps nearly as large as the blade, and it completely sheaths the stem (Fig. 5.2B). In many species, such as corn, the sheath extends over at least one complete internode. If the union between the blade and the sheath is examined carefully, a small flap of delicate tissue extending upward from the sheath may be seen, closely enveloping the stem; this is called the **ligule** (Fig. 5.2C). It may, in some cases, serve to keep water and dirt from sifting down between stem and sheath. In many species, of which barley (*Hordeum*) (Fig. 5.2C) is a good example, the base of the blade, at its union with the sheath, is carried around the stem in two earlike points, the **auricles.** Ligule and auricles may both be present.

THE BLADE

In most leaves, the blade performs most of the photosynthesis. Its broad, flat surface provides maximal surface area for capturing light and carbon dioxide, and its internal cells are close to the surface so that carbon dioxide reaches the photosynthetic tissues after traveling only a short distance.

Leaf blades vary widely in form. These variations can be useful in classifying plants, and technical terms have been developed to describe the shape of the leaf apex, base, and margin. For instance, the leaf margin may be *entire* (smooth) (Fig. 5.1F), *dentate* (toothed) (Fig. 5.1A), or *lobed* (deeply indented) (Fig. 5.1C). Some leaves are so deeply lobed that the blade forms several distinct units, or **leaflets** (Fig. 5.1D, 5.1E, and 5.1F). Such a leaf is said to be **compound,** whereas a leaf that has a single blade is **simple** (Fig. 5.1A, 5.1B, and 5.1C). The leaflets of a com-

Figure 5.1 Different kinds of leaves. *A,* poplar (*Populus deltoides*); *B,* castor bean (*Ricinus communis*); *C,* oak (*Quercus lobata*); *D,* rose (*Rose odorata*); *E,* Virginia creeper (*Parthenocissus quinquefolia*); *F,* faba bean (*Vicia faba*), × ½.

pound leaf may themselves be subdivided into still smaller leaflets, in which case the leaf is **twice compound.** If these leaflets, in turn, are subdivided, the leaf is **thrice compound.**

How does a plant benefit by having its leaf blades divided into leaflets rather than forming a single large blade? The spaces between leaflets allow better air flow over the leaf surface. This may help to cool the leaf and reduce evaporation while improving carbon dioxide uptake. However, there must be compensating disadvantages, for plants exist with both compound and simple leaves. Very likely, the balance of advantages and disadvantages depends on the environment.

Leaflets resemble small simple leaves. How, then, do we distinguish between a compound leaf and a branch with several simple leaves? Look for axillary buds: buds occur in the axils of leaves but not in the axils of leaflets.

Leaves may be **pinnately** or **palmately compound.** In the pinnate arrangement, the leaflets occur along an extension of the petiole called a **rachis,** as in the pinnae of a feather (hence the term *pinnate*). Examples are found in the garden pea and the rose (Fig. 5.1*D*). If all the leaflets arise from a common point, the leaf is palmately compound; Vir-

ginia creeper (Fig. 5.1*E*) is an example of the palmate arrangement.

VENATION

Leaves also differ in the arrangement of veins, or **venation.** Monocots and dicots usually exhibit quite different venation patterns. Most monocot leaves have **parallel venation** (Figs. 5.2*A* and **5.3*A*, Color Plate 8**), in which many veins of approximately equal size are arranged in parallel along the leaf. The large veins may branch from one or a few main veins, or they may already be distinct when they enter the leaf directly from the stem. The large veins in turn may give off many fine branches that connect to larger veins **(Fig. 5.3*A*).** By contrast, dicot leaves usually show **netted venation (Fig. 5.3*B*).** Here one or a few large veins enter the leaf and branch repeatedly to form a conspicuous network in which the veins are of graded sizes. The smallest veins may end freely **(Fig. 5.3*C*).** With these differences, monocots and dicots can often be distinguished on the basis of their leaf venation. However, such identifications can sometimes be wrong, because sometimes monocots have netted venation and dicots have parallel venation.

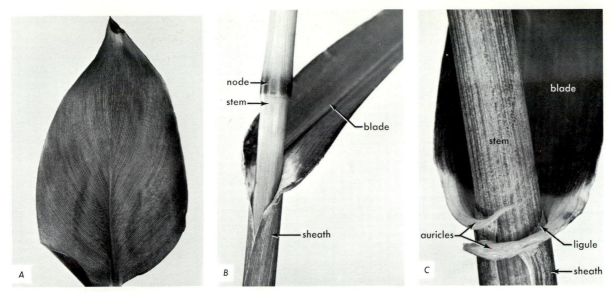

Figure 5.2 Leaves of monocotyledonous plants. *A,* lily (*Canna* sp.) leaf showing parallel venation with secondary veins extending from the midvein toward the leaf apex. *B,* crabgrass (*Digitaria sanguinatis*), showing sheath and blade. *C,* barley (*Hordeum vulgare*), showing ligule and auricles.

ANATOMY OF THE FOLIAGE LEAF

Just as the overall shape of the leaf contributes to photosynthetic efficiency, the internal anatomy also has features that promote photosynthesis. Leaves arise from the shoot apex, and like the primary shoot, the leaf is composed of three primary tissues: the *epidermis* which surrounds and supports the *ground tissue* in which *vascular tissue* is embedded. The arrangement of tissues is shown in Fig. 5.4.

The epidermis usually consists of a single layer of cells that covers the entire leaf surface. It protects the leaf from drying out and from mechanical injury, and it provides a framework for the delicate photosynthetic tissues within the leaf. The ground tissue, or **mesophyll,** is composed of parenchyma cells that contain chloroplasts and perform photosynthesis. The veins have xylem and phloem elements which conduct water, inorganic salts, and foods. Fibers and collenchyma may be associated with conducting elements of the midrib and the larger lateral veins. The veins stiffen the leaf and hold the blade outstretched, serving in much the same way as the ribs of an umbrella.

The petiole has its own specialized structures that enable it to support the leaf blade, conduct food, water, and salts, and to abscise from the stem when the leaf is injured or the growing season comes to an end.

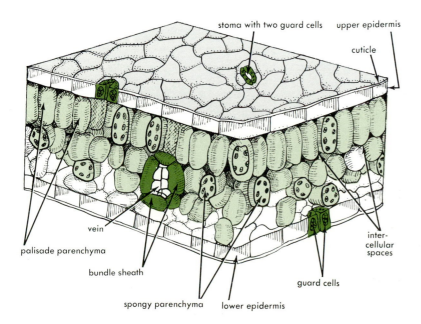

Figure 5.4 Diagram of a generalized leaf showing its three primary tissues: epidermis, mesophyll consisting of palisade and spongy parenchyma cells and vascular bundle (vein).

EPIDERMIS

The epidermis covers the entire leaf surface and is continuous with the surface of the stem to which the leaf is attached. In most leaves the epidermis is a single layer of cells that consists of **ordinary epidermal cells, guard cells,** and **hair cells** (Figs. 5.5 and 5.6). Ordinary epidermal cells show a variety of shapes, depending on the species, and may resemble tiles or jigsaw puzzle pieces when viewed from the leaf surface. They fit tightly together, forming a tough layer that resists rupture. A thickened outer wall adds to the strength of the epidermal layer. The epidermal cells secrete a waxy **cuticle** that coats the leaf and sharply limits the loss of water (and the entry of carbon dioxide) (Fig. 5.6).

Since the cuticle blocks gas exchange, special openings in the epidermis—called **stomata** (singular, **stoma**)—are needed to permit the entry of carbon dioxide. The stomata are scattered over the leaf surface (Fig. 5.6*B*). There may be hundreds or thousands of stomata per square cen-

Table 5.1. Average Number of Stomata per Square Centimeter

Plant	Upper Epidermis	Lower Epidermis
Alfalfa (*Medicago*)	16,900	13,800
Apple (*Malus*)	0	29,400
Bean (*Phaseolus*)	4,000	28,100
Cabbage (*Brassica*)	14,100	22,600
Corn (*Zea*)	5,200	6,800
English Oak (*Quercus*)	0	45,000
Nasturtium (*Tropaeolum*)	0	13,000
Oat (*Avena*)	2,500	2,300
Potato (*Solanum*)	5,100	16,100
Tomato (*Lycopersicon*)	1,200	13,000

timeter of leaf area, as shown in Table 5.1. Usually there are more stomata on the bottom of the leaf than on the top; this probably helps to limit water loss through the stomata because the leaf is cooler on the bottom side.

A stoma is a compound structure composed of two **guard cells** separated by a **pore** (Fig. 5.7). The guard cells open and close the stoma according to environmental conditions. They operate so that the leaf is sealed against water loss except when carbon dioxide is required for photosynthesis or when the leaf needs evaporative cooling. The mechanism of stomatal action is discussed in Chapter 7.

With the exception of a few submerged aquatic plants, stomata are present in all angiosperms and gymnosperms. Functional stomata have been found in liverworts, mosses, horsetails, club mosses, ferns, and cycads. In the angiosperms, they occur on stems, petals, stamens, and pistils as well as on leaves. The major group of green plants that lack stomata is the algae. Stomata are structurally similar in the many different groups in which they are found.

Guard cells occur in pairs; each is crescent-shaped or semicircular in form, as seen in a surface view (Fig. 5.7*C*). Chloroplasts occur in guard cells but are lacking in ordinary epidermal cells. Both kinds of epidermal cells have long-lived protoplasts. The outer wall of epidermal cells, including guard cells, has a cuticle (Fig. 5.7*A*) like that of the epidermis of stems. It is effective in limiting both the inward or outward movements of water vapor and other gases. The cuticle is usually thicker on the upperside of the leaf, where temperatures are higher.

Several different types of **hairs** grow out from the epidermis of leaves and resemble hairs from the epidermis of stems. They may be unicellular or multicellular, simple or branched, scalelike or glandular. The unicellular hair, the simplest kind, may be branched or unbranched. Multicellular hairs may consist of a single row of cells but may also be branched (Fig. 5.5). Glandular hairs bear a single large cell or a group of cells; some of these cells excrete ethereal oils, which often impart a stickiness to leaves and young shoots **(Fig. 5.8, Color Plate 8).**

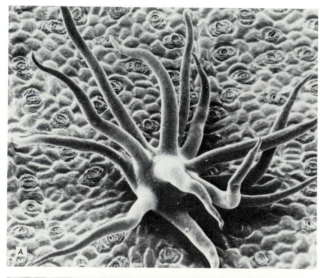

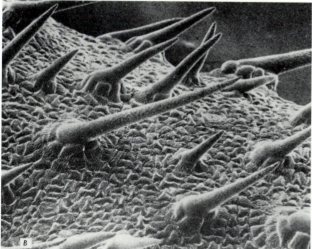

Figure 5.5 Trichomes: *A, Aleurites* leaf—branched; *B, Croton* leaf—simple.

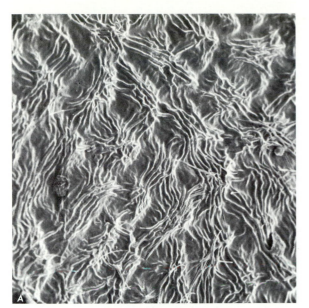

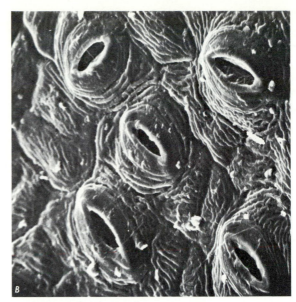

Figure 5.6 Cuticle on the epidermis of mala mujer (*Cnidoscolus* sp.) as seen with the scanning electron microscope. *A,* upper surface, showing cuticular ridges, ×650. *B,* lower surface at higher magnification, showing cuticular ridges between stomata, ×1100.

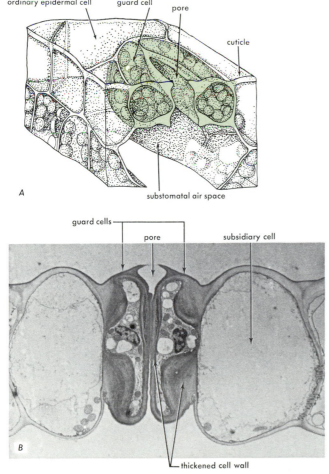

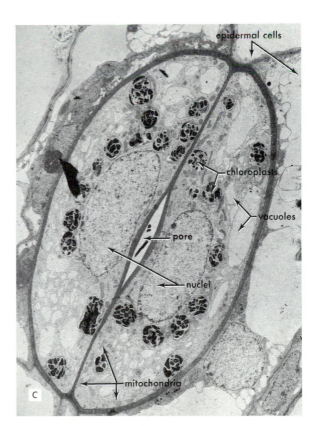

Figure 5.7 Views of stomata. *A,* diagram; *B,* cross section in corn (*Zea mays*), ×1000; *C,* surface view in rice (*Oryza sativa*), ×900.

MESOPHYLL

Mesophyll is the photosynthetic tissue between the upper and lower epidermis. It is parenchyma tissue and is traversed by veins. Chloroplasts are present in the mesophyll cells, which, in some species, are differentiated into two distinct layers: **palisade parenchyma** and **spongy parenchyma** (Figs. 5.4 and 5.9). The palisade parenchyma is just below the upper epidermis and usually consists of from one to several layers of narrow cells with their long axes at right angles to the leaf surface. The spongy parenchyma extends from the palisade parenchyma to the lower epidermis. Cells of the spongy parenchyma are irregular in shape and loosely arranged. Mesophyll cells have thin, moist cell walls that allow a rapid diffusion of carbon dioxide into the cells.

Intercellular spaces are much larger in spongy parenchyma than in palisade parenchyma. Since the air spaces between mesophyll cells are interconnecting, many cells are in contact with an intercellular space. Thus, most food-making cells have free access to carbon dioxide and oxygen. To facilitate gas exchange, large air spaces, called **substomatal chambers,** are generally present beneath each stoma (Fig. 5.7A).

The leaves of some plants, especially those that grow in very dry habitats, have a modified leaf structure. For example, their mesophyll cells are tightly packed together and lack the differentiation between a palisade and spongy layer. Such plants may also have a thick cuticle and sunken stomata. All of these modifications are believed to reduce transpiration.

VASCULAR SYSTEM

The veins, or vascular bundles, form a network that extends throughout the leaf. The conducting elements are xylem and phloem. Veins conduct water, mineral salts, and foods, and also mechanically support the mesophyll tissue. In addition to the midrib and larger lateral veins that are visible to the naked eye, innumerable minute branch veins can be seen with the aid of a microscope. The large veins contain vessels, tracheids, sieve tubes, companion cells, and also some mechanical tissue (Fig. 5.9). Such veins, in some leaves, may have both primary and secondary vascular elements.

In larger veins that have both xylem and phloem elements, the xylem is toward the upper surface of the leaf, and the phloem is toward the lower surface (Fig. 5.9). Smaller veins have few vascular elements and few or no mechanical elements. The very end of a vein is usually a single tracheid with a spiral wall **(Fig. 5.3C, Color Plate 8).** Veinlets are usually surrounded by the **bundle sheath** (Fig. 5.10)—one or more layers of parenchyma cells, which may or may not possess chloroplasts. Water and solutes must pass from conducting elements of the veinlet through the bundle sheath in order to reach the mesophyll. The smallest veinlet has an unbroken connection with vascular elements of the midrib, petiole, and stem to which the leaf is connected.

In some plants, the photosynthetic system shows a distinct division of labor between the bundle sheath cells and the mesophyll cells. These plants are called C_4 plants. The mesophyll cells in C_4 plants are very efficient at capturing carbon dioxide but cannot form sugar. Instead, they bind

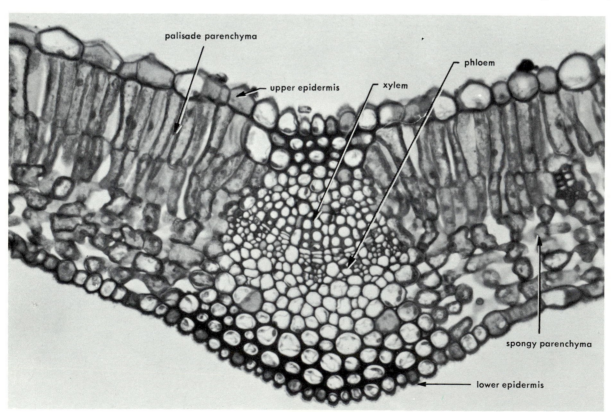

Figure 5.9 Photomicrograph of a cross section of the midrib and mesophyll of lilac (*Syringa vulgaris*) leaf, ×150.

the carbon dioxide to a carrier compound. The resulting molecules travel to the bundle sheath, where the carbon dioxide is released and used for sugar synthesis. These plants have very enlarged bundle sheath cells (Figs. 5.10*B* and **5.11, Color Plate 8**). The C$_4$ process is discussed more fully in Chapter 8.

THE PETIOLE

In the petiole, phloem and xylem maintain their relative positions; as in the stem, phloem is on the underside of the petiole (and leaf blade) and xylem is on the upperside. One or more vascular bundles are embedded in parenchyma. Fibers may be associated with vascular tissues of bundles and, not infrequently, groups of collenchyma cells occur beneath the epidermis. Collenchyma cells are also commonly present in leaves. Usually they form part of the epidermis but may be formed in the mesophyll near the leaf surface adjacent to the vascular bundles.

LEAF DEVELOPMENT

Leaf development is intimately associated with the differentiation of the young shoot apex. The leaf is initiated on the flanks of the apex, as a slight bulge resulting from the enlargement and division of several cells in the outer layers of the apical meristem (Fig. 5.12). This bulge continues to enlarge until it forms a thin, elongated **leaf primordium** (Fig. 5.13*A*). The flanks of the primordium (the marginal meristem) will initiate the formation of the leaf blade (Fig. 5.13*B* and 5.13*C*). This region of growth remains active until the leaf has fully formed. Most cell divisions are completed early in leaf development; cell enlargement accounts for most of the increase in leaf size.

During the early stages of leaf formation, the cells immediately beneath each primordium also become meristematic. These cells, the procambium, will form the vascular tissue that connects the leaves to the stem's vascular system (Fig. 5.12).

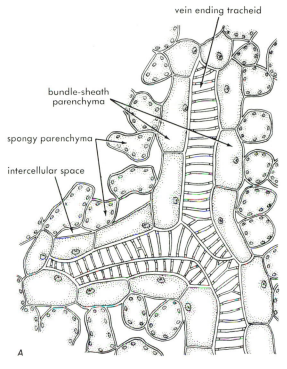

vein ending tracheid

bundle-sheath parenchyma

spongy parenchyma

intercellular space

A

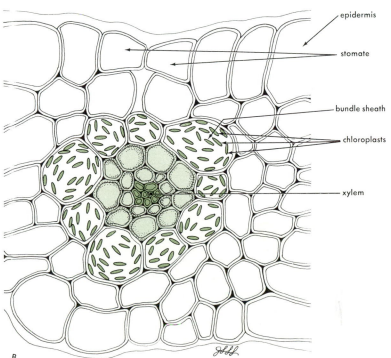

epidermis

stomate

bundle sheath

chloroplasts

xylem

B

Figure 5.10 Diagrams of leaf vascular bundles. *A,* view from leaf surface; note the parenchyma tissue bordering the vein and the presence of only annular wall thickenings in the tracheids at the vein tip. *B,* cross section of a leaf with C4 photosynthesis; note the prominent bundle sheath cells.

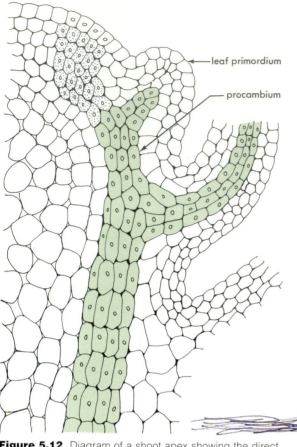

Figure 5.12 Diagram of a shoot apex showing the direct relationship between the initiation of a leaf and procambium strand. The second leaf shows a definite primordium, and the third leaf is well-formed. The procambium strands associated with these leaves are shown.

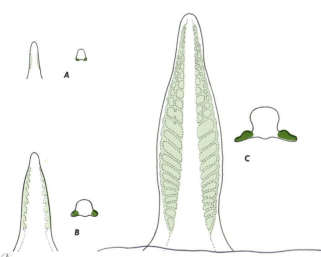

Figure 5.13 Growth of a young leaf. *A*, pencil stage, formation of marginal meristems, shown in green; *B*, activity of marginal meristems initiates lateral expansion of a young leaf; *C*, continued lateral growth through the activity of marginal meristems.

One problem that has continually intrigued plant biologists is how the plant regulates the site of new leaf initiation. The leaves of vascular plants are distributed on the stem in four distinct patterns **(phyllotaxy)—alternate, opposite, whorled,** and **spiral** (Fig. 5.14). The position of a new leaf primordium is closely correlated with the position of the next older primordium. This has been proven by the experimental removal or isolation of the youngest visible primordium on a shoot apex, causing the next formed primordium to develop closer to it. It is also known, at least in ferns, that if the procambium beneath a suspected leaf primordium is severed, the primordium will not develop.

LEAF SHAPE

Leaves vary in shape at different times of the plant's life cycle. The first leaves to form in the embryo are the cotyledons, which are usually quite unlike foliage leaves. In the seedling, the first foliage leaves sometimes differ from later leaves. In buds, modified leaves called bud scales protect the apex. All of the parts of the flower—sepals, petals, stamens, and carpels—are modified leaves.

Leaf shape may vary considerably with the environment or the physiological age of the plant. This variation is called **heterophylly.** Thus, some plants (ivy—*Hedera helix*, and some species of *Eucalyptus* are examples) have distinctive juvenile leaves for the first few years of growth. Subsequent adult leaves of a different form are characteristic of the older or adult plant (Fig. 5.15). The age-related response is usually associated with flowering; that is, only adult plants are capable of producing flowers and subsequently fruits and seeds.

ENVIRONMENTAL FACTORS

Leaf shape is also influenced by environmental factors such as light, moisture, and temperature. The total intensity, wavelength, and daily duration of light have separate but interrelated effects. Plants growing in intense sunlight **(sun leaves)** usually have thick leaves with a thick palisade tissue and a dense, spongy parenchyma **(Fig. 5.16A, Color Plate 8).** Their intercellular spaces are small, and the epidermis is heavily cutinized and generally glossy, with the stomata confined to the lower epidermis. These plants may also have wooly epidermal hairs. Leaves of the same species growing in shade **(shade leaves)** have contrasting traits **(Fig. 5.16B).** They are thin, possessing a single palisade layer and a spongy parenchyma with many intercellular spaces. The epidermis has a thin cuticle.

Light is required for the normal development of leaves, including the differentiation of chloroplasts to a state in which they are capable of carrying on photosynthesis. In darkness, the leaves of a typical dicot remain small and pale yellow, while the stems grow long and slender. This condition is called **etiolation.** Regular daily illumination that is intense enough to support photosynthesis is required for the continued healthy existence of leaves. Excessive shading usually results in the death of a leaf.

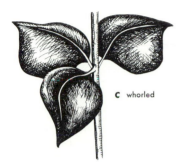
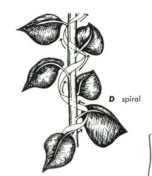
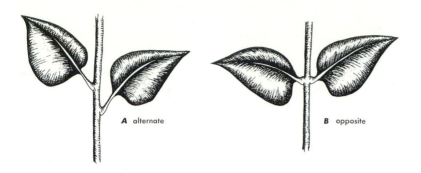

Figure 5.14 Leaf arrangement on stem axis. *A,* alternate; *B,* opposite; *C,* whorled; *D,* spiral.

Figure 5.15 *A,* adult form of *Eucalyptus* sp. leaves $\times \frac{1}{4}$. *B,* juvenile form, $\times \frac{1}{2}$.

Submerged leaves of semiaquatic plants may be vastly different in shape from aerial leaves of the same plants (Fig. 5.17). Similar changes can be induced in the aerial portion of the shoot by reducing the CO_2 content of the air and lowering the temperature.

Nitrogen starvation or water stress may also induce the plant to acquire leaf traits similar to those of sun leaves.

LEAF MODIFICATIONS
Leaves may be considerably modified to perform functions other than photosynthesis.

Bud scales are short, thick, sessile, often covered with dense hairs on the outer surface, and sometimes waxy or

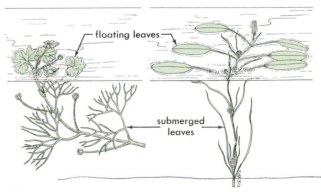

Figure 5.17 Two examples of aquatic plants with both air and water leaves.

resinous. They protect the delicate meristematic tissue of the shoot tip and the rudimentary leaves from drying out.

The **spines** of various species of cacti and those of ocotillo *(Fouquieria)* represent entire transformed leaves. In the black locust *(Robinia Pseudoacacia,* Fig. 5.18*A),* the spines are stipules rather than leaves.

In some species of vetchling *(Lathyrus),* the **tendrils** are transformed leaflets; in other species, the whole leaf is transformed into a single tendril, and leaflike stipules perform the normal functions of the leaves. In trumpet flower *(Bignonia capreolata)* the third leaflet is transformed into a tendril (Fig. 5.18*B).* Both leaf tendrils and stem tendrils serve to attach the plant to a support.

Leaves are sometimes modified as food or water storage organs. The thick, fleshy bases of leaves that comprise much of the daffodil *(Narcissus)* bulb (see Fig. 4.38*D)* accumulate large quantities of food. Succulents found in deserts and saline soils have thick, fleshy leaves with special water-storage tissue.

A striking adaptation of leaves to a special function occurs in insectivorous plants. In these plants, the leaves have taken on forms and various structural features that enable them to capture insects and obtain food from their bodies. Well-known insectivorous plants are sundew *(Drosera,* **Fig. 5.19*C*** and **5.19*D,* Color Plate 9),** pitcher plants *(Darlingtonia,* **Fig. 5.20, Color Plate 9),** and the Venus flytrap *(Dionaea muscipula,* **Fig. 5.19*A*** and **5.19*B).***

Leaves sometimes function effectively in vegetative reproduction. In certain species of *Kalanchoe* (Fig. 5.18*C),* patches of tissue located in notches along the leaf margins remain meristematic. This tissue will eventually develop small, new plants while the parent leaf is still active. The little plants later drop from the leaf to the ground, where under favorable conditions they develop into new individuals. Leaves of *Begonia* produce plantlets by the differentiation of small clusters of epidermal cells on the upper leaf surface.

LEAF ABSCISSION

The separation of plant parts from the parent plant is a normal, continually occurring phenomenon. Leaves fall, fruits drop, flower parts wither and fall away, and even branch tips or whole branches may be separated from the parent plant as a normal part of its life. The fall of leaves from woody dicotyledons in the autumn is a common example of this phenomenon. In practically all cases, separation, or **abscission,** is the result of differentiation in a specialized region known as the **abscission zone,** which is located at, or close to, the base of the petiole (Fig. 5.21). Parenchyma cells comprising the abscission layer may be smaller and lack lignin, compared to the cells of adjacent tissues. Even vascular elements may be shorter, and fibers may be absent from the bundle in the abscission zone. These anatomical features make this zone an area of weakness.

Previous to the fall of leaves, changes normally occur in the abscission zone. Cell divisions, though apparently not necessary, frequently take place and produce a layer of brick-shaped cells across the petiole. Actual separation of the leaf can be brought about in several ways. In some species, the middle lamella is dissolved away and cells separate. In other plants, walls and cells are dissolved. In a third, small group of plants, a layer of cork forms across the petiole so that the leaf simply withers in place and is blown away. In all cases, a protective, corky layer of cells develops

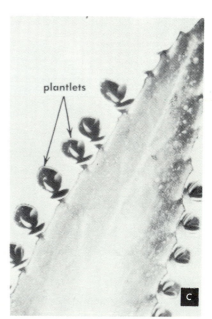

Figure 5.18 Examples of leaf modifications. *A,* black locust *(Robinia Pseudoacacia)* stipular spines; *B, Bignonia capreolata* tendrils; *C,* foliar plantlets on a *Kalanchoe pinnata* leaf.

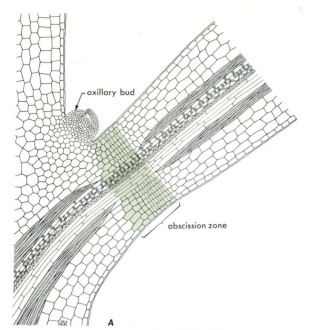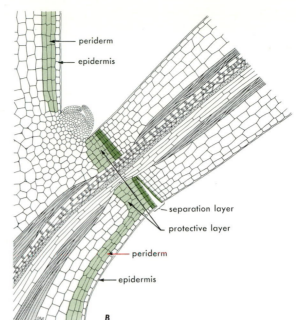

Figure 5.21 The formation of the abscission zone.

across the leaf scar; this layer of cells is continuous with the stem cork. Furthermore, the vessels are likely to become plugged with tyloses or gums. These devices prevent the fall of a leaf from leaving an open wound or a point of entrance for organisms that might cause disease.

The abscission zone has two functions: (1) to bring about the fall of the leaf or other plant part and (2) to protect the region of the stem from which the leaf has fallen against insect damage or rot caused by bacteria or fungi.

Environmental cues are important in synchronizing abscission with the seasons. These cues are most commonly cold temperature or short days that in turn induce hormonal changes that affect the formation of the abscission zone (see Chapter 9).

SUMMARY

1. The chief functions of leaves are photosynthesis and transpiration.

2. Angiosperm leaves are generally supplied with flat, thin blades that are attached to the stem by petioles or sheaths. Veins strengthen the blades and transport food and water. Leaf blades may be simple or compound; leaf margins may be entire, dentate, or lobed.

3. A cross section through the blade of a typical leaf shows the following tissues: upper epidermis; mesophyll, dif- ferentiated into palisade parenchyma and spongy parenchyma; and lower epidermis. Generally, a waxy cuticle coats the epidermis. Guard cells of the epidermis form stomata that control gas exchange.

4. Mesophyll tissue is comprised of palisade and spongy parenchyma tissues and veins. Chloroplasts are present in both parenchyma tissues. Mesophyll is adapted for photosynthesis. Some plants do not differentiate mesophyll into spongy and palisade regions.

5. Leaves may become modified, serving as bud scales, spines, tendrils, the source of propagules, food or water storage organs, and insect traps. Environmental factors such as strong light intensity, nutrient deficiency, and water stress can affect leaf morphology and anatomy. This phenomenon is called heterophylly.

6. Leaf primordia develop in a definite spatial sequence on the flanks of the shoot apex.

7. The young leaf lamina is produced by marginal meristems that are also responsible for the shape of the leaf.

8. Cell divisions are completed while the leaves are enclosed in the bud; leaf expansion results mainly from cell enlargement.

9. The formation of a definite abscission zone across a petiole or fruit stem is responsible for leaf fall or fruit drop. Environmental cues and hormonal changes affect formation of the zone.

6 THE STRUCTURE OF ROOTS

The functions of the root system are **anchorage, storage, conduction,** and **absorption.** You need only walk through a streambed and observe the exposed root systems of large trees, or attempt to pull weeds, to get a firsthand idea of the anchorage function of roots. Plants with unusual roots provide an anchorage function in atypical ways. Adventitious roots on ivy and some viny figs, for example, develop from the stems; these short roots have a flat end-pad that anchors the vines to their growing surface **(Fig. 6.1C, Color Plate 10).** The sticky substance that is secreted for this purpose tends to erode building surfaces, often requiring the plant to be pulled down to prevent the building from crumbling. Parasitic plants, like dodder, anchor themselves by sinking haustorial roots into their host's vascular tissue in order to tap its water and nutrient supply.

Contractile roots provide a more dynamic form of anchorage (Fig. 6.17A). These roots occur in species such as dandelion (*Taraxacum* sp.) and ginseng *(Panax quinquefolius)*. Contractile roots have a cortex of loosely packed cells that shrink when they lose water. Thus, a contractile root shortens when the soil around it dries out, which can help to pull a bulb underground or to hold the aboveground parts of the plant tightly against the soil.

In anchoring the plant, roots help to hold the shoot system erect against the forces of wind and gravity. This function becomes very clear in the case of **prop roots,** such as those of maize (Fig. 6.2D) and the banyan tree **(Fig. 6.1A, Color Plate 10).** Banyan *(Ficus bengalensis)* trees grow in saline mud in tropical lagoons and tidal marshes. Branches of these trees form adventitious prop roots that extend down into the mud where they enlarge and actually prop up the large branches.

The storage function of roots is seen in the carrot, which has a large taproot that undergoes secondary growth (Fig. 6.2A). Its mass of secondary xylem, however, is composed mostly of storage parenchyma cells filled with water and carbohydrates. Root storage usually assists the plant at some developmental stage, such as providing nutrients during flower production. Sugar beets, for instance, are biennial plants that begin to fill their storage root with food at the end

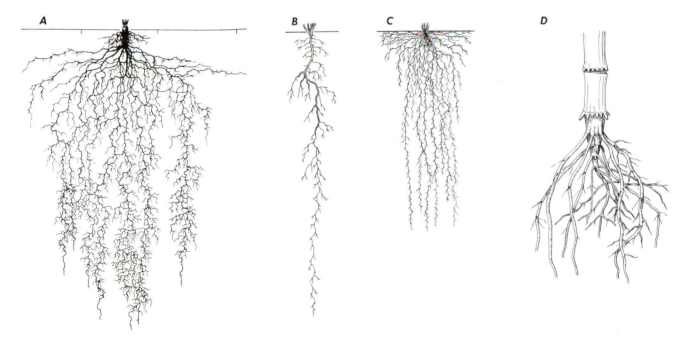

Figure 6.2 Types of root systems. *A,* tap root system of carrot (*Daucus carota*); *B,* tap root system in which the main root penetrates the soil but few secondary roots develop; *C,* fibrous root system penetrating the soil evenly; *D,* type of fibrous root system composed of adventitious roots growing at lower nodes of the stem of corn (*Zea mays*).

of the first growing season. This stored material is partially used to assist the growth of early shoots during the following season, but flowering during summer uses most of the stored energy.

In herbaceous plants, water is responsible for about 95% of root weight. Water is needed for all root processes and for every metabolic reaction. Soil water contains dissolved salts and minerals, such as potassium, sulfur, phosphorous, calcium, and magnesium, which are needed by the plant (see Chapters 2 and 7).

Roots extract water and minerals from the soil by means of an elaborate absorption and conduction system, which is described on page 81 and in Chapter 7.

TYPES OF ROOT SYSTEMS

When pulled up, many grasses and small garden plants take with them a massive clump of soil. This happens because the **fibrous root system** of these plants consists of several main roots that branch to form a dense mass of intermeshed lateral roots (Fig. 6.2C).

Vegetables with a large storage root, like the carrot, have a **taproot system** that consists of one main root from which lateral roots radiate (Fig. 6.2A and 6.2B). Some desert plants have a rapidly growing taproot system that enables them to penetrate the soil quickly to reach deep sources of water. The depth of root penetration of various plants is quite remarkable. A University of Nebraska botanist, John Weaver, exhumed the entire root systems of 45 species of native

Table 6.1. Rye Root-System Measurements

Root	Number	Length (m)
Main roots	143	65
Secondary roots	35,600	5,181
Tertiary roots	2,300,000	174,947
Quarternary roots	11,500,000	441,938
Total	14,000,000	609,570

plants from the Midwest prairie. Only 4 species maintained root systems primarily in topsoil; most of them had roots that extended to 4 m deep. Even in very hard soils, the depth of root penetration well exceeded the height of the aboveground plant. A typical annual plant, such as corn or rye, will build an immense fibrous root system in one growing season. In a study on rye, a single plant 50 cm tall, with 80 tillers (shoot branches), had a root system amounting to 210 m² of surface area compared to only 4 to 5 m² of the aboveground portion (Table 6.1).

In trees and shrubs, most functioning roots are localized in the upper 1 m of soil, with the majority of the feeder roots in the upper 15 cm (Fig. 6.3). However, these plants may extend lateral roots well beyond the expansion of overhead branches. In areas of closely packed trees, or on sandy soils, competition and low surface moisture may reduce the amount of area covered and encourage deeper root penetration.

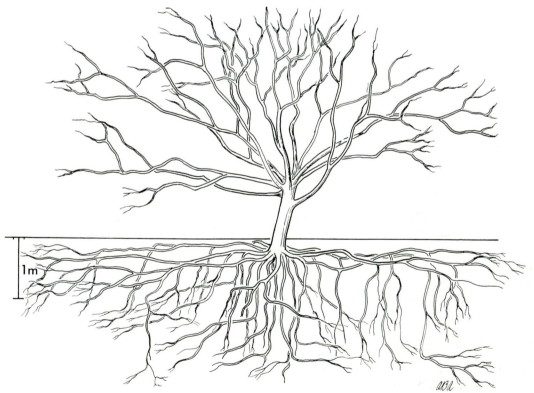

Figure 6.3 Diagram of excavated root system of a 16-year-old tree to show the shape and size of the root system compared to the above ground shoot system.

INITIATION OF ROOTS

The seeds of higher plants contain a small undeveloped plant, the embryo. When the seed germinates, the embryonic root, or **radicle,** extends by the division and elongation of its cells to form the **primary root** (Fig. 6.4). Taproot systems develop from only one enlarged primary root, which then forms **lateral,** or **secondary, roots.** Further branching of secondary roots gives rise to succeeding orders of roots, for example, tertiary, quarternary, and so on. Fibrous root systems develop in a different way. The embryos of some grasses, such as barley, consist of not only one radicle but also of several additional embryonic roots called **seminal roots.** The seminal roots emerge soon after the radicle during germination (see Fig. 13.11, p. 195). They elongate rapidly, and soon it is not possible to distinguish the primary root. Usually the seminal roots do not persist, but in some grasses they may function throughout the plant's life.

Roots that develop from organs other than roots are called **adventitious roots.** In most instances, adventitious roots develop at the nodes of the stem. In a young corn plant, soon after the emergence and development of a rudimentary root system, **prop roots** develop from the shoot node nearest the soil level (Fig. 6.2D). These prop roots function as roots but also assist in the support of the plant in the soil. The prop roots of the banyan, mentioned on page 79, are adventitious; they arise from large aerial branches.

Often adventitious roots can be induced to form artificially. Pieces of stem (cuttings), such as a cane from a blackberry plant, may form roots at their cut ends if they are placed in moist soil. Leaves from *Begonia* and several other species also can be rooted simply by soaking them in water. Many commercially important ornamentals, in fact, are reproduced by root propagation from leaves or stems (see Chapter 9).

THE ABSORBING REGION OF THE ROOT

The terminal few millimeters of the root tip is the region that absorbs water and minerals from the soil. In this function the root is so efficient that it dries the soil in its vicinity. In well-drained soils, roots must continually grow into new areas if they are to maintain their absorptive function. Growth is accomplished in two ways: First, a growth region occurs just behind the tip of the root (Fig. 6.4). Cells in this **region of elongation** grow in length and push the root tip through the soil. A **root cap,** described below, protects the root tip from abrasion against soil particles. Secondly, **root hairs** arise from epidermal cells just behind the region of elongation

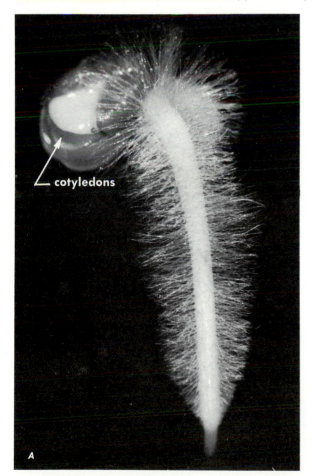

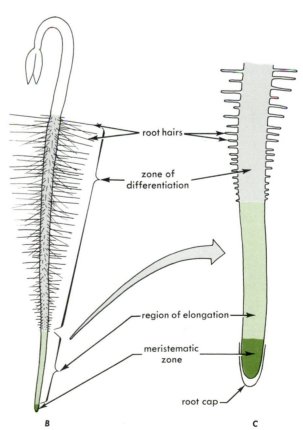

Figure 6.4 Radish seedling (*Raphanus sativus*). *A,* photograph showing root hair growth. × 2; *B,* and *C,* diagrams showing growth regions.

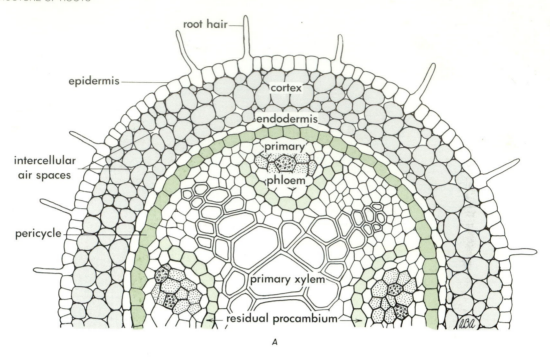

A

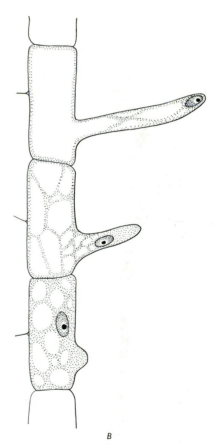

B

Figure 6.5 *A,* cross section of a root showing the primary tissues. *B,* development of a root hair.

(Fig. 6.4). Each root hair arises from a single epidermal cell and grows between soil particles for a distance as great as 1 to 2 cm.

Root hairs penetrate between small soil particles and are in intimate contact with them and the water film that surrounds them. The cell walls of the root hairs and epidermal cells are made of cellulose and pectin. The pectic coat on the outside of the cell wall makes the root hairs gummy, facilitating their adherence to soil particles. Water is absorbed by the root hairs and epidermal cells and is then passed on by diffusion and active selective processes to the conducting elements. Root hairs dry the soil around themselves and usually wither and die in a few days. However, new root hairs continually form as old ones die. In one rye *(Secale)* plant, 100 million new root hairs are estimated to be formed each day.

INTERNAL ANATOMY

A CROSS SECTION OF THE ABSORBING REGION

A precise inner anatomy gives the root an ability to select and accumulate nutrients from the soil. The structure is best viewed in a cross section taken through the root-hair zone (Fig. 6.5*A*). The root shows the same three primary tissues that occur in shoots: epidermis, ground tissue, and vascular tissue. The epidermis is a single layer of cells, some of which form hairs. Beneath the epidermis is the cortex, a cylindrical zone that is composed chiefly of parenchyma cells and water-filled spaces. The innermost cells of the cortex are brick-shaped and packed tightly together, comprising a layer called the **endodermis.** The core of the root, enclosed by the

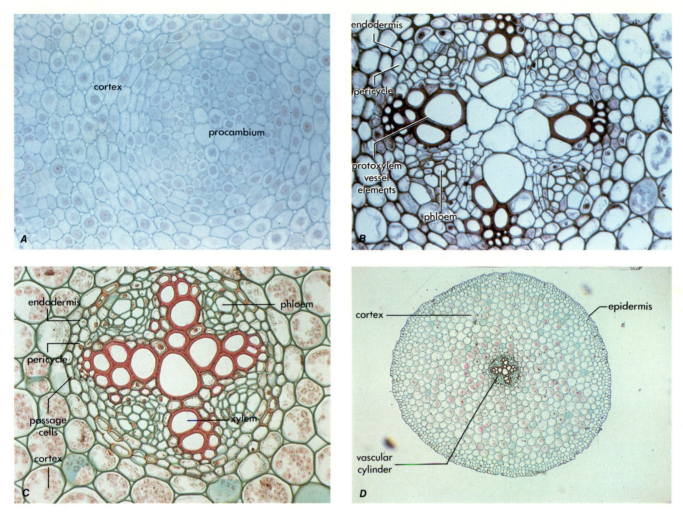

Figure 6.11. Buttercup (*Ranunculus* sp.) roots showing the stages of differentiation in primary growth. *A*, immature section taken from near the meristematic region. The layers of tissues and regions can already be seen. The metaxylem and protoxylem cells in the central stele are clearly formed. *B*, the cells of the protoxylem have developed secondary walls (shown here stained red with safranin). This particular root is tetrarch with four protoxylem points with phloem between. *C*, fully mature root with all primary tissues differentiated. Note that the endodermis cells adjacent to the protoxylem poles are lacking secondary walls (passage cells). *D*, this is the same root section as *C*, but shows all tissues.

Figure 6.14. Willow (*Salix* sp.) root transverse section showing a large lateral root soon after its emergence.

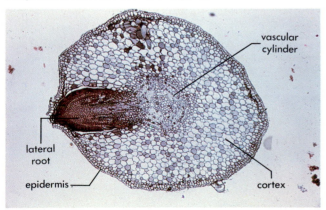

Figure 6.16. Alfalfa (*Medicago sativa*) root transverse section with some secondary tissues and periderm (cork) on its periphery.

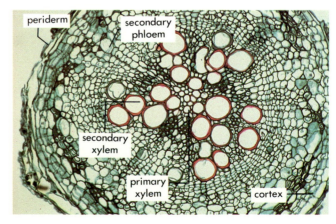

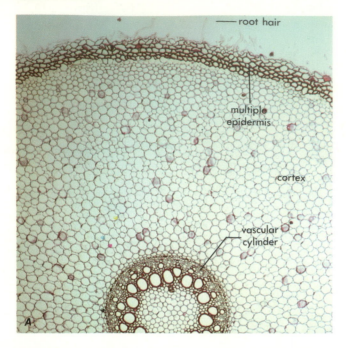

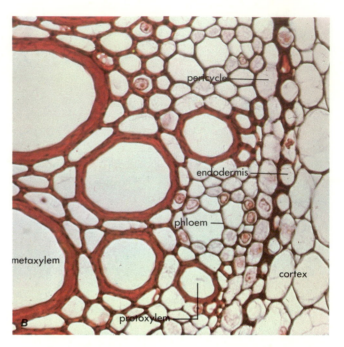

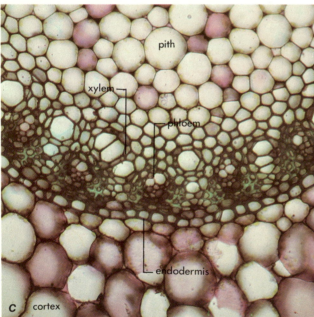

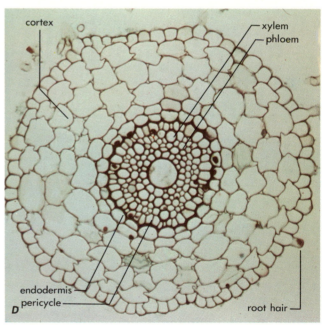

Figure 6.12. *A, Asparagus officinale* root in transverse section showing all tissues and root regions. The vascular cylinder (stele) has a pith of all parenchyma cells. *B, Asparagus* root at higher magnification to view the cell layers in greater detail: the endodermis, pericycle, xylem, and phloem of the polyarch stele are clearly defined. *C, Epidendron* sp. (orchid) root in transverse section to show details of the stele. *D, Triticum* sp. (wheat) root; this is an example of a monocotyledonous root lacking a parenchymatous pith.

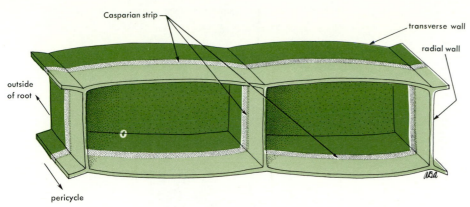

Casparian strip

transverse wall

radial wall

outside of root

pericycle

Figure 6.6 Three-dimensional view of two endodermal cells. The suberized Casparian strip is shown in gray. Water from solution outside the root wets the cellulose walls until it reaches the Casparian strip. Since water cannot wet and cross the Casparian strip, the cell walls inside the Casparian strip are wet by water that has passed through the protoplast of the endodermal cell.

endodermis, is the **vascular cylinder,** or **stele.** In dicot roots, the stele has xylem at the center, with two or more ribs of xylem that radiate toward the root surface. Masses of phloem lie between the xylem ribs. A meristematic layer, the **pericycle,** surrounds the vascular tissues. The pericycle is usually inactive in the absorbing part of the root, but in older regions it may give rise to branch roots and secondary tissues.

The endodermis plays a special role in controlling nutrient accumulation by the root. Water and dissolved solutes from the soil can travel freely through the porous walls of the cortex and epidermis, without entering the living cells. Likewise, water and solutes in the stele can escape from the dead xylem vessels and, entering the walls, can travel freely between the cells. Thus, if there were no barrier, solutes could pass without hindrance between the soil and the xylem. But the endodermis provides a barrier against such free movement of materials. The lateral walls of the endodermal cells contain a band of waxy material (suberin) called the **Casparian strip** (Fig. 6.6). Electron micrographs show that the plasmalemma of the endodermal cell is fused with the Casparian strip. This arrangement creates a barrier that prevents materials from passing via the walls between the stele and the cortex. To pass, molecules can only travel through the living protoplasts of the endodermal cells; and to enter the protoplasts, they must cross the selectively permeable plasmalemma. This enables the root to regulate the movement of solutes into and out of the root.

THE FORMATION OF PRIMARY TISSUES

As mentioned before, the root continually grows into new areas of soil. Older regions of the root gradually lose their absorptive role and undertake other functions, while new absorbing regions form at the root tip. For a realistic picture of the functioning root, we must look at the way new tissues arise and mature.

Figure 6.7 shows a three-dimensional drawing of a root with the internal regions visible. As you read, you may find it helpful to look at this diagram for orientation.

The **root apical meristem,** which occurs just behind the root cap (Fig. 6.8), is composed of small, regularly shaped cells that are capable of dividing. The emergence of the radicle during germination is dependent on the initiation of cell division in this region. After the radicle emerges, the cells of the root apical meristem continue to divide; on the average, one daughter cell of each division remains as part of the meristem and divides again, while the other daughter cell differentiates into a mature, specialized cell.

The central portion of the apical meristem is composed of a pellet-shaped region of cells known as the **quiescent center** (Fig. 6.9). These cells divide at an extremely slow rate. They release dividing cells, **initials,** just fast enough to continuously maintain the root's shape and to keep pace with its growth. The function of the quiescent center is unknown; one hypothesis, however, is that it may be the location where growth regulators are synthesized and released to control development of the root as new cells are made.

Some initials divide and produce cells in front of the apical meristem to form the **root cap** (Figs. 6.8 and 6.10). The root cap is a thimble-shaped region of cells that surrounds the apical meristem and precedes it as the root elongates and forces its way through the soil. The root cap protects the apical meristem and lubricates its passage through the soil; it is also the site that perceives gravity and controls the direction of root growth.

Initial cells behind the quiescent center divide to supply the cells for the primary root tissues. These cells continue to divide, so that the region of cell divisions may extend as far as 1.5 mm from the tip of the root.

The apical meristem provides tissues for the root proper by supplying the same three primary meristems that occur

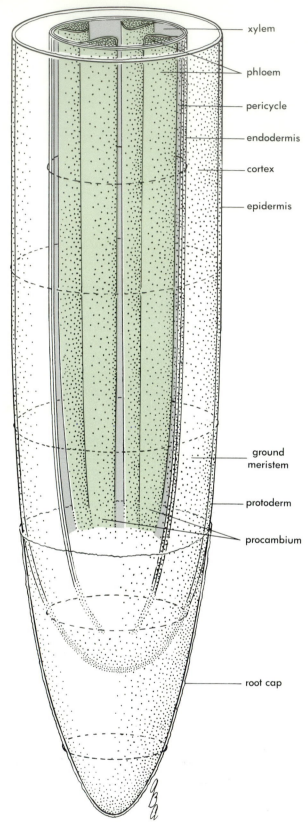

xylem

phloem

pericycle

endodermis

cortex

epidermis

ground meristem

protoderm

procambium

root cap

Figure 6.7 Three-dimensional diagram showing the primary tissues of a root.

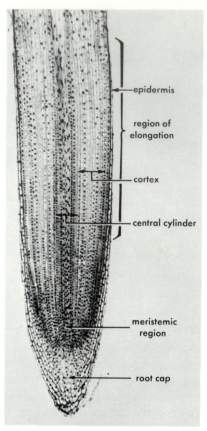

epidermis

region of elongation

cortex

central cylinder

meristemic region

root cap

Figure 6.8 Median longitudinal section through an onion (*Allium cepa*) root tip.

in the shoot: *protoderm,* which gives rise to epidermis; *ground meristem,* which forms the cortex; and *procambium,* which forms the stele (Fig. 6.9).

The protoderm consists of uniformly shaped meristematic cells. At some distance from the root apex, some cells of the protoderm develop into **root hairs** by the extension of their cell wall into the surrounding soil (Fig. 6.5*B*).

Ground meristem cells produce the cells of the cortex. The cortex consists mostly of storage parenchyma and occasionally of sclerenchyma cells. The innermost cell layer of the cortex is the **endodermis** (Figs. 6.5 and 6.6).

Endodermal cells in mature regions of the root often develop partial secondary cell walls. Passage cells adjacent to xylem points often do not form secondary walls, thus ensuring the continued passage of water.

The vascular cylinder, or stele, develops from the procambium and forms the central core of the root. In dicotyledonous roots, the xylem occupies the middle of the stele, with the periphery consisting of alternating radii of xylem and phloem (Figs. 6.5 and **6.11, Color Plate 11**). In monocotyledonous roots, the central core contains parenchyma cells **(Fig. 6.12, Color Plate 12).** The first matured xylem-conducting elements, the **protoxylem,** develop at the points of the radiating xylem. Dicot roots with two protoxylem points are called diarch roots, those with three points are triarch, and so on; most monocot roots have more than five proto-

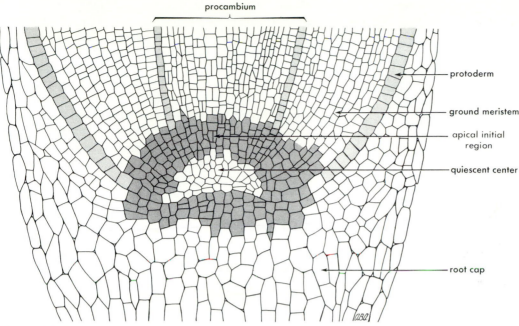

Figure 6.9 Diagram of a median longitudinal section through the meristem of a hypothetical root. The quiescent center cells divide very slowly in the center and more rapidly near the periphery. All three primary meristems (promcambium, ground meristem, and protoderm) plus the root cap are derived from the apical initial cell region.

Figure 6.10 Scanning electron micrograph of a corn (*Zea mays*) root with the root cap lifted off but still held by a strip of epidermis. (Used by permission of Dr. Peter Barlow.)

xylem points and are therefore called polyarch. **Metaxylem** cells form within the mass of xylem to complete the central core.

The protoxylem is modified so that it is capable of transporting water while the root is elongating; that is, protoxylem cells must be able to withstand the forces that move water and still be flexible enough to elongate. This is accomplished by the formation of a secondary cell wall in the shape of annular rings or as a spiral (see Fig. 4.12A). When the root has completely elongated, the metaxylem cells mature. They no longer are required to elongate, and consequently, they form thick secondary cell walls with pitted areas through which lateral exchange can take place (see Fig. 4.12A). Protoxylem cells often become crushed while the metaxylem develops. Xylem of roots consists of vessel members (in flowering plants), tracheids (in most ferns and conifers), parenchyma, and fibers.

Phloem cells form in the spaces between the protoxylem arms (Figs. 6.5, **6.11,** and **6.12, Color Plates 11 and 12**). The protophloem is the first part of the vascular system to become functional. These cells form at the periphery of the phloem and function primarily during root elongation. Metaphloem cells develop toward the inside, and they function during the plant's adult life. Phloem of roots consists of parenchyma and fibers; sieve-tube members and companion cells (in flowering plants); and sieve cells (in conifers and ferns).

THE ORIGIN OF BRANCH ROOTS

One major difference between shoots and roots is in the way branch organs arise. You may recall that lateral shoots

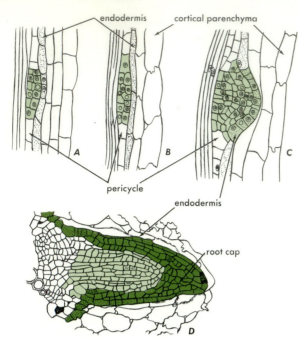

Figure 6.13 Branch roots, *A,* initiation of branch root in the carrot (*Daucus carota*) starting from meristematic cells in pericycle; *B* and *C,* later stages; *D,* young root pushing through cortex, ×50.

arise from axillary buds, which form at the base of each leaf from tissues near the surface. By contrast, lateral roots arise from the pericycle, which lies deep inside the root, beneath the cortex (Figs. 6.5, 6.13, and **6.14, Color Plate 11**).

Initiation of lateral roots at particular locations is controlled by chemical growth regulators that cause pericycle cells to begin to divide (Fig. 6.13). The **lateral root primordia** that result continue to form new cells which in turn elongate. Endodermal cells subtending this region often also divide for a short time, contributing cells to the tip of the new lateral root. As it expands, the lateral root pushes its way through and destroys the cortical cells and the outer epidermis. As it emerges, the new lateral root becomes organized into an apical meristem with primary meristems identical to its parent root **(Fig. 6.14, Color Plate 11)**. This system of lateral root formation may present a hazard in the plant's development because the lateral root forms a wound in the cortex and epidermis of the primary root. If the wound doesn't heal rapidly, it is a good point of entry for plant pathogens.

SECONDARY GROWTH IN ROOTS

As the root system grows and branches, the primary vascular tissues may not be able to supply the needs of the many root tips. In long-lived dicots, the older regions of the root form secondary vascular tissues. Two secondary, or lateral, meristems are required to form these tissues, the **vascular cambium** and the **cork cambium.**

In roots, secondary growth is initiated by the division of (residual) procambium cells between the arcs of xylem and phloem. These cells form secondary xylem to the inside and phloem to the outside (Fig. 6.15). After a time, the crescent-shaped region of dividing cells joins with the pericycle, which in turn also begins to divide. This connection forms a complete ring of vascular cambium around the entire vascular cylinder (stele). Secondary tissues are produced during the growing season to form annual growth rings, just as in the stem. Continued growth expands the root and finally causes the splitting, sloughing off, and destruction of the cortex and epidermis. However, early stresses of expansion, plus the activity of growth hormones, stimulates the pericycle between the xylem arms to resume division and form the cork cambium. This meristem forms cork cells to the outside and parenchyma to the inside, just as in the stem **(Fig. 6.16, Color Plate 11).**

ROOT SYMBIOSES

Many roots form symbiotic associations with soil microorganisms. The associations fall into two classes: **mycorrhizae** and **bacterial nodules.**

MYCORRHIZAE

Mycorrhizae are short roots in association with a soil-borne fungus. Two types of mycorrhizal roots are found, depending on whether the fungus penetrates into the root cells or not. **Ectotrophic** types are found in roots of such trees as pines (*Pinus*), birches (*Betula*), willows (*Salix*), and oaks (*Quercus*). This type causes a drastic change in the root shape (Fig. 6.17*B*, 6.17*C*, and 6.17*D*). These mycorrhizal roots are about 0.5 cm long, have no root cap, and exhibit a simple monarch stele. In addition, the fungus penetrates between the cell walls of the cortex and forms a covering sheath, or mantle, of fungal hyphae (see Chapter 15) around the entire root. Mycorrhizal roots are short and forked and sometimes are borne as tight clusters. **Endotrophic** mycorrhizae do not form a mantle over the root, and the fungus actually enters the cortex cells. Mycorrhizal roots are more efficient in mineral absorption, but they are apparently not absolutely essential for the growth of the usual host plants. This is known because plants that are artificially fed adequate nutrients can grow without mycorrhizae. Mycorrhizae also may be beneficial to their host plants by secreting hormones or antibiotic agents that reduce the potential of plant disease. Mycorrhizae are discussed further in Chapter 15.

BACTERIAL NODULES

Some plants, like certain legumes, are capable of fixing nitrogen, that is, changing N_2 in the soil into the NH_4^+ (ammonia) form of nitrogen that is usable by the plant. This is done by an unusual relationship between the bacterium *Rhizobium* and the roots of legume plants. Root cells are infected by the passage of a thin infection thread of the bacteria into the root hair cells of roots. The infection thread passes through the root hair by causing the cell membrane to enfold as it penetrates the cell. The infection thread grows

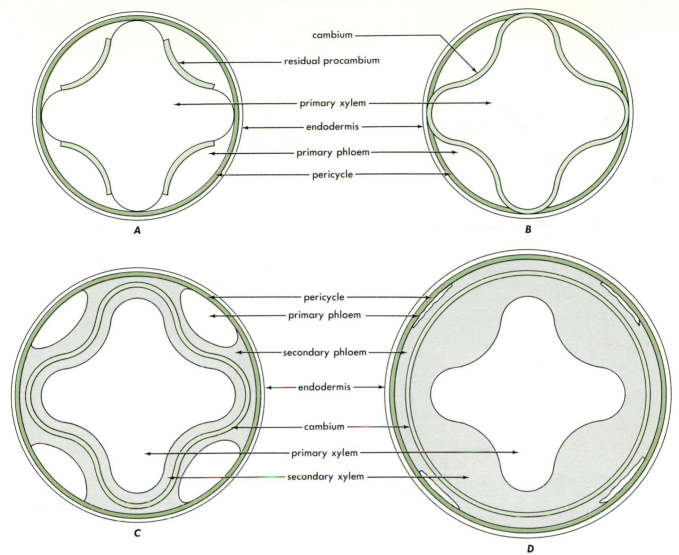

Figure 6.15 Diagrammatic representation of the development of the secondary plant body in a root. *A,* at the completion of primary growth, arcs of residual procambium cells remain (light green) and a complete circle of pericycle is present (dark green). *B,* the residual procambium (light green) joins with the pericycle cells outside the xylem arms to initiate a continuous cylinder of vascular cambium (light green). *C,* the vascular cambium forms secondary xylem internally and secondary phloem externally (light gray). The primary phloem is being pushed outward and a complete ring of pericycle (dark green) is still associated with it. *D,* a smooth circle of vascular cambium forms, producing secondary xylem and phloem. The primary xylem remains in the center of the stem, the primary phloem has been crushed, and the pericycle that remains will form the cork cambium.

through the epidermal cell into the cortical cells. The bacteria then divide and also stimulate the cortical cells to divide, thereby forming the root nodule (Fig. 6.17*E* and 6.17*F*). The bacteria are the actual agents for fixing the nitrogen.

SUMMARY

1. The principal functions of roots are absorption of water and nutrients, conduction of absorbed materials and food, and anchorage of the plant in the soil.

2. There are two general types of root systems: a fibrous root system and a taproot system.

3. Roots differ from stems because they lack nodes and internodes.

4. The root tip is divided into four zones of specialization: (1) the root cap, which protects (2) the meristematic region as it moves through the soil; (3) a region of elongation; and (4) a region of differentiation characterized externally by root hairs and internally by the formation of primary vascular tissues.

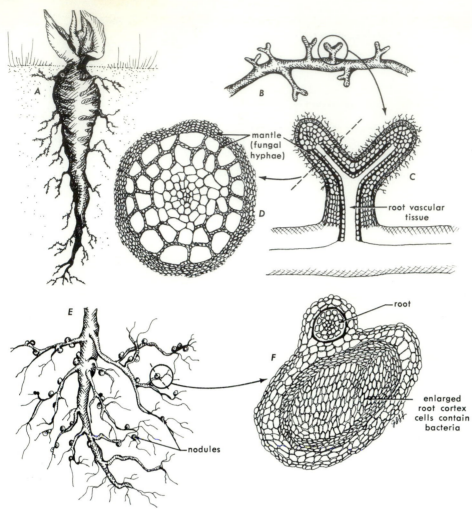

Figure 6.17 Views of root modifications. *A*, contractile root; *B*, branched habit of mycorrhizal root; *C*, longitudinal section through one of the Y-branched mycorrhizal roots; *D*, cross section through one root showing the fungal mantle; *E*, soybean (*Glycine max*) root with bacterial nodules; *F*, cross section of root and nodule.

5. Soil water and nutrients move easily through epidermal and cortical tissues.

6. The endodermis forms the innermost cell layer of the cortex. The Casparian strip, a suberized band in radial and transverse walls, completely encircles endodermal cells.

7. Water cannot move across the Casparian strip. Therefore, all water with dissolved nutrients inside the endodermis has passed through the protoplasts of endodermal cells.

8. The pericycle, a row of cells internal to the endodermis, represents the outermost row of cells of the vascular cylinder; these cells differentiated from procambium. Cells of pericycle may eventually form part of vascular cambium or cork cambium or give rise to branch roots.

9. In cross section, the primary xylem is star-shaped and generally triarch to pentarch. Primary phloem arises between the arms of primary xylem. There is no pith in the roots of most dicot plants.

10. Water and inorganic nutrients enter primary xylem from parenchyma cells without passing through cells of phloem.

11. Water and inorganic nutrients enter root-hair cells, pass through the cortex, are filtered by protoplasts of endodermal cells, cross the pericycle, and move directly into the vessels or tracheids of primary xylem.

12. In roots having secondary growth, a vascular cambium is formed by pericyclic cells over xylem arms and procambial cells remaining between primary xylem and phloem. This vascular cambium at first forms a wavy line in cross section. It eventually becomes circular, and there is little difference in the appearance of wood from root or stem.

7 THE ABSORPTION AND TRANSPORT SYSTEMS

The living plant collects materials from the environment and assembles them into plant cells and organs. It therefore requires systems for absorbing materials, manufacturing new substances, and transporting materials within the plant. The manufacturing systems—metabolism and photosynthesis—are discussed in Chapters 2 and 8. The present chapter deals with the absorption and transport systems.

WATER ABSORPTION AND TRANSPORT

Green plants require only a few simple materials from the environment: water, carbon dioxide, oxygen, and several minerals. Of these materials, water is the substance that plants absorb and transport in the greatest quantities. Some of the absorbed water is consumed in metabolic reactions such as photosynthesis, and some of it is retained in the vacuoles as cells grow. But as much as 98% of the water that enters the plant is lost through the shoot system by evaporation. This evaporative loss is termed **transpiration**.

Plants transpire a large amount of water. A single corn plant in Kansas transpired 196 liters (l) between May 5 and September 8. It has been estimated that the water lost by a grove of red maple trees in a growing season may be equivalent to a sheet of water 72 cm deep over the entire grove.

Plants draw their major water supplies from the soil solution, the bulk water that lies between soil particles. Water molecules enter the plant (Fig. 7.1) through the epidermis of the root tip. They move through the cortex, the endodermis, the pericycle, and finally into the xylem vessels. The liquid in the xylem, **xylem sap,** moves up the plant to growing organs and to leaves, where water is used in photosynthesis and lost by transpiration. In leaves, water passes from the xylem to the mesophyll cells, where it evaporates into the intercellular spaces and escapes from the leaf through stomata.

WATER POTENTIAL AND WATER MOVEMENT

What causes water to move through the plant? Two answers apply at different times. Sometimes water is pushed up the plant as the roots absorb materials from the soil solution. More often, however, water is *pulled* up the plant by forces that arise in the leaves and growing shoots. To understand both the push and the pull, we must first look into the principles that govern water movement in general.

Physiologists take two approaches to the study of water movement. One approach speaks in terms of *energy,* while the other deals with the *forces* that affect water molecules. Every mass of water has a certain amount of stored energy, which is expressed as the **water potential.** Water potentials can be measured precisely by means of instruments. The measurements show that *water always flows from regions of high potential to regions of low potential.*

Several factors affect water potentials, and their effects are additive. Therefore, the water potential (symbolized by the Greek letter ψ) can be expressed as a sum of terms:

$$\psi = \psi_\pi + \psi_p + \psi_m$$

The most important factors, summarized in Table 7.1, are:

1. *Temperature:* Temperature is a measure of heat, which increases the water potential. The equation has no separate term for temperature because the heat effect is incorporated into the other terms.

2. *The presence of solutes:* Solutes reduce the water potential. The contribution of solutes is symbolized as ψ_π, and is called the **osmotic potential.** In pure water, ψ_π is zero. Therefore, ψ_π is negative whenever solutes are present.

3. *Pressure:* Physical pressure increases the water potential. Pressure can result from the weight of the atmosphere, from the restraint imposed by cell walls, or from the weight of overlying tissues. The **pressure potential,** symbolized as ψ_p, is zero at room temperature and normal atmospheric pressure. Higher pressures generate a positive ψ_p, while pressures below that of the standard atmosphere result in negative values of ψ_p.

4. *Solid particles:* Solid materials, such as sand grains or cell-wall microfibrils, decrease the water potential. The **matric potential,** symbolized as ψ_m, expresses this effect on

Table 7.1. Factors That Affect Water Potential in a Solution

Water Potential Is Increased by:	Water Potential Is Decreased by:
Pressure Heat Elevation (gravity)	Solutes Polymers (adhesion)

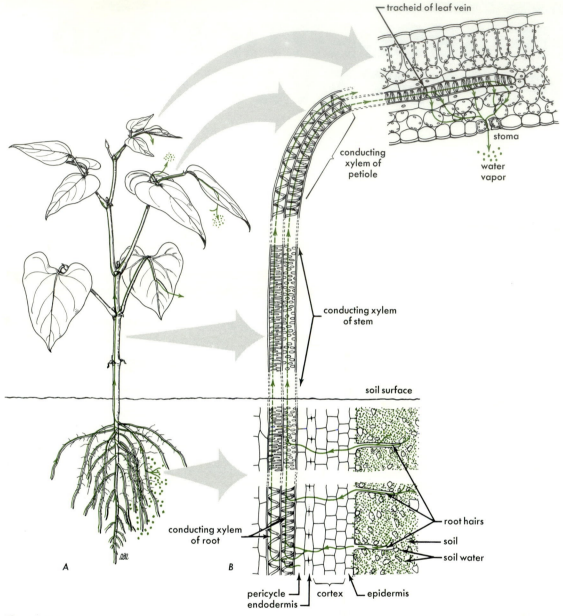

Figure 7.1 *A,* diagram showing the water movement in a whole bean plant. *B,* details of the xylem pathway.

the water potential. The matric potential is zero in pure water. The addition of solid materials leads to negative values of ψ_m.

5. *Gravity:* Any material gains potential energy when it is lifted above the ground. This is as true of water as it is of a ball that has been thrown into the air. The gravitational potential rises as water moves up the trunk of a tree. However, in small plants the gravitational potential can be neglected because the other factors are much more prominent.

Which of the preceding factors is most important? The answer varies from one point to another. The cell wall and the protoplast offer an instructive contrast (Fig. 7.2). The

protoplast has a high osmotic potential (ψ_π) because it is rich in solutes. Its internal pressure can also be high, giving a substantial ψ_p. The wall, by contrast, has fewer solutes and little or no pressure. Instead, it has microfibrils that create a matric potential (ψ_m). The matric potential can become high enough, in a drying wall, to pull water from any protoplast.

Students are sometimes confused by the fact that water potentials are usually negative. We have said that water moves from high to low potential. In this respect, a potential of -50 is lower than a potential of -10. To avoid confusion, think of water potential as a vertical scale, with zero in the middle and negative numbers below zero (Fig. 7.3). If two regions differ in water potential, water will always move to the region that is lower on the scale.

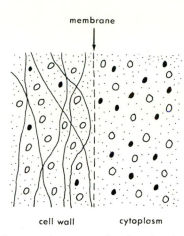

membrane

cell wall   cytoplasm

Figure 7.2 Contrasts between cell wall and protoplast, affecting water movement. The cell wall contains fewer solutes, but has polymers that create a matric potential.

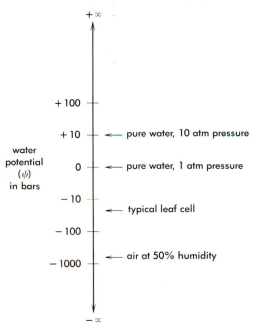

water potential (ψ) in bars

+∞

+100

+10 ←— pure water, 10 atm pressure

0 ←— pure water, 1 atm pressure

−10 ←— typical leaf cell

−100

−1000 ←— air at 50% humidity

−∞

Figure 7.3 Scale of water potential. The standard unit is the bar. By convention, pure water at 25°C and 1 atmosphere pressure is assigned the potential of zero bars. If two parts of a plant differ in water potential, water will move toward the part that is lower on the potential scale.

THE FORCES AND EVENTS BEHIND WATER POTENTIAL

Water potentials allow us to predict the direction of water movement, but they do not show *why* water obeys the rules we have stated. To develop an understanding of this, we must examine the forces and events that occur between molecules.

Two factors must be considered to understand water movement: the effect of *thermal motion* and the effect of *forces between molecules*.

Thermal motion is a ceaseless, random wandering of molecules that occurs because the molecules carry kinetic energy. Thermal motion has no intrinsic direction; molecules collide with one another millions of times per second, and they rebound at widely varying angles. Thus, a molecule in a fluid will wander throughout the whole liquid mass in the course of time. Such wandering is called **diffusion.** Thermal motion gives water (and all other molecules) an *escaping tendency*—a tendency to wander away from the present location. Water potential is indirectly a measure of the escaping tendency of water.

Forces between molecules can influence the escaping tendency. For example, suppose a container is filled with water molecules, and then additional water is forced into the container. Crowded molecules repel one another and tend to push their neighbors away; the escaping tendency is increased. This description can help you visualize why water potential increases with pressure.

Molecules generally cling together when they are not artificially compressed. The molecules of organic solvents such as gasoline are only slightly attracted to one another; hence gasoline evaporates easily. By contrast, water molecules attract one another quite strongly. The attraction between like molecules is called **cohesion.** Water molecules are cohesive because each molecule has positive and negative regions:

$$\delta^{(+)} H \diagdown \underset{\delta^{(+)} H \diagup}{O} \delta^{(-)} \cdots\cdots \underset{H}{\overset{\delta^{(+)}}{H}} - \underset{H \delta^{(+)}}{\overset{\delta^{(-)}}{O}}$$

The negative part of one molecule attracts the positive parts of neighboring molecules, as shown by the dotted line. These attractions are called **hydrogen bonds** (see Appendix).

Water molecules may also cling to other kinds of molecules that have charged regions. Such an attraction between unlike molecules is called **adhesion.** As an important example, water is strongly attracted to the cellulose of cell walls. As described in Chapter 2, cellulose polymers have many hydroxyl groups. These groups have positive and negative regions like water molecules, and they give cellulose a powerful tendency to acquire a coating of water. The matric and solute components of water potential are partly due to adhesive forces between water and other molecules. These forces reduce the mobility of water and decrease its escaping tendency. Correspondingly, they reduce the water potential.

OSMOSIS AND TURGOR PRESSURE

The concept of water potential is useful in explaining osmosis and turgor pressure, two phenomena that are important in the mechanisms of water movement. **Osmosis** is the diffusion of water across a differentially permeable membrane (Fig. 7.4). It occurs when the solutions on the two sides of the membrane differ in water potential. If the two solutions are at the same pressure and temperature, the one with the higher solute concentration will have a lower water potential. Therefore, water will move into the solution that contains the most solute particles. This water flow across a membrane—osmosis—brings the two solutions closer to the

membrane

Figure 7.4 System that would show osmosis. A membrane separates two solutions that contain water (○) and solute (●) molecules at the same pressure. The pores in the membrane permit water but not solutes to pass. The water potential will be lower in the left-hand solution because water is less concentrated there; water will move into that solution.

same water potential. Unless an outside force intervenes, osmosis will continue until the two solutions achieve equal water potentials. Then equilibrium prevails.

In cells that have walls, the osmotic entry of water creates pressure because the walls resist being stretched. Pressure increases the water potential, and water stops entering when the potentials inside and outside the cell are equal. Actually, *two* pressures arise from osmosis: the crowded molecules inside the cell create an outward-directed **turgor pressure,** which tends to stretch the wall; and the wall exerts an opposing force called **wall pressure.** The two pressures are equal and opposite when the cell is neither gaining nor losing water. In a growing cell, by contrast, turgor pressure exceeds wall pressure.

Turgor and wall pressures are important in giving cells a firm shape, just as air pressure gives shape to a rubber tire. The soft tissues of a plant become limp or **wilted** when the cells lose turgor pressure.

A cell may lose much water if it is bathed in a solution of low water potential, such as strong brine. In that case, the vacuole becomes markedly smaller and the protoplast pulls away from the cell wall. This phenomenon is called **plasmolysis.** Plasmolyzed cells can regain their turgor if they are placed in pure water or a very dilute solution.

A cell may die if osmosis causes severe and prolonged dehydration. For example, if a heavy application of ordinary salt is placed on the soil where weeds are growing, so that the root cells are surrounded by a solution of high concentration, water diffuses from the cells and they become severely dehydrated. If this state is prolonged, the roots die.

In parts of the western United States where rainfall is low and the evaporation rate is high, salts may accumulate on the surface and form what are known as "alkali flats." In such soils, the salt concentration of the soil solution may be so high that many crop plants cannot grow; only species that are especially adapted to tolerate high salt concentration are able to survive. These plants are known as **halophytes.**

HOW WATER IS PULLED THROUGH THE XYLEM

As mentioned earlier, water is usually pulled, rather than pushed, through the xylem. The pulling forces can arise from either transpiration or osmosis, and they tend to occur in regions that are photosynthesizing, growing, or transpiring. Photosynthesis consumes water and produces solutes, giving the cells a lower water potential than other regions. Consequently, water moves from the xylem into the leaf mesophyll and other photosynthetic tissues.

Growing cells take solutes from the phloem. This gives the growing cells an exceptionally low solute component of water potential. At the same time, the walls of growing cells continuously stretch under turgor pressure. Therefore, the pressure does not rise as high in a growing cell as in a mature cell. With low solute and pressure potentials, the growing cells have a strong tendency to absorb water from the vascular system.

In transpiration, water is moved by another means. Here, the primary event is the diffusion of water vapor from the humid air inside the leaf to the dry air outside the leaf. This reduces the water potential of the air channels within the leaf. Water evaporates from the moist walls of the leaf cells, obeying the rule of movement toward regions of low water potential. The walls are left in a drier state, and their matric potential becomes very negative. The walls soon have a potential low enough to pull water from the xylem.

Figure 7.5 illustrates how the drying cell wall pulls in water from the xylem. The diagram represents the upper end of a vessel and shows one of the many small channels by which water can reach an air space in the leaf. For clarity, the diagram greatly exaggerates the pore and the water molecules, compared to the size of the vessel. Also, the walls of parenchyma cells would normally stand between the vessel and the air space. However, these differences do not affect the principles that we wish to illustrate.

Water molecules (dots in the diagram) adhere very strongly to the cellulose molecules of the cell wall. This anchors the edges of the water surface to the top of the pore. Also, the water molecules cohere to one another. The bonds generate a tension that keeps the free water surface as small

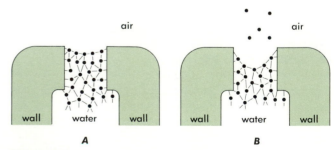

Figure 7.5 How evaporation from the cell wall generates a pulling force (tension) in the xylem. The diagram represents the top end of a vessel with a greatly exaggerated channel that leads through the wall to an air space. Dots are water molecules. Forces of adhesion and cohesion are shown as short lines. *A,* before evaporation; *B,* after evaporation. Note the increased curvature and area of the water surface in *B*. This raises the surface tension, which pulls water toward the surface.

as possible, much as if the water surface were a stretched sheet of rubber. But when water molecules evaporate away, the surface recedes and becomes more deeply curved. This leads to a larger surface area and a higher surface tension. The surface tends to contract, and it can only do so by pulling more water up from below. In brief, then, the tension needed to lift water up the xylem is based on the combination of cohesion between water molecules and adhesion between water and the cell wall.

Once a pull is established in the walls adjoining the xylem water, the remaining question is how the motive force is transmitted throughout the plant and into the soil solution from which the water is obtained. Cohesion is a major part of the answer. The water mass within the plant-soil system is a continuous whole that is bound together by the forces of cohesion. When the walls of the leaf cells are drying and the water potential is reduced, the forces that attract nearby water molecules pull on a chain of molecules that extends all the way down to the soil solution and into each cell of the plant.

The forces that move water up the plant must work against gravity as well as the drag (resistance) exerted by cell walls. The taller the tree, the greater the weight of the water columns that must be drawn up the xylem and the greater the force that must be sustained to keep up a given flow rate. Therefore, a tall tree must be able to survive with drier leaf cell walls than a short plant.

Plant physiologists refer to this picture of water movement as the **cohesion-tension theory** (among other names). In this theory, water is pulled, not pushed, and the water column may be under considerable tension. It has often been asked whether water columns really can sustain the tensions needed for lifting water as high as the tops of tall trees, which may exceed 100 m in height. Laboratory experiments have shown that thin columns of water do indeed have the required tensile strength and can withstand pulling forces great enough to raise water up to a height of 300 m.

But the presence of a single, tiny gas bubble destroys the tensile strength of a water column. With a bubble in the column, tension can raise water no higher than 9.75 m (at sea level). Beyond that point, further increases in tension merely expand the bubble with no further rise in the water column.

Under sufficient tension, a water column may spontaneously produce bubbles. This is called **cavitation.** With sustained tension, bubbles will expand until they meet a barrier such as a cell wall. Bubbles cannot cross cell walls because the wall polymers exert too great an attraction for water. In the plant, the conducting area of the xylem is divided into a large number of microscopically narrow vessels and tracheids. Cavitation may occur in many of these during transpiration, but the bubble will be confined to the vessel or tracheid of its origin, and the rest of the xylem continues to function. When transpiration stops (e.g., at night when stomata close), bubbles formed by cavitation are sometimes resorbed so that the blocked conducting cell can function again.

ROOT PRESSURE AND GUTTATION

The rooted stump of a freshly cut plant often "bleeds" xylem sap from the cut end. Such "bleeding" represents the movement of water up the xylem without transpiration as a driving force. The xylem sap is pushed rather than pulled upward in this case. The pushing force is termed **root pressure.** It is an osmotic phenomenon: the xylem sap has a higher concentration of ions than the soil solution because of the action of ion-transport systems in the plasma membranes of root cells. Thus water moves osmotically from the soil into the xylem. This raises the volume of liquid in the xylem and forces it up the stem.

Root pressure does not exist when transpiration is occurring, but it may be a significant water-moving force in small plants when transpiration is not taking place. When a well-watered, vigorously growing plant is placed under a bell jar, transpiration ceases as the atmosphere in the jar becomes saturated. Continued absorption of water by osmosis in the roots may then result in a slow exudation of water from the tips of the leaves (Fig. 7.6). This loss of liquid from leaves is termed **guttation.** Many plants have specialized openings called **hydathodes** at the tips of their leaves through which guttation liquid passes outward.

Figure 7.6 Guttation from tips of barley (*Hordeum vulgare*) leaves.

HOW ROOTS "LOCATE" WATER

Without extensive irrigation or frequent rainfall, plants usually remove water from the soil faster than it can be replaced by movement from neighboring regions. Hence, for continued water absorption, the roots must continually grow into new areas that have not yet been dried.

Geotropism (see Chapter 9) gives the growth of many roots a general downward orientation, increasing the probability that the root will contact untapped reserves of water. Another form of geotropism causes some roots to grow laterally, while branch roots proliferate in any moist region that a root may pass through.

Experiments show that roots do not "go in search of water." However, roots only grow in moist soil. In practice, this means that if only the upper layers of soil are moist, the roots will be confined to the upper layers. In irrigating a garden or lawn with a hose, one is easily misled into believing, because the soil surface is wet, that sufficient water has been applied. Examination may reveal that at depths of 12 cm or more the soil is very dry—too dry for the growth of roots. The loss of water from soils at levels below the first few centimeters is essentially due to plants. Thus one of the principal reasons for removing weeds from a growing crop is that they take water from the soil.

CARBON DIOXIDE ABSORPTION

Carbon dioxide is often the factor that most limits plant growth: in most plants it is the sole source of carbon for building organic compounds, and it is only a minor component of the air (about 0.03% of air is carbon dioxide). Correspondingly, there is an advantage in being able to trap the largest possible fraction of the carbon dioxide that passes the plant in currents of air.

When air that contains carbon dioxide meets the plant surface, the only reason it enters the plant is that photosynthesis keeps the carbon dioxide concentration in the tissues very low. This causes CO_2 to enter by diffusion along the concentration gradient. The steeper the gradient, the faster the flow of CO_2 into the plant and the larger a fraction of the available CO_2 the plant will be able to trap from a moving airstream. The concentration gradient is steepest when the photosynthetic cells are directly in contact with the air.

In line with these principles, the plant body tends to be structured in a way that brings many photosynthetic cells near to the outside air. The broad, flat shape of the leaf greatly increases the area of plant tissue that is close to the air. Within the leaf, air spaces form an extensive system of channels, giving almost all the photosynthetic mesophyll cells direct contact with an air space that leads to the outside of the leaf.

CONTROL OF CARBON DIOXIDE ABSORPTION AND WATER LOSS

Cells that are directly exposed to dry air quickly lose too much water by evaporation. Thus, the same air exposure that promotes the capture of CO_2 also creates dangerous problems of water loss. The balance between these two factors has doubtless played a great part in the evolution of leaf structure.

In land plants, fatal water loss is prevented by impermeable cork on older stems and by an impermeable waxy cuticle that covers the young shoots and leaves. These coatings limit gas exchange to special apertures, the stomata, which can be opened and closed to establish the best possible balance between CO_2 capture and water loss. A typical leaf may have as many as 6 million stomata (see Chapter 5).

Stomata function in the following way (Fig. 7.7): Each pore is lined by a pair of special epidermal cells known as guard cells. These cells are sausage-shaped; they also differ from ordinary platelike epidermal cells in having thicker walls and an ability to perform photosynthesis. The walls of the guard cells are thickest on the side that faces the pore. This feature allows the stoma to be opened and closed by changing the turgor of the guard cells. When turgor pressure is low, the guard cells press together and the stomatal pore is closed. An increase in turgor pressure stretches the walls of the guard cells slightly, the thinner portions stretching the most. This causes the guard cells to become curved, opening the stomatal pore.

The turgor in the guard cells is controlled by adding and removing solutes. If solutes are added, the water potential decreases; water enters the guard cells from adjacent cells, and the turgor pressure rises. The reverse happens if the guard cells lose solutes.

Guard cells take up potassium (K^+) ions from adjacent cells as the stomata open. They release the K^+ ions as the stomata close. This ion transfer seems to depend on an active transport "pump" in the plasma membrane. Without some compensating factor, a one-way movement of charged ions would soon lead to charge imbalances across the membrane, and the process would stop. A charge balance is achieved by compensating movements of other ions and by the synthesis of organic ions, such as malate. Protons (H^+) and chloride (Cl^-) have also been implicated in the-

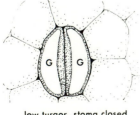

low turgor, stoma closed

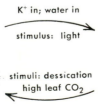

K^+ in; water in
stimulus: light

stimuli: dessication
high leaf CO_2
K^+ out; water out

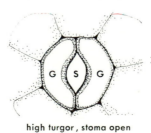

high turgor, stoma open

Figure 7.7 Diagram summarizing the mechanism by which guard cells open and close stomata. The gain or loss of K^+ involves active transport; water follows by osmosis. G = guard cell; S = stomata l pore.

ories of stomatal action. Together with K$^+$, these ions provide much of the solute potential for opening the stomata.

Starch and sugars may also be important in this mechanism. Some stomata have starch grains that are partly converted to free sugar when the stomata are induced to open. The reverse occurs when the stomata are caused to close. These facts show that the solute relations of the stomata are likely to be complex.

Elaborate control systems govern the opening and closing of stomata. Under extreme dehydration, stomata automatically close because the guard cells lose turgor pressure. In some leaves, moderate water stress causes cells to produce the hormone abscisic acid, which signals the stomata to close before dehydration is severe. Later, the hormone is gradually destroyed.

In addition, environmental factors act as control cues for the stomata. Carbon dioxide inside the leaf causes the stomata to close, and light usually causes stomata to open. The result is that the stomata open when there is light for photosynthesis and when the carbon dioxide has been reduced to a point where more is needed.

Some plants have their light signals reversed, so that the stomata open and take in CO$_2$ at night but close in the daytime. As CO$_2$ enters these plants, it is stored by attachment to an organic carrier molecule. In daylight, with the stomata closed, the stored CO$_2$ is gradually released by the carrier and used in normal photosynthesis. This reversed control system is especially common in succulent plants of dry regions. Its value is that less water is lost with the stomata open at night because the lower night temperatures yield slower evaporation.

The rate of transpiration and its effect on the plant depend on the humidity and temperature of the air. Low humidity and high temperature promote evaporation, increasing the rate of transpiration and the extent of drying in the walls of leaf cells. At moderate transpiration rates, the mesophyll cells of leaves have a solute concentration that is high enough to enable them to maintain turgor pressure even though the walls are partially dried. But with increased rates of evaporation, as in the midday heat of a dry summer

day, the leaf cell walls may develop such a low water potential that they extract water from the protoplasts as well as from the xylem. This removes the turgor pressure needed to maintain leaf shape and, consequently, the leaves wilt. Midday wilting is common in big-leafed herbs such as squash plants. It is a temporary condition; normal turgor returns when the rate of transpiration decreases in the evening and the walls within the leaf can regain moisture. Plants that are native to hot, dry areas often have reinforcing sclerenchyma in their leaves, a feature that makes the leaf less sensitive to drying.

Heat promotes transpiration, so it is not surprising that stomata often close as the temperature rises during a summer day. But paradoxically, a still larger increase in leaf temperature may cause stomata of some plants to remain open. This seems to be the case when the leaf is in danger of overheating. Transpiration cools the leaf at the risk of dehydration.

Certain plants called xerophytes (see Chapter 10) are able to survive in extremely dry habitats. They often have roots that penetrate deeply into the soil; or, as in many desert cacti, they have an extensive surface root system that can rapidly absorb limited rainfall from spring or summer showers. Many desert plants have special water storage tissue, as in the plants with fleshy stems and leaves that are commonly known as succulents (see Fig. 10.30).

Anatomical features that decrease water loss are (a) the cuticle of leaves, young stems, and fruits, (b) sunken stomata (Fig. 7.8), (c) distribution of stomata, and (d) reduction of the transpiring surface. In addition, stomatal behavior is an important factor in controlling water loss.

Most of the water lost from a plant passes out through the stomata. Some water, however, is lost through the cuticle. Various modifications in leaf structure reduce cuticular transpiration; for example, thickening of the outer wall of the epidermal cells and the presence of a waxlike material, cutin, in this wall. Most plants of arid and semiarid climates have a thicker cuticle than do those of humid climates.

Most plants have their stomata chiefly on the lower surface, where the temperatures are lower. This reduces the

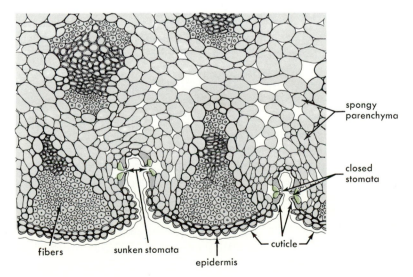

spongy parenchyma

closed stomata

fibers   sunken stomata   cuticle   epidermis

Figure 7.8 Sunken stomata in crypts of a yucca leaf. Note also the thick epidermal cuticle and the large bundles of fibers.

rate of transpiration. In some plants (Fig. 7.8), the stomata occur in pits or folds below the general leaf surface. Here the rate of transpiration is slowed because the water vapor must diffuse for a long distance to reach the dry air.

In most plants, the principal organs of transpiration are leaves. Therefore, any decrease in leaf surface will reduce transpiration and conserve the water absorbed by roots. In corn and some other monocots, the leaves roll up during drought, thus exposing less surface to the air than fully expanded leaves. Cacti and some other desert-adapted plants have no foliage leaves; in these plants the transpiring surface is restricted to the stem surface.

The leaves of many plants are clothed with hairs. The hairs reflect light, reducing leaf temperature. In some plants, such as common mullein, the hairs reduce transpiration. But experiments with other plants have shown that water loss from the leaves is greater when the hairs are present than when they are lacking.

MINERAL ION ABSORPTION

PRINCIPAL MINERAL ELEMENTS

A variety of mineral elements are needed by the plant. These and the typical quantities withdrawn from the soil by wheat plants are shown in Table 7.2. The mineral elements are largely absorbed in the form of ions dissolved in the soil water.

Nitrogen (N) is an essential component of proteins, nucleic acids, and many other molecules that play important metabolic roles. The atmosphere is rich in N_2 gas; it comprises 78% of the air. Some 8000 kilograms of N_2 lie above every square meter of land. However, ordinary green plants obtain their nitrogen mainly as nitrate ions (NO_3^-) from the soil. To a lesser degree, ammonium ions (NH_4^+) also serve as a nitrogen source, and some plants apparently can derive nitrogen from certain organic compounds in the soil.

Nitrogenous fertilizers, both natural and commercial, are usually the most important fertilizers applied to growing plants. The chief commercial nitrogenous fertilizers contain nitrogen (a) in the nitrate form; (b) in the form of ammonia or its compounds; (c) in organic compounds, such as cottonseed meal, and (d) in the amide form, such as urea and calcium cyanamide. Organic nitrogen in complex molecules cannot be used directly by plants but must first be converted into available forms through the action of microorganisms in the soil.

The rate of growth of plants is influenced greatly by available nitrogen. Nitrogen is very mobile in the plant and can be translocated from mature to immature regions. An early symptom of nitrogen deficiency is a yellowing of leaves, particularly the older leaves; a stunting in the growth of all parts of the plant then follows.

An excess of available nitrogen results in vigorous vegetative growth and a suppression of food storage and of fruit and seed development.

Phosphorus (P) is a component of nucleic acids, energy-carrying molecules such as ATP, and the important phospholipids of cellular membranes. Some proteins also contain phosphate groups. The highest concentrations of phosphorus occur in rapidly growing plant parts, such as maturing fruits, seeds, and shoot tips.

Applications of phosphorus to deficient soils may promote root growth and hasten maturation, particularly in cereals. Phosphates (potash) are the principal source of phosphorus for plants.

Sulfur (S) comprises a small but vital fraction of the atoms in many protein molecules. As disulfide bridges (SS), the sulfur atoms aid in stabilizing the folded protein, while sulfhydryl (SH) groups participate in the active sites of some enzymes. Some enzymes require the aid of small molecules that contain sulfur. Finally, the machinery of photosynthesis includes some sulfur-containing compounds (e.g., ferredoxin).

Ordinary green plants obtain sulfur in the form of sulfate ions. When plants deficient in sulfur are given sulfur fertilizers, the most noticeable response is increased root development and a deeper green color of the leaves.

Potassium (K) serves as an enzyme activator. Over 40 enzymes have been found to require potassium for maximum activity. In addition, potassium ions are important in controlling the stomata. This is a mobile element, which will migrate from older tissues to meristematic regions. For example, during the maturing of a fruit crop, potassium moves from leaves into fruits.

Any water-soluble inorganic compound of potassium, such as potassium sulfate, phosphate, or nitrate salts, can be used by plants as a potassium source.

Calcium (Ca) is required by all ordinary green plants. It is one of the constituents of the middle lamella of the cell wall, where it occurs in the form of calcium pectate. Calcium also affects the permeability of cytoplasmic membranes. Calcium may be found in combination with organic acids in the plant. Oxalic acid, for example, is a by-product of metabolism. It is a soluble substance and is toxic to the protoplasm if it reaches a high concentration in the cell. When united with calcium, however, the soluble oxalic acid is converted into the highly insoluble calcium oxalate, which does not injure the protoplasm. There is also evidence that calcium favors the translocation of carbohydrates and amino acids and encourages root development. Calcium deficiency is frequently characterized by a death of the growing points because it is not readily translocated in the plant from mature to immature regions. General disorganization of cells and tissues also results from calcium deficiency. This effect

Table 7.2. Amounts (in kilograms per hectare) of Macronutrients and Micronutrients Removed from the Soil in One Growing Season by a Wheat Crop

	Macronutrients						Micronutrients				
Element	N	K	P	Ca	S	Mg	Fe	Mn	B	Zn	Cu
Amount	85	47	17	13	12	9	0.8	0.6	0.3	0.2	0.03

is consistent with one of the key roles of calcium—maintaining the normal structure of cell membranes.

Magnesium (Mg) is a constituent of chlorophyll; it occupies a central position in the molecule. Many enzyme reactions, particularly those involving a transfer of phosphate (energy metabolism), are activated by Mg^{+2} ions. Magnesium is also vital to the function of ribosomes in protein synthesis.

A deficiency in magnesium results in **chlorosis**—a lack of chlorophyll and, therefore, foliage that is pale yellowish rather than green. This disease is common in cultivated plants, and in many cases, applications of magnesium effect a cure. However, chlorosis can also be caused by deficiencies in other elements such as iron.

The materials listed above provide elements that are known as **macronutrients** (macro = large) because plants require them in large quantities. The macronutrient elements are C, H, O, N, P, S, K, Ca, and Mg.

In addition, plants require tiny amounts of several other elements. These elements, known as **micronutrients,** include at least those listed below. It is difficult to study the plant's need for an element that is required only in trace quantities, because the element may be hard to eliminate from the nutrient medium or the air around the plant. Thus physiologists anticipate that in the future more essential elements will be discovered. (There is already evidence that some plants may require small amounts of sodium (Na) and nickel (Ni).

Iron (Fe) ions are components of several electron-carrying compounds (cytochromes and ferredoxin) essential in respiration and photosynthesis. In picking up and giving off electrons, the iron ion is alternately reduced and oxidized by other compounds.

Although iron is not a component of chlorophyll, it is essential for chlorophyll synthesis. Iron deficiency may be responsible for chlorosis. The quantity of iron required by plants is very small. For example, chlorosis of pineapples in Hawaii is cured by spraying the leaves with iron salt solutions. Orchard trees suffering from iron chlorosis may be cured by injecting iron compounds into the trunk or applying iron salts to the soil.

Boron (B) deficiency causes varied and complex symptoms, such as decreased root and shoot elongation, inhibition of flowering, and darkening of tissues. The metabolic basis of boron action is not clear, though boron is known to play a part in regulating carbohydrate breakdown. Deficiency diseases may be cured by applying very small quantities (10 to 25 kg) of sodium tetraborate (borax) per acre. However, excessive amounts of boron in the soil or water may be toxic; in fact, borates are used as weed killers.

Zinc (Zn) is an activator or a part of several enzymes. One of the earliest effects of zinc deficiency is a drop in production of the growth hormone auxin (see Chapter 9), resulting in inhibition of growth. Deficiency diseases such as "little-leaf" of deciduous fruit trees (Fig. 7.9) can be cured by spraying the trees with zinc salts, by injecting dilute so-

Figure 7.9 Disease of peach (*Prunus persica*) known as "little leaf," caused by a deficiency of zinc. Branch at left untreated; branch at right cured by driving in zinc-coated nails.

lutions of zinc salts into the trunks, or by driving zinc brads into the trunks. These correctives show how small are the quantities of zinc needed by the plant.

Manganese (Mn) activates several enzymes and also plays an essential role in the oxygen-liberating steps of photosynthesis. The most striking deficiency symptom is chlorosis (Fig. 7.10), but this chlorosis is somewhat different from that caused by iron deficiency. In iron chlorosis the young leaves become white or yellow with prominent green veins, while manganese chlorosis gives the leaf a mottled appearance. Spraying or dusting crops with as little as 22 kg of manganese sulfate per hectare often cures deficiencies.

Figure 7.10 Chlorosis of tomato (*Lycopersicon esculentum*) leaf caused by a deficiency of manganese in the nutrient solution.

Figure 7.11 Tomatoes (*Lycopersicon esculentum*) growing in solution with all essential chemical elements except copper. Leaves at left were sprayed with solution containing copper.

Copper (Cu) is a constituent of certain enzyme systems such as ascorbic acid oxidase and cytochrome oxidase, and of the compound plastocyanin, which is part of the electron transport chain in photosynthesis. Deficiency causes abnormalities of growth (Fig. 7.11), especially in plants growing in marsh and peat soils.

Molybdenum (Mo) is important in enzyme systems involved in nitrogen fixation and nitrate reduction. Plants suffering from molybdenum deficiency can absorb nitrate ions but cannot then metabolize this form of nitrogen. If nitrogen in the form of ammonium is supplied to these plants, the effects of molybdenum deficiency are less severe. For some Australian soils that are low in molybdenum, only 140 gm of MoO_3 per hectare, applied once every 10 years, increased pasture yield by six to seven times.

Chlorine (Cl) participates in the oxygen-evolving steps of photosynthesis. This element is present in the soil solution as the very soluble Cl^- ion. Because of the very small quantities required by plants and its almost universal occurrence in soils and air (salt spray travels long distances), chlorine is never intentionally added as a component of fertilizer. In fact rainfall may contain enough chlorine to satisfy the requirements of plants.

THE ABSORPTION OF MINERALS

As mentioned above, the minerals required by plants are normally absorbed by roots from the soil solution. However, there are exceptions: the pitcher plant and the Venus flytrap, for example, capture insects that, on decomposition, release minerals that are absorbed by the specialized leaves that serve as traps. **Epiphytes** (plants that use the aerial parts of other plants as supports for growth and are not themselves rooted in the soil) may obtain minerals by trapping airborne dust and debris. Highly specialized parasitic plants, such as dodder and mistletoe, sink modified stems known as **haustoria** into the vascular tissues of host plants and obtain their minerals in this fashion.

Finally, many plants form symbiotic associations with soil microorganisms. In these associations, called mycorrhizae and bacterial nodules, fungi and bacteria bring in mineral nutrients in exchange for vitamins and carbohydrates. (They are discussed in Chapter 6.)

Aside from the special cases mentioned above, minerals are absorbed from the soil by roots. Appropriately, the root is selective in absorbing minerals, and it can accumulate higher concentrations of minerals than are found in the soil. The plasma membrane and the endodermis play key roles in absorption. Their action is described in the following paragraphs.

The soil solution enters the plant through the epidermis, chiefly the root hairs. Epidermal cells have porous walls that readily allow the solution to pass. The absorbed materials then follow two paths into the interior of the root (Fig. 7.12).

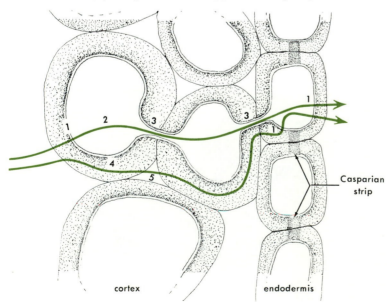

Figure 7.12 Paths of water and ion movement in outer root tissues. Two alternative pathways are shown. Water and solutes move similarly, except at membranes, where transport systems may affect the movement of solutes. 1 = plasma membranes; 2 = protoplasm; 3 = plasmodesmata; 4 = cell walls; 5 = intercellular space.

One path leads through the walls and intercellular spaces of the cortex, and ends abruptly at the endodermis. The brick-shaped cells of the endodermis fit tightly together like tiles. Furthermore, in their lateral walls a waxy material (the Casparian strip) fills the spaces between the microfibrils. The plasma membrane of each endodermal cell is fused to the Casparian strip. As a result, molecules cannot enter (or leave) the vascular cylinder by way of the endodermal cell walls.

The second path through the root is by way of the living protoplasts. Molecules can cross the plasma membrane of any cell, from the epidermis to the endodermis. Once inside the cell, a molecule travels freely from one cell to the next by way of the channels called **plasmodesmata.** By this route, molecules travel between the cortex and the vascular cylinder.

The function of the endodermis, then, is to guarantee that no molecule can enter or leave the vascular cylinder without first crossing a plasma membrane. The membrane allows water molecules to pass freely in both directions. However, the membrane is selectively permeable to solutes and thus will only allow certain molecules to pass. Moreover, the passage is often unidirectional. Such one-way transport involves an active "pump" in the membrane that consumes ATP as an energy source. The result is that scarce minerals such as nitrate and potassium can be accumulated in the xylem in a concentration many times that in the soil.

To appreciate the value of the endodermis, consider what would happen if there were no Casparian strip. Solutes would be actively accumulated in the outer root tissues, as before, and would be deposited in the xylem vessels. But the vessels are dead; they have no membranes, and they readily allow solutes to escape through pits in the walls. If there were no Casparian strip, the solutes would be able to leak back to the soil by way of the cell walls and intercellular spaces.

OXYGEN ABSORPTION AND TRANSPORT

Oxygen is required for respiration by all the cells in the plant. Photosynthesis produces more than enough oxygen to meet the needs of nearby cells during the daylight hours, but much of this oxygen leaks away to the atmosphere and little of it is transported as far as the root tips. Consequently, leaves absorb oxygen from the air at night, and roots usually require a supply of oxygen from air trapped in the soil.

The soil contains a reservoir of atmospheric gases in the spaces between solid particles. This trapped air may exchange with the open air above ground if the soil has a loose texture. Cultivation of heavy and compacted soils improves the air supply. Roots may absorb oxygen directly from the soil gases, or oxygen may dissolve in soil water, from which it is taken up along with the water and minerals. There seems to be no active mechanism for trapping soil oxygen; entry is by diffusion along the concentration gradient that results from the consumption of oxygen in respiration.

The amount of air in the soil depends not only on the pore space but also on the water content of the soil. If water occupies the pore space, air is forced out. Water-soaked (saturated) soil contains practically no air except the amount dissolved in water. If the soil around the roots is continuously water-soaked, plants die because of insufficient oxygen and possibly as a result of the accumulation of carbon dioxide. Most species of land plants will grow normally with their roots in a water solution if it is well aerated by bubbling air through it (Fig. 7.13). Inadequate soil aeration reduces the rate of water and mineral absorption.

Most land plants, including agricultural plants, will not survive long with the root system submerged in unaerated water or surrounded by a soil that is water-soaked. But some plants flourish under such conditions: among them are rice,

Figure 7.13 Effect of aeration on root growth in tomato (*Lycopersicon esculentum*). Left, plants growing in complete nutrient solution through which air was bubbled; right, plants growing in same solution without aeration.

Figure 7.14 Aerial "stump roots" or "knees" of bald cypress (*Taxodium distichum*).

various swamp and marsh plants, and the bald cypress (*Taxodium distichum,* Fig. 7.14). Almost all such plants contain large communicating air spaces in the stem and roots. Some swamp-dwelling trees, such as the bald cypress and certain mangroves, develop special roots that grow upward until their ends are above the water level (Fig. 7.14). These branch roots have a central core of loose tissue through which air moves downward to the submerged organs.

TRANSPORT OF ORGANIC MATERIALS

Within a living plant, substances are constantly moving from one place to another. Four types of movement can be detected: (1) slow diffusion of molecules and ions; (2) moderate movement of materials carried by protoplasmic streaming in living cells; (3) more rapid flow of material in the sieve tubes; and (4) very rapid conduction of water and mineral solutes in the xylem.

DIFFUSION AND PROTOPLASMIC STREAMING

Water and solutes move by diffusion between cells and through cell walls. Diffusion is slow because the numerous collisions between molecules lead to an erratic path of motion.

Once a molecule is inside a cell, its rate of movement increases many times as it is picked up by the protoplasmic stream. Cytoplasm streams at rates of a few to several hundred millimeters per hour. The highest rate of protoplasmic streaming, observed in a slime mold, amounted to 486 mm per hour.

Protoplasmic streaming may carry a molecule the length of a cell in a matter of seconds. In terms of a whole plant, however, this is a very slow process. It would take days for a molecule in the leaf to be carried upward to the growing shoot tip or downward to the roots. Figure 7.15 shows the way in which one solute may be moving by diffusion and protoplasmic streaming in one direction through a series of cells while another solute may be moving in another direction. Diffusion and protoplasmic streaming are the chief methods by which organic solutes move through living plant tissues, except in the sieve tubes.

THE MECHANISM OF PHLOEM CONDUCTION

The phloem sieve tubes move organic materials rapidly throughout the plant. The phloem is the major path of food transport, as is shown by the results of ringing experiments. When a ring of bark is removed from a stem down to the cambium (thus removing the phloem), the soluble carbo-

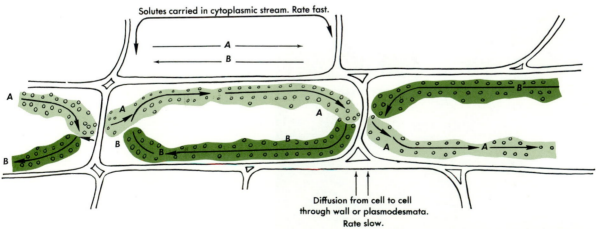

Solutes carried in cytoplasmic stream. Rate fast.

A

B

A

A

B

A

B

B

Diffusion from cell to cell
through wall or plasmodesmata.
Rate slow.

Figure 7.15 Diagram illustrating how two solutes can move in opposite directions by diffusion and protoplasmic streaming in the same tissue.

hydrate content in tissues above the ring increases after a few hours. Below the ring, the concentrations decrease. These conditions would not prevail if foods were moving downward in the xylem tissues.

Translocation of material in the phloem may reach a rate of 100 cm per hour. When this is compared to the much lower rates of protoplasmic streaming, it is evident that normal protoplasmic streaming is insufficient to cause this rapid transport. In fact, the mature functional sieve-tube members have highly modified cellular contents (see Chapter 3) in which protoplasmic streaming does not seem to occur.

The pores in the sieve plates connect the sieve-tube members into a continuous tube. According to one widely accepted theory—the **mass flow theory**—water and solutes move together as one mass along the sieve tube.

The mechanism of mass flow can be illustrated by the flow of photosynthetic products from the leaves to the roots or to other sites of storage and consumption (Fig. 7.16). The process begins with the production of glucose molecules in the mesophyll cells of the leaves. Glucose diffuses into the cells around the leaf veins, and is converted into sucrose. Using energy from ATP, these cells actively transport sucrose into the sieve tubes. Water follows by osmosis, so that the sieve tubes develop high turgor pressures. Meanwhile, roots and growing organs are taking sucrose from other points along the sieve tubes. Water follows by osmosis, and the pressure in those areas decreases. The result is a difference in pressure along the sieve tube. The pressure difference causes the sieve tube contents to flow out of the leaf and along the stem, to regions where sucrose is being used.

The direction of phloem transport depends on the activities of the plant. For example, during its first growing season the sugar beet transfers a great deal of sucrose from the leaves to the roots via the phloem. Next spring the process is reversed; sucrose moves from the roots to the young, growing shoot system.

Although sucrose is usually the principal organic substance carried in the sieve tubes, other organic materials such as hormones and amino acids, and even inorganic ions, may also be carried.

The rate of phloem transport may be controlled by the rates at which solutes are added and removed. But at least two other mechanisms can also affect the rates of movement. They are discussed below.

Callose, a special polysaccharide with unusual linkages between the glucose subunits, is deposited on the sieve plates by the action of enzymes. Other enzymes remove callose from the same locations. The thickness of the callose layer on the sieve plates is therefore dynamically maintained. The thicker the callose, the smaller the pores in the sieve plates. To the extent that pore size influences the rate of flow in sieve tubes, regulation of the callose system may affect the flow rates. The enzymes of the callose system are highly sensitive to mechanical stimuli such as vibrations or crushing. Such stimuli cause an extensive buildup of callose in a matter of seconds, a response that is slowly reversed when the stimuli cease.

A network of protein strands also occurs in the conducting space of each sieve-tube element. Under normal conditions of smooth flow in the sieve tubes, the protein network seems to be firmly anchored to the sieve plates and

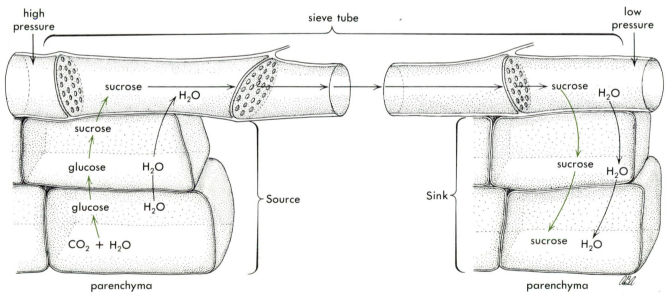

Figure 7.16 The mass flow theory of transport in the phloem. Sucrose is actively transported into the sieve tubes at food source regions in the plant (e.g., leaves or storage organs) and removed at sink regions (e.g., regions of food utilization or storage). Water follows by osmosis and raises the pressure in the sieve tubes at the source region and lowers the pressure at the sink region. The sieve tube contents flow from high-pressure to low-pressure regions.

...fect the transport of materials. But abrupt changes ...ow rate, such as occur when a stem is cut, cause ...rotein network to rip loose and pile up against the near-... sieve plate, blocking the sieve pores. Both the callose and the protein network may be significant in limiting the loss of phloem sap from stems and petioles that have been injured.

SUMMARY

1. Water is absorbed chiefly from the soil by roots and moves throughout the plant in the xylem.

2. Some water is consumed in metabolism or retained in growth, but most of it is lost by evaporation (transpiration).

3. Water moves from regions of high water potential to regions of low water potential.

4. The water potential of a solution is raised by adding pressure and is lowered by adding solutes or materials such as wall fibrils that absorb water.

5. The water potential of air increases as its humidity increases.

6. If two solutions are separated by a membrane that will permit water but not solutes to pass, water will move through the membrane from a high to a low water potential. This water movement is called osmosis.

7. Because plant cells are enclosed in walls, water uptake creates turgor pressure, which gives the cell a rigid shape. Wilting results from the loss of turgor.

8. The cohesion-tension theory states that evaporation of water from the leaf pulls water upward through the xylem.

9. Cohesive and adhesive forces between molecules prevent the xylem water columns from breaking under tension.

10. When there is no transpiration, water enters the xylem by osmosis, creating positive root pressure and pushing water up the plant.

11. Roots do not seek water, though growth is generally directed downward. Roots grow only in moist soil.

12. Carbon dioxide is absorbed from the air, reaching leaf cells through stomata and intercellular spaces.

13. Gas exchange is limited by cork and cuticles and is controlled by stomata.

14. Stomata are opened and closed by raising and lowering the turgor pressure of guard cells. The changes in pressure are achieved by pumping potassium and other ions into and out of the guard cells, and by producing or removing organic molecules. Water follows by osmosis. Light exposure, CO_2 depletion, and heat stimulate the stomata to open.

15. Minerals enter the plant through the roots, moving passively with the water flow except when they meet membranes, which may present a barrier. Carriers control ion movements through membranes.

16. The impermeable Casparian strip, in the walls of the endodermis, prevents water and solutes from using the cell walls as an avenue for moving between the vascular system and outer root tissues. This permits membranes to control traffic.

17. Leaves absorb oxygen from the air at night. Roots usually absorb O_2 from air in the soil.

18. Organic solutes move through the phloem between regions of production, storage, and use.

19. The mass flow hypothesis states that the solution in the sieve tubes moves from regions of high turgor pressure to regions of low pressure. The high source pressures are thought to result from active pumping of sucrose into sieve tubes, followed by osmotic entry of water. Low pressures occur where sucrose is being removed, because water leaves the sieve tubes by osmosis at those points.

20. The rate of movement in the phloem may also be affected by a protein network that can plug the sieve pores when sieve tubes are injured; it is also affected by enzymes that control the size of the sieve pores by depositing and removing callose.

8 PHOTOSYNTHESIS

Cells are constantly expending energy, whether they are growing, transporting materials, or repairing damage. The fuels utilized by cells are molecules such as sugars and fats, which are degraded into low-energy wastes as they are used. The quest for additional energy is a central theme in the life of an organism.

In this respect, organisms can be divided into two categories: **autotrophs** ("self-feeders"), which absorb energy from the sun and make their own fuels; and **heterotrophs** ("other-feeders"), which depend on fuels made by other organisms because they cannot use the energy of light.

As heterotrophs, humans have a vital interest in the welfare of the green plants and algae, which are the world's principal autotrophs. In this chapter we will investigate the process of **photosynthesis,** by which plants convert light to chemical energy.

THE DISCOVERY OF PHOTOSYNTHESIS

Though humans have always depended on plants, the question of where plants obtain their nourishment was not seriously studied until the mid-1600s. Until then, scholars drew their ideas from Aristotle, who had written that roots obtain nutrients from decaying materials (humus) in the soil. This view agreed with the common knowledge that plants grow better if manure is added to the soil.

Perhaps the first real challenge to Aristotle came in 1450 when Nicolas of Cusa pointed out that crop after crop can be planted and yet the soil remains undiminished. His view was that most of the plant material must come from water. The first person to test this idea was a Belgian chemist and physician named Jan van Helmont, whose work was published in 1648. He planted a willow branch in a tub of soil and added rainwater as needed. After 5 years the willow weighed about 76,000 grams (g), while the soil had lost only about 60 g. Clearly most of the plant's weight came from something other than soil, and water seemed the most logical source.

Most of a century passed before a new idea emerged. The English scientist Stephen Hales suggested in 1727 that part of the plant's substance might come from the air. Hales experimented extensively with plants and is regarded today as the father of plant physiology. He observed that wood emits gases when heated, and it seemed logical that leaves might do the opposite, absorbing gases from the air. However, the idea could not be tested because at that time the nature of gases was still unknown.

Further insights were closely connected with the origin of modern chemistry, in which the study of gases played a large part. It was long known that animals will suffocate in a confined space where a candle has been allowed to burn out. In 1772, the Englishman Joseph Priestley announced the surprising discovery that a plant can replenish the exhausted air. The result was disputed; others could not always duplicate it. Then, in 1779, the Dutch physician Jan Ingenhousz settled the dispute with the most fundamental discovery of all: plants restore air only in the light, and only the green parts are effective. The nongreen parts exhaust the air just as animals do. This discovery emerged in a time when the composition of air was still unclear. But within a few years the Frenchman Antoine Lavoisier (the founder of modern chemistry) had established the idea of oxygen as a substance, and by 1784 it was clear that green plants produce oxygen in the light.

About the same time, the Swiss Jean Senebier concluded that plants not only purify exhausted air, but they also require something that is put into the air by respiration and combustion. By 1804 the required material was known to be carbon dioxide, and another Swiss scientist, Theodore de Saussure, showed that half the dry weight of the plant consists of carbon that is taken from carbon dioxide. These discoveries, stretching over 150 years, could be summarized in the simple equation:

$$\text{carbon dioxide} + \text{water} \xrightarrow[\text{green plants}]{\text{light}} \text{oxygen} + \text{organic matter}$$

What is the "organic matter" that arises in photosynthesis? Later studies showed that plants accumulate sugars when exposed to light. Assuming that glucose is the direct product, the equation for photosynthesis becomes:

$$6CO_2 + 6H_2O \xrightarrow[\text{green plant cell}]{\text{light energy}} C_6H_{12}O_6 + 6O_2$$

With this equation it is easy to see that light energy has been converted to chemical energy. Water and carbon dioxide are stable and low in energy. By contrast, anyone who has burned paper (which consists largely of cellulose, made of sugar) knows that great amounts of energy are released when oxygen reacts with sugar. In short, the equation suggests that photosynthesis stores energy by separating oxygen from other atoms.

Here matters rested through the nineteenth century, for the workings of photosynthesis are too intricate to be plumbed without such modern tools as radioisotopes and electron microscopy.

...ne last triumph of traditional methods came in ...when the Dutch physiologist C. B. van Niel made ...able deduction about the source of the oxygen that ...uced in photosynthesis. Previously, the equation for ...osynthesis had suggested that the O_2 comes from CO_2. ...y examining photosynthesis in bacteria, van Niel drew quite a different conclusion: that the oxygen comes from water. In various bacteria, hydrogen sulfide or molecular hydrogen or various organic compounds take the place of water in photosynthesis. The green and purple sulfur bacteria produce sulfur instead of oxygen:

$$CO_2 + 2H_2S \xrightarrow{\text{light}} \underset{\text{carbohydrate}}{(CH_2O)} + H_2O + 2S$$

When H_2 is used in place of water, the following equation results:

$$CO_2 + 2H_2 \xrightarrow{\text{light}} (CH_2O) + H_2O$$

Comparing these cases, van Neil concluded that a general equation for photosynthesis should be written as follows:

$$\underset{\substack{\text{carbon} \\ \text{dioxide}}}{CO_2} + \underset{\substack{\text{hydrogen} \\ \text{donor}}}{2H_2A} \xrightarrow{\text{light}} \underset{\text{carbohydrate}}{(CH_2O)} + \underset{\text{water}}{H_2O} + 2A$$

The hydrogen donor H_2A can be H_2O, H_2S, H_2, or any other substance capable of donating hydrogen to CO_2 in the process of photosynthesis.

To test van Niel's idea, a group of biochemists supplied plants with water that contained the unusual oxygen isotope, ^{18}O instead of the common isotope ^{16}O. The oxygen liberated in photosynthesis contained ^{18}O, as van Niel would have predicted.

These findings have an important bearing on the way light energy is trapped and stored. We said before that energy is stored by separating oxygen from other atoms. Now we can refine this view. In green plants, light energy is used to separate oxygen from the hydrogen of water.

TWO PHASES OF PHOTOSYNTHESIS

Although the overall equation of photosynthesis seems simple, the discovery that O_2 comes from H_2O rather than from CO_2 makes it clear that there are hidden complexities. A good understanding of photosynthesis requires that the process be taken apart and studied piece by piece. Plant biochemists are still working on this task.

One successful approach has been based on the examination of chloroplasts, the organelles that perform photosynthesis. Chloroplasts are abundant in the cells of leaves and other green parts of plants; the green color is due to light-absorbing compounds, or **pigments,** within the chloroplast.

Chloroplasts are very fragile, but if properly treated, they can be removed in good working condition from the cell. In the test tube, their operation can be explored without the

confusion that occurs when other kinds of organelles are in action at the same time.

The procedure for isolating chloroplasts or other organelles (fractionation) begins with a method for breaking open cells. Fragments of tissue are placed in a dilute solution of salts and organic compounds. The precise composition of the medium is important and must be determined by trial and error. The cells are broken by grinding with sand in a mortar, or with a kitchen blender. This releases some of the chloroplasts intact, though others are broken. The ground material is filtered to remove large solid fragments. Then the suspension of organelles is spun in a centrifuge to separate the organelles according to their density or mass. Again, experiments are needed to determine the most effective procedure. The final result is a solution in which chloroplasts are suspended almost free of other organelles. The isolated chloroplasts can produce carbohydrate and oxygen if they are given light, carbon dioxide, and water.

The electron microscope has shown that a chloroplast has three principal parts: (1) a surrounding envelope consisting of a pair of membranes; (2) a liquid called **stroma** which is enclosed by the envelope; and (3) a set of flattened membrane-lined sacs called **thylakoids** which are embedded in the stroma (see Figs. 3.7 and 8.1). Chloroplast suspensions contain both intact chloroplasts that have all their membranes, and damaged chloroplasts that have lost their

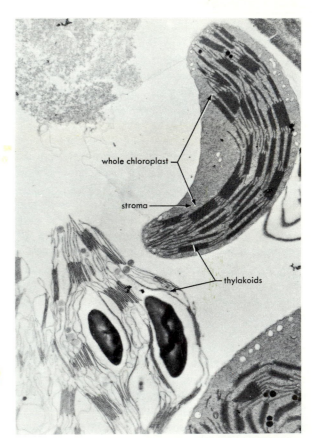

whole chloroplast

stroma

thylakoids

Figure 8.1 Electron of micrograph of isolated chloroplasts of *Vicia faba,* showing one plastid with stroma and one plastid that has lost its outer envelope and stroma, ×2500.

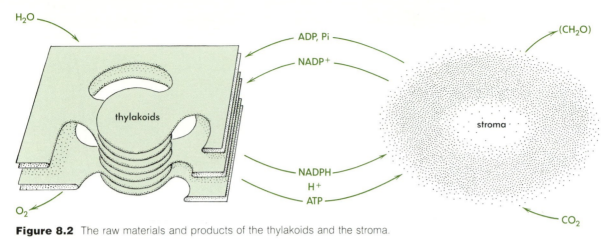

Figure 8.2 The raw materials and products of the thylakoids and the stroma.

envelope and stroma and retain only the thylakoids. This important fact makes it possible to isolate thylakoids and to study their activity apart from the stroma.

These experiments revealed a division of labor within the chloroplast (Fig. 8.2). The green thylakoids capture light and produce O_2; the colorless stroma captures carbon dioxide and produces sugar. Several compounds pass between the thylakoids and stroma, carrying energy and materials for sugar synthesis. Because they depend on light, the reactions of the thylakoids are often called the **light reactions.** The reactions of the stroma are called the **dark reactions** because they do not directly involve light; however, they normally occur in daylight.

In the following sections we will examine the thylakoid and stroma reactions in detail.

THE THYLAKOID REACTIONS

Between thylakoids and stroma, the thylakoids perform the more noteworthy task: they capture light and convert it to chemical bond energy. Some of the energy is used to make ATP from ADP and phosphate. More demanding, however, is the task of separating electrons and protons from the oxygen of water, for use in forming sugar. This requires a great deal of energy, for the oxygen atoms have a strong affinity for electrons. Two electrons and one proton from a water molecule are transferred to a molecule of $NADP^+$, while the other proton is released as H^+. The oxygen atoms from two water molecules join in a pair to form the electron-hungry compound O_2:

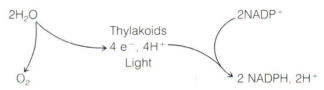

To accomplish their task, the thylakoids present a very complex system indeed. It is still only partly understood, though we can offer an outline of its nature. First we will investigate how thylakoids absorb light energy.

THE LIGHT-CAPTURING SYSTEM

Thylakoids capture light by means of photosystems—**photosystems I** and **II**—which are embedded in the membrane (Fig. 8.3). The photosystems appear as lumps scattered over

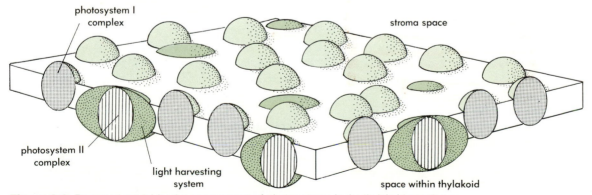

Figure 8.3 Proposed model for the arrangement of photosystems in the thylakoid membrane. Each complex is a group of proteins to which chlorophyll and carotenoid molecules are bound. The Photosystem I and Photosystem II complexes include reaction centers. Adapted from P. A. Armond, L. A. Staehelin, and C. J. Arntzen, *J. Cell Biol.* 73:400–418 (1977).

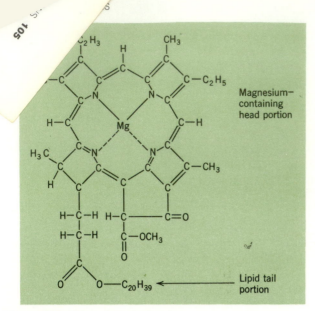

Figure 8.4 Chlorophyll *a* molecule.

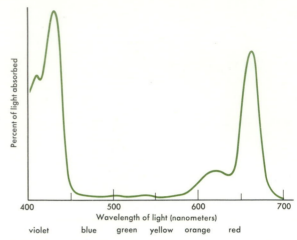

Figure 8.5 Absorption spectrum of chlorophyll. The graph shows the fraction of received light that is absorbed when the pigment is exposed to various wavelengths of light. The relation between wavelength and color of light is also shown.

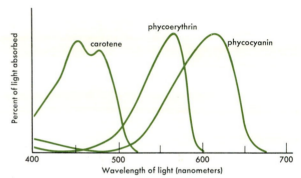

Figure 8.6 Absorption spectra of three accessory pigments.

the membrane. Each photosystem II unit is coupled with secondary light traps called **light-harvesting complexes.**

Biochemists have isolated the photosystems and analyzed their contents. Each one contains several protein molecules plus many molecules called **pigments.** The pigments absorb light energy, while the proteins hold the pigments in place and channel the flow of energy. Each photosystem has an "antenna" of light-collecting pigments, plus a **reaction center** where the trapped energy is converted to an electrical form. The accessory light-harvesting complexes have pigments and proteins, but they lack reaction centers. They pass energy to photosystem II.

Chloroplasts contain several kinds of photosynthetic pigments. Most important is **chlorophyll a** (Fig. 8.4), which occurs in every photosystem and is essential for photosynthesis. It has a flat, light-absorbing "head" made of four linked rings, with a magnesium (Mg^{2+}) ion in the center; and a hydrocarbon "tail" that probably helps to bind the pigment to the photosystem. There are many molecules of chlorophyll *a* in each photosystem; two of them occupy the reaction center.

Other pigments are also associated with the light-harvesting complexes: chlorophyll *b*, which differs from chlorophyll *a* in the small groups that line the absorbing ring; and **carotenoids,** which are related to the compounds that color tomatoes and carrots. These **accessory pigments** absorb light and pass the energy to chlorophyll *a*.

Accessory pigments allow the chloroplast to capture light that would have been missed by chlorophyll *a*. These pigments are needed because chlorophyll *a* does not absorb all colors of light equally well. The absorption behavior of chlorophyll is best shown by means of an **absorption spectrum** (Fig. 8.5). In this figure the color is related to a measurable property of light called the **wavelength.** Chlorophyll *a* absorbs red, blue, and violet light very well but is

a poor absorber in the green and yellow regions of the spectrum. This is why leaves are green: sunlight is a mixture of many colors, and chlorophyll removes the red, blue, and violet portions. The green is reflected or passed through the leaf, where it can be seen. The accessory pigments absorb some of the light that escapes chlorophyll (Fig. 8.6).

How do pigments absorb light? A beam of light acts as if it is composed of many fast-moving light particles called **photons.** Each photon carries a fixed amount, or **quantum,** of energy. In light absorption, a pigment molecule captures a photon. The photon disappears, and its energy causes the electrons in the pigment to change their pattern of motion. The molecule is said to enter an **excited state,** and the absorbed energy is called **excitation energy** (Fig. 8.7). Before absorbing light, the low-energy molecule is said to be in its **ground state.** In equations, the excited state is symbolized by placing an asterisk beside the name of the pigment.

Absorption is only the first step in capturing light energy. Excitation energy is very unstable and stays in the molecule only for a fraction of a second. To be stabilized and put to use, the excitation energy must make its way

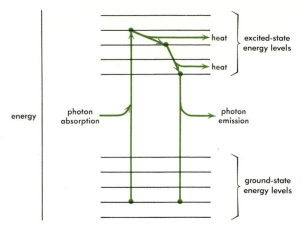

Figure 8.7 Absorption of a photon by a pigment causes an electron to move from a low-energy, ground-state condition to a high-energy excited state. The energy difference is the excitation energy, equal to the photon energy. The excited electron may give up energy to other molecules as heat or light (fluorescence).

from the absorbing pigment to a reaction center. This is accomplished by a process called **resonance transfer,** in which the energy passes from one pigment molecule to another:

$$\text{pig}_1{}^* + \text{pig}_2 \xrightarrow{\text{resonance transfer}} \text{pig}_1 + \text{pig}_2{}^*$$

Resonance transfer is poorly understood, but we do know that efficient transfer requires the pigments to be close together and precisely positioned. The proteins that bind pigments are probably important in providing the right conditions.

If the excitation energy does not reach a reaction center quickly enough, it may be lost in two ways (Fig. 8.7). Part of the energy is converted to **heat** through collisions between molecules. (The conversion of light to heat is familiar to anyone who has worn black—highly absorbing—clothing on a sunny day.) Alternatively, some or all of the energy may be converted into a new photon, which escapes from the chloroplast; such light emission is called **fluorescence.** From the standpoint of photosynthesis, heat and fluorescence are a waste of energy. Therefore, photosynthesis depends on putting the energy to use before these events take place.

Upon reaching the reaction center, the excitation energy causes chlorophyll *a* to pass an electron to a carrier compound:

$$\begin{array}{c} \text{chl}^* \\ \\ \text{chl}^+ \end{array} \xrightarrow{\ \ e^-\ \ } \begin{array}{c} \text{acceptor} \\ \\ \text{acceptor}^- \end{array}$$

With this event, the excitation energy is converted into more stable electrical energy. The nature of the electrical energy is easy to understand if you recall that positive and negative

charges attract one another. The electron transfer results in a positive chlorophyll ion and a negative carrier ion. The electron is strongly attracted to the chlorophyll ion but cannot return to it because of the structure of the photosystem. Therefore, energy has been stored in the separation of electrical charges. From this point on, the thylakoid performs useful work by controlling the way electrons move back to the positive chlorophyll ions. This process is described in the next section.

FROM ELECTRICAL ENERGY TO CHEMICAL BOND ENERGY

Figure 8.8 is a simplified view of the way thylakoids produce energy-rich chemical bonds. Two kinds of working units are involved: ATP-producing units called **ATPases,** and light-driven **electron transport systems.** Hundreds or thousands of these units are embedded in the membrane; for simplicity, only one of each is illustrated. A summary of their operation follows:

1. Light drives electrons from water through the electron transport system to $NADP^+$. The product, NADPH, is unstable and donates electrons and hydrogen for sugar synthesis. NADPH contains much of the energy that was derived from light.

2. The electron-transport process subtracts hydrogen ions from the stroma and adds them to the fluid inside the thylakoid. The result is a **proton gradient**—a difference in H^+ concentration across the membrane. This is similar to the proton gradient that occurs in the respiring mitochondrion, as described in Chapter 2. The gradient stores energy, which is released when protons escape from the thylakoid.

3. ATPase enzymes are embedded in the membrane, and they draw energy from the proton gradient to make ATP.

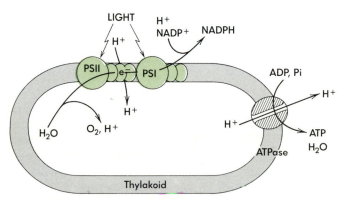

Figure 8.8 How thylakoids handle energy. An electron transport system (green) uses light energy to drive electrons from water to $NADP^+$ in a way that produces a proton (H^+) gradient across the membrane. Energy is stored in the proton gradient and in the production of O_2 and NADPH. An ATPase enzyme (oval at right) draws energy from the proton gradient to produce energy-rich ATP. Many ATPase units and electron transport systems are present in a thylakoid. PSI and PSII are photosystems I and II.

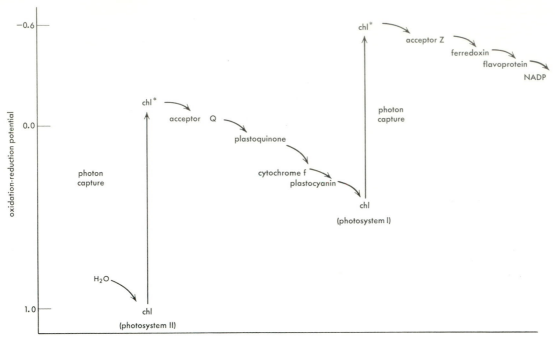

Figure 8.9 Oxidation-reduction potential diagram of the electron carriers and photosystems I and II. The arrows show the path of electrons in photosynthesis. Carriers high on the scale tend to give up electrons to lower carriers; electrons lose energy as they move to carriers that are lower on the potential scale. Symbols: chl = chlorophyll *a* in the ground state; chl* = chlorophyll *a* in the excited state.

With the help of Fig. 8.9, we can follow the changes in energy that occur during light capture and electron transport. The vertical scale (the oxidation–reduction potential) can be interpreted in two ways: it shows the energy that is associated with having an electron on a given carrier, and at the same time, it measures the tendency of each carrier to give up the electron. A carrier high on the scale will transfer an electron to a carrier lower on the scale, which results in a release of energy.

Tracing the events in Fig. 8.9, observe what happens when photosystem I absorbs a photon. At the reaction center, excitation gives chlorophyll a strong tendency to transfer an electron; in effect, the excitation energy is concentrated in the electron. The electron jumps to acceptor Z and then quickly moves from one carrier to another, losing a little energy at each step. Arriving at $NADP^+$, the electron still retains much of the original energy.

Meanwhile, the chlorophyll at reaction center I was left as a positive ion. Its remaining electrons are in the ground state. The chlorophyll ion holds these electrons very strongly and will not absorb another photon until the lost electron is replaced. In this condition, photosystem I must await events in photosystem II.

Meanwhile, a photon strikes photosystem II and drives an electron from the reaction center to acceptor Q. This electron migrates down the chain of carriers until it meets and joins the chlorophyll ion of photosystem I. Now photosystem I can absorb more light.

But now photosystem II has a positive chlorophyll ion, which has a powerful tendency to replace its lost electron.

By a route that is still poorly understood, an electron passes from water to the chlorophyll ion.

In summary, it is as if two photons have pushed an electron from water to $NADP^+$. The electron has moved from a molecule with a high attraction for electrons (water) to a molecule with a low attraction for electrons ($NADP^+$). The result is a storage of energy.

Cyclic versus Noncyclic Electron Transport

The events described above involve a one-way flow of electrons from water to NADP, with oxygen, NADPH, and H^+ as products. This is called **noncyclic electron transport.** Many biochemists believe that light also drives electrons in a closed circuit that produces neither O_2 nor NADPH (Fig. 8.10). Here, photosystem II is not involved. Light strikes photosystem I and drives electrons onto acceptors. The electrons pass through a chain of carriers and return to photosystem I. This is called **cyclic electron transport.** It is useful because, as in noncyclic electron transport, it drives protons through the membrane and feeds the proton gradient.

ATP SYNTHESIS (PHOTOPHOSPHORYLATION)

As described, electron transport leads to a difference in hydrogen ion concentrations across the thylakoid membrane. In effect, light has caused an electrical current that drives an active "proton pump." The resulting proton gradient stores energy, just as energy is stored when water is pumped to a reservoir at the top of a hill. Exactly the same situation occurs in the mitochondrion; the basis of its energy storage was described in Chapter 2.

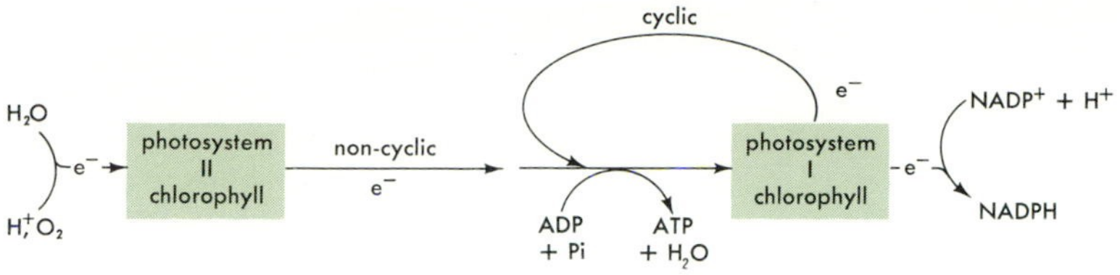

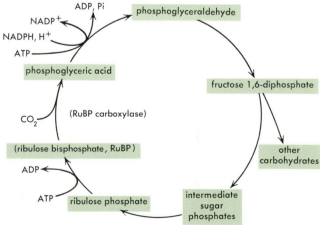

Figure 8.10 Noncyclic photophosphorylation (ATP formation) draws energy from the flow of electrons between photosystems I and II. In cyclic photophosphorylation the electrons start and end in photosystem I.

As in the mitochondrion, ATPase enzymes draw energy from the proton gradient to make ATP (Fig. 8.8). For simplicity, the ATPase is pictured as having a central channel through which protons pass to the stroma. For every two or three protons that pass, a molecule of ATP is produced. In this way, much of the energy in the proton gradient is replaced by the more stable bond energy of ATP. (Some of the energy is lost as heat.) Biochemists are currently seeking the precise way in which the ATPase causes this energy transformation.

CARBON FIXATION: MAKING ORGANIC MOLECULES FROM CO_2

While some investigators have been exploring the thylakoids, others have been examining the process by which carbon dioxide is incorporated into organic molecules such as sugars. This process occurs in the *stroma*, the fluid that surrounds the thylakoids.

Six molecules of carbon dioxide are needed to make one molecule of sugar. Eighteen molecules of ATP are split in the process, and hydrogen is taken from a dozen molecules of NADPH. With so many molecules involved, one can easily imagine that sugar formation is a stepwise process with many enzymes and intermediate products. Tracing the steps was a difficult research problem, for many sugar molecules are being made at once in the chloroplast, and at any moment there are molecules in many stages of completion. In addition, the chloroplast produces numerous substances that are not involved in photosynthesis at all—molecules such as lipids that are being built for incorporation into membranes. These other metabolic activities interfere with studies of photosynthesis.

Biochemists overcame these problems by using carbon dioxide molecules labeled with the radioactive isotope of carbon, ^{14}C. In chemical reactions, ^{14}C (called "C-fourteen") acts almost exactly like ordinary carbon (^{12}C). However, ^{14}C atoms occasionally "decay." The carbon atom becomes an atom of nitrogen, and a high-energy particle is released. The particles can be detected, and this makes it possible to identify molecules that have taken up ^{14}C.

To analyze the steps of carbon fixation, biochemists introduced radioactive carbon dioxide into a vessel that contained photosynthetic cells. At intervals thereafter, samples of cells were removed for analysis. The cells were quickly killed to stop their metabolism, and the cell contents were divided into fractions containing different kinds of molecules. Each fraction was then checked for the presence of radioactivity.

In the early 1950s, Melvin Calvin and his colleagues at the University of California used this method to work out the early carbon pathways of photosynthesis. Calvin eventually received the Nobel Prize for this work. Calvin's group used cells of a unicellular alga, which appear to follow the same steps of photosynthesis as the leaves of higher plants. When they allowed photosynthesis to proceed for 1 minute in the presence of $^{14}CO_2$, they found radioactive carbon in a wide variety of compounds. But when photosynthesis was stopped after only a few seconds, most of the labeled carbon atoms were found in molecules of phosphoglyceric acid (PGA). PGA, then, was the first identifiable product of carbon fixation. By allowing photosynthesis to continue for progressively longer times, the biochemists found additional radioactive products, which must have been made from PGA.

THE C_3 PATHWAY

Through such experiments, a total picture of carbon fixation and sugar formation gradually emerged (Fig. 8.11). This complex pathway is called the **Calvin cycle,** or the **C_3 pathway.** The latter name relates to the fact that the first product, PGA, has three carbon atoms. The Calvin cycle occurs in all green plants.

The chief steps of the Calvin cycle are:

1. CO_2 combines with ribulose bisphosphate (RuBP), a five-carbon sugar with two phosphate groups. The product immediately breaks into two molecules of the three-carbon compound phosphoglyceric acid (PGA). These events require the enzyme RuBP carboxylase, which is probably the most abundant enzyme on earth.

2. Two molecules of PGA are reduced to form two mole-

Figure 8.11 Diagram showing some steps in the carbon cycle of photosynthesis, the Calvin, or C_3, cycle.

cules of a triose sugar phosphate, phosphoglyceraldehyde. The energy that drives this reaction comes from NADPH and ATP, and energy from these molecules is stored in the newly formed triose sugar. ADP and $NADP^+$ are regenerated.

3. Two triose phosphates combine to form a six-carbon sugar phosphate, fructose diphosphate. By this process the plant has essentially added one CO_2 molecule to a five-carbon sugar to produce one molecule of a six-carbon sugar.

4. As this process continues, some of the fructose phosphate may be transformed through other reactions into other carbohydrates, including sucrose and starch.

5. Some of the fructose phosphate molecules are used to form new molecules of RuBP, the compound that accepts CO_2 in step 1. Thus the whole process forms a cycle of reactions, with CO_2 from the air and hydrogen from water entering the cycle, and various sugars being produced.

THE C₄ PATHWAY AND PHOTORESPIRATION

In addition to the Calvin cycle, an auxiliary path of carbon fixation—the **C₄ path**—occurs in many plants that have evolved in regions of high light intensity, high temperature, and drought. The C₄ path is so named because its first stable product is a compound with four carbon atoms, rather than PGA.

Figure 8.12 illustrates the C₄ path and how it relates to the Calvin, or C₃, cycle. In the C₄ path, carbon dioxide is first taken up by an enzyme called PEP carboxylase. This enzyme has a stronger affinity for carbon dioxide than does RuBP carboxylase. It adds CO_2 to the three-carbon compound **phosphoenolpyruvate** (PEP), to form the four-carbon product **oxaloacetate** (OAA). Later, other enzymes may convert OAA into two more C₄ compounds, malate and aspartate. The C₄ products may accumulate to high levels.

C₄ plants have a division of photosynthetic labor within the leaf (Figs. 5.10 and 8.12). The ordinary mesophyll cells perform C₄ photosynthesis and are very efficient at extracting carbon dioxide from the air spaces. The C₄ products then move to the cells that line the smaller leaf veins (the **bundle sheath cells**) for further processing. Entering a bundle sheath cell, the C₄ compound is broken down into pyruvate and CO_2. Then the carbon dioxide is made into sugar by the Calvin cycle, while the pyruvate returns to the mesophyll for reuse. This arrangement delegates the capture of carbon dioxide to the efficient C₄ system and leaves sugar formation to the Calvin cycle in the cells that line the veins. The sugars can immediately be passed to the phloem for export from the leaf.

C₄ photosynthesis may have its chief value in avoiding **photorespiration,** a wasteful process that occurs in C₃ plants. Like true respiration, photorespiration oxidizes a sugar and produces CO_2. But despite this similarity, photorespiration has no relation to the energy-yielding mitochondrial process; photorespiration occurs in different organelles and does not generate ATP.

Photorespiration begins with the enzyme RuBP carboxylase, which normally joins carbon dioxide to ribulose bisphosphate (RuBP) in the Calvin cycle. At times, this enzyme

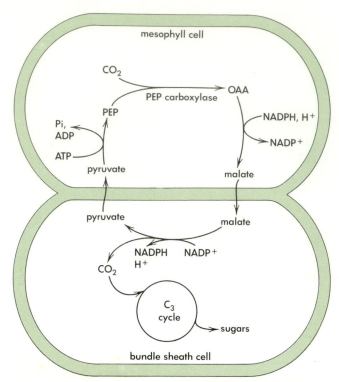

Figure 8.12 C₄ photosynthesis. In the mesophyll cells, the enzyme PEP carboxylase captures CO_2; malate and other four-carbon products (not shown) diffuse to the bundle sheath, where they release CO_2 for the production of sugars in the Calvin C₃ cycle.

reacts with O_2 instead of CO_2. When this happens, RuBP is cleaved into a three-carbon compound and the two-carbon molecule **glycolic acid.** Glycolic acid diffuses to an organelle called the **microbody,** where it is oxidized with the release of CO_2. The result is a loss of previously captured carbon, with no compensating benefit. Biochemists currently believe that photorespiration is an accident that results from the ability of O_2 to substitute for CO_2 in the enzyme's active site.

Photorespiration is particularly intense when oxygen is plentiful and carbon dioxide is scarce, because in this situation, oxygen is most likely to enter the active site of RuBP carboxylase. These conditions prevail on hot, bright days, when the chloroplasts are most active. Then photorespiration may release up to 50% of the captured carbon.

The C₄ system avoids photorespiration because the mesophyll cells capture CO_2 by means of the enzyme PEP carboxylase, which does not react with O_2. Therefore, the mesophyll cells have no photorespiration. The Calvin cycle is restricted to the bundle sheath cells, which continuously receive C₄ acids from the mesophyll cells. The acids break down and maintain a high CO_2 concentration in the bundle sheath, which keeps photorespiration to a minimum. Any CO_2 that does escape from the bundle sheath is immediately recaptured by mesophyll cells. Since the mesophyll cells remove nearly all the CO_2 from the air spaces in the C₄ leaf, carbon dioxide diffuses in from the air much faster than it would enter the leaf of a C₃ plant. With such speed and efficiency, a C₄ plant may produce two to three times as much sugar as a C₃ plant during the hot, bright days of the summer. This is part of the reason why such C_4 crops as sugarcane and maize are so productive. Under milder conditions, when photorespiration is less prominent, the C_3 plants are more efficient than C_4 plants because they expend less energy to capture CO_2.

FACTORS AFFECTING PRODUCTIVITY

The photosynthetic leaf is a highly evolved unit with many features that promote an efficient capture of light and carbon dioxide. Nevertheless, only about 0.3 to 0.5% of the energy that strikes a leaf is stored in photosynthesis. The yield may be increased by a factor of 10 by providing ideal conditions. In a hungry world, there is much to be gained by exploring the conditions that limit plant productivity and how they might be overcome.

Productivity is determined partly by the environment and partly by the hereditary traits of the plant. One avenue toward greater productivity is to breed more efficient plants. This approach has already been quite successful in the case of cereal grains. For instance, the productivity of maize has been increased over 200% since the 1960s (see Chapter 11). Recently, Norman Borlaug received the Nobel Prize for developing rice strains so productive as to constitute a "green revolution" in tropical countries.

Photorespiration is one example of a hereditary trait that reduces yields, and the C_4 system offers a compensating hereditary advantage. Now that research has uncovered these traits, perhaps breeding programs or the more advanced recombinant DNA technology (see Chapter 11) will lead to new C_4 varieties as well as C_3 plants that are less prone to photorespiration.

THE ENVIRONMENT AND PHOTOSYNTHESIS

Besides affecting photorespiration, the environment has other, more general influences on plant productivity. The main factors are temperature, illumination, carbon dioxide supply, and mineral elements of the soil.

Temperature affects the rate of transpiration, so that stomata may close as the temperature rises during the day. This limits the supply of carbon dioxide. Also, temperature affects the rates of enzymatic reactions. Cold molecules collide with little energy and are unlikely to react. On the other hand, enzymes are damaged by high temperatures. The result is that each enzyme works best at an intermediate temperature called the **optimum.** Since various enzymes have different optima, the entire balance of metabolism varies with temperature. Plants are most productive when the temperature favors photosynthesis over respiration. For instance, potatoes are most productive in northerly latitudes because the potato plant *(Solanum tuberosum)* respires too rapidly to form food-storing tubers in warmer climates.

Light and **carbon dioxide** have interacting effects. The carbon dioxide content of the open air is more or less constant at 0.03%, while the intensity of light changes during the day and from one season to another. Thus, photosynthesis is sometimes limited by the carbon dioxide supply and sometimes by the amount of light. Under greenhouse conditions, productivity can sometimes be raised by providing artificial light and additional carbon dioxide; however, these practices require careful handling. To be effective, light must be of the wavelengths that pigments can absorb and must be regulated to prevent overheating. Carbon dioxide levels can only be moderately increased without damaging the plants. In some cases, concentrations of 0.5% are known to be injurious.

Mineral nutrients can affect productivity by influencing the production of photosynthetic leaves. Some minerals are directly involved in photosynthetic molecules; for instance, chlorophyll contains magnesium, the water-splitting system requires manganese, some electron carriers contain iron, and all proteins contain nitrogen. Thus, poor soils can result in plants with poorly developed photosynthetic systems. In these cases, yields can be greatly increased by effective fertilizing programs.

SUMMARY

1. Photosynthesis is the primary energy-storing process of life. In it, light energy is stored as chemical energy in organic compounds.

2. The raw materials of photosynthesis are carbon dioxide and water; the products are sugar and oxygen (O_2).

3. Chloroplasts are organelles that perform photosynthesis in green plants.

4. The energy of photons (light) removes electrons from chlorophyll. Some of the energetic electrons, plus hydrogen ions from water, are taken up by NADP to form NADPH. The rest of the photoelectrons return to chlorophyll ions after passing along chains of electron carriers. These events result in the formation of a proton gradient across the thylakoid membrane. The proton gradient provides energy for producing ATP.

5. The electrons lost by chlorophyll in making NADPH are replaced by electrons from water. This splits water molecules to release H^+ and O_2.

6. The reactions that require light to form ATP, NADPH, and O_2 are called the light reactions. They occur in thylakoids of the chloroplasts.

7. In the stroma of the chloroplast, the H^+ and electrons of NADPH are transferred to organic compounds. These compounds are used in a series of reactions that incorporate CO_2 and produce molecules of sugar. ATP is used as an energy source. These steps are called the dark reactions. Two major pathways of carbon fixation are the C_3 and C_4 pathways.

8. In C_3 plants, photorespiration releases up to 50% of the previously captured CO_2. This does not happen in C_4 plants, which are therefore more productive under certain conditions.

9. The photosynthetic process captures only a small fraction of the available light energy.

10. Photosynthesis may be limited by the CO_2 supply, light, temperature, minerals, and the hereditary efficiency of the plant.

9 THE CONTROL OF GROWTH AND DEVELOPMENT

Which are more alike: cattle in a herd or oaks in a woodland? The cattle are decidedly more uniform than the trees in terms of the number of limbs and appendages they have. The systems that control growth and development in the higher animal body are tightly programed and conservative; they resist influences from the environment far more than they adjust to them.

Plants are the opposite. They have stimulus-response systems that allow the environment to modify their path of development. This is the principal way in which plants adjust to the environment. If in a poor location, plants detect the situation by means of environmental signals, and physiological systems adjust their growth in a way that permits them to persist. And whereas animals meet the threat of predators by running away, the stationary plant suffers the attack and later replaces the lost parts.

ENVIRONMENTAL ADAPTATION BY THE YOUNG PLANT

The plant's use of environmental signals can be illustrated by tracing the behavior of a seedling as it starts the life of an independent plant.

When a seed germinates, the survival of the young plant usually depends on rapidly establishing a water supply to support growth. Appropriately, the primary root senses gravity and directs its growth downward, in the direction that usually leads to water.

The seed may germinate in soil or leaf litter, and the seedling must reach light before its food supply has been exhausted. Several mechanisms help to make the best use of stored energy. First, the shoot senses gravity and grows upward, toward the usual location of light. Second, a sensory pigment detects any light that reaches the shoot, and it causes growth to be directed toward the light. Finally, all aspects of growth except stem elongation are suppressed until the stem reaches the light **(Figs. 9.1, 9.2, 9.3, and 9.4, Color Plate 13).** The leaves remain small and yellow; energy is not expended on photosynthetic machinery in leaves that might never be exposed to the sun. In the stem, the internodes grow rapidly and form little lignin. The shoot is weak and spindly, but it can rely on support from the surrounding soil or litter. A shoot is said to be **etiolated** when it grows in this way. Etiolation is easy to demonstrate in the laboratory; almost all shoots (including the twigs of trees) adopt the etiolated type of growth when kept in darkness or in very dim light.

When the seedling grows through soil or litter, there is a risk of damage to the delicate shoot tip and young leaves. In dicot plants, damage is avoided by a hairpin bend, or **hook,** in the stem just below the cluster of young leaves **(Figs. 9.1** and **9.2).** The stem elongates below the hook, so the tip and the young leaves are pulled, rather than pushed, through the soil. When the shoot breaks into the light, the hook straightens, presenting the leaves to the sun; the leaves then expand in preparation for photosynthesis. In grasses (which are monocots), damage is avoided in a different way: the young leaves are rolled up inside a protective tube, the **coleoptile** **(Figs. 9.3** and **9.4).** The spearlike coleoptile pushes between soil particles until it reaches the surface; then the coleoptile stops growing. The enclosed leaves split the tip of the coleoptile, emerge, and unroll **(Fig. 9.4).**

Emergence into the light initiates widespread changes in growth. The principal light sensor is a pigment called *phytochrome.* When a shoot is illuminated, phytochrome triggers all the changes mentioned above. In addition, it causes more thickening, more lignification, and less elongation. The result is a stem with the strength needed to stand erect against the pull of gravity.

The greening of the leaves is one of the most striking changes that occurs when an etiolated seedling reaches the light **(Figs. 9.2** and **9.4).** The mesophyll cells of a leaf and the cortex of a stem are programed to delay the formation of the photosynthetic apparatus until they are exposed to light. In darkness, the immature plastids contain a small amount of **protochlorophyll,** an incomplete form of chlorophyll, which is attached to a special catalytic protein. Upon illumination, the protochlorophyll absorbs light and is converted to chlorophyll. More protochlorophyll is formed in its place. As chlorophyll accumulates, membranes in the plastids are organized into thylakoids, and the full battery of chloroplast enzymes is assembled. The leaf can be fully green within 36 to 48 hours. Phytochrome is also involved in these events.

As this introduction has shown, the systems that control development in the plant can respond to signals from the environment. Physiologists have long felt that a study of these signals and the responses might lead us to valuable insights about the plant's developmental systems. Charles Darwin, for example, held such a belief; and the book that he and his son Francis published in 1881, *The Power of Movement in Plants,* helped to launch a revolution of knowledge about developmental control. But to understand this revolution, we must first shift our focus to the level of the cell.

HORMONES AND THE CONTROL OF GROWTH AND DEVELOPMENT

The typical plant begins its life as a single cell, the zygote. Many cell divisions occur, and the organism becomes a colony of cells that cooperate to form and maintain an integrated plant body. The cells in one region form a leaf; those in another region form a root or a flower. Some cells become green and perform photosynthesis; others lose their nuclei, form sieve plates, and join the phloem. What causes the cells to develop so differently? What tells each cell the proper path for its development? How do the cells achieve cooperation?

Acorns always produce oak trees; corn kernels always produce corn plants. This shows that development depends on hereditary information. At the outset, all the information is contained in the zygote. Thus one might suppose that cells develop differently because each cell receives only a fraction of the zygote's original store of hereditary information.

But experiments have shown that mature parenchyma cells can give rise to whole plants, if the cells are removed from their original location and are given suitable stimuli and raw materials. This demonstrates that mature parenchyma cells contain all the hereditary information that the zygote contained. Evidently, then, each cell uses only a part of its hereditary information. A common idea is that each cell follows a particular reading program as it refers to the hereditary information. This selective pattern of reading guides the cell through a sequence of developmental changes. Cells differ because they follow different reading programs, but if so, how is the program determined?

The simplest suggestion is that cells follow a path of development in response to signals from other cells. Such internal signals were postulated by Charles and Francis Darwin, and much of modern plant physiology has been concerned with identifying these signals and tracing their action. Today we know of five classes of molecules that act as developmental signals in plants (Fig. 9.5). These compounds, called **hormones,** act at very low concentrations and function in the plant only as signals.

We do not yet know in detail how the hormones initiate their effects. But even small changes in their structure tend to change their activity. This suggests that hormones are precisely shaped to fit receptor sites, much as modulators attach to regulatory enzymes (see Chapter 2). Proteins are likely candidates for receptors since other types of molecules generally lack the structure-recognition capabilities that proteins have.

With a hormone attached, a receptor might alter the cell by several mechanisms. The receptor might interact directly with DNA to stimulate or inhibit the reading of a particular

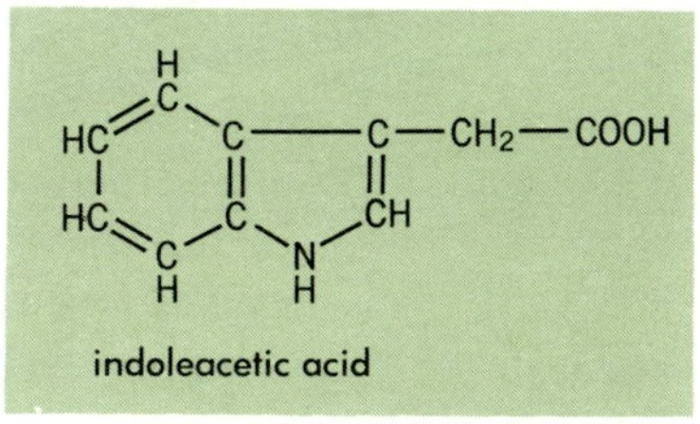

A: Molecule of the auxin, indoleacetic acid (IAA), made up of two rings and a side chain

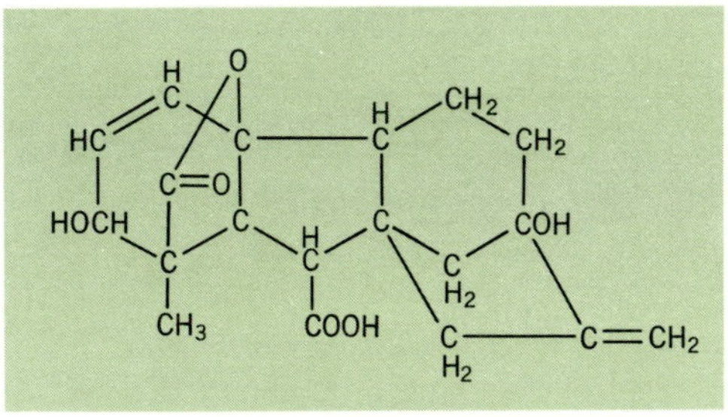

B: Molecule of gibberellic acid

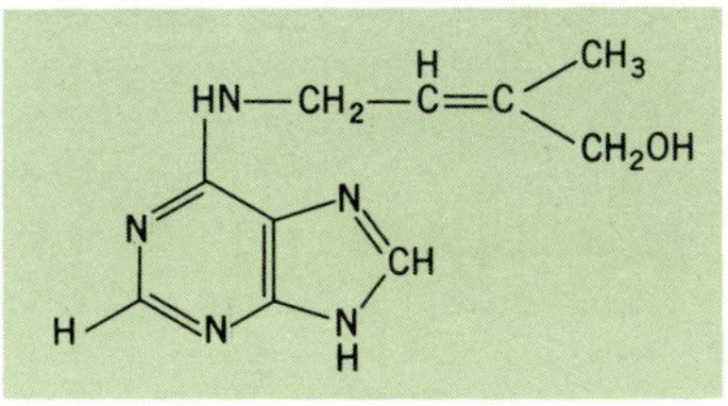

C: Zeatin

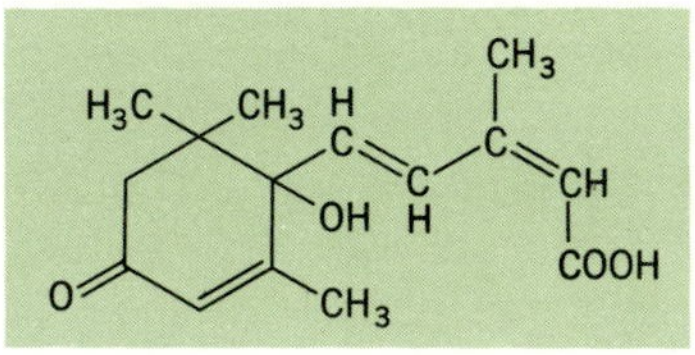

D: The structure of abscisic acid (also called dormin or abscisin II)

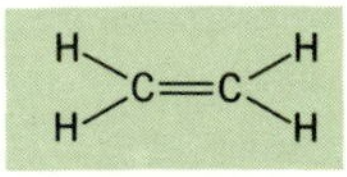

E: Ethylene

Figure 9.5. Plant hormones. *A,* indoleacetic acid, IAA, the most common natural auxin; *B,* gibberellic acid, GA₃, one of more than 40 very similar natural gibberellins; *C,* zeatin, one of several natural cytokinins; *D,* abscisic acid, ABA; *E,* ethylene, a gaseous growth regulator.

Figure 9.1. Pea seedlings grown in light and in darkness. Note minimum stem elongation and maximum leaf development in light. The center two plants have had only one day in light, which caused straightening of the stem (hook-opening) and start of leaf expansion and greening, as compared to the fully etiolated seedlings on the left.

Figure 9.2. Pea leaf development as a function of light. Fully developed green leaves from plants in continuous light; early stage of development after one day in light (center); etiolated leaf from dark-grown plant protected by plumular hook. All eight days old.

Figure 9.3. Corn seedlings grown in light or in darkness showing the long mesocotyl growth that pushes the coleoptile and its enclosed first leaf and shoot apex to the soil surface. In the light-grown seedlings the mesocotyls are only a few millimeters long.

Figure 11.25. Tissue culture propagation. *A*, direct method from shoot tips of jojoba *(Simmondsia chinensis)*. *B*, jojoba culture showing new root initiation.

Figure 9.4. Corn *(Zea mays)* leaf development. A series showing the leaf enclosed in the coleoptile, emergence from the coleoptile in darkness where unrolling is prevented, an artificially unrolled leaf, and a similar leaf grown in light, fully unrolled with greening completed.

Figure 15.19. *A*, apothecia of the brown rot fungus *(Monilinia fructicola)*, ×½; *B*, brown rot on cherry compared with a healthy fruit, ×½.

Figure 15.35. Various types of lichens. *A*, the conspicuous green foliose lichen is *Peltigera aphthosa* and it is surrounded by the white branches of the fruticose lichen *Cladonia* sp.; *B*, crustose lichen *Acarospora flava* colors a rock on the skyline of a Sierra Nevada ridge; *C*, *Ramalina reticulata*, a fruticose lichen, clothes the branches of an oak in the foothills of the Coast Ranges along the Pacific Coast.

gene, as in the operon mechanism (see Chapter 2). Alternatively, the receptor might alter the concentration of another compound (a secondary messenger) that in turn acts on the DNA. Investigators have found that hormones sometimes bind to membranes. This suggests that some receptors are membrane proteins. If so, then on binding a hormone, the receptor might change the permeability of the membrane, or a part of the receptor itself might detach within the cell to serve as a secondary messenger.

Although we cannot define exactly how hormones work, we can describe the kinds of developmental processes that they help to control, and we can illustrate the hormones in action. Development combines the results of three major processes: **cell division,** which produces new cells; **growth,** which increases the size of cells and organs; and **differ-**

entiation, the changes by which cells become specialized in structure and function. Each of these processes can vary in speed and direction. And all of these variables must be controlled in each part of the plant body if development is to be normal. The following passages outline what we know about the part that hormones play.

AUXINS

Our knowledge of hormones began with the discovery of the **auxins,** a class of hormones that were first detected because they stimulate the elongation of grass coleoptiles. The kind of experiment that led to their discovery is shown in Fig. 9.6.

The only naturally occurring auxin to be identified so far is **indoleacetic acid (IAA),** which is shown in Fig. 9.5A. But

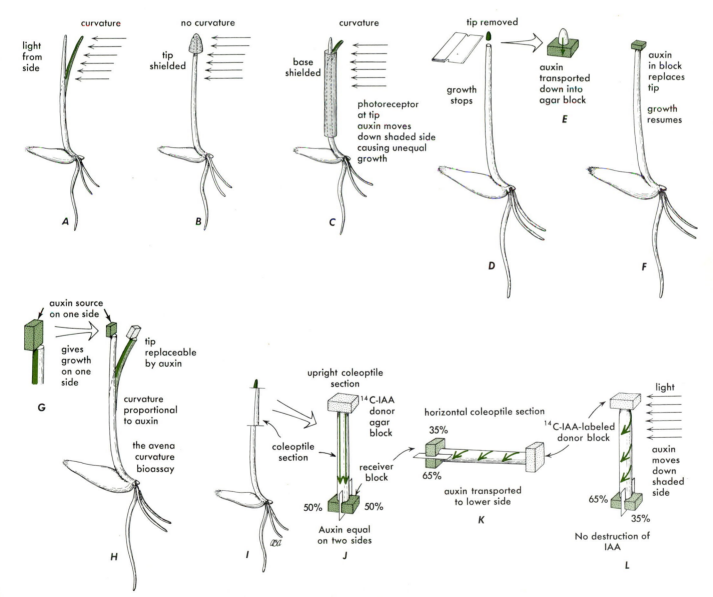

Figure 9.6. Diagram of experiments showing how regulation of auxin transport can result in phototropic curvature (*A, B, C, L*) or in geotropic curvature (*I, J, K*), and the *Avena* curvature bioassay used in estimating minute amounts of auxin (*D, E, F, G, H*). Experiments in which indoleacetic acid (IAA) labeled with radioactive carbon was used (*I, J, K, L*) show how geotropic and phototropic stimuli cause auxin to be transported laterally. Unequal amounts of IAA accumulate in the receiver blocks.

many artificial compounds have been developed that affect plants in the same way as IAA, and these are also called auxins.

Control of Cell Enlargement by Auxin

Cell enlargement is a process critical to the life of almost every cell in a plant, and the ability of the plant to control this process precisely is central to plant development. A plant cell may be thought of as an inflatable bag (the protoplast) surrounded by a cover (the cell wall) that may be rigid or may be expanded under pressure, depending on its makeup. By using respiratory energy and carrier systems, a variety of solutes are pumped into the vacuole, and water follows osmotically, creating turgor pressure. The turgor pressure in a typical cell might be maintained at five times the pressure of the surrounding air (for comparison, automobile tires are usually inflated to about three times atmospheric pressure).

Whether growth occurs will depend on both the pressure and the extensibility of the wall. As the cell grows, the wall is forced to stretch. Stretching hardens the wall and makes it less extensible. To continue growth, the cell wall must be continually loosened.

Auxin stimulates growth by causing the walls to loosen. However, auxin does not act directly on the wall but, rather, causes events inside the protoplast that ultimately lead to wall loosening.

To learn how the wall is loosened, physiologists have looked very closely at the way cell walls are constructed. The wall is built of polysaccharide polymers such as cellulose and pectin, as well as proteins. These framework polymers are thought to be joined into a network by cross-links. Many physiologists believe that the cross-links limit growth and that enzymes loosen the walls by breaking the cross-links. This would allow the polymers to slide into new positions. Then the same enzymes would form new cross-links. All these events would happen at the same time in different parts of the wall. Wall-loosening enzymes have not yet been isolated, but indirect evidence suggests that they may be short-lived. Thus growth will continue only as long as new enzyme molecules are added to the wall. If so, auxin might control growth by stimulating the release of wall-loosening enzymes. In support of this idea, physiologists have found that auxin binds to the endoplasmic reticulum, where enzymes are made. In addition, auxin may cause cells to secrete small molecules or ions that stimulate enzyme action in the wall. For example, auxin causes many cells to release hydrogen ions, and experiments have shown that hydrogen ions can cause walls to loosen.

In addition to its wall-loosening effect, auxin stimulates metabolism. It speeds respiration, promotes solute transport, and stimulates the synthesis of new cell wall material. This contributes to growth in several ways. First, the cell wall tends to become thinner as it stretches, and new material must be added or else the wall could become thin enough to break under turgor pressure. Secondly, the active uptake of solutes is needed both to build new protoplasm and to promote the osmotic uptake of water. Faster respiration is needed to support all these activities.

Controlling Auxin Concentration

Elongation rates respond within minutes if auxin is added or removed. This emphasizes the importance of factors that control the concentration of auxin.

Auxin is formed in regions of the plant that are separate from the zones of rapid elongation (Fig. 9.7). The tip of the coleoptile and the stem tip with its young leaves make IAA from the amino acid **tryptophan.** The auxin is then moved down the organ through the elongating region (Fig. 9.6). The movement is called **polar transport** because it is a one-way, energy-requiring movement away from the tip. Typical rates of movement are 10 to 15 mm per hour, much slower than movement of materials in the xylem and phloem, which may exceed 1 m per hour. Furthermore, polar transport in the youngest stem regions is opposite to the direction of flow in the phloem and xylem and has been shown to occur even in stem tissue from which all vascular elements were removed.

Thus, auxin moves down from the tip through the growing zone, and the concentration in this stream establishes the growth rate of the tissue. If the tip of a coleoptile—the auxin source—is removed, the remaining auxin is rapidly drained out of the growing zone. Within a few minutes the growth rate drops. Growth can be restored if the tip is put back or the auxin supply is replaced with a block of gel containing IAA (Fig. 9.6). If, however, the auxin source is placed on only one side of the cutoff stump, the cells that line up below the auxin supply grow more than those on the other side. This leads to a bending of the coleoptile or stem.

One might suppose that continued production of auxin at the tip and its transport downward should lead to an accumulation somewhere. But no region of accumulation has been found. Auxin is altered by enzymes along the way. Some enzymes break auxin molecules down into inactive products, whereas other enzymes inactivate auxin by tying it to another molecule, forming an inactive compound:

$$\text{indoleacetic acid} \longrightarrow \text{indoleacetylaspartic}$$
$$\text{+ aspartic acid} \qquad \text{acid (inactive)}$$

Supplying the plant with high concentrations of auxin frequently stimulates the synthesis of the enzymes responsible for inactivation.

Tropic Curvatures

Plants can orient the direction of growth of organs in response to environmental cues. These growth responses are called **tropisms.**

When the shoot and root emerge from the seed, it is of obvious advantage to direct shoot growth upward and root growth downward. The direction of gravitational force is a reliable cue. When growth is directed by gravity, the plant is said to show **geotropism** (Fig. 9.8).

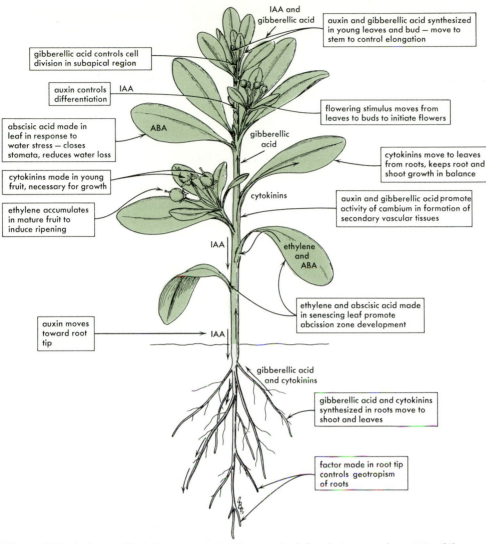

IAA and
gibberellic acid

auxin and gibberellic acid synthesized
in young leaves and bud — move to
stem to control elongation

gibberellic acid controls cell
division in subapical region

auxin controls
differentiation

IAA

flowering stimulus moves from
leaves to buds to initiate flowers

ABA

abscisic acid made in
leaf in response to
water stress — closes
stomata, reduces water loss

gibberellic
acid

cytokinins move to leaves
from roots, keeps root and
shoot growth in balance

cytokinins made in young
fruit, necessary for growth

cytokinins

auxin and gibberellic acid promote
activity of cambium in formation of
secondary vascular tissues

ethylene accumulates
in mature fruit to
induce ripening

IAA

ethylene
and
ABA

ethylene and abscisic acid made
in senescing leaf promote
abcission zone development

auxin moves
toward root
tip

IAA

gibberellic acid
and cytokinins

gibberellic acid and cytokinins
synthesized in roots move to
shoot and leaves

factor made in root tip
controls geotropism
of roots

Figure 9.7. A diagram illustrating some typical hormonal relations between various parts of the plant.

Figure 9.8. Apical dominance and negative geotropism in shoots of *Photinia* sp. Right: upright shoot showing apical dominance and suppression of lateral bud growth. Left: shoot that has been bent. Note development of lateral shoots and their negative geotropic response.

We do not yet know how the plant senses gravity, but many physiologists think the answer may involve **statoliths**—particles in the cell that are heavy enough to sink to the bottom, comparable to the stones in the inner ear that give human beings a sense of the vertical. Colorless plastids, heavy with starch grains, may act as statoliths; when a plant is laid on its side, these **amyloplasts** tumble to the lowermost side of the cell.

Once the shoot has sensed its orientation in space, the direction of auxin transport changes, so that by the time the auxin has traveled from the tip of the stem to the main growing zone, there may be twice as much auxin moving through the lower side of the stem as through the upper side (Fig. 9.6). The greater growth on the lower side turns the tip upward. The amyloplasts settle back to their original position, and auxin transport becomes uniform on both sides again. The plant requires a few minutes to sense its original change in position, and within 15 minutes there is enough auxin concentration difference to cause a curvature. In some rapidly growing plants, the stem turns upright within an hour after being placed horizontally.

The geotropic curvature of roots has many aspects in common with the shoot response, except that there is considerable doubt that auxin is involved. The root cap produces a growth inhibitor—perhaps the hormone abscisic acid—that moves away from the tip to control curvature.

Phototropism is a change in the direction of growth in response to light (Fig. 9.6). The response depends on a pigment that absorbs blue and violet light. The pigment is thought to be a flavoprotein, which is a different kind of molecule than chlorophyll or phytochrome. In an unevenly illuminated shoot, there is more pigment activation on the lighted side than on the shaded side. Careful experiments have shown that IAA labeled with radioactive carbon and applied to the tips of coleoptiles moves symmetrically down the coleoptiles in the dark. Light treatments that cause curvature result in twice as much auxin moving down the shaded side as down the lighted side, but the total amounts transported are the same as in dark controls. Thus, somehow the light-sensing pigment causes a lateral diversion of the auxin flow, as in geotropism. The result is a curvature toward the light.

Apical Dominance

In the typical pattern of shoot growth, the active shoot tip inhibits the sprouting of lateral buds below the apex. This phenomenon is called **apical dominance.** Plants with strong apical dominance (e.g., many sunflowers) produce few branches. Weak apical dominance leads to a bushy appearance, as in tomato plants or oak trees. Even if lateral buds do sprout, the main shoot tip influences the rate and direction of branch growth. This is part of the reason why lateral branches usually do not grow vertically.

The dominance of apical growth over lateral growth decreases with distance from the tip of the stem and varies with the age of the plant, the genotype, nutrition, and other environmental factors. For example, when a plant is laid on its side, buds on the upturned flank may sprout and become active shoots (Fig. 9.8). Generally, applications of rich fertilizers to the soil will also weaken apical dominance and promote branching.

Because the control of branching is important in the human use of plants, physiologists have done many experiments on the basis of apical dominance. If the main shoot tip is removed, one or more of the lateral buds (usually the uppermost) will sprout and take over the role of the main shoot tip, asserting dominance over the other buds. However, none of the lateral buds sprout if the main shoot tip is removed and auxin is immediately applied to the stump. This suggests that the main shoot tip exerts its dominance by producing auxin. However, physiologists are divided on the question of how auxin maintains control. Since auxin is known to travel down the stem, one school of thought is that high auxin concentrations at the bud can directly inhibit sprouting. Another idea is that lateral buds are inactive because

the nutrients needed for growth are sent primarily to the main shoot tip in response to the high auxin concentration there. This idea is based on many observations, including the fact that auxin stimulates metabolism, transport, and growth.

Apical dominance is an example of a **growth correlation,** an interaction between the growth of two plant parts. Other examples of growth correlations, mediated in part by hormones, are shown in Fig. 9.7.

Cell Differentiation

When xylem elements differentiate, the process is easily observed because the changes in the cell are extreme. The wall develops heavy thickenings with characteristic patterns, and eventually the protoplast dies. These visible signs of change have made xylem a favorite tissue for studying cell differentiation.

One experimental approach has been to perform delicate operations on the shoot tip to study how xylem forms in the young leaf primordia and the nearby stem. These studies show that a leaf primordium stimulates differentiation in the procambial strand leading to it (Fig. 9.9A). If the leaf primordium is sliced away, the same differentiation of vascular tissue in the stem can be induced by applying auxin (Fig. 9.9C).

In *Coleus* stems, a wound that severs a vascular bundle is followed by cell divisions and differentiation of xylem elements from parenchyma cells around the wound. The new xylem elements reconnect the injured bundle. Observed more closely, it has been found that new phloem cells are formed even earlier than the xylem cells in this regeneration process. These events require a supply of auxin that is normally transported out of the leaves above the wound and then down the stem. If the leaves are removed at the time of wounding, regeneration does not occur. The effect of the missing leaves can be replaced by applying auxin to the petiole stumps.

Farther down the stem where elongation has ceased, even an herbaceous plant like the tomato may have some secondary growth taking place from a vascular cambium. This cambial activity depends on a supply of auxin that is received from the shoot tip. In woody plants that have been dormant over winter, the buds that sprout in the spring give off auxin, which stimulates the cambium. A wave of cambial divisions can be traced down the stem, following the flow of auxin.

Further advances concerning the differentiation process have come with **tissue cultures.** Under suitable conditions, blocks of parenchyma tissue from stems can be induced to continue cell divisions and growth without differentiation. This produces a uniform mass of unspecialized cells—a **callus.** Blocks of callus tissue made from lilac stems show a remarkable response to auxin and sucrose: if these two substances are applied to the top of the block, a discontinuous ring of vascular tissue develops within the tissue block, at a distance below the surface. The diameter of the ring and its distance from the droplet can be increased by raising the concentration of auxin. If auxin spreads by diffusion, its

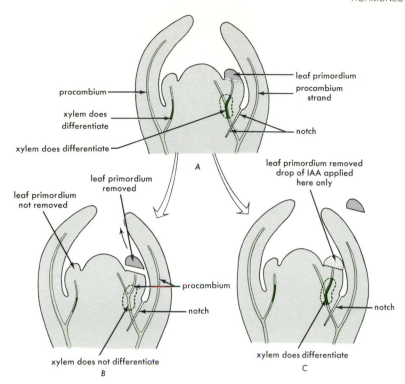

Figure 9.9. Diagram of an experiment showing how a leaf primordium provides a stimulus for xylem differentiation in the procambium. The notch severing the procambium isolates the tissue of interest above it. *A,* control with intact leaf primordium; *B,* leaf primordium removed; *C,* leaf primordium removed and a drop of auxin substituted. Auxin is an effective replacement for the stimulus from the primordium.

concentration must decrease with distance from the source. If we put these facts together, it seems that the signal for cells to become vascular elements must involve their exposure to a particular auxin concentration.

Whether the differentiating cells become xylem or phloem elements in lilac callus depends on the concentration of sucrose that is applied. High concentrations induce nothing but phloem; low concentrations induce only xylem; and intermediate levels stimulate both xylem and phloem to differentiate. Significantly, when both xylem and phloem are formed, their arrangement is concentric; the xylem is toward the center, as in the cross section of an intact stem.

GIBBERELLINS

While European scientists of the 1920s were occupied with the auxins, Japanese workers discovered another class of hormones. These hormones were first identified in studies of a disease of rice plants, the *bakanae* (foolish seedling) disease. Afflicted seedlings grow very tall and eventually fall over. The disease is caused by a fungus, *Gibberella fujikuroi.* The abnormal growth was duplicated by treating normal seedlings with the liquid in which the fungus was grown. The active agent in the liquid proved to be a group of substances that were given the name **gibberellins.** They form an extensive group of related compounds, variations on the basic structure shown in Fig. 9.5*B.* Although these compounds were first discovered in a disease, they have since been found as natural hormones in a wide array of flowering plants. The *bakanae* disease results from an oversupply of gibberellin.

Gibberellins and Stem Growth

Gibberellin promotes stem growth in many plants besides rice. The response of some **rosette** plants is especially striking (Fig. 9.15). A rosette plant has a very short stem with a cluster of leaves at ground level. As a prelude to flowering—often in response to seasonal stimuli such as changes in night length—the stem elongates rapidly, or bolts. Gibberellin treatments can substitute for the stimuli, causing early bolting. Gibberellin is thought to govern natural bolting as well, because the plants cannot bolt if they are prevented from making gibberellin. Bolting seems to involve a stimulation of cell division just below the shoot tip, in the region that provides cells for growing internodes. This region is much more active in long-stemmed plants than in rosettes. Gibberellin causes these cells to divide much more rapidly.

In other cases, gibberellin promotes growth by causing walls to soften, or by causing solute formation and water uptake.

Auxins and gibberellins often act simultaneously to control stem elongation. In some cases, gibberellin may stimulate cell division, thus providing more cells on which auxin can act. In the coleoptile, gibberellins act at an early stage of development and the auxin-sensitive stage shown in Fig. 9.6 comes later in life. Woody stems need both auxin and gibberellin to maintain cambial activity.

In Europe, commercial use is made of the growth retardant CCC, or Cycocel (2-chloroethyltrimethyl ammonium chloride), to inhibit stem elongation in wheat plants. Cycocel inhibits the production of gibberellin in the plant. Shorter, stronger stems result, and the plants are much more resistant to lodging, that is, to being knocked down by wind and rain. Lodging makes harvesting difficult.

"Dwarf" forms in a variety of plants are often due to a diminished gibberellin synthesis or to an enhanced production of compounds that oppose the action of gibberellin. Dwarf forms of corn and peas are used in estimating the concentrations of gibberellins in extracts of plants or plant parts. Such methods are called **bioassays.** Figure 9.10 shows a bioassay in which quantities of gibberellin are estimated by comparing their effect on growth to that of known hormone concentrations. Bioassays have been important in studying all the plant hormones because plant materials are often sensitive to quantities of hormone that would be far too small to detect chemically.

Gibberellins and Enzyme Synthesis

Besides affecting growth, gibberellin promotes the germination of many seeds. Studies of this effect have shown much about the early steps by which gibberellin alters cells. When a grain such as barley or corn starts to germinate, the embryo begins to grow but has limited food reserved in itself. The main reserves are in the starchy endosperm—a tissue containing cells that are loaded with starch, reserve proteins, and some nucleic acids. A special layer of living cells, the aleurone layer, surrounds the main endosperm tissue and is instrumental in digesting the stored materials to soluble forms that can diffuse to the embryo. Early in germination, the embryo synthesizes gibberellin (GA), which diffuses to the aleurone layer and triggers the synthesis of enzymes to digest stored materials in the endosperm. If the embryo is removed from the seed prior to germination, very little digestion of the remaining endosperm takes place. Adding gibberellin in minute quantities to the embryoless endosperm induces the synthesis and secretion of enzymes just as if the embryo had provided the stimulus (Fig. 9.11).

The barley aleurone layer can be isolated and studied independently of the embryo and storage endosperm. It is a collection of cells with no growth activities and no division; however, it can synthesize and secrete several enzymes. In this connection the aleurone layer of barley has been used to study the mechanism of action of gibberellin.

About 6 hours after gibberellin is added, the aleurone cells start to make enzymes that break down starch, proteins, and nucleic acids. Prior to the appearance of the enzymes, new ribosomes and endoplasmic reticulum membranes are formed, as well as enzymes involved in membrane lipid synthesis. We have much to learn about the way in which gibberellin unleashes all these changes, though it does appear that an activation of genes (new enzyme synthesis) is involved.

In the manufacture of beer, the starch stored in the grain endosperm must be hydrolyzed to a soluble form before its conversion into alcohol by the glycolytic enzymes of yeast. The natural production of amylase occurs in the malting barley. The addition of extra GA speeds up amylase synthesis by the aleurone enough to be used commercially.

CYTOKININS

Hormones of a third major class, the **cytokinins,** were discovered through work with tissue cultures. Physiologists have found that keeping cells alive and active after they have been removed from the plant depends on the exact constitution of the medium to which the cells are transferred. Finding the best medium is a matter of trial and error. Such trials led to the discovery, in the mid-1950s, that cell division could be stimulated by adding adenine, one of the nucleic acid bases. Since then a variety of related compounds have been found to work much better than adenine. These are the cytokinins, one of which is shown in Fig. 9.5C. Some cytokinins

Figure 9.10. Dwarf peas (*Pisum Sativum*) showing the promotion of stem elongation by gibberellic acid applied 7 days prior to photograph. This response is used to bioassay the gibberellin contents of plant extracts.

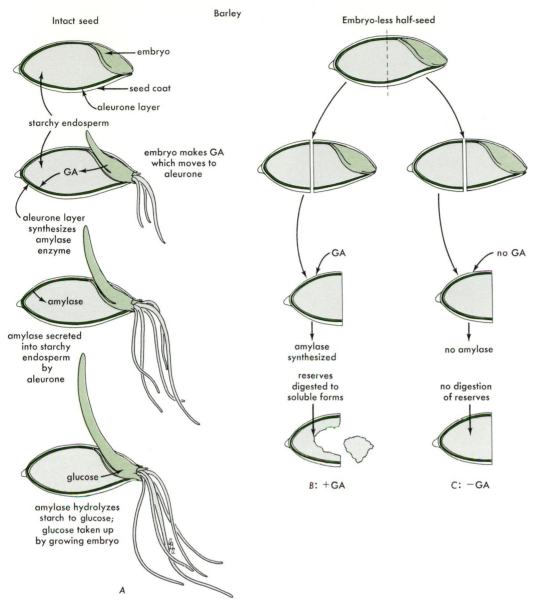

Intact seed

Barley

Embryo-less half-seed

— embryo

— seed coat

— aleurone layer

starchy endosperm

embryo makes GA which moves to aleurone

GA

aleurone layer synthesizes amylase enzyme

amylase

amylase secreted into starchy endosperm by aleurone

glucose

amylase hydrolyzes starch to glucose; glucose taken up by growing embryo

A

GA

no GA

amylase synthesized

no amylase

reserves digested to soluble forms

no digestion of reserves

B: +GA

C: −GA

Figure 9.11. Diagram illustrating how gibberellin from the embryo induces the synthesis of the starch-degrading enzyme, α-amylase, in the aleurone layer.

are synthetic, while others occur naturally, especially in young developing fruits.

Besides affecting cell divisions, the cytokinins play a part in controlling the initiation of plant organs. In tissue cultures, for example, cells of tobacco pith enlarge if supplied with nutrients and auxin, but they divide only if small amounts of a cytokinin are added. Moreover, by varying the balance between auxin and cytokinin, it is possible to selectively initiate the development of roots and shoots (Fig. 9.12). High auxin-to-cytokinin ratios cause root initials to differentiate. A low ratio of auxin to cytokinin causes clumps of cells to become apical meristems that grow into shoots. Intermediate concentration ratios cause callus to form. Shoots can be removed, rooted, and grown to mature plants. These

methods are becoming increasingly important in genetic engineering, in which geneticists are striving to develop novel crop plants with the use of cultured cells (see Chapter 11).

Cytokinins also contribute to apical dominance. If a cytokinin is applied directly to a lateral bud that is being suppressed by an active shoot tip, in some plants the bud will grow out. The applied cytokinin promotes the formation of vascular connections between bud and stem, a process that is inhibited by the dominant shoot tip. Cytokinins appear to be made in the tips of roots, and they travel upward in the xylem. They may accumulate at the cut end of a rooted stump, where adventitious buds often form. The opposed movements of auxin and cytokinin can give each region of the plant a unique balance between quantities of these two

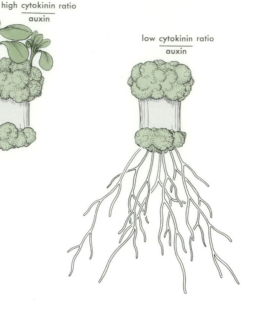

high cytokinin ratio / auxin

control

callus

pith

low cytokinin ratio / auxin

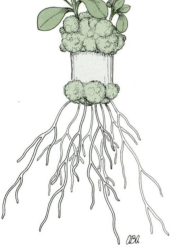

intermediate cytokinin ratio / auxin

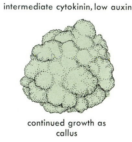

intermediate cytokinin, low auxin

continued growth as callus

Figure 9.12. Interaction between auxin and cytokinin in the control of development. Pieces of tobacco pith tissue were grown aseptically on nutrient medium, supplemented with various levels of the hormones.

hormones, and their opposing effects on bud growth may help to determine the pattern of branching as the plant grows.

Plants normally maintain a close balance between shoot and root growth. The upward movement of cytokinins from the root may help to maintain this balance. Increases in root growth cause more abundant cytokinin supplies, which cause a corresponding increase in shoot growth. Leaves also contribute to the shoot/root balance by producing carbohydrates and vitamins (especially vitamins of the B group, needed as cofactors in respiration) that roots cannot build for themselves.

Cytokinins also prevent leaf deterioration, or **senescence.** If a leaf is removed from a stem, senescence is initiated. Protein synthesis stops, and storage carbohydrates, proteins, nucleic acids, and chlorophyll are broken down. If the leaf is put under conditions in which it forms adventitious roots, the senescence is halted. Applications of cytokinin have a similar effect in halting the deterioration of detached leaves. The maintenance of active RNA and protein synthesis seems to be an important part of cytokinin action in delaying senescence.

ETHYLENE

Some years ago it was the practice among lemon growers to pick lemons green and allow them to ripen in heated boxcars as they were shipped across the United States. This system worked nicely, until the leaky kerosene heaters in the freight cars were replaced by more efficient steam heaters. The shippers were dismayed to find that the lemons no longer ripened in time for marketing. Investigators found that the kerosene stoves had been leaking small quantities of a simple gas, **ethylene,** which acted as a ripening stimulus. This is one of many early observations that ethylene as a pollutant affects plant development. More recently, ethylene has been found to be a potent natural hormone that contributes to normal development in plants. Its structural formula is shown in Fig. 9.5E.

Ethylene is built in the plant from the amino acid **methionine,** a constituent of all cells. Since ethylene has only limited solubility in the aqueous phase of a cell, it diffuses into the atmosphere. Thus, the ethylene concentration depends on the balance between synthesis and diffusion. Enclosing plants—especially fruits—in a tight container can cause the concentration of ethylene in the plant to reach high levels, with drastic effects on growth and development. Effective levels are often in the range of 0.1 to 1 part per million (ppm) of air. Even the levels of ethylene present in urban air pollution are sometimes enough to cause biochemical changes in plants.

The action of ethylene in controlling vegetative growth can be illustrated by its effect on the pea seeding **(Fig. 9.1, Color Plate 13).** As the plumule emerges from the seed, the tip of the shoot has a hooked curvature that protects the apex. The young plumule synthesizes ethylene rapidly in darkness. The ethylene maintains the hook and prevents the leaves from enlarging. When the shoot breaks through the soil, light is absorbed by phytochrome, which in turn triggers a decrease in ethylene synthesis. The leaves are released from ethylene inhibition and expand. The hook "opens" to present the leaves to sunlight, which further stimulates leaf development.

While growing underground, the shoot encounters obstacles such as clods. The pressure of the obstacle against the shoot causes a dramatic increase in ethylene production. The results can be simulated by gassing exposed seedlings with ethylene (Fig. 9.13). Ethylene induces the stem to swell and inhibits elongation. The thickened stem can exert more upward force against the obstacle. If the stimulus is prolonged, the stem's geotropic response is modified; it starts to grow almost horizontally. These responses increase the likelihood that the shoot will penetrate or grow around the obstacle to reach the open air.

Some of these effects involve changes in the direction of cell growth, and they raise one of the most fundamental unsolved problems in plant development: What controls the direction of cell growth? What causes some cells to grow equally in all directions while others become cylindrical or spindle-shaped? Studies of growing cells have suggested that the direction of growth is determined when new cellulose microfibrils are added to the wall. If the microfibrils are laid down at random, the cell expands uniformly. If they are deposited around the cell like the hoops around a barrel, the cell will grow chiefly in length, because the hoop arrangement prevents a large increase in cell diameter.

What does this have to do with ethylene? In some of the responses in which ethylene induces a thickening, as in the shoot that is trapped beneath an obstacle, the effect seems to depend on a change in the direction of cell growth. Ethylene causes some of the cells to deposit their new cellulose microfibrils more randomly, resulting in the consequences mentioned above.

We do not know how the protoplast, *inside* the plasma membrane, controls the orientation of wall polymers that are laid down *outside* the plasma membrane, or how an information storehouse such as DNA can specify such a directional process. But the effect of ethylene seems to offer a step toward understanding these problems.

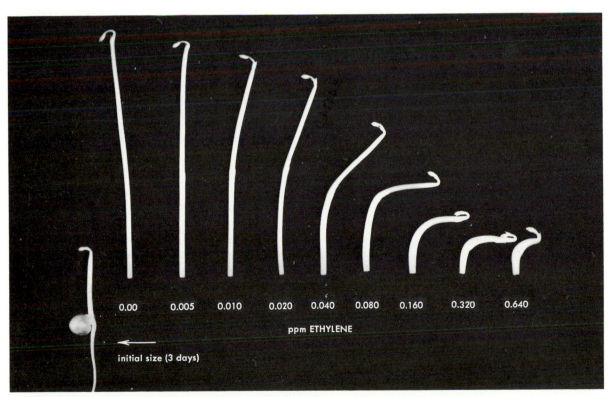

Figure 9.13. The response of dark-grown pea seedlings to various levels of ethylene during 4 days. Internally generated ethylene, formed in response to stress, can induce similar responses. ppm = parts per million ethylene in air.

Another effect of ethylene is seen in the growth of young trees, in which the relationship between height and diameter of the trunk is very much a function of the motion it undergoes as a result of wind. If a young tree stem is tied to a rigid stake, as is frequently the case in container-grown trees, or is supported by other plants in a crowded nursery, the usual response is a tall, slender, weak stem. If the stake is flexible enough to allow some movement in response to gentle winds, the plant has a sturdier form resembling one grown without support. The plant that grows without support from a young stage typically has a shorter stem, larger in diameter, and a tapered form coming from a wide base at the soil level. Routine exposure to high winds can greatly increase this response, as can be seen in trees on windy sea coasts or mountain ridges. Controlled gassing of tree trunks with ethylene stimulates enlargement, and it is thought that stress-induced ethylene synthesis in the trunk contributes to this growth response.

Ethylene and Fruit Ripening

During the early development of a fruit, the seeds are immature and delicate. Dispersal at this time would be premature. Many fruits have a programed pattern of changes that convert the fruit from a seed-manufacturing organ to a seed-dispersal organ. These changes are called **ripening.**

Often the fruit changes its color from green to yellow, orange, red, or blue. This increases visibility to potential animal dispersal agents. To make this change, chlorophylls are broken down, and other pigments such as the red and blue anthocyanins are synthesized. Another change is the conversion of starch and organic acids to sugar so that the fruit becomes sweet. Some of the materials in the cell walls are broken down; the cells now become more loosely bound to each other, resulting in a softer fruit. Volatile flavor components are synthesized, and these contribute to much of the flavor we value. One important link in this grand plan is a change in ethylene relations within the fruit. Either ethylene rises to a critical concentration, or the fruit becomes more sensitive to the existing ethylene. Then ripening is initiated. Usually this is signaled by, or coordinated with, the maturation of the seeds.

Ethylene may cause ripening when it is not wanted. When fruit is held in cold storage, elaborate precautions are used to prevent ethylene from accumulating in the atmosphere in quantities that would trigger ripening. When a fruit goes through the phase of rapid ripening, the production of ethylene may increase many times beyond the critical triggering level, and much ethylene can be released into the air. Thus, "one bad apple in a barrel" producing ethylene may trigger the ripening of all the apples in the barrel. Or the ethylene produced by the mold on an orange may be enough to trigger undesirable changes in the rest of the oranges.

ABSCISIC ACID

Many plants of the temperate zone become inactive, or *dormant,* during the cold season. In addition, seeds are usually dormant at the time of release. Even in active, photosynthesizing trees in midsummer, the terminal buds on the young twigs may go dormant after producing a sufficient number of leaves to clothe the tree. These phenomena illustrate the importance of dormancy and its release in the plant's repertoire of control capabilities.

Although the mechanisms that control dormancy are not fully understood, one factor that is often involved is the hormone **abscisic acid.** Abscisic acid (often abbreviated **ABA**) (Fig. 9.5D) is so named because it is plentiful in leaves that have recently been dropped, or **abscised.** Applications of abscisic acid to some plants will cause leaf and fruit abscission.

Some dormant buds and seeds are also rich in ABA, and their release from dormancy is correlated with a loss of ABA. In some seeds the ABA may be lost by leaching as seeds are exposed to rains. In other seeds and dormant buds, ABA may be gradually broken down by enzymes during the winter.

Just how ABA causes dormancy is not entirely clear. In some instances it seems that dormancy is actually due to a lack of gibberellin. Where this is true, dormancy may be overcome by adding gibberellin. If ABA is removed (e.g., by leaching), the level of gibberellin in the tissue increases as a prelude to the release from dormancy.

ABA also helps to control transpiration. In many plants, the stomata close when transpiration causes mild dehydration. There is good evidence that ABA governs this response: water stress causes many leaves to accumulate high levels of ABA, and stomata close in response to ABA treatments. ABA seems to prevent the guard cells from taking up or retaining the potassium ions that are needed to maintain turgor pressure. With the stomata closed, transpiration stops and the leaf can recover its water balance. Then the ABA is slowly metabolized away, and the stomata reopen.

INTERACTIONS BETWEEN HORMONES

For each class of hormone, there are examples in which one hormone alone seems to control a particular process: for instance, the control of growth by auxin in the oat coleoptile. But it is more usual for the various hormones to interact in complex ways to control development. This has already been seen in the control of root and shoot formation in tissue cultures by auxin and cytokinin. Also, the effects of ethylene and auxin are often interrelated. An experimental treatment of tissues with high auxin levels often induces the synthesis of ethylene. Ethylene may then inhibit the transport of auxin. These results suggest that auxin and ethylene levels may mutually control each other in many tissues.

The control of flower morphogenesis in cucurbits (cucumbers, squash, melons) is a complex case of hormonal interaction. A typical cucurbit plant first forms male flowers; at later nodes it forms bisexual flowers. Some strains produce plants with only female flowers. It has been shown that applied auxin raises the internal ethylene level and that ethylene stimulates the initiation of female parts on flowers that would normally lack them, or it tends to suppress the formation of male parts on flowers that would normally be bisexual. Gibberellin, either in naturally high concentrations or artificially applied, promotes maleness. There appears to be an ethylene–gibberellin balance that acts at the time of flower initiation to determine which organs the flower will have.

Ethylene and auxin may interact in controlling the abscission of leaves and fruits. Ethylene causes an abscission layer to form and mature at the base of the petiole. Auxin both promotes ethylene formation and blocks ethylene action. If the auxin supply drops, as it does in an aging leaf, ethylene can act, the abscission layer forms, and the leaf soon abscises.

THE PRACTICAL USE OF HORMONES

Natural hormones and synthetic compounds that act like them are used in growing and marketing plant products. Several examples are mentioned below. Bear in mind that these are only a sampling of the compounds and methods used.

Fruit Ripening

Ripe bananas store poorly. Today it is common practice to pick, ship, and store bananas while they are still green. Gassing with ethylene triggers ripening, and the bananas are ready to eat in 5 days.

Bud Sprouting

Stored potato tubers are prevented from sprouting by treatment with auxins. This is a case of bud inhibition, comparable to natural apical dominance.

Fruit Drop

Fruits such as apples, pears, and citrus may drop from the trees before harvest time. The fruits can be induced to cling to the trees for several days longer by spraying with synthetic auxinlike compounds, chiefly naphthyleneacetic acid and 2,4-dichlorophenoxyacetic acid (2,4-D). These compounds apparently retard the formation of the abscission layer at the base of the petiole.

Flowering

Pineapple plants can be induced to form flowers and fruits on a precise schedule by applying auxin, ethylene, or a compound that breaks down to release ethylene. The auxin acts by inducing ethylene production in the plant. This aids the pineapple industry in coordinating harvesting, processing, and shipping. The same principle can be used at home; an ornamental bromeliad houseplant can be induced to flower by putting it in an airtight container with a ripe apple as an ethylene source.

Rooting of Cuttings

In many commercially valuable plants the preferred method of propagation is by means of cuttings—because the seeds are hard to germinate or valuable hybrid properties are lost in seed formation or because cuttings produce a large plant

Figure 9.14. The promotion of root initiation by the synthetic auxin, indolebutyric acid, in cuttings of holly (*Ilex opaca*). Row *A*, cuttings that stood in a solution of 0.01% indolebutyric acid for 17 hours before being placed in sandy rooting medium. Row *B*, untreated controls. Photographed after 21 days.

faster than a seedling. A "cutting" is a detached shoot. To form a new plant, it must form roots, usually at the cut end (Fig. 9.14). In some species, cuttings spontaneously form roots if they are kept moist. Other species are more difficult to root, and for them various methods have been devised as rooting aids. Treatment of the cut end with auxins such as naphthaleneacetic acid or indolebutyric acid is often effective. The bases of the cuttings are immersed in the growth regulator solution for 12 to 24 hours or dusted with a dry powdered preparation; they are then placed in a bed of planting compound or moist sand. The conditions required for rooting (e.g., hormone concentration and treatment time) vary with the species. Interested amateur gardeners can buy these chemicals at retail nurseries.

Growth Regulators as Herbicides

Although auxin is essential in small quantities, an excess can cause serious physiological disturbances. Some synthetic auxinlike compounds have proven to be very potent weed killers (herbicides) for this reason. The so-called "phenoxy" compounds are especially effective: for instance, 2,4-D and 2,4,5-trichlorophenoxyacetic acid (2,4,5-T). These compounds are selective, killing broad-leafed plants and leaving the grasses relatively unharmed. This makes them useful for destroying dicot weeds in grain fields, dandelions in lawns, and brush in rangeland.

LIGHT AND THE CONTROL OF DEVELOPMENT

Plants have at least three pigment systems that initiate developmental changes in response to light. *Phototropism,* the directional control of growth by light, involves a flavoprotein pigment and has been described previously. The *greening* of plants when they break out of the ground is initiated by a pigment called *protochlorophyll* which was discussed at the beginning of the chapter.

The third and by far the most versatile pigment system that regulates development is the **phytochrome** system. Phytochrome is a pigment that exists in two stable forms, active and inactive, each having a different color. The inactive form absorbs red light most strongly and is called P_{red} or P_r; the active form is called $P_{far-red}$ or P_{fr} because it most strongly absorbs the far-red light that lies just at the extreme of human visual sensitivity. When either form absorbs light, it is converted to the other form.

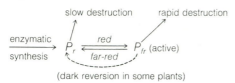

(dark reversion in some plants)

P_r is continuously synthesized from precursors and is slowly destroyed by enzymes. P_{fr} may exist and exert its effect for many hours, but it is unstable and is more rapidly destroyed by enzymes than is P_r. In some plants, P_{fr} is spontaneously converted back to P_r in darkness.

Continuous sunlight with its broad spectrum of wavelengths constantly converts the available phytochrome molecules back and forth between the P_r and P_{fr} forms, so that a steady-state mixture results with the P_r/P_{fr} ratio determined by the spectral composition of the light.

Although the mechanism of action of phytochrome has not yet been established, some work suggests that it can bind to membranes and may exert some control over membrane function.

Phytochrome has a wide variety of effects on plant development, of which the following are important examples.

SEED GERMINATION

Some seeds do not germinate in darkness. After a brief soaking, a short exposure to red light will induce them to germinate. Red light acts by converting P_r to P_{fr}, which in turn relieves some unknown block in metabolism and permits germination. If far-red light is given immediately after the red treatment, P_{fr} is converted back to the inactive P_r form and the seeds will not germinate. For this latter cancelation effect, the far-red treatment must occur before P_{fr} has had time to act.

Responses to light that are canceled by a second dose are said to be **photoreversible.** Red/far-red photoreversibility is a unique phenomenon that is seen only in events governed by phytochrome.

SHOOT GROWTH

The main signal to a seedling that it has emerged from the soil is the formation of P_{fr} by light. The complex response of the seedling has already been described. In these de-etiolation responses the action of P_{fr} may involve such factors as the release of active gibberellin from an inactive form and the activation of genes to form new enzymes.

When a plant finds itself almost continuously in the shade of other plants, the quality of light falling on the plant is changed in a way that can markedly affect the phytochrome status of the tissue. As light passes through successive layers of leaves, the red wavelengths are filtered out by chlorophyll, while the far-red wavelengths pass through. In the shade created by other plants, light rich in far-red tends to markedly lower the P_{fr} concentration. The result is more rapid stem elongation and an increased probability of growing out from under the shade that is cast by other plants.

PHOTOPERIODISM AND FLOWERING

Over most of the earth's surface, there are pronounced seasonal differences in temperature, water supply, and illumination. The developmental system of the flowering plant is equipped to prepare for these changes. Well before the weather is harsh, deciduous trees begin to shut down the anabolic systems in the leaves. There is an orderly withdrawal; the leaves are stripped of some nutrients, abscission layers are formed, and the leaves are abandoned to the elements. If the plant can make underground storage organs, they are filled up before there is a great danger of the loss of food. In addition, the metabolic system gradually shifts its composition to protect against frost damage before ice appears. Then the tree becomes dormant. After the cold season, the plant waits until it is well into the fair season before stirring to active life and does not respond to the occasional day of false spring that occurs in late winter before the weather is reliable.

Not all plants are so good at predicting seasons. But on the other hand, the preceding comments were scarcely a sampling of the seasonal adjustments that plants can show. Some desert plants prepare themselves for a dangerously hot spell rather than a dangerously cold one. And more subtly, some plants time their activities during the growing season in a way that minimizes competition for light, moisture, and pollinators.

Of all the earth's seasonal variables, the most reliable is the annual cycling of day and night lengths. Plants have been found to contain a system that measures the lengths of the nights. The control of development by this timing system is called **photoperiodism.**

All forms of photoperiodism are based on the system that monitors the day/night cycle. This system is located in the leaves. Given suitable light/dark cycles, photoperiodically sensitive leaves send out stimuli through the phloem to other parts of the plant: to the shoot apex, where flowers or winter buds may be induced; or to underground stems, which may be induced to form tubers.

To explore photoperiodism, then, the principal focus of attention is on the leaves and their timing system. But to illustrate its operation, we will use the control of flowering as our example in the following paragraphs.

Figure 9.15. Effect of day length on behavior of the long-day plant henbane (*Hyoscyamus*). Plants were grown with an 8-hour photoperiod until they were about a month old and were then subjected to 24 photoperiods of 10, 11, 12, 13, 14, or 16 hours. Initiation of flowers and accompanying stem elongation (bolting) occurs on days longer than 12 hours.

In regard to flowering, some plants achieve their timing without reference to light stimuli, relying instead on factors such as maturation. These plants are said to be **day-neutral.** Other plants, however, use photoperiodism to time the flowering response. These plants fall into two major groups that are traditionally designated as **long-day** and **short-day** plants. Actually, current research has shown that the plants are really measuring the lengths of nights rather than days. But the names are too firmly embedded in the language of physiologists to be changed.

Long-day plants usually flower in the spring when nights are growing shorter (Fig. 9.15). In these plants the signal for flowering is sent from leaves to shoot tips when the nights are less than a particular critical length. The critical night length is characteristic of the species and, in general, is set so that the plant has a suitable period of vegetative growth in the spring before flowering. Thus plants in the more northerly regions, where winter ends later, tend to have shorter critical night lengths; they therefore flower later in spring.

Short-day plants, such as many chrysanthemums, typically time their flowering for fall or winter, when the nights are progressively lengthening (Fig. 9.16). Here again there is a critical night length, but in short-day plants the signal for flowering occurs when nights become *longer* than the critical length. These plants fail to flower in the shorter days of early spring because at that time they are immature.

Presumably the advantage of flowering in the fall is that the plants have been able to accumulate photosynthetic products all summer for better seed production. A well-set, accurate timing system should choose the optimum time to initiate flowering so that the plant has the longest possible period of active growth while still allowing time for seeds to mature before the cold sets in.

It is not entirely understood how the leaf measures lengths of night. We do know that phytochrome signals the end of the day. During the daylight hours there is always some P_{fr} present, but at night the P_{fr} gradually disappears. It seems that the disappearance of P_{fr} somehow "tells" the plant that night has begun. The next light exposure results in new P_{fr}

Figure 9.16. Effect of day length on behavior of *Chrysanthemum*, a short-day plant. *A,* plant that received light of natural short days of autumn and blossomed at the usual time. *B,* plant that received an hour of light near the middle of each night for several weeks beginning just before flower buds would normally have been initiated; thus, each long dark period was divided into two short ones. This interruption was sufficient to delay flowering.

being formed; this acts as a signal that the night is over. Since P_{fr} formation requires only a little light, even a brief flash of red light during the night can seriously interfere with the timing of night length. As would be expected with a phytochrome-mediated response, the effect of a red light flash at night can be canceled when it is immediately followed by a flash of far-red light.

Though phytochrome signals the beginning and end of night, it appears that in darkness some other component in many plants actually measures the time. This timing system has been called the **biological clock.** Its nature is unknown, but it is highly accurate and insensitive to such disturbing factors as differences in temperature and nutrition. Some plants can discriminate between night lengths that differ by as little as 10 minutes, a degree of precision that allows flowering to occur within a week of the same date every year.

The biological clock is a metabolic system that completes a cycle approximately once each 24 hours. It synchronizes many physiological functions on a daily cycle and is present in organisms ranging from humans to algae.

In photoperiodic plants, the disappearance of P_{fr} after nightfall signals the clock to begin measuring time. At dawn, P_{fr} is re-formed and the timing is stopped. For the control of flowering, the important point is whether the clock has run long enough to exceed the critical night length. The answer determines whether or not the leaf will send a flowering stimulus to the shoot apex.

In considering the flowering stimulus we have only mentioned that it is made in the leaves and moved to the apices. We do not yet know its chemical nature. Some experiments strongly suggest that the same flowering stimulus serves long-day, short-day, and day-neutral plants. They apparently differ only in the conditions required to bring about its synthesis.

Whatever the nature of the flowering stimulus, it must be produced by a metabolic system that we might call a *stimulus generator.* In short-day plants the stimulus generator is activated if the clock passes its critical point during one or more successive nights (Fig. 9.17). To produce a short-day flowering habit, by contrast, the stimulus gener-

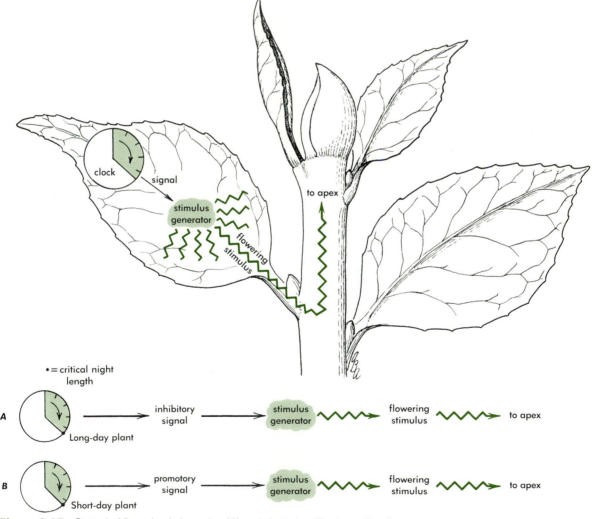

Figure 9.17. Control of flowering in long-day (*A*) and short-day (*B*) plants. Details are hypothetical and diagrammatic. In both cases the timing of night length is started by the disappearance of P_{fr} after nightfall. In both, the same kind of flowering stimulus originates from a metabolic ''generator'' in the leaves; the stimulus moves to the apex. Whether a stimulus is sent depends on whether the clock reaches the critical night length measurement before being interrupted by light. The plants act as if the clock sends a signal on reaching the critical length; short-day and long-day plants differ as to whether the signal promotes or inhibits the stimulus generator.

ator must be set to produce its flowering stimulus automatically if the clock does *not* pass the critical point for one or more nights in succession. That is, long-day and short-day plants differ according to whether the clock activates or inhibits the stimulus generator when the critical night length is exceeded.

TEMPERATURE AND DEVELOPMENT

Photoperiodism is not the only system that times flowering. Some plants require a prolonged exposure to winter cold as a prerequisite to flowering.

Plants having this type of control include both winter annuals and biennials. Winter annuals typically germinate just prior to or during the winter so that the young plant passes through weeks of low temperatures before reaching the warmer weather of spring and summer when flowers form.

Biennials (plants with a 2-year life cycle) generally have a full season of vegetative growth the first year, exposure to cold weather during the next winter, and then they flower in the second spring and summer. Without exposure to the low temperatures, flowering will be delayed or prevented (Fig. 9.18). Gibberellins can sometimes substitute for the cold requirement. This promotion of flowering in response to cold

Figure 9.18. Substitution of gibberellic acid for the cold requirement (winter) in the flowering of the biennial carrot (*Daucus carota*). *A,* control plant under long days only; *B,* long days plus gibberellic acid, no cold treatment; *C,* long days plus cold treatment, no gibberellic acid.

treatment is called **vernalization.** The cold treatment generally requires temperatures of 10°C or less for several weeks. Often, the cold requirement is followed by a requirement for long days. This prevents the plant from flowering too early in spring. The growing tip itself, rather than the leaves, may perceive the cold stimulus in vernalization.

"Winter" cereals, wheat and rye, are normally planted in the fall so that they germinate before the onset of winter and flower promptly in the following spring and summer after making a minimum number of leaves. If they are planted in the spring and are not vernalized, flowering is delayed until many more leaves are formed. "Spring" cereal strains are normally planted in the spring and flower in response to long days. Winter cereal strains can be artificially vernalized prior to a spring planting date by moistening the seeds and holding them at about 1°C for a period of weeks.

In the discussion of photoperiodism, we indicated that many woody perennial plants use the photoperiod to induce a dormant condition in the bud prior to the onset of unfavorable weather. The bud typically goes through a period of increasing dormancy until it can no longer be rapidly reactivated by favorable environmental conditions or a shock treatment such as defoliation. By fall it may be in a state of rest in which no treatment will activate it other than the passage of time at low temperature, a process called **chilling.** The chilling requirement generally requires temperatures a little above freezing but below 10°C. Normally, the requirement is met by midwinter and at that time, a branch brought into the greenhouse will have buds bursting in a few weeks. The buds are no longer in a condition of rest; they can respond to the warmer days of spring with renewed growth. The photoperiod induced the dormant condition, and the chilling process signaled that it was safe to respond to springlike weather.

Fruit trees typically need 250 to 1000 hours of chilling to produce good growth and fruit production in spring and summer. The lack of sufficient cold weather sets the southern limit for peaches, apricots, cherries, plums, and apples, and for effective flowering of ornamentals such as lilac (*Syringa vulgaris*).

Another temperature-related phenomenon is the development of **frost hardiness.** In the summertime, temperatures that are a few degrees below freezing will kill leaves and buds. During the fall, many plants gradually develop resistance to temperatures as low as 50°C below freezing. Short photoperiods and exposure to temperatures just above freezing are the stimuli that induce frost resistance.

The seeds of some species will not germinate until they have been exposed to several weeks of temperatures near freezing. Such a cold, wet treatment of seeds to promote germination is called **stratification.**

SUMMARY

1. Plant development can be modified by environmental stimuli such as light, gravity, and temperature. These responses contribute to the survival of the plant.

2. Plant hormones are natural compounds that act in small quantities as signals to regulate development. Auxin, the gibberellins, the cytokinins, abscisic acid, and ethylene are the major plant growth hormones.

3. Auxins promote cell enlargement by altering the cell walls, making them more extensible. Growth is controlled by factors that affect the synthesis, transport, and inactivation of auxin. Phototropic and geotropic curvatures in the shoot are regulated by changing the transport of auxin to the growing cells.

4. Auxins also help to control other developmental processes such as apical dominance, cell division, and differentiation of specific cell types in the vascular system.

5. Gibberellins control stem elongation through effects on cell elongation and division. They also affect seed germination and cambial activity. In certain instances they may induce the synthesis of enzymes.

6. Cytokinins control cell division in specific tissues; they interact with auxin to control organ initiation and bud growth; they help to prevent aging in leaves; and they help to coordinate root and shoot growth.

7. Ethylene interacts with auxin to control stem elongation and organ abscission. It also helps to maintain the etiolated growth habit in darkness and to regulate ripening in many fruits.

8. Abscisic acid helps to regulate the leaf's response to water stress through stomatal closure and contributes to seed and bud dormancy.

9. Light acts as a developmental signal when it is absorbed by specific photoreceptor pigments. The chief receptor pigments are phytochrome, the phototropic pigment, and protochlorophyll.

10. Phytochrome exists in active and inactive forms that are interconvertible by red and far-red light. When stimulated by light, it causes seedlings to adopt the aboveground mode of growth. It may inhibit stem elongation, promote leaf expansion, promote the germination of seeds, and participate in many other developmental events.

11. Photoperiodism, the control of development in response to the lengths of days and nights, is the basis for many seasonal adjustments in plants. The photoperiodic timing system occurs in the leaves and seems to involve a "biological clock" that is governed by phytochrome.

12. There are two main types of photoperiodically controlled plants, as regards flowering: long-day and short-day plants, which respond respectively to night lengths shorter than critical or longer than critical. Many plants are day-neutral; in them, flowering is not photoperiodically controlled.

13. Buds may be induced to become dormant in autumn by a photoperiodic system. Some dormant buds must be exposed to a prolonged chilling treatment, normally provided by winter, before they will resume their activity.

14. A prolonged period of low temperature may be needed to vernalize certain plants, that is, to remove a block that prevents the formation of flowers by the shoot apex. Some seeds also require a chilling treatment before they will germinate.

10 PLANT ECOLOGY

Through evolution, existing plant species have achieved a balance with their environment. In each type of habitat, certain species group together. Fossil records indicate that similar groups of species have lived together for millions of years. The species of these groups share incoming solar radiation, soil, water, and nutrients to produce a relatively constant amount of living matter. They recycle nutrients from the soil to living tissue and back to the soil again; and they divide the environment's resources. Plant ecology attempts to explain this balance between plants and their environment. For example, what stresses does the environment put on plants, and how do plants respond to them? Plant ecology also attempts to determine how this natural balance can be artificially imitated or manipulated to improve human life.

Plant ecology deals with levels of biological organization beyond the organism level: the population, the community, the ecosystem. A **population** is a group of closely related organisms that normally interbreed. A **community** is composed of all the populations in a given habitat. An **ecosystem** consists of the living community plus the nonliving factors of its environment.

COMPONENTS OF THE ENVIRONMENT

The environment of an organism includes all of the living and nonliving things around it. Some important environmental components are moisture, temperature, light, soil, and living organisms, and all of them work separately and jointly to influence plant distribution and behavior. Species differ in their range of tolerance to these environmental factors. For example (Fig. 10.1), coast redwood *(Sequoia sempervirens)* is restricted to a narrow strip of California and adjacent southwestern Oregon, about 9500 km², that experiences heavy summer fog. Coast redwood is not tolerant of variations in moisture or temperature. Red maple *(Acer rubrum)*, however, occurs over about 4,000,000 km² of eastern North America in wet, dry, cold, or warm environments. It is tolerant of wide variations in moisture and temperature.

There are two types of environment. The **macroenvironment** is influenced by the general climate, elevation, and latitude of the region. Weather Bureau data on rainfall, wind speed, and temperature are measures of the macroenvironment. Measurements are taken at a standard height of 1.5

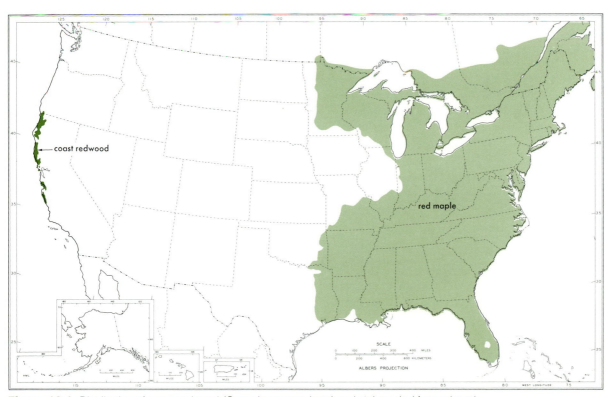

Figure 10.1 Distribution of coast redwood *(Sequoia sempervirens)* and red maple *(Acer rubrum)* in the United States.

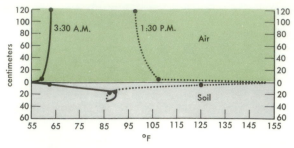

Figure 10.2 Air and soil temperature at several positions just above and just below the soil surface. Temperatures were measured during the warmest and coolest parts of the day.

m above ground in a clear area away from buildings or trees.

The **microenvironment** is the environment that is close enough to the surface of an organism or object to be influenced by it. For example, bare soil tends to absorb heat; consequently, the temperature just above or below the soil surface is much higher than air temperature (Fig. 10.2). Light quality and quantity are much different for herbs beneath a forest canopy than for the leaves of the canopy itself (Fig. 10.3). Air as far as 10 mm from the surface of a leaf on a still day is less turbulent and higher in humidity than free air further away from the leaf. In marshy areas, where the water table is close to the surface, minor dips or rises in the topography greatly influence the root microenvironment. Plants growing in shallow depressions are often subject to more frequent frost than those growing on higher ground, because cold air settles in the depressions. Ecologists are convinced that the microenvironment is as important for plant growth as the macroenvironment.

MOISTURE

Water is a most important factor in determining both the distribution of plants over the earth's surface and the character of an individual plant. Probably no single factor is so largely responsible for the abundance of plants in a habitat as is the supply of water. So important are the water relationships of plants that various attempts have been made to classify plants on the basis of these relationships. One such classification divides plants into (a) **xerophytes,** which are able to live in very dry places, (b) **hydrophytes,** which live in water or in very wet soil, and (c) **mesophytes,** which thrive best with a moderate water supply.

Plants with xerophytic characteristics, which limit transpiration or in other ways closely control water balance, occur in different climatic zones, especially deserts. Xerophytic plants do not necessarily have a lower transpiration rate than do mesophytes when water is ample, but they do possess one or more characteristics that enable them to survive periods of drought. The most effective of them are a thick cuticle, early or daytime stomatal closure, reduction of the transpiring surface, and water storage tissue.

The time of precipitation is as important to plant growth as the total annual amount. For example, tropical rain forests and tropical savannahs may both receive the same total yearly rainfall—perhaps 200 cm—but the forest receives an equal share of the total each month, while the savannah has pronounced dry and wet seasons. The result is that the two vegetation types are quite different: tropical rain forests support tall, evergreen trees and vines in profusion, but tropical savannahs support grasses, shrubs, or short trees, which are often deciduous.

Plants adapt to dry climates in several ways. Cactus plants (see Fig. 10.30) develop water storage tissue in the stem and reduce transpiration by the loss of leaves; all their photosynthetic activity is performed in stem tissue. Mesquite shrubs *(Prosopis)* and salt-cedar trees *(Tamarix)* possess long roots that tap groundwater, sometimes at depths below 53 m. Epiphytes, growing on tree trunks above the soil (Fig. 10.4), trap falling rain in leaf axils or in dead cells on the surface of roots. These epiphytes are not parasites; they depend on trapped water and debris for all growth requirements. Annual herbs adapt to drought by completing a brief life cycle only during periods of sufficient soil moisture. *Boerhaavia repens* of the Sahara Desert is a small annual that can go from seed to seed in 10 to 14 days, but most annuals have a life span of 3 to 8 months.

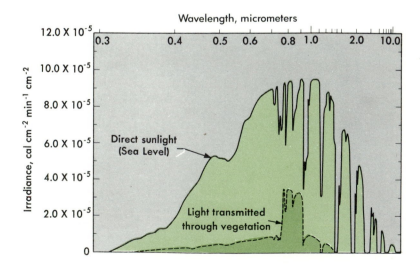

Figure 10.3 Distribution of solar radiation energy by wavelength. Energy was measured in direct sunlight and beneath a forest canopy.

There is some evidence to show that plants can absorb dew through their leaves, but the amount of water obtained in this way is not great. Usually, dew, fog, or high humidity aid the plant more by reducing the rate of transpiration than by physically entering the plant. (Exceptions include small epiphytes, certain annual plants, mosses, and lichens.)

Plants sometimes utilize fog by condensing the mist into large drops that trickle down the stem or drip from the branches, increasing soil moisture considerably. The amount of water added to soil in this way by trees on the San Francisco Peninsula was measured as 5 to 74 cm in one summer, the exact amount within that range depending on the type of exposure. Normal precipitation (rain) for the area is only 64 cm a year, with almost no rain in summer.

TEMPERATURE

Temperature influences moisture availability and the rate at which chemical reactions occur. At freezing temperatures, water changes to ice and is unavailable for plant growth; consequently, in cold climates one of the principal factors limiting growth may be drought. Absorption of water by roots is slower in cold soil than in warm soil. At high temperatures, evaporation may remove much of the soil moisture before the new growth of shallow roots can reach it, and transpiration will also be stimulated, resulting in further rapid depletion of soil water.

Brief extremes in temperature may be more important in determining plant distribution than are long-term moderate temperatures. Palms and many cactus plants seem ex-

cluded from areas that experience a specific frequency of frost. California lilac (Ceanothus megacarpus) seeds germinate poorly unless they are exposed briefly to high temperatures. Many temperate zone plants are not frost-resistant until they are exposed to moderately low temperatures for a time and are thus "hardened." Other plants require an alternation of day and night temperatures (a **thermoperiod**) for best growth.

Topography greatly influences soil and air temperature. In mountains, latitude and elevation combine to restrict species to certain belts (as shown for mountain hemlock in Fig. 10.5). Even at the same latitude and elevation, minor differences in the direction of slope **(aspect)** create different microenvironments. A gorge running east–west through an Indiana forest was studied to illustrate the effect of aspect on plant growth. The gorge was 65 m wide at the top and 45 m deep; its sides supported scattered trees, shrubs, and herbs. Meteorological instruments were placed 15 cm above and below the soil surface midway down each side. During spring, the south-facing slope was found to exhibit a greater daily range of topsoil temperature, a higher mean air temperature, a higher rate of soil water evaporation, and lower relative air humidity than the north-facing slope. Of nine spring-flowering species present on both banks, the average flowering time was 6 days earlier on the south-facing bank. A similar difference in flowering time could be achieved on level ground only over a distance in latitude of 180 km.

Temperature is an estimate of the heat energy available from solar radiation. Solar radiation at the limits of our atmosphere is equivalent to about 2 cal/cm^2/min. Much of this radiation is absorbed, scattered, or reflected within the atmosphere, and only half may reach the ground and heat it. In turn, the warm earth reradiates energy back to space—terrestrial radiation. The difference between solar (incoming) radiation and terrestrial (outgoing) radiation is termed **net radiation.** During a clear day at the equator, solar radiation is greater than terrestrial radiation, and thus, net radiation is positive; during the night, the reverse is true and net radiation is negative. But, over a 24-hr period, a month, or a year, net radiation may be positive or negative, depending on the location. For example (Fig. 10.6), at Aswan, Egypt—a warm desert climate—net radiation is positive each month, with a peak in June. By contrast, net radiation is positive only 5 months of the year at Turukhansk, Siberia—a cold, subarctic climate—and the annual net radiation is very close to zero.

LIGHT

Figure 10.3 shows how solar radiation is changed in quality and quantity by its passage through green leaves. Light intensity on the forest floor may be only 1 to 5% that of full sunlight. Some herbs of the deciduous forest complete their life cycle in early spring, before the trees above them develop their complete foliage and thus reduce light intensity. Leaves of plants that develop in shade exhibit different morphology, anatomy, and physiology than those that develop in sunlight, even when both are attached to the same plant

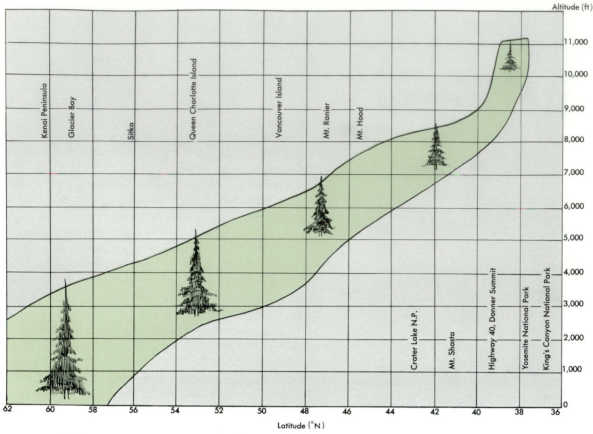

Figure 10.5 Distribution of mountain hemlock (*Tsuga mertensiana*), showing the relationship between altitude and latitude.

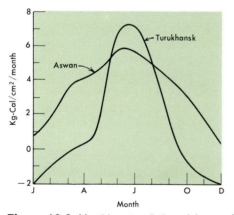

Figure 10.6 Monthly net radiation of Aswan, Egypt, and Turukhansk, Siberia.

(see Chapter 5). Shade leaves are larger, thinner, contain less chlorophyll per gram of tissue, have less well-defined palisade and spongy mesophyll layers, and achieve maximum rates of photosynthesis at much lower light intensities than do sun leaves. Biochemical evidence has shown that the difference in photosynthetic rates is due to a difference in the activity of the enzyme responsible for CO_2 fixation.

When a large tree dies in a mature forest, the canopy is opened and more light reaches the ground beneath. This increased light radiation induces seedlings there to begin rapid growth. Seedlings must be able to survive many years of slow growth in shade before such an opening "releases" them. Eastern hemlock *(Tsuga canadensis)* is remarkably shade-tolerant. Mature trees of this species live for 1000 years. They can retain the capacity for rapid growth when released from shade to an age of 400 years. Saplings only 2 m tall and 2 to 3 cm in diameter may have a ring count indicating an age of 60 years.

Trees that are not shade-tolerant may become established and grow to maturity if some disturbance such as fire or logging first removes the shade. Seedlings of shade-tolerant species grow beneath them, however, and replace them as they die, while their own seedlings do not remain alive. For this reason, it is difficult to maintain pure stands of valuable shade-intolerant species such as teak, mahogany, Douglas fir, or the southeastern yellow pines. These species only invade disturbed sites and do not long maintain themselves.

Light is also reduced in intensity and quality by passage through water. Algae are not commonly found below a depth of 60 m except in very clear water. A record depth for algae may have been found in Lake Tahoe, California. A sample of water and bottom mud from a depth of 140 m contained the alga *Chara*. Light intensity at that depth, despite the water clarity, is only 0.1% of full sunlight.

Brief exposure to light may be as important for some plants as long-term exposure. As discussed in Chapter 9, seeds of some species can be stimulated to germinate in the dark if they are first exposed briefly to full sunlight; this is done by simply shaking them for a few seconds in soil held in a glass container. This treatment could be duplicated in nature by a plow turning the soil and seeds in it, briefly exposing the seeds to sunlight, then reburying them. Plowing aerates the soil, makes it more permeable for root growth, and removes plants at the surface—all conditions of advantage to an emerging seedling.

SOIL

Soil is the part of the earth's crust that has been changed by contact with the living and nonliving parts of the environment. It is a weathered, superficial layer typically 1 to 3 m thick, made up of decomposed and partly decomposed parent rock material with associated organic matter in various stages of decomposition. There are many kinds of soils and many different soil conditions. The character of natural plant covering and the behavior of crops depend on soil conditions as well as climatic conditions. Environmental influences that operate through soil are called **edaphic factors;** those that act on the plant through the atmosphere are called **climatic factors.**

Each environment creates a unique soil. These soils have their own history of development, morphology, and chemical attributes. These soils are described in the technical literature, just as plant species are described, and the basic unit is a **soil series.** Examples include Miami, Yolo, Auburn, Fargo, and Mesa series. Series names are place names, identifying the type locality where the series was first described. There are more than 12,000 soil series in the United States alone.

Each soil series may have several **types** which differ from each other only by the texture of the uppermost layer of soil. Thus we have Miami silt loam (Table 10.1), Miami loam, and Miami sandy loam soil types.

Soil Formation

Most soils consist primarily of mineral particles that are released by a slow, continual process of weathering of the parent rock. Mineral matter is derived from fragmented rock. The kind of parent rock (e.g., granite, lava, sandstone, limestone, and shale) and the degree of weathering determine the nature of the mineral or inorganic components of the soil.

Weathering may involve simply **mechanical breaking** of the parent rock. For instance, the action of strong winds or wave action may hurl sand against rock outcroppings and wear them down, or water collecting in pores of the rocks may expand at freezing temperatures to create fissures that ultimately fragment the rock. **Chemical weathering** results in more profound changes in the mineral matter itself. Atmospheric gases, such as CO_2 or SO_2, become dissolved in rainwater and produce acidic solutions that dissolve the

Table 10.1. Sample Description of a Soil Series[a]

Miami series

The Miami series consists of well-drained soils developed on highly calcareous glacial till of Late Wisconsin age.

Soil profile (Miami silt loam):

A_0 Partly decomposed forest litter from deciduous trees. ¼ to 2 inches thick.

A_1 0 to 3 inches, very dark-gray to dark grayish-brown (10YR 3/1 to 4/2, moist) silt loam; moderate fine crumb structure; friable when moist, soft when dry, and nonsticky when wet; organic content relatively high; gradual and wavy horizon boundary; slightly acid. 2 to 4 inches thick.

A_2 3 to 12 inches, light yellowish-brown to brown (10YR 6/4 to 5/3, moist) silt loam; weak thin platy structure; friable when moist; slightly hard when dry, and slightly sticky when wet; gradual lower horizon boundary; slightly to medium acid. 7 to 11 inches thick.

B_1 12 to 16 inches, light yellowish-brown to yellowish-brown (10YR 6/4 to 5/4, moist) heavy silt loam to light silty clay loam; moderate fine subangular blocky structure; friable to slightly firm when moist, slightly hard when dry, and slightly sticky when wet; medium to strongly acid; irregular lower horizon boundary. 3 to 5 inches thick.

B_{21} 16 to 24 inches, brown (7.5YR 5/4, moist) to yellowish-brown (10YR 5/4, moist) silty clay loam; strong medium to coarse subangular blocky structure; firm when moist, hard when dry, and sticky when wet; variable quantities of small partly weathered rock fragments; wavy lower horizon boundary; strongly to medium acid. 6 to 14 inches thick.

B_{22} 24 to 28 inches, dark-brown (7.5YR 3/2 to 10YR 4/3, moist) silty clay loam; strong coarse to very coarse subangular blocky structure; firm when moist, hard when dry, and sticky when wet; variable quantity of partly weathered rock fragments; wavy lower horizon boundary; neutral to slightly acid. 2 to 6 inches thick.

C 28 inches +, light yellowish-brown (2.5YR 6/4 to 10YR 6/4, moist) mixed clay loam till with coarse fragments and stones; massive to weak very coarse subangular blocky structure; varying mineralogical composition with sufficient lime to produce general effervescence with dilute acid; usually has a high percentage of limestone rock fragments.

Range in characteristics.—The color of the cultivated surface soil is grayish brown to yellowish brown when moist. The reaction of the A and B horizons in local areas is slightly acid. Silt loam, loam, sandy loam, and fine sandy loam types have been mapped. The sandier types usually have correspondingly sandier B and C horizons.

Distribution.—South-central Michigan; central-western and western Ohio; central, northern, and eastern Indiana; southeastern Wisconsin; and northeastern Illinois.

Series established: Montgomery County, Ohio, 1900.
Source of name: Named for Miami River in western Ohio.

[a]Source: *The Soil Survey Manual,* U.S. Department of Agriculture Handbook No. 18 (Washington, D.C., 1951).

Table 10.2. Classification of Soil Mineral Matter According to Size of Particles (International System)

Type of Particles	Range in Diameter of Soil Particles (mm)
Coarse sand	2.0–0.2
Fine sand	0.2–0.02
Silt	0.02–0.002
Clay	0.002 and smaller

Table 10.3. Composition of Three Soils According to the Size of the Mineral Particles

Soil Type (texture)	Course Sand (%)	Fine Sand (%)	Silt (%)	Clay (%)
Sandy loam	67	18	6	9
Loam	27	32	21	20
Clay	1	9	22	68

parent rock material. Plant roots secrete weak acids. Certain algae, bacteria, and lichens hasten the decomposition of the least resistant mineral materials.

Weathering of the parent material, under the influence of climate and organisms, proceeds along a definite series of stages from a young soil in which the processes of soil development are continuing to a mature soil that has approached a steady state.

Soil formation may proceed rapidly if the type of parent material and type of climate are favorable. Soils form rapidly where there is a warm, humid climate, forest vegetation, flat topography, and a parent material that is easily broken down. For example, Fort Kamenetz, built of limestone slabs in the Ukrainian part of the Soviet Union, was abandoned in 1699. Today, a mature soil 10 to 40 cm thick has developed from the limestone. Soils form slowly in regions with a cold, dry climate, grass vegetation, steeply sloping topography, and parent rock material that is not easily broken down. It has taken from 1000 to 10,000 years for mature soils to develop in northern areas scoured during the Wisconsin glaciation.

Plants influence soil development in several ways. They move ions through the soil by absorbing them through their roots. These ions accumulate in the leaves and are returned to the topsoil after leaf abscission. Plant roots and shoots decay and add nitrogen and sometimes acids to the soil. In addition, plant canopies shelter the soil surface, reducing erosion and water loss.

The **texture** of soil is determined by the mineral fractions of a soil; we customarily speak of coarse or fine **sand, silt,** and **clay.** The size of soil particles decreases in the order given (Table 10.2). Clay is not only the smallest mineral soil particle, but because of chemical weathering it has also undergone the most change from the parent rock.

Although most soils contain both mineral and organic matter, the organic matter they contain constitutes only 2 to 5% of their weight. These predominantly **mineral soils** can be classified according to the relative amounts of the sand, silt, and clay they contain. Soils that contain roughly equal amounts of sand, silt, and clay are called **loams.** Depending on the relative amounts of each component in the soil, we speak of clay, clay loam, silt loam, loam, sandy loam, loamy sand, and sand textures (Table 10.3). The suitability of a soil for plant growth is greatly influenced by its sand, silt, and clay content. **Organic soils** are defined as soils having a layer, at least 30 cm thick, that is more than 30% organic matter.

Soil Water and Its Dissolved Substances

An important difference among the various soils is their ability to hold water: it is greatest in clay and organic soils, and least in coarse sand.

If water is applied to a soil in the field, the spaces between the soil particles **(pore spaces)** become filled only for a short time to the depth wetted. With drainage, the water begins to move downward under the influence of gravity. After a while, this movement downward stops; the soil particles hold a certain amount of water against the pull of gravity. The amount of water held by the soil after drainage is called the **field capacity** of that soil. When a soil is at field capacity, or even at a moisture content well below this amount, a film of water completely surrounds each soil particle, and water also exists in the form of wedges between soil particles. Clay particles and organic matter in the soil are **colloidal** particles, which hold additional water by **imbibition.** Water imbibed by soil particles is much more difficult to remove from the soil than water that exists as a film or as wedges.

Soil particles themselves, particularly finely divided clay and organic matter, act as giant negatively charged ions that attract a cloud of positively charged ions such as Ca^{2+} and K^+. Thus, colloidal soil particles, while not in solution themselves, serve as reservoirs to which many of the various ions essential for plant growth are loosely attached and from which these ions may be released into solution by a process known as **cation exchange** (Fig. 10.7).

Acidity influences the physical properties of the soil, the availability of certain minerals to plants, and the biological activity in the soil; consequently, it strongly influences plant growth. Certain plants such as azaleas, camellias, and cranberries grow best in acid soils, but most plants do best in soils near neutrality, and a few plants grow satisfactorily in slightly alkaline soils.

Acid soils, such as those beneath northern coniferous trees that shed acidic foliage, range from pH 3.5 to 5.0; agricultural soils in the humid areas range from pH 5.0 to 7.0; and soils in arid or saline regions may reach pH's as high as 11.0, but a range of 8.0 to 9.0 is more common. Acidity per se—that is, H^+ or OH^- ions—is not directly responsible for limiting plant growth. Rather, soil pH affects the availability of plant nutrients in two ways. In an acid soil, hydrogen ions (H^+) replace other **cations** (positive ions, such as K^+, Ca^{2+}, and Mg^{2+}) on the negatively charged clay particles, allowing the nutrient ions to float free in the soil water and to be easily leached from the soil as the water

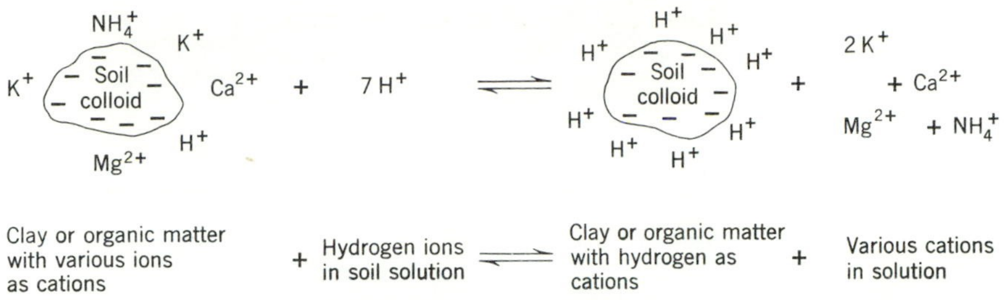

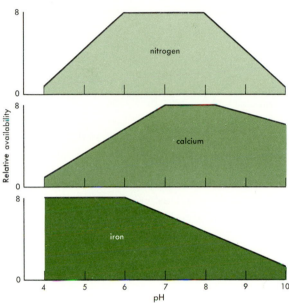

Clay or organic matter with various ions as cations $+$ Hydrogen ions in soil solution $\rightleftharpoons$ Clay or organic matter with hydrogen as cations $+$ Various cations in solution

Figure 10.7 Cation exchange on soil colloids.

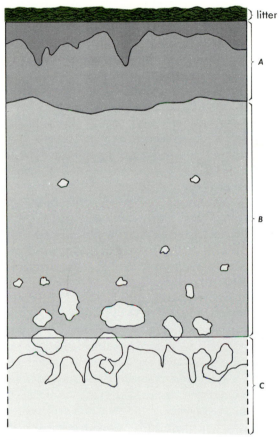

Figure 10.8 Relationship of relative nutrient availability to soil pH.

moves downward. Soil pH also affects the solubility of plant nutrients. As shown in Fig. 10.8, some nutrients (e.g., iron, manganese, and aluminum) increase in solubility as pH decreases. Aluminum may even reach toxic levels in some acid soils. Other nutrients (e.g., calcium and magnesium) increase in solubility as pH increases. Extremes of soil pH may also affect plant growth indirectly by suppressing bacterial growth.

Soil Profile Development

In wet temperate regions, such as those that support dense coniferous forests, soil development is principally caused by percolating rainwater. The rainwater continuously dissolves nutrient salts in the upper portion of the soil and carries them, along with particles of finely divided mineral and organic matter, downward into the lower levels of the soil. This process is called **leaching.** In time, the soil consists of a series of superimposed layers, or **horizons,** that differ in color, texture, and chemical attributes (Fig. 10.9). At present, ten soil orders are recognized: *spodosol, entisol, inceptisol, aridisol, mollisol, alfisol, ultisol, oxisol, vertisol, and histosol.* Five of the soil orders are discussed below.

Spodosols (old name = podzols) develop beneath conifer forests. The upper, or *A*, horizon of a spodosol is

Figure 10.9 Profile of a typical soil in a wet, temperate region. The *A* horizon, except for a thin region just beneath the litter, is leached of organic matter, clay, and salts. These leached substances accumulate in the *B* horizon. The *C* horizon, removed from weathering processes, represents parent material from which the *A* and *B* horizons have been formed.

rich in organic matter only in the few uppermost centimeters, where freshly deposited organic litter is decomposing. The lower portion of the *A* horizon is sandy and light-colored, has a pH of 3 to 5, and may be deficient in nutrients, organic matter, and clay. Rather abruptly, about 30 cm below the surface, the *B* horizon begins. The *B* horizon is often reddish-brown from the oxidized iron that has accumulated in it; it is relatively rich in mineral nutrients and clay and usually has a more neutral pH than the *A* horizon. The *B* horizon, then, is a layer of accumulation. Finally, 100 to 130 cm below the surface, the *C* horizon, made up of slightly weathered parent material, appears. The relative thickness of the *A* and

B horizons depends on the climate and amount of plant cover.

Alfisols and **ultisols** lie beneath the deciduous forests of the eastern United States. They are related to spodosols in that they are leached and acidic, but not as extremely as spodosols. Ultisols have reddish-orange B horizons, showing iron oxides.

Mollisols (old name = chernozem) develop beneath grasslands that experience moderate rainfall and high summer temperatures. Moisture is no longer sufficient to leach the soil severely; the pH is 7 to 8. Fibrous root systems of grasses thoroughly permeate this soil, and their decay evenly enriches it with organic matter. An *A* horizon, containing up to 10% organic matter by weight, occupies the top 60 cm. It may be almost black in the upper part but becomes dark brown below. Just beneath the *A* horizon, calcium carbonate accumulates, leached from above. The layer is sometimes called the A_c horizon. Below it is the parent material; there may be no *B* horizon.

Oxisols (old name = laterites) form in regions of high rainfall with continuously warm temperatures, such as those that support tropical rain forests. Mineral cycling is rapid because of fast plant growth and fast litter decomposition. Soil is acidic in pH. Removal of the forest canopy quickly leads to soil deterioration because litter decomposition no longer compensates for leaching. Large amounts of fertilizer must be added if the land is farmed. Silicates (sand) are leached, but clay and iron are not; the iron in clay is oxidized to Fe_2O_3, giving the topsoil a brilliant reddish color.

FIRE

In the past 40 years, fire has been rediscovered as a major ecological factor. We say "rediscovered" because primitive human beings were well aware of its effect and learned to use it for their own purposes. It may well be that their most important food crops, their domestic animals, their routes of migration, and some of their cultural attributes were molded by natural and human-made fires. Certainly, many animals are today attracted to fire and exhibit behavioral patterns in relation to fire that imply thousands of years of evolution in response to it.

Most natural fires are started by lightning, and in North America many vegetation types, including grassland, chaparral, and at least some conifer forests, owe their distribution and community structure in large measure to lightning fires. On U.S. National Forest land, lightning fires account for over 60% of all fires, an average of 5000 fires each year. In addition, early explorers and settlers invariably commented on the frequency of fires in forests and grasslands. In the words of E. V. Komarek, a fire ecologist: "Lightning fires are an integral part of our environment and though they may vary in both time and space, they are rhythmically in tune with global weather patterns. Our environment can be called a fire environment."[1]

[1]E. V. Komarek, Sr., "The Nature of Lightning Fires," in *Proceedings of the 7th Tall Timbers Fire Ecology Conference* (Tallahassee, Florida: Tall Timbers Research Station, 1968), pp. 5–41.

Figure 10.10 The pine savannah of the southeastern coastal plain. The pine shown is longleaf (*Pinus palustris*); the area is in North Carolina.

One of the first vegetation types shown to be dependent on frequent fire for its maintenance in this country is the pine savannah along the southeastern coastal plain (Fig. 10.10). Tall, scattered loblolly, slash, shortleaf, and longleaf pines dominate the area, and a thick growth of grasses (mainly *Andropogon* and *Aristida*), with some herbs, cover the ground beneath. During the nineteenth century, when this vegetation was protected from fire, it was noticed that the pines gave way to oak; in time the result was a thick oak forest with little grass. This was undesirable, as the pines were valuable for lumber and turpentine, and the grass was valuable for forage. However, the role of fire was not fully appreciated until the present century.

The importance of fire was conclusively demonstrated in 1930, with longleaf pine *(Pinus palustris)*. Longleaf pine not only tolerates fire but depends on it. The seeds germinate in the fall soon after they drop to the ground from the cones. During their first year of growth, the seedlings are very sensitive to even the slightest fire. However, during the next two to four years, longleaf pine seedlings are in the "grass" stage (Fig. 10.11). Most of the growing is done by the roots, and the stem apex remains close to the ground. The terminal bud is covered with a dense mat of hairs that protects it from surface fires that may sweep the area during this time. Fire is even beneficial at this stage, because it kills a fungus that parasitizes the long needles. (*Scirrhia acicola*, which causes brown spot needle blight).

By the time the seedling is 3 to 5 years old, a surface fire is desirable to remove grasses that may have covered the seedlings and shaded them from the sun. The seedling will then make a spurt of stem growth so rapid that, by the age of 8 to 9 years, the young tree's canopy is beyond the reach of surface fires. In addition, its thick bark can endure the heat of a fire without permitting damage to the cambium. If fire is kept from an area for 15 years, grasses and young oak and pine saplings of other species become so dense

Figure 10.11 Close-up of longleaf pine seedling in the "grass" stage.

Figure 10.12 Thick understory of shade-tolerant tress that has developed in a Sierran mixed-conifer forest as a result of fire exclusion.

that reproduction of longleaf pine is suppressed. If fire continues to be kept out beyond that point, oak trees gradually replace the pines and form a closed forest.

The standard practice today is to burn the grass and pine debris about every 4 years. The grass is quickly able to regenerate itself, and pine reproduction is favored, while oak seedlings are killed. If longer periods go by without fire, too much dry litter collects on the ground, increasing the hazard of setting a fire so hot that mature trees are damaged. Of course, no matter how "cool" the fire, great care is taken to control it. The cost of controlled burning amounts to considerably less than one dollar per acre.

The value of controlled burning was recognized sooner on the East Coast than on the West Coast. In California, the absence of fire in some areas has produced potentially catastrophic conditions, especially in the mixed-conifer forest of the Sierra Nevada Mountains. Before settlement by Europeans, natural fires swept these forests about once every 8 years (as revealed when fire scars in tree trunks are dated by ring count). These natural fires were not raging, rampaging, extremely hot fires, because there was only a relatively small amount of litter on the ground. The resulting fire-adapted forests were stately and open, according to early reports by travelers. As shown in Fig. 10.12, however, much of the mixed-conifer belt of the Sierra Nevada (about 1500-m elevation) today is no longer stately and open. We now have evidence that in the absence of fire, these forests are transformed into crowded fir *(Abies)* and incense-cedar *(Calocedrus)* forests, and that the longer fire is excluded, the more catastrophic is the eventual, inevitable fire.

California experiences a high frequency of lightning strikes during dry, late summer, and it is impossible to prevent fires from starting. A considerable amount of litter—needles, twigs, and bark—is shed each year. So long as fires move through an area frequently, say every 8 to 25

years, the fire does not become hot enough to reach the upper canopy and significantly damage the trees. The native trees are adapted to such frequent, light burns. The seedlings of many of them require contact with mineral soil in order to become established because litter does not retain moisture. Fire, of course, clears the ground of litter. In controlled experiments, seedlings of ponderosa pine, Jeffrey pine, and big tree are much more abundant on burned plots than on unburned plots. If the seedlings are protected from fire for 10 to 15 years, they become tall enough and develop a bark that is thick enough to withstand a light surface fire. However, should fire be excluded for much longer periods, so much brush and litter collects that even mature trees cannot withstand the flames that eventually come. Many magnificent stands of big tree *(Sequoiadendron)*—the most massive tree in the world, found nowhere else—are now, for this reason, in great danger of being destroyed by fire.

BIOLOGICAL FACTORS

It is typical for several plant and animal species to coexist in a given habitat. Very rarely does a single species, to the exclusion of all others, control a habitat. If all the species in a group are individually examined, they will be seen to utilize different parts of the environment or alternate with each other in time. Green plants are producers of carbohydrates, but parasitic plants and fungi are consumers or decomposers of it; some animals feed on plants, others feed on insects, and others feed on small mammals; some plants are trees that utilize full sunlight, others such as ferns utilize weak light; some animals are nocturnal, others are diurnal; some plants are evergreen, others are deciduous; and so on. The portion of the environment utilized by each species is called its **niche.** The niche has often been called the "occupational address" of a species. In a particular area and time, it is thought that only one species can persist in each niche. Two

Table 10.4. Some Categories of Biological Interaction in Terms of the Effect the Relationship Has on Each Partner[a]

| | Effect | | | |
| | On | | Off | |
Name of Interaction	#1	#2	#1	#2
Neutralism	0	0	0	0
Competition	−	−	0	0
Mutualism	+	+	−	−
Protocooperation	+	+	0	0
Commensalism	+	0	−	0
Amensalism	0	−	0	0
Parasitism/herbivory	+	−	−	+

[a]Entries under the heading "On" show the effect that occurs when the two partners, #1 and #2, are in contact; entries under the heading "Off" show the effect of breaking off the contact. 0, no effect; +, stimulation; −, depression.

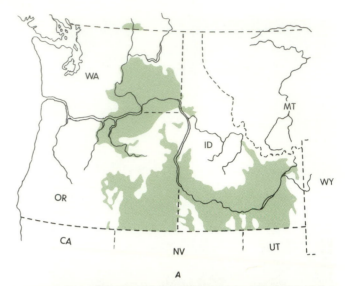

Figure 10.13 *A*, present distribution of cheatgrass (*Bromus tectorum*) in the intermountain region. Range extensions to the south, into California, Nevada, and Utah are not shown. From Mack, Agroecosystems 7:145–165 (1981). *B*, cheatgrass in flower.

species that occupy the same niche compete strongly with each other.

The organisms in a given area can interact in various ways, as shown in Table 10.4. Some of the interactions will be discussed below.

Competition is one form of biological interaction. It occurs when two plants exploit the same limited resource, such as light, water, minerals, or animals that disperse pollen and seeds. Competition leads to decreased growth of the plants when the resource is insufficient. Sometimes only a single factor, such as a mineral element, is lacking, but usually two or more factors are limiting, and it is difficult to separate them and determine which one creates the greater competitive stress. Competition is very important in determining plant distribution. Many species restricted to saline, dry, or nutritionally poor soils would actually grow better on "normal" soil if other plants were removed. These species are restricted to their niches because they are poor competitors in comparison with the numerous plants that populate more moderate niches. Sometimes, introduced plants become widespread pests and actually replace native species because they are better competitors than the native species. This seems to be the reason for the enormous increase of the introduced annual plant cheatgrass *(Bromus tectorum)* in the intermountain region of the United States (Fig. 10.13).

In general, plants respond to competition in one of two ways: by thinning, during which smaller, weaker individuals die and larger individuals survive to reproductive age; or by plastic (nongenetic) reduction in size of all individuals and survival of all to reproductive age. The thinning response is typical of long-lived perennial plants, and the plastic stunting is typical of short-lived plants such as desert annuals. Cheatgrass, for example, exists in natural densities of from 10 to 1000+ individuals per square meter; survival rates at both density extremes are similar, and the only difference is that plants in dense plots are small at reproductive maturity and each produces fewer seeds compared to plants in open plots.

Amensalism, another form of biological interaction, is the inhibition of one species by another. Whereas competition results from the removal of a resource, amensalism results from the addition of something to the environment. In the coastal hills of southern California, sage shrubs *(Salvia)* cover the slopes, and grasses carpet the valleys. Occasional pockets of shrubs occur in the grassland. Figure 10.14 is an aerial view of these pockets of shrubs. The ground beneath and around the shrubs is bare of grass, and the grass is stunted as far as 9 m from the shrubs. Since the zone of stunting is well beyond the limits of shrub root growth, competition between shrubs and grass for soil moisture is not the cause of stunting. If clumps of grass are transplanted to the bare zone, their growth is severely retarded. Analysis of the soil beneath the shrubs, the bare zone, or the grasses,

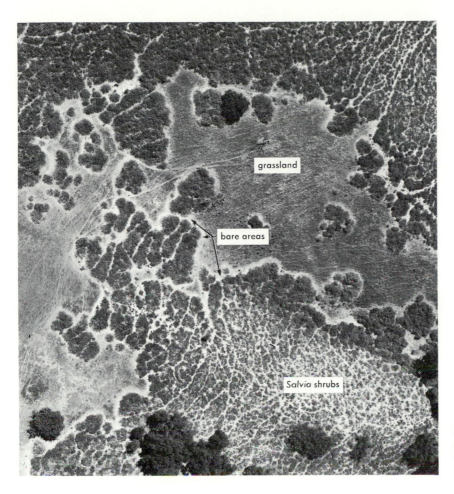

grassland

bare areas

Salvia shrubs

Figure 10.14 Aerial photo of *Salvia* shrubs adjoining grassland. The light bands between and around the shrubs are bare of most plant cover.

shows no major differences in physical or chemical attributes. The reason for this distribution is that many volatile oils are emitted from sage leaves, and two of the oils in particular—cineole and camphor—are very toxic to grass seedlings. When both fresh sage leaves and grass seeds are placed in a closed chamber, germination is often reduced to zero, and radicle growth is considerably reduced. These same toxins can be isolated from the air around sage shrubs, from the natural soil, and from seed coats of grasses in the soil. The conclusion of some ecologists is that the bare areas are caused by toxins emitted by *Salvia* that inhibit the germination and growth of grasses. More recent studies indicate that a high intensity of foraging by birds and small mammals, which nest in the shrubs, also contributes to maintaining the bare zones.

Amensalism by chemical means may be very common in nature, and it has been recognized for 2000 years by farmers. Pliny, the Roman scholar and historian (23–79 A.D.), wrote that walnut *(Juglans nigra)* inhibits the growth of nearby plants. It is now known that a yellow water-soluble pigment, juglone, is leached from all walnut organs, enters the soil, and is the toxic principle that damages plants. Plants are leaky systems, passively contributing all sorts of substances to their environment. One investigator grew seedlings of 150 species in a nutrient culture that included several radioactive elements as markers. The plants took up the isotopes through their roots and transported them to all parts of the

plant, including the leaves. The plants were then exposed to a mist, and the water that condensed and ran off the leaves was collected for analysis. He found that 14 elements, 7 sugars, 23 amino acids, and 15 organic acids—all radioactive—had been leached from the plants. In nature, these substances would have been leached by rain from the leaves and would have accumulated in the soil. Other studies have shown that roots are similarly leaky.

Herbivory, another form of interaction, is the consumption of plant parts or seeds by another organism. The consuming organisms could be parasitic plants or microbes; browsing and grazing animals, which consume only a minor part of the plant; or animals that consume enough of the plant to cause death.

Herbivory may also be a part of other interactions. For example, metabolic by-products (secondary compounds) in plant tissue can repel some herbivores. This is an example of amensalism. Other herbivores enter into **mutualistic,** or mutually beneficial, relationships with their food sources: one herbivore may pollinate flowers, while another disperses fruits or seeds. In some tropical trees, such as acacia, ants dwell in the bases of hollow thorns and actually guard the tree against other herbivores and competing plants. In return, the ants feed on various nutrients that are concentrated in the leaves. In addition to these effects, a moderate intensity of grazing can sometimes result in higher rates of plant photosynthesis and a resulting biomass that surpasses ungrazed control plots. The enhancement can be due to an opening up of the canopy and a change in plant hormone balance that leads to the initiation of new shoots and vigorous growth. These mutualistic effects reflect the fact that plants and herbivores have coevolved for thousands and millions of years. An active area of current ecological research is the study of plant–herbivore interactions: the nature of plant defenses and the nature of herbivore adaptations that penetrate the defense (for an example, see Table 10.5). These interactions illustrate the process of **coevolution.**

Herbivory plays an obvious part in plant distribution. A glance along pasture fences reveals that normally abundant but palatable plants are absent from the pasture, yet they are common outside it; weedy species, which are not palatable, are common in the pasture, yet they are rare outside it. Grazing animals are responsible for this difference. Plant defenses against herbivores include: (1) a spatial distribution that makes it difficult for herbivores to locate and damage host plants; (2) a life-cycle timing, such as seasonal leaf drop, that makes the host plant unavailable to herbivores; (3) the production of a physical barrier such as cuticle, thick pubescence, thorns, spines, or stinging hairs; and (4) the manufacture of substances that are distasteful or otherwise inhibit herbivores from feeding on host plants.

Many insects feed exclusively on plants, and some plant species have evolved sophisticated chemical defense mechanisms against insect predation. Such chemicals include tannins, alkaloids, or volatile terpenes and oils. In some cases the deterrents are concentrated in glandular hairs or trichomes on leaf surfaces; this is true of a wild relative of the cultivated tomato, *Lycopersicon hirsutum* forma *glabratum*. Our selective breeding of tomatoes has diminished the concentration of the toxin in commercial strains.

Some plant metabolites attract animals instead of repelling them. Flower scents are a good example. Some volatile substances emitted by flowers are chemically similar or identical to insect pheromones (animal-generated volatiles that affect animal behavior). Thus, the oriental fruit fly (*Dacus dorsalis*) responds to blossoms of several plant species that produce a volatile compound identical to the insect's sex pheromone.

Table 10.5. Apparent Plant Defenses of Some Tropical Legumes to Seed Weevil Herbivores and Corresponding Adaptations by the Seed Weevil Species to Circumvent These Plant Defenses

Plant Defense	Weevil Adaptation
Gum production by seed pods following penetration of first larva from egg mass; this may push off remaining eggs or drown or otherwise obstruct young larvae.	A period of quiescence in embryonic development in the egg until seed maturation is completed. Resistance to these gummy fluids. Eggs laid singly instead of in clusters.
Dehiscence, fragmentation, or explosion of pods, scattering the seeds to escape from larvae coming through the pod walls.	Oviposition on seeds only after they have been scattered. Attachment of seeds to one another or to the pod valve.
Production of a pod free of surface cracks, which prevents eggs from adhering. Also, indehiscent pods, excluding weevils which oviposit only on exposed seeds.	Oviposition into the soft, fleshy exocarp. Chewing holes in the pod walls in which eggs are then laid. Gluing the eggs to the pod surface.
A layer of material on the seed surface that swells when the pod opens and detaches the attached eggs.	Attachment of the eggs by anchoring strands to allow for substrate expansion. Entry of the larva through the pod into the seed before the pod opens.
Production of allelochemic substances.	Mechanisms for avoidance and detoxification.
Flaking of the seed pod surface, which may remove eggs laid on that surface.	Oviposition beneath the flaking exocarp. Rapid embryonic development or feeding on immature seeds so that entry occurs before flaking begins.
Immature seeds remaining very small throughout the year and then abruptly growing to maturity just before being dispersed.	Entry and feeding upon immature seeds. Delay of embryonic maturation until seeds are mature.
Seeds so small or thin that weevils cannot mature in them.	Utilization of seed contents more completely. Feeding on several seeds.

Source: T. D. Center and C. D. Johnson. 1974. Coevolution of some seed beetles (Coleoptera:Bruchidae) and their hosts. Ecology 55:1096–1103.

ECOLOGY AND PLANT POPULATIONS

Ecologists would like to use species as indicators of particular environments. They would like to know, for example, that a certain species only grows in a certain range of temperature or moisture. Then the temperature and moisture patterns of an area could be accurately predicted by just noting whether that plant species grows there. Unfortunately, most species are tolerant of a wide range of environments, and they are not genetically uniform. Individuals of species X that grow in one environment are slightly different from others of species X that grow in another environment. Their genetic variation makes it possible for them to compete successfully in several different environments.

ECOTYPES

In the early twentieth century, the Swedish botanist Göte Turesson collected seeds of species that had a wide range of habitats—from lowland, southern, and central Europe to northcentral Russia in the Ural Mountains. When he germinated the seeds and grew the plants in a uniform garden in Åkarp, Sweden, he found that members of widespread species were not uniform. Plants that grew from seeds collected

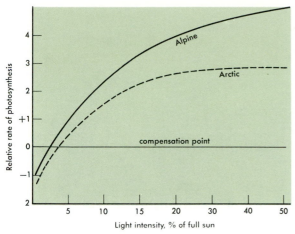

Figure 10.15 Photosynthetic response to different light intensities for alpine and arctic ecotypes of alpine sorrel. (*Oxyria digyna*).

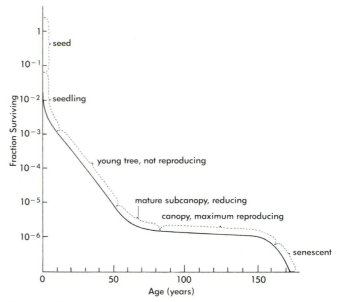

Figure 10.16 Number of plants in various age stages, for a population of the tropical palm *Euterpe globosa*. If the top left "1" equals 10 million, then for every 10 million seeds present in a given area, there are 100,000 seedlings, 10,000 young trees, and so forth. Redrawn from Van Valen, Biotropica 7 (1975): 260–269. [Van Valen. 1975. Life, death, and energy of a tree. Biotropica 7:260–269.]

in warm, lowland sites often were taller and flowered later in the year than those that grew from seeds collected in cold, northern sites. Yet these variants had very similar flower morphology and were interfertile, traits indicating that the variants were all part of one species. Turesson called these variants within species **ecotypes.** He hypothesized that they were ecologically and genetically adapted to particular environments.

Evidence that has since accumulated indicates that most wide-ranging species are composed of a continuum of ecotypes, each differing slightly in morphology and/or physiology. For example, alpine sorrel (*Oxyria digyna*) is a small perennial herb that grows in rocky places above timberline in mountains and north of timberline in the arctic. Both the alpine and arctic environments experience long periods of freezing weather, but alpine areas receive much higher light intensities during the growing season than do arctic areas. Physiological ecotypes have resulted. For example, the optimum light intensity for photosynthesis in alpine plants is much higher than that for arctic plants (Fig. 10.15). The two ecotypes also differ morphologically in leaf color, leaf shape, and the frequency of rhizome production.

Applications of the ecotype concept have been especially useful in forestry. The U.S. Forest Service conducts tests to determine the best ecotype of a species to be used for reforestation in a given area. Seed is collected throughout a species' range and is germinated in a uniform garden; seedling growth is carefully watched, and the ecotype best suited for the garden climate is selected for nearby reforestation programs.

DEMOGRAPHY

Demography (or demographics) is the study of vital statistics: births, deaths, reproductive rates, and ages of individuals in populations. Every plant population has its own particular birth (germination) rate, age class distribution, and average life span.

Plants present some unique problems to demographic

studies, compared to animals. Perennial plants have primary and secondary meristems that theoretically permit unlimited growth in length and girth. Some, in addition, can reproduce asexually by runners, layering, or rhizomes so that the same genetic "individual" can continue life endlessly. (In such a case, each member of the clone is called a **ramet,** but they all belong to the same, single genetic entity called a **genet.**) The longest-lived organisms on earth are plants, including lichens 4500 years old, clones of shrubs 4000 years old, and conifer trees nearly 5000 years old (see Fig. 19.7). In addition, the seeds of some species lie dormant for as long as 1000 to 10,000 years. Finally, plants are more plastic than animals, so that demographic events for the same species can vary enormously from place to place or from year to year.

Every vigorous, healthy plant population should contain the following age states: viable seeds (or other propagules), seedlings, juveniles, immature vegetative plants, mature vegetative plants, initial reproductive plants, maximally reproductive plants, and senescent plants (Fig. 10.16). If a population shows only the first four or five age states, it is probably invading a plant community or area; if it shows only the last four states, it may be declining and will ultimately disappear from the community or area.

ECOLOGY AND PLANT COMMUNITIES

Some environments support only one species. Dense clusters of cattail (*Typha*) grow in water-filled ditches, and all other vascular species are excluded; widely spaced spinifex (*Triodia*) covers hundreds of square kilometers of arid Aus-

Figure 10.17 A community composed of essentially a single species: spinifex, or porcupine grass (*Triodia basedowii*) in Central Australia.

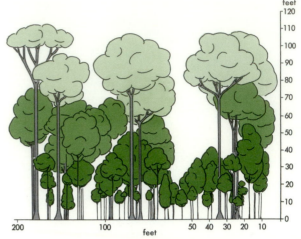

Figure 10.18 Profile diagram of the architecture of a tropical rain forest in British Guiana. There are three tree strata, at approximately 30, 18, and 9 m. Shrubs and ground herbs are uncommon, but tree sapings are included in the 9-m stratum.

tralia, and few other perennial species are present (Fig. 10.17). Most environments, however, support groups of species associated together in **communities.**

Provided that distances are not too great, a given community repeats itself wherever the environment is suitable. For example, high peaks in the Smoky Mountains of North Carolina all exhibit red spruce *(Picea rubens),* mountain ash *(Sorbus americana),* Fraser fir *(Abies fraseri),* and a ground carpeted with several moss and fern species. The spruce, fir, ash, mosses, and ferns are important members (but not the only ones) of this high-elevation forest. High elevations at different latitudes or on different continents support other species, so there are other types of high-elevation forest communities. There are hundreds of nonforest communities as well. Along slopes of the California Coast Range at moderate elevation, a shrubby community of scrub oak *(Quercus dumosa),* manzanita *(Arctostaphylos),* and California lilac *(Ceanothus)* repeats itself. Black spruce *(Picea mariana),* *Sphagnum* moss, and the shrub Labrador tea *(Ledum palustre)* are repeatedly associated together in virtually every bog across Canada.

Communities are named after their most common large species—the plants that receive the full force of the macroenvironment. The spruce and fir of the high-mountain community in the Smoky Mountains occur in greater numbers than does the mountain ash, and their leaf canopies influence the microenvironment in which the carpet of ferns and mosses grow. The spruce and fir are said to **dominate** the community, and the community name is **spruce-fir.** For similar reasons, we have **black spruce** communities and **scrub oak** communities.

Communities exhibit a unique architecture or structure, which is produced by the leaf canopies of their dominant and subordinate species. In the tropical rain forest of British Guiana, one group of species forms a canopy at 30 m, a second group at 18 m, and a third at 9 m (Fig. 10.18). The ground floor is almost bare of shrubs and herbs. In decid-

uous forest communities of the United States, there is one tree canopy layer at about 18 m, a layer of short trees (such as dogwood) at 3 m, a rich shrub layer at 1 m, and a seasonally present layer of herbs on the ground. Some desert communities exhibit only a single (shrub) layer.

Fossil impressions millions of years old sometimes reveal communities very similar to those that exist today. These fossil communities are so similar to modern communities that we are able to reconstruct past climates from them. A fossil community near the town of Goshen in central Oregon, dated to the Eocene epoch (50 million years ago), included the following genera: *Drimys, Magnolia,* holly *(Ilex),* oak *(Quercus), Ocotea,* and fig *(Ficus).* The closest living approximation of this fossil community today occurs 2400 km south of it in the temperate uplands of Central America. Apparently the climate in North America is now much colder than it was in the Eocene. The reconstruction of past vegetation types and climates by the use of fossil evidence is called **paleoecology.**

Another form of fossil evidence is pollen that has been trapped in lake sediments. As pollen is shed near a lake, some sinks to the bottom and is incorporated with silt and organic matter that forms sediment. A chronological sequence of pollen is thus preserved; the deeper the pollen occurs in the sediment, the older the pollen. Pollen of many species is resistant to decay under such anaerobic conditions and may remain intact for thousands of years. Cores of sediments can be examined under the microscope and the pollen identified as to genus, and sometimes even to species. Within limits, paleoecologists assume that the abundance of pollen in the core is proportional to the abundance of that species in the past.

HOW CAN COMMUNITIES BE MEASURED?

If communities are to be useful as tools with which to estimate the nature of the environment, they must be described more completely than just by their name, because minor variations in species composition indicate different environments. For example, in areas where spruce and fir are both dominant and very abundant, the environment is probably different from where they are still dominant but less abundant. How can the abundance of a species or the composition of a community be measured?

1. One way to characterize a community is to measure the amount of ground covered by each species. The amount of ground cover is important because it affects the microenvironment of the community. To sample the ground cover, one could stretch a tape measure across the ground; the length of tape covered by each plant canopy is proportional to the total ground cover of each species. This method of determining ground cover is called a **line transect.** Numerical information from line transects often contradicts first impressions. In desert shrub communities, for example, a person looking across the landscape in Fig. 10.19 would think that the shrubs covered much of the ground, but such an estimate is in error because of the low perspective. If seen from above, when the canopy edges can be projected perpendicularly to the ground, it is apparent that actually only 10 to 20% of the ground is covered. Forest communities, on the other hand, exhibit more than 100% cover because of overlap in canopy layers at different levels.

2. Another method of estimating cover, which is especially useful when plants are low to the ground, is by the use of **quadrats.** Quadrats are variously shaped frames that can be placed on the ground directly over the plants. The amount of ground covered by each species can be estimated as a percentage of the area enclosed by the quadrat frame. A community can be sampled by placing quadrats at random or at regularly spaced locations throughout it. Usually not more than 1-10% of the total community area is included in the quadrat samples.

3. Other methods of sampling, based on mathematical assumptions, can be used to determine whether individuals are distributed at random, regularly (like trees in an orchard), or in clumps (Fig. 10.20). Nonrandom distribution (regular or clumped) may result from propagation by asexual means, the method of seed dispersal, competition, or some pattern (like terracing) in the microenvironment. Most temperate zone species are distributed in clumps.

SUCCESSION

As we have already mentioned, the microenvironment beneath a plant community is very different from that in the open. Temperature, humidity, soil moisture, and light are all affected by the canopy. A stable community consists of spe-

Figure 10.19 Shadscale (*Atriplex conferifolia*) shrubs dominating a desert community in southern Nevada. Plant cover is less than 20%.

cies whose seedlings can survive in its unique microenvironment; seedlings of other species do not survive. But if the stable community is removed by some catastrophe, leaving the soil exposed to full sunlight, the first species that colonize the site are not those of the old community. They are seedlings of species adapted to growth in full light intensity.

In parts of the southeastern United States where oak–hickory communities are stable, the first species to invade following disturbance are annual and perennial herbs that grow best in high light intensity, such as horseweed (*Conyza canadensis*), *Aster,* and broomsedge (*Andropogon virginicus*). These make up the **pioneer** community. Pine seeds blow in from neighboring areas during the first several years after the disturbance, and within 5 years there are

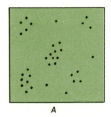
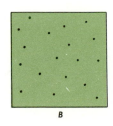
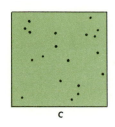

A   B   C

Figure 10.20 Diagrammatic representation of three types of plant distribution: *A,* clumped; *B,* regular; *C,* random.

many pine seedlings. As the seedlings grow, they compete with the herbs for moisture, and their expanding canopies begin to change the microenvironment. Within 30 years of the disturbance, a tall stand of pine results. Examination of the forest floor, however, shows many oak and hickory seedlings and very few pine seedlings. Pine seedlings grow poorly in shade and under conditions of root competition, but those of oak and hickory do well. Within 50 years of the disturbance, the hardwood seedlings form a well-defined understory beneath the pines (Fig. 10.21). Whenever a pine tree dies, it is replaced in the upper canopy by an oak or a hickory. Within 200 years of the disturbance, the forest again consists only of mature oak and hickory trees, and the forest floor is covered with many seedlings of the same trees. The shrubs and herbs associated with an oak–hickory community are also present. Such a sequence of change in plant communities is called a **succession,** or **sere,** and the stable community that is capable of indefinitely maintaining itself is called a **climax community.** In this case, oak–hickory is the climax community. All the intermediate, temporary communities are called **seral stages.**

Sometimes, climax communities are prevented from becoming established because of a recurring disturbance. There is some evidence that part of the prairie of the central United States that greeted pioneers could have been a forest were it not for recurring prairie fires. The fires destroyed slow growing tree seedlings, but grasses quickly regenerated or germinated. When fire is controlled, prairie near the forest edge soon supports many tree seedlings.

The first ecologists to elaborate successional pathways—such as Josef Paczoski in Europe and Frederick Clements in the United States—had no experimental evidence for documentation. They could not record changes as time passed, for most successions take hundreds of years to complete. They determined the existence of successional pathways by comparing a climax community with adjacent small areas that had been disturbed at known times in the past. They examined freshly disturbed or denuded areas (as a result of, for example, ice, landslides, fire, logging, and wind-throw), which revealed the likely pioneer community, or the first stage of succession. Sites that had been disturbed at some time earlier (verified from records of fires, grazing, or logging) exhibited later successional communities because the passage of time was greater. Complete sequences of succession were sometimes represented as bands of different communities about a gradually filling bog or behind a retreating glacier (Fig. 10.22). The community closest to the lake or glacier was the most recently established and represented the pioneer community. The next

Figure 10.22 Succession of communities about a pond near Juneau, Alaska. Standing water in the pond supports leaves of cow lily (*Nuphar*); mats of *Sphagnum* moss and sedge (*Carex*) encroach on the pond along its edge; beyond is a shrub-dominated community of *Empetrum, Cassiope,* and *Kalmia*, with some mountain hemlock trees (*Tsuga mertensiana*); finally, Sitka spruce (*Picea stichensis*), a tall, dark-green tree, dominates the rest of the area.

Figure 10.21 A stand of longleaf pine (*Pinus palustris*), about 50 years old, showing a well-developed understory of broad-leaved species of oak and hickory. In time, the pines will die and be replaced by maturing members of this understory.

closest community occupied soil that had been exposed for a longer time; at some earlier time it had supported the pioneer community, but enough time had passed for the next successional stage to appear. Beyond the second community was a third, representing the third successional stage, and so on until the climax community was reached. Ring counts of trees closest to the pond or glacier, coupled with estimates of the rate of silting or of glacial movement, gave clues as to the time necessary for each successional stage to be reached.

The conclusions of these early workers have yet to be experimentally or empirically verified, and only recently have some of the causes of succession (such as competition, amensalism, and growth requirements of seedlings) been experimentally dealt with.

VEGETATION TYPES OF THE WORLD

The term **vegetation** refers to the life form of plants dominating a community: trees, shrubs, or herbs; deciduous or evergreen; coniferous or broad-leafed. A community dominated by evergreen angiosperms belongs to a different vegetation type than one dominated by shrubs or one dominated by conifers. On the other hand, many different communities dominated by different conifers belong to the same vegetation type. We will briefly examine six major vegetation types: tundra, taiga, deciduous forest, tropical rain forest, savannah–prairie, and scrub.

TUNDRA

If one were to travel north or south toward the poles, it would be evident that the trees gradually become stunted and less common, and finally disappear completely. At the same time, low shrubs, perennial herbs, and grasses become dominant. The resulting meadowlike vegetation is called **tundra** (Fig. 10.23). Annual plants are very rare, and the reproduction of perennials is chiefly by vegetative means such as rhizomes. Shrubs are dwarfed, gaining normal height only

in the protection of the lee of boulders or small hills. The winter wind, carrying ice, acts like a sandblast and prunes back stems wherever they project beyond the boulder or hill.

The warmest month has an average daily mean temperature of 10°C or less, and the growing season is short. Only 2 to 4 months of the year have average daily temperatures above freezing. Approximately the top 30 cm of soil thaws during the growing season and roots may freely penetrate it, but below this level soil water may be permanently frozen; this soil is called **permafrost.** Annual precipitation is about 25 cm, very little falling as snow. The terrain is generally flat, but small depressions are common, and water tends to stand in them during the growing season. Day length fluctuates from 24 hours in June (Northern Hemisphere) to none at all in December.

A similar tundra vegetation occurs in mountains at elevations above treeline. The climate, however, is slightly different. Solar radiation is more intense, day length does not fluctuate so extremely, and frosts are more common during the growing season. Many plants are tinged with red from abundant anthocyanin. This substance absorbs some of the high light intensities that may damage the plant's photosynthetic apparatus.

The lower limit of these alpine areas depends on latitude. At 60°N in Alaska, treeline is at 900 m; at 45°N in the Rocky Mountains, it is at 3000 m; at 20°N in central Mexico, it is at 4000 m; at 5°S in the Andes of South America, it is back to 3600 m; and at 50°S in Chile, it is back to 900 m. Because summer temperatures are cooler near the coast, timberline also depends on distance from the ocean. Thus, timberline is lower in the Cascades than in the Rockies.

TAIGA

Just south of the tundra is the **taiga,** a broad belt of communities dominated by conifers. A similar belt occurs in mountains, just below the alpine zone (Fig. 10.24). Trees are slender, short (15 to 20 m), and relatively short-lived (less than 300 years), but the forests are dense. The taiga is made up of a patchwork of forests on upland sites that alternate

Figure 10.23 Alpine tundra, Beartooth Plateau, Wyoming, 3300 m elevation.

Figure 10.24 Taiga, near Canadian-United States border.

Figure 10.25 Moss-draped maple in the spruce and Douglas fir forest of lowlands in Olympic National Park, Washington.

with bogs formed in depressions. Soils are acidic and relatively infertile spodosols; in northernmost or uppermost parts of the taiga the soil has permafrost below a surface layer.

The growing season is 3 to 5 months long, and temperatures above 30°C are occasionally reached. Winter temperatures are very severe; the mean daily temperature is below freezing for 6 months of the year, and at Yakutsk, Siberia, the average January daily temperature is −43°C. Annual precipitation is less than 50 cm, and the air is exceptionally dry. The taiga is monotonous and silent; animal life is not abundant.

At lower elevations in some mountains, and along the coast (where temperature extremes are not so severe), coniferous forests are much more luxurious. Trees are taller and the undergrowth is denser. Many favorite recreation areas of the western United States are situated in these richer forests. The Olympic Peninsula of Washington receives over 200 cm of annual precipitation and supports the most spectacular coniferous forest in the world (Fig. 10.25). Douglas fir, the most heavily cut lumber species of the United States, reaches its best growth in this area.

DECIDUOUS FOREST

In contrast to the taiga, the **deciduous forest** supports a diversity of plant life and is vibrant with animal activity. Deciduous angiosperm trees dominate communities of this forest. When day length becomes shorter and nights become cooler in late summer, hormone concentrations in such trees

change, and the destruction of chlorophyll follows, allowing other pigments to show through in brilliant fall colors of red, yellow, and orange. In spring, before new leaves have fully expanded from buds, beautiful and distinctive herbaceous perennials come into flower. Shrubs are common but not dense.

These forests occur in areas with a cold (though not severe) winter and a warm, humid summer. Snowfall may be heavy, but most of the annual precipitation falls as summer rain. Annual precipitation ranges from 75 to 125 cm. Since the growing season may be as long as 200 days, it is not surprising that much of the deciduous forest in Europe, Asia, and North America has been cut and replaced with cropland.

Soils are alfisols in the north and west, ultisols in the south. (The latter have striking yellow or red colors in the *B* horizon, which is often exposed if plowed.)

TROPICAL RAIN FOREST

No other vegetation can equal the **rain forest** in diversity of species and complexity of community structure (Figs. 10.18 and 10.26). The deciduous forest exhibits 5 to 25 tree species per acre, but the tropical rain forest supports 20 to 50. The tree canopies typically form three layers, and beneath this dense overstory there are very few shrubs and herbs. Light intensity on the ground is less than 1% of full sun. Most of the herbaceous plants grow as epiphytes on trunks and branches of trees. Vines cling to tree trunks and wind their way up to the canopies. Seeds of a few species, such as the well-named strangler fig, germinate in the canopy and

Figure 10.26 *A,* tropical rain forest in northwestern Argentina; note epiphyte on trunk in foreground. *B,* interior of *Euterpe* palm montane rain forest, 1200 m elevation, central mountains of Puerto Rico.

grow down to the soil. Many of the tree trunks flare out at the base in peculiar buttresses.

The species are evergreen. A given leaf may be shed or a given bud may break at any time of the year. The leaves tend to be very large, with long, tapering tips that may serve to drain the surface of moisture. Rainfall is high, often over 250 cm a year, but it is evenly distributed throughout the year. Temperature is also uniform throughout the year, averaging 27°C. Warm temperatures and high humidity couple to produce a climate oppressive to humans. Afternoon showers, which reduce the temperature, offer only short relief.

Farming peoples of the rain forest have for centuries practiced agriculture on a cut–burn–cultivate–abandon plan. The vegetation is removed in a small area by cutting and burning, cultivation follows for a few years, and the plot is abandoned. Abandonment follows because crop growth declines each year. The rain leaches nutrients from the exposed topsoil, and the nutrients are not replaced by litter as they are beneath the normal forest canopy, as mentioned earlier in this chapter. Much of the rain forest of Asia and Africa has been disturbed in this way. More recent disturbance involves bulldozing and clear-cutting larger areas, especially in South America; the rain forest may be incapable of reinvading these areas. The vegetation that invades these abandoned plots is shrubby, dense, and often includes bamboo. It is this successional stage, which eventually will give way to rain forest, that serves as the "jungle" of television and films. Soils are typically in the oxisol order.

SAVANNAH AND PRAIRIE

Forests give way to grassland as rainfall decreases and becomes more seasonally distributed. At first, trees remain close enough to retain a closed canopy, but gradually they become more widely spaced, and eventually they drop out altogether except along river banks and canyons. **Savannah** is the term applied to grassland with widely spaced trees whose canopies cover less than 30% of the ground (Fig. 10.27); **steppe** or **prairie** refers to grassland without trees (Fig. 10.28).

Grasslands of the United States are heavily used for agriculture and grazing purposes, and little of the original vegetation or animal life remains today. Only from the records of explorers and early settlers are ecologists able to reconstruct in their imagination what these thousands of square miles once looked like.

Figure 10.27 African savannah in Kenya. The flat-topped trees are *Commiphora,* which is the source of the biblical myrrh.

Figure 10.26 (*cont.*)

Figure 10.28 Short-grass prairie, South Dakota.

The eastern part, with relatively high rainfall, supported a dense "sea" of grass some 1 to 2 m tall. Here is a description from a journal dated 1837:

The view from this mound . . . beggars all description. An ocean of prairie surrounds the spectator whose vision is not limited to less than 30 or 40 miles. This great sea of verdure is interspersed with delightfully varying undulations, like the vast swells of ocean, and every here and there, sinking in the hollows or cresting the swells, appears spots of trees, as if planted by the hand of Art for the purpose of ornamenting this naturally splendid scene.

This is now the corn belt and is one of the most agriculturally fertile areas of the world. Productivity is so high that it permits us to export grain to many countries around the world. Its mollisol soils are dark brown in color from the rich amount of organic matter left by decaying fibrous grass roots.

The drier, western part supported shorter and sparser stands of grass, but there was still enough growth to allow graziers to cut and bale it as hay. Today, the western prairie supports a shorter, sparser cover of grass. Reasons for such a decline include too much grazing pressure by herds of cattle and a slight drying trend in the climate of the area over the past 75 years. Shrubs from the desert and junipers from the foothills have invaded (possibly because of fire suppression), and they have further reduced grass cover.

Grasslands are the home of the largest terrestrial herbivores and carnivores. Vast herds of bison that roamed the North American prairie are gone, but animal preserves in African savannahs still excite visitors by a spectacular animal cast that includes eland, wildebeest, giraffe, elephant, and lion.

SCRUB

Scrub is vegetation dominated by shrubs, and it occurs in areas with hot, dry summers and moderate winters. In areas with high but seasonal precipitation, the shrubs are tall and thorny, but if rainfall is only 40 to 50 cm a year, scrub can be a nearly impenetrable thicket of shrubs with evergreen, hard, spiny, small leaves (Fig. 10.29). Scrub of this sort is found around the Mediterranean Sea, on coastal hills of California and Chile, in southwestern Australia, and at the tip of South Africa. Even though the species of plants in these widely separate areas are entirely different, the vegetation has come to look strikingly similar. In the western United States, this type of scrub is called **chaparral.**

Desert scrub (Figs. 10.19 and 10.30) grows in areas of rainfall below 25 cm a year. Shrubs are widely spaced and have far-reaching root systems (much of which runs just beneath the soil surface). The shrubs may be associated with **succulents,** plants that store water in stems or leaves, such as the common cactus of the southwestern United States. Some shrubs have deep roots that tap permanent supplies of groundwater, while others reduce transpiration by dropping their leaves every dry period and developing new ones every rainy period. Seeds of desert annuals may remain viable for long periods in the soil until a favor-

Figure 10.29 Chaparral in the foothills of California.

Figure 10.30 Desert scrub in southern Arizona with succulents (cacti).

able year, when rainfall is sufficient to trigger germination. A colorful show of flowers follows in a few weeks or months. Although desert scrub seems to be made up mainly of perennial species, an actual count of all species appearing throughout the year reveals nearly as many annuals as perennials.

A FINAL NOTE

We have described major vegetation types as if they have sharp boundaries with, and differ greatly from, all other types. Actually, vegetation types typically grade into one another or form complex mosaics that may extend for thousands of hectares. Zones where two or more vegetation types meet are called **ecotones.**

ECOLOGY AND ECOSYSTEMS

ENERGY FLOW

We have examined plant ecology at the population, community, and community, and vegetation levels. A final level of organization is the **ecosystem.** The ecosystem of a particular habitat includes the plant and animal communities, the physical (nonliving) environment, and all the interactions between them.

The organisms of an ecosystem can be divided into three categories: producers, consumers, and decomposers. **Producers** are green plants and certain bacteria that perform photosynthesis. **Consumers** are parasitic, herbivorous, or carnivorous plants and animals that feed on other plants or animals. **Decomposers** are saprophytic bacteria, fungi, and certain animals that obtain nutrition by breaking down dead organic matter. The three groups form a complex web of interdependence: consumers feed on producers, and decomposers feed on the organic remains of both and convert them to lower-energy, smaller molecules that are taken up again by producers.

Caloric energy is also passed through the ecosystem, and in this transfer, organisms are linked to each other by a **food chain.** Food chains are short in the arctic tundra: meadow plants trap light energy and convert it to chemical energy (e.g., proteins or carbohydrates); caribou transfer some of this energy to themselves, retaining some of it, excreting some, and respiring some; and human beings transfer some of the caribou caloric energy by eating a fraction of the total caribou. Aquatic ecosystems exhibit longer food chains: algae (plankton especially) produce the caloric energy; some energy is passed to the small animals that feed on plankton, then to the small fish that eat the plankton-feeders, then to larger fish, and then to shore birds or land animals such as bears or humans.

There is tremendous variation in net productivity from one ecosystem to another. **Net productivity** equals calories produced in photosynthesis minus calories lost in respiration. Desert ecosystems, for example, exhibit less than 0.5 g of dry matter produced per square meter of land per day; grasslands, 0.5 to 3.0; deciduous and coniferous forest, 3 to 10; agricultural areas utilized all year (as in sugarcane production), 10 to 25.

Energy transfer through the ecosystem involves losses each time energy passes from one type of organism to another. Figure 10.31 summarizes a very simple food chain consisting of California rangeland on which steers are grazed. The range receives 700,000 kcal of sunlight energy per square meter per year. After deducting the energy lost by reflection and heating, the fraction spent by plants on their own respiration, and other energy channeled into roots and other unconsumable parts of the plants, only 800 kcal are taken in by the grazing animals. The animals in turn spend nearly all their energy unproductively (from the point of view of

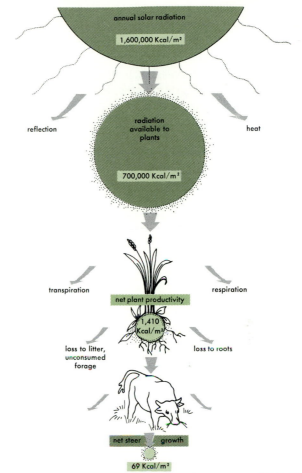

Figure 10.31 Loss of energy in western rangeland. Of 700,000 kcal of radiant energy that reach each square meter of surface, only 0.1% becomes available as plant food for the cattle grazing the rangeland. Only 0.01% of the plant food consumed during an 8-month period is converted to meat available for human consumption, that is, only 0.004% of the original incident solar energy.

human beings) in respiration, waste matter, and inedible bones and hide, leaving a net of only 69 kcal per square meter, of the original 700,000, that goes into meat. This represents about 139 kg of edible beef on the hoof per hectare per year.

POLLUTION

Human beings add to the environment many substances whose effect on plants and animals is poorly understood. If organisms in food chains or ecosystems shared by humans are damaged by these substances, humans may be indirectly, but nevertheless severely, damaged.

These substances are called **pollutants.** Some are liquids or solids that have been introduced to waterways and may affect the metabolism of aquatic plants, such as algae (see Chapter 16). Others are gases released to the atmosphere; they may affect the growth of wild or cultivated terrestrial plants. Among the most important of these airborne pollutants are sulfur dioxide (SO_2), ozone (O_3), and peroxyacetyl nitrates ($C_2H_3O_5N$, also called PAN).

Sulfur is present in iron, copper, lead, and zinc ores; during refining it is released as SO_2. It is also released from the burning of coal. The SO_2 enters leaves through stomata and goes into solution on the wet mesophyll cell walls as sulfurous acid. Inside the cells, such vital metabolic processes as protein synthesis are disrupted because of the reducing nature of this acid. Ultimately, cells shrink and collapse, producing visible chlorotic lesions on the leaves. Chronic exposure of such sensitive crops as alfalfa, barley, and lettuce to only 0.1 to 0.5 ppm of SO_2 causes visible damage and reduced yield. The effect of SO_2 on natural vegetation is dramatically revealed in Copper Basin, Tennessee, where early in this century a refinery freely released SO_2 before its toxic nature was understood. Over an area of perhaps 40 km^2 the original forest was destroyed (Fig. 10.32). The barren land was then so severely eroded that subsequent attempts to revegetate the area have been unsuccessful.

SO_2 can also become dissolved in rain water as a strong acid, creating **acid rain.** Acid rain has acidified entire ecosystems such as forests in New England and Sweden, causing a significant decline in plant productivity. Efforts to correct the problem are complex because the source of pollutants can be across international boundaries.

Ozone and PAN are important components of urban smog. In large part they are a result of the incomplete combustion of automobile fuel (Fig. 10.33). Our chemical understanding of smog and its effect on plants started less than 30 years ago. Ozone is naturally present in air, but smog contains 10 to 100 times the natural concentration, and this concentration is sufficient to cause millions of dollars of crop damage a year. Ozone enters through the stomata, and its oxidizing nature causes many metabolic malfunctions, including chloroplast breakdown. In the mountains near Los Angeles, air circulation takes damaging quantities of ozone to a height of 1700 m; this causes needle drop, reduced growth, and the ultimate death of trees in ponderosa pine forests.

Other experiments in the Los Angeles Basin have shown the effect of PAN on citrus. Established lemon and orange trees were encased in miniature greenhouses. Some greenhouses circulated normal, smoggy, outside air; others circulated air that had first been filtered of PAN. Over the period of 6 years, the average yearly fruit yield of lemon trees was reduced 39% by PAN, and the orange tree yield was reduced 64%. Evidently, PAN interferes with the light reactions of photosynthesis and also with respiration.

It is hoped that ecologists will eventually show us how ecosystems can be rationally manipulated to simultaneously satisfy all our needs: industry, urbanization, agriculture, conservation, and recreation. Knowledge must be gained and compromises made between our needs and what the environment can support.

Figure 10.32 Copper Basin, Tennessee. The normal eastern deciduous forest was destroyed by sulfur fumes from a smelter in the early twentieth century, and revegetation attempts have been unsuccessful.

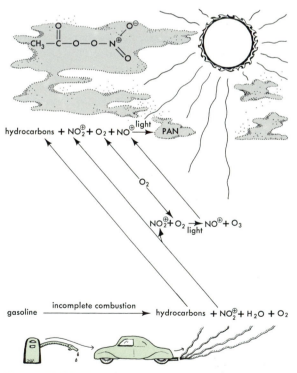

Figure 10.33 The formation of ozone (O_3) and peroxyacyl nitrates (PAN) in smog. The formula for PAN is given at the top of the diagram.

SUMMARY

1. Plant ecology is the study of how the environment influences plant distribution and behavior. It is also the study of the structure and function of nature. Ecology deals with the population, community, and ecosystem levels of organization.

2. The components of the environment include moisture, temperature, light, soil, fire, and living organisms. Each species utilizes a different part of the environment: its niche.

3. The seasonal timing of rainfall may be as important as the annual total in determining the kind of vegetation present.

4. Slope aspect affects temperature in the microenvironment; it also affects the amount of net radiation that reaches the plants.

5. Light quantity and quality is changed as light passes through a leaf canopy or through water.

6. Soil formation proceeds through mechanical and chemical weathering and through plant activity. Colloidal soil particles are important as nutrient reservoirs, and soil pH affects their ability to retain nutrients. Through time, soils in different regions develop their own characteristic profiles. Soils can be classified by profile characteristics into 10 orders and thousands of soil series. The ultimate taxonomic unit is the soil type.

7. Fire is an important, natural environmental factor that affects the distribution and architecture of many vegetation types.

8. Forms of biological interaction include competition, amensalism, herbivory, and mutualism. Plant defenses against herbivores are varied and complex, but herbivores appear to have coevolved mechanisms to penetrate each defense.

9. A taxonomic species includes genetically similar plants that are interfertile. Within any wide-ranging taxonomic species, however, are many ecotypes that have become adapted to different environments.

10. Plant demography is the study of the vital statistics of plant populations: germination rate, death rate, reproductive capacity, and the distribution of ages within the population. Reproduction includes vegetative production of genetically identical plants (ramets), as well as sexual production of genetically unique plants (genets).

11. A plant community is composed of a group of associated species or populations that occupy a particular type of environment. One or more of the species dominate the community, but all of the species together combine to form a unique architecture of canopy layers. The structure and composition of communities can be objectively determined by the use of sampling methods such as line transects or quadrats.

12. Climax communities perpetuate themselves, but successional communities are transitory.

13. Vegetation types of the world include tundra, taiga, deciduous forest, tropical rain forest, savannah–prairie, and scrub.

14. An ecosystem consists of the plant and animal communities of a habitat, their abiotic environment, and all the interactions between these components. Organisms of an ecosystem are tied together by food chains. Damage done to one component of a food chain (as through the action of pollution) may cause indirect but serious damage to the entire ecosystem.

15. Important atmospheric pollutants include sulfur dioxide, ozone, and peroxyacetyl nitrate. The latter two are components of smog and cause millions of dollars of crop damage a year as well as ecological damage to natural vegetation. SO_2 also causes acid rain.

11 GENETICS AND THE LIFE CYCLE

The preceding chapters have dealt with the growth and survival of the individual plant. Now we turn to matters of reproduction, where the issue is survival beyond the life of the individual. This chapter will introduce the general principles of reproduction and heredity and the related sciences of genetics and plant breeding.

By understanding the mechanisms of heredity, we can breed plants that are more beautiful, more productive, and less prone to disease. To perceive the importance of genetics, consider the case of wheat and its chief pest, the wheat-rust fungus. Growing in the stems of wheat plants, wheat rust costs millions of dollars per year in lost harvests. Plant breeders combat the fungus by developing new strains of wheat that resist attack. Every few years a new wheat variety is developed that not only resists the fungus but also produces more and better grain. However, the rust fungus adapts rapidly, and in 10 years it can attack and devastate the newest and most resistant kind of wheat. Thus the breeders must always produce new varieties to stay ahead of the fungus.

The science of plant breeding is based on the fact that plants of the same species vary in ways that can be passed on to their offspring. Therefore, we will begin by seeing what causes the variation.

SOURCES OF VARIETY

As background, let us briefly review the way information is stored in the cell. (A more complete discussion can be found in Chapter 2.) The hereditary information consists of distinct units (equivalent to messages) called *genes*. Each gene affects specific traits: one gene might influence flower color; another might affect leaf shape; and so on. Nearly all the genes reside in the nucleus, where they are carried on *chromosomes*. Every chromosome has hundreds or thousands of genes. Within a chromosome, the genes are segments of a very long DNA polymer.

Variations in hereditary information can arise in two ways: by *mutation* and by *recombination*. Though DNA is copied very precisely, once in a while the copying process makes an error or the DNA is changed by environmental chemicals and radiations. These changes are called **mutations;** the altered plants are called **mutants.** In future rounds of cell division the mutant DNA is copied as faithfully as was the old DNA.

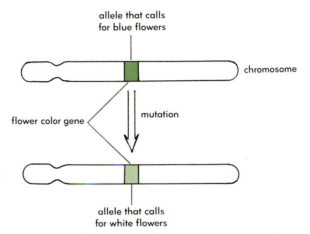

Figure 11.1. The relation between alleles, genes, and chromosomes. Alleles are alternative forms of the same gene, produced by mutation. The size of the gene in relation to the chromosome has been greatly exaggerated for clarity. A chromosome has hundreds or thousands of genes, each occupying a very small fraction of the chromosome's length.

Most mutations are barely noticeable, but some can be spectacular. For example, a plant with white flowers may occur in a field where all the other plants have blue flowers. The mutant gene is a new version of the instruction for flower color. The two versions—blue and white—are said to be **alleles,** or alternative forms of the same gene (Fig. 11.1). They are related like two copies of the same book, one of which has a typographical error. There are many ways to change a gene, and correspondingly there are many possible alleles for each gene.

Mutations are important because they are the source of new alleles. However, they occur so rarely that we can ignore the possibility of new mutations during an ordinary breeding program. The important thing is that mutations have altered the DNA in many places in the past, so natural populations have two or more alleles for many genes.

Sexual reproduction creates varied offspring through **recombination,** by mixing genes from two parents. This is of major importance because most plants are produced by sexual matings. In the next section we will examine sexual reproduction in detail.

REPRODUCTION AND LIFE CYCLES

There are two general types of reproduction: **sexual** and

asexual. The difference between the two is that *sexual reproduction involves a step in which two nuclei fuse into one nucleus*, whereas asexual reproduction does not involve such a step. The difference is important because nuclear fusion can bring the genes of two individuals together, creating offspring with new combinations of hereditary information. Asexual reproduction yields no such novelty; the offspring are identical to the parent.

ASEXUAL REPRODUCTION

In asexual reproduction, a new individual arises directly from a fragment of the parental body. If the reproductive unit is a single cell, it is called a **spore.** Most algae and fungi produce asexual spores as their principal means of dispersal. In the higher plants, asexual reproduction involves larger, multicellular units. Rhizomes and stolons (see Chapter 4) can form roots and green shoots at the nodes; when the connection with the parent dies, the rooted portions become independent plants. In this way, a single plant can grow into a large colony of genetically uniform individuals (a *clone;* see Chapter 14). Even leaves, which are determinate in most plants, can sometimes serve for reproduction. For example, the leaves of *Bryophyllum* (Fig. 11.2C) retain patches of residual meristem at notches along the margins, and these areas give rise to plantlets with a few leaves and roots. The plantlets drop to the ground and grow independently.

When an organ is severed, the injury often activates a regenerative system that replaces the lost organs. This is routinely used in the nursery industry, where many plants are produced from one plant by inducing roots to form on stem cuttings (see Chapter 9). Such crops as bananas and potatoes have long been grown by cutting storage organs (tubers and corms, respectively) into pieces that are planted. In the jade plant *(Crassula argentea)* and its close relatives, detached leaves can produce a new plantlet from a vascular bundle near the wound surface (Fig 11.2A). Leaves with this ability can be cut into segments, each of which will form a plantlet (Fig. 11.2B). Such methods are called **vegetative propagation** because they produce plants from vegetative organs.

SEXUAL REPRODUCTION

Sexual reproduction is more complex (Fig. 11.3). It begins with the formation of **gametes,** which are cells that are specialized for fusion. The animal sperm and egg are examples of gametes. Each gamete carries one full set of chromosomes. In an event called **fertilization,** or **syngamy,** two gametes fuse to make a cell called a **zygote.** The zygote is the first cell of the offspring, and it has twice as many chromosomes as a gamete. Thus, to make another generation of gametes, a special form of cell division must occur later in the offspring so that the chromosome number is cut in half. The reduction division is called **meiosis.**

Because the number of chromosomes varies during the sexual process, it is convenient to have terms for the number of chromosomes. A cell with one set of chromosomes is said to be **haploid** (from the Greek *haploos,* single). The symbol **1n** is used to designate a haploid cell. A cell with two sets is **diploid,** or **2n.**

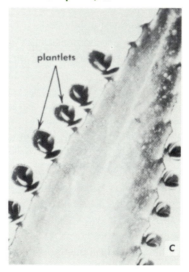

Figure 11.2. Vegetative propagation from leaves. *A,* detached leaf of the jade plant *(Crassula argentea)*, which has formed shoots and roots. *B,* plantlets produced on segments of *C. argentea* leaves in sterile culture. *C,* plantlets at the margins of a *Bryophyllum* leaf.

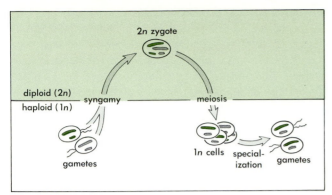

Figure 11.3. Simplified sexual life cycle showing only the essential steps. Bars inside cells represent chromosomes.

THE ALTERNATION OF PHASES

As the generations pass, sexual reproduction carries the species back and forth between a diploid phase and a haploid phase. Organisms vary widely in the extent of the diploid and haploid phases. This is important in the study of heredity because gene expression depends on how many sets of genes are present when the body is being built.

Some common variations in the life cycle are shown in Fig. 11.4. They differ in the phase of life when cells multiply by mitotic divisions. In humans and animals, the zygote and its diploid products undergo mitosis to produce a diploid body (Fig. 11.4A). Only a small proportion of the diploid cells go through meiosis, and their haploid products immediately become gametes without first building a haploid body. This is called a **gametic life cycle** because meiosis directly leads to gametes. These organisms show complex patterns of heredity because diploid cells have two rival sets of alleles that take part in controlling the development of the body.

Some of the fungi and algae are the opposite (Fig. 11.4C): their only diploid cell is the zygote, which goes through meiosis without first dividing mitotically. This is called a **zygotic life cycle,** which reminds us that the meiosis occurs in the zygote. The haploid nuclei of algae and fungi multiply by mitosis, and they often build a haploid body. Later, some of the haploid cells specialize to become gametes. These organisms show the simplest pattern of heredity because each cell has only one set of genes at the time when the body is developing.

The most complex life cycle includes both haploid and diploid bodies, which is the case in the higher plants. The seaweed *Ulva*, or sea lettuce (Fig. 11.4B), is a relatively simple case: haploid cells divide mitotically to build a sheet-like body. Later, some of the cells specialize to become gametes. The gametes fuse and form a zygote, which forms another sheetlike body by mitotic divisions. Some of its cells later go through meiosis to produce swimming reproductive cells, or **spores.** This is called a **sporic life cycle** because meiosis produces not gametes but spores that directly grow into a plant. Because both haploid and diploid bodies are formed, these plants are said to go through an **alternation of phases** or **generations.**

Flowering plants have an alternation of diploid and haploid phases, but like animals, their life cycle emphasizes the

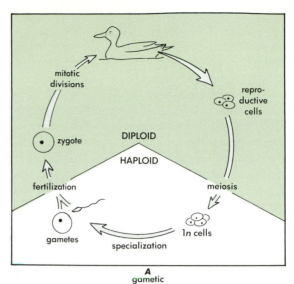

A
gametic

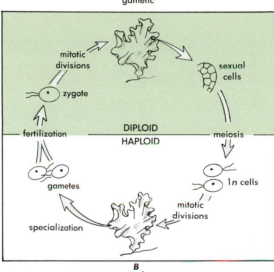

B
sporic

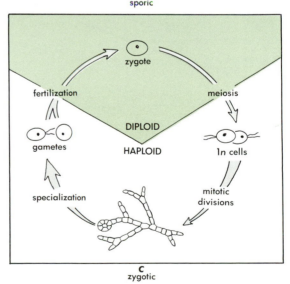

C
zygotic

Figure 11.4. Three sexual life cycles. They differ as to whether mitotic divisions occur in the haploid phase, the diploid phase, or both. *A*, gametic life cycle, typical of animals. The haploid phase has been exaggerated for clarity; it occupies only a small fraction of the cycle. *B*, sporic life cycle, as in *Ulva*. *C*, zygotic life cycle, found in many algae and fungi. The diploid phase has been exaggerated for clarity.

diploid phase. To demonstrate this, we will first briefly describe the structure of the flower (Fig. 11.5). The flower is made of diploid cells. Within it, meiosis occurs in structures called **anthers** and **carpels.** Cells inside the anther go through meiosis to form haploid cells, and each haploid cell divides once or twice by mitosis to make a very small haploid plant. These tiny plants build a thick outer wall and become **pollen grains,** which are specialized for transport through dry air to reach another flower. Meanwhile, some cells in the carpel also go through meiosis, and some of these haploid cells then divide three times by mitosis to make a haploid stage called the **embryo sac.** The embryo sac remains inside the carpel, and one of its cells serves as an egg. When a pollen grain arrives at the carpel, it grows a **pollen tube** that (among other things) deposits a nucleus in the egg. This **sperm nucleus** fuses with the egg nucleus to make a diploid zygote. Finally, by mitotic divisions, the zygote grows into a diploid embryo and ultimately into a plant.

MEIOSIS AND ITS IMPORTANCE

As mentioned earlier, sexual reproduction is important in studies of heredity because it produces varied offspring. Therefore, the steps that lead to variety demand special attention. Both fertilization and meiosis add variety. Fertilization creates varied diploid plants by combining genetic material from two plants.[1] In a way that is less obvious, *meiosis* generates new kinds of *haploid* cells.

Meiosis consists of two rounds of cell division: first, one cell becomes two, and then the two become four (Fig. 11.6). The two divisions are called Meiosis I and Meiosis II.

---

[1] In many cases, pollen fertilizes eggs in the same flower where the pollen was made; this is **self-fertilization.** The alternative—a transfer of pollen between plants—is called **cross-fertilization.** Self-fertilization is useful when partners are scarce, but it loses the advantages of bringing together traits from varied individuals. Thus most wild species have barriers against self-fertilization.

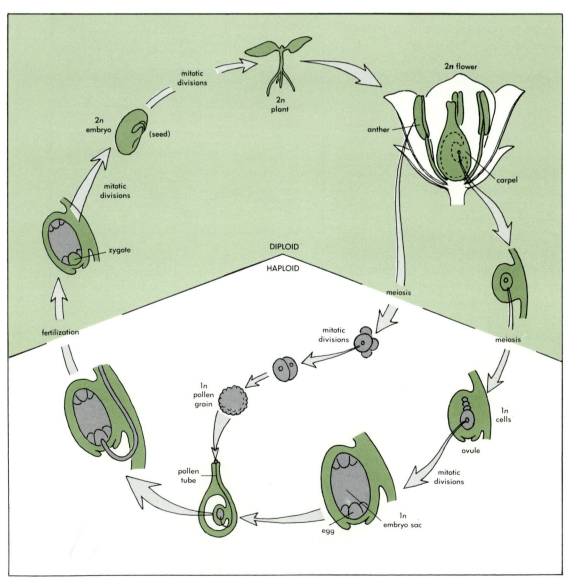

Figure 11.5. Sexual cycle of the flowering plant.

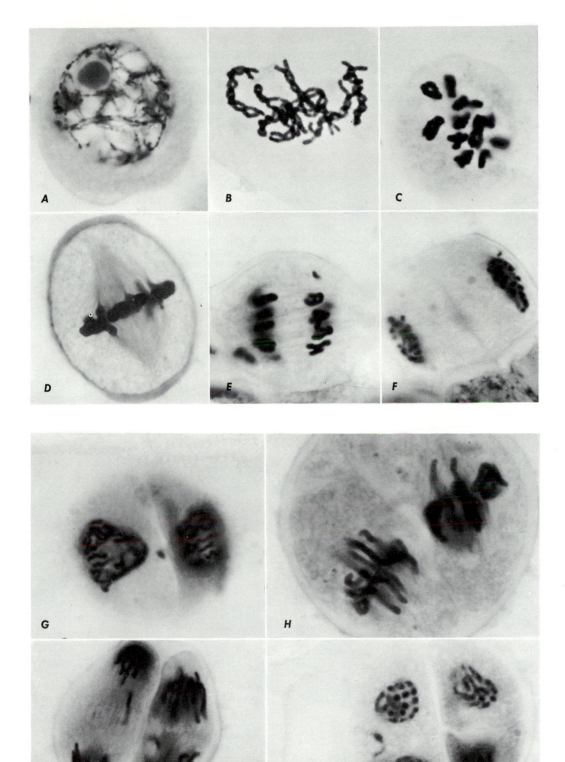

Figure 11.6. Meiosis in a lily anther. Photomicrographs showing: *A*, early Prophase I—note paired chromosomes. *B*, late Prophase I—each body represents two paired chromosomes. *C*, later in Prophase I. *D*, Metaphase I—note spindle fibers extending from chromosomes. *E*, Anaphase I. *F*, Telophase I. *G*, Prophase II. *H*, Metaphase II. *I*, Anaphase II. *J*, Telophase II.

Each division in meiosis looks superficially like mitosis: a spindle forms, chromosomes are separated in four stages **(prophase, metaphase, anaphase,** and **telophase),** and the cell divides **(cytokinesis).** But on closer inspection, *Meiosis I* differs from mitosis in important ways.

Figure 11.7 illustrates Meiosis I in more detail. The cell that approaches meiosis is diploid; it has two copies of each chromosome, one from each parent (Fig. 11.7*A*). To em-

phasize their different origins, one member of each pair is colored green and the other is gray. The two matching chromosomes of each kind are said to be **homologous**—that is, they carry similar, if not identical, genetic messages. If one chromosome has a gene that governs flower color, the homologous chromosome also has a gene for flower color.

Not only is the cell diploid, but it also duplicates its chromosomes before meiosis (Fig. 11.7*B*). The copies stick

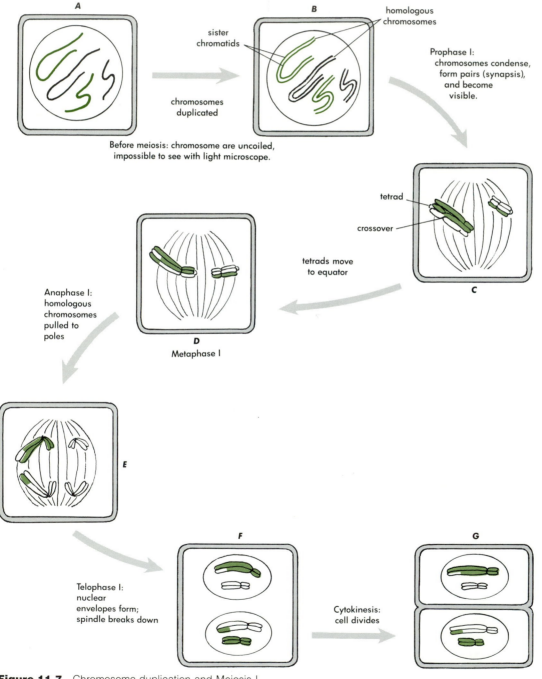

Figure 11.7. Chromosome duplication and Meiosis I.

together and appear as a single chromosome, but as meiosis proceeds they begin to separate, and one can finally see that each chromosome is a double structure. The two copies are called **sister chromatids.**

Before meiosis, the chromosomes are uncoiled and are too slender to be visible. But in prophase of Meiosis I, they coil and thicken, and then they can be seen (Fig. 11.7C). As they coil, the homologous chromosomes come together in pairs, lined up side by side. This process, called **synapsis,** occurs only in Meiosis I; there is nothing like it in mitosis or Meiosis II. In synapsis all four copies of the chromosome make up a unit called a **tetrad** (Fig. 11.8). The number of tetrads equals the haploid number of chromosomes.

While in the tetrad arrangement, the homologous chromatids may exchange parts. This process, called **crossing-over,** is important because it mixes, or **recombines,** the genes of the two parents. The result of a crossover is indicated by the shading in Fig. 11.7C. In effect, the two chromatids break, and the fragments are switched and then rejoined. Several crossovers can occur in each tetrad, and they are more or less randomly placed along the tetrad. Thus each chromatid becomes a patchwork made of parts from both parental chromosomes. Note, however, that each chromatid emerges with a complete set of genes.

While synapsis is taking place, a spindle forms. Microtubules connect the tetrads to the spindle, and the tetrads are pulled to the equator of the cell. They pause at the equator during what is known as **metaphase** (Fig. 11.7D). Then the spindle pulls each tetrad apart **(anaphase).** Half of each tetrad goes to each pole (Fig. 11.7E). This creates two equal sets of chromosomes, each with the haploid number. In **telophase** the spindle vanishes, and nuclear envelopes form (Fig. 11.7F). Finally, the cell divides and Meiosis I is finished

(Fig. 11.7G). The original diploid cell has been converted into two haploid cells, and each cell has a mixture of genes from the two parents.

Meiosis II is shown in Fig. 11.9. By steps that resemble mitosis, it separates the sister chromatids so that each cell has two haploid nuclei. Cell division follows, yielding four haploid cells.

As indicated by the shading in the figures, the haploid cells that emerge from meiosis carry a mixture of genes from the green and white sets of chromosomes. Two processes help to mix the genes: first, the crossing-over during Prophase I exchanges parts between homologous chromosomes; and second, the tetrads line up independently of one another when they reach the equator at Metaphase I. Thus for each tetrad, each cell is as likely to receive the green homolog as the white homolog. Together, crossing-over and independent segregation produce haploid cells with a variety of allele combinations.

GENETICS AND PLANT BREEDING

Now that we have examined the process of sexual reproduction, we are ready to discuss the problems that breeders face and the strategies they use to shape the heredity of plants. We shall concentrate on diploid genetics because the diploid phase predominates in the life of domestic plants.

Since sexual matings alter heredity, the first stage in any breeding program is to bring matings under control. The possibility of accidental pollination by wind or insects must be eliminated, usually by placing bags over the flowers during their receptive period. Also, the possibility of self-fertilization must be taken into account. It can be controlled by cutting off the anthers before the pollen have matured.

Success in a breeding program also depends on the choice of **traits** to be studied. Traits are observable characteristics such as color of petals and shape of leaves. Some traits are easier to study than others. For example, the weight of seeds produced by a plant is determined partly by heredity and partly by the environment. Seed weight, oil or starch content, growth rates, and so on are **quantitative traits.** Such traits commonly show **continuous variation,** in which the differences between plants are distributed widely and evenly about a mean (see Fig. 11.21). Traits with continuous variation are important because they include many features of economic importance. However, they are difficult to study because the progeny of matings are difficult to classify, and patterns of heredity are complex.

By contrast, some traits show **discontinuous variation.** Thus a population of garden pea plants might contain some individuals that form red flowers and some that form purple flowers, but no intergrades. Traits that show discontinuous variation are easy to study because the progeny can be assigned to definite classes and the various types occur in simple ratios. Historically, studies of discontinuous traits gave us the key to understanding heredity. Today, it is still easiest to learn genetics by starting with discontinuous traits.

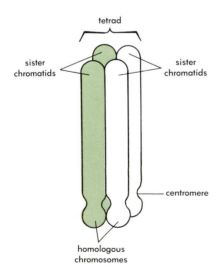

Figure 11.8. Diagram to illustrate terms that refer to chromosomes. The centromere is the location where the spindle fibers attach.

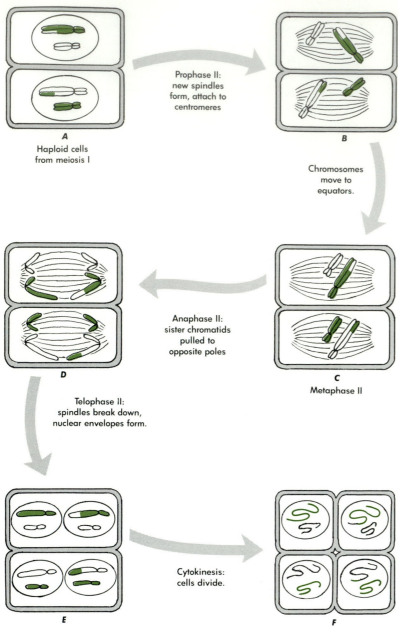

A
Haploid cells
from meiosis I

Prophase II:
new spindles
form, attach to
centromeres

B

Chromosomes
move to
equators.

Anaphase II:
sister chromatids
pulled to
opposite poles

D

C
Metaphase II

Telophase II:
spindles break down,
nuclear envelopes form.

E

Cytokinesis:
cells divide.

F

Figure 11.9. Meiosis II.

TRAITS THAT SHOW DISCONTINUOUS VARIATION

The first systematic breeding program was done in the mid-nineteenth century by Gregor Mendel. He chose the garden pea, partly because many varieties of peas were available and partly because crossing could be controlled. The carpels mature before the anthers while the flowers are still closed. At this stage Mendel pried the flower open, removed the immature anthers, and dusted the inside of the flower with pollen from another plant of his own choice. Fertilization occurred inside the closed flower, without risk of exposure to stray pollen.

Each variety of pea that Mendel studied had its own set of unique hereditary features. In a field sown with one vari-ety, all the progeny for generation after generation would have the features typical of that variety. The plants in such a field are said to "breed true" for the traits under study; collectively, the plants form a **true-breeding strain.** Most breeding programs start with true-breeding strains because they provide an especially simple background against which to judge the effects of matings.

The Monohybrid Cross

Given two true-breeding strains that differ in some interest-ing trait, the next step is to see what will happen when plants of the two strains are mated. This will establish how many genes are involved in determining the trait and how the al-ternative alleles interact with one another. The test is easiest

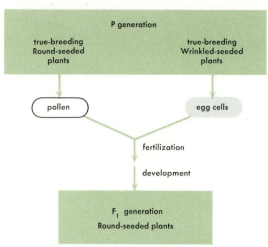

Figure 11.10. Results of a monohybrid cross with garden peas.

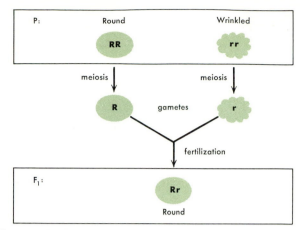

Figure 11.12. Interpretation of Mendel's results from the monohybrid cross.

if we pay attention to just the one trait and ignore any other differences between the parents. For example, if we are studying flower color, we ignore differences in height. Matings with this one-trait focus are called **monohybrid crosses.**

Figure 11.10 shows the kind of results that Mendel obtained in monohybrid crosses with peas. The original parents are called the **P generation.** One parental plant in Fig. 11.10 came from a strain that always produces round seeds; the other came from a strain that always forms wrinkled seeds. The progeny are the **First Filial,** or **F₁ generation.** In this cross, Mendel found that all of the F₁ plants formed round seeds. This seems surprising at first because since they received hereditary information for making both round and wrinkled seeds, one might expect some of the progeny to form wrinkled seeds. Either the hereditary information for wrinkled seeds was suppressed, or it never entered the zygote during fertilization.

An additional cross (Fig. 11.11) proved that the information for making wrinkled seeds was indeed present in the F₁ plants. Here the F₁ plants were mated with one another. In such a cross, the F₁ plants are said to be **self-crossed,** or **selfed,** and their progeny are the **F₂ generation.** Although all the F₁ plants formed round seeds, some of the

F₂ plants made wrinkled seeds. This showed that the F₁ plants did contain the information for wrinkled seeds, but the information was suppressed.

To explain the patterns of inheritance that he saw, Mendel proposed the scheme shown in Fig. 11.12. He suggested that the hereditary information for seed shape is carried by two kinds of particle: one kind (call it **R**) carries information for making round seeds, and the other **(r)** carries information for making wrinkled seeds. Mendel proposed that each gamete carries one of these particles, so that the zygote contains two of them. To account for the suppression of the "wrinkled" trait, he proposed that the factor **R** is **dominant** over **r** when the two are both present in the same organism. That is, the presence of **R** prevents **r** from affecting the shape of seeds.

In modern language, we would rephrase Mendel's proposal by saying that allele **R** (round seeds) is dominant over allele **r** (wrinkled seeds); allele **r** is said to be **recessive** to **R.** This pair of alleles is said to show **classical dominance.** We have not explained *how* **R** suppresses **r;** we have merely developed a picture that helps to predict the progeny in matings. Classical dominance has been found between many pairs of alleles, affecting many traits in many plant and animal species.

The idea of dominance between alleles explains why some of the F₂ plants produced wrinkled seeds (Fig. 11.11). The analysis is shown in Fig. 11.13. Each nucleus in the parental plants contains the allele combination **Rr.** Meiosis separates the alleles, so that half of the gametes contain allele **R** and half contain allele **r.** Gametes with these alleles meet at random in matings, so that zygotes are formed with all the possible combinations. These are shown in the four boxes in Fig. 11.13, along with the seed shapes that the progeny will form. Three of the four combinations result in round-seeded plants; the fourth combination, **rr,** yields plants with wrinkled seeds.

To be useful, genetic analyses should predict not only the kinds of progeny that will occur but also their relative

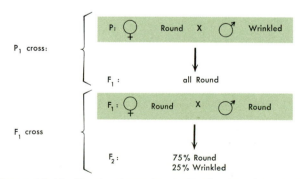

Figure 11.11. Results of crossing the F₁ offspring with one another in a monohybrid cross.

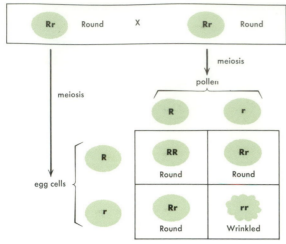

Figure 11.13. Why a recessive trait reappears in the F₂ generation.

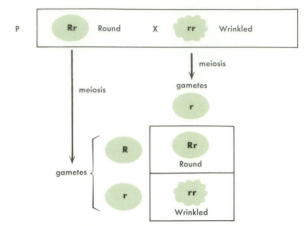

Figure 11.14. The testcross, to determine the genotype of a parent that has the dominant phenotype.

abundance, or **frequency.** Looking back to Fig. 11.13, one can see that 25% of the progeny formed wrinkled seeds. Why 25%? To understand this, one must be aware that chance plays a major part in bringing the gametes together. The contained alleles do not affect a gamete's chances of meeting and fusing with another gamete. Since meiosis forms equal numbers of **R** eggs and **r** eggs, the laws of chance assure that the two types of eggs will be involved in matings equally often. Thus half the progeny should contain an **r** egg nucleus. By chance, half of these will combine with pollen that carry allele **r,** since meiosis also assures that **r** pollen are just as abundant as **R** pollen. Half of half is 25%—hence 25% of the F₂ plants are the **rr** type, which form wrinkled seeds. The frequency of a given progeny type is set partly by the mechanics of meiosis, which fixes the frequency of gamete types; and partly by the laws of chance, which govern the random combination of the various gamete types.

As anyone knows who has played games with dice or cards, the laws of chance leave room for variation. With many tosses of a coin, "heads" will come up very nearly 50% of the time. But with just a few tosses, the percentage may be quite far from the ideal 50% figure. Likewise in counting the progeny of a mating, we do not expect the numbers to be exactly as predicted; deviation is likely to be greater, the smaller the number of progeny. Geneticists have adopted statistical methods that help to estimate how far the real results of a cross are likely to deviate from ideal predictions.

A note on terminology: the combination of alleles present in a plant is known as the **genotype,** while the observable trait itself—seed shape in this case—is the **phenotype.** True-breeding plants have two copies of the same allele, **RR** or **rr.** They are said to be **homozygous** for this gene. Plants that carry two different alleles, such as **Rr,** are **heterozygous.**

The Testcross

The dominance mechanism makes it easy to recognize plants that are homozygous for the recessive allele, because they are the only plants that show the recessive phenotype. But plants with genotypes **RR** and **Rr** are identical in appearance; both produce round seeds. How can we deduce the true genotype of a round-seeded plant, so as to make accurate predictions in matings that use the plant?

The easiest approach is to use a **testcross.** In a testcross, one parent has only recessive alleles for the trait being studied; the other parent has the dominant phenotype, but its genotype is unknown. The testcross for seed shape in peas is diagrammed in Fig. 11.14. There it has been assumed that the unknown parent has the genotype **Rr.** The recessive parent forms only gametes with the allele **r.** The other parent produces gametes with allele **R** and an equal number with allele **r.** Only two progeny genotypes are possible; they are shown in the boxes. If the unknown parent is heterozygous, half the progeny form round seeds and half form wrinkled seeds. If the unknown parent is homozygous **RR,** all the progeny in the testcross form round seeds. The testcross is useful because the alleles contributed by the recessive parent do not interfere with the expression of the other parent's alleles. The progeny phenotypes directly reveal the unknown parent's gamete types. From this information we can easily deduce the unknown parent's genotype.

Incomplete Dominance

Though many traits show classical dominance, some do not. One exception is shown in Fig. 11.15. When plants from a red-flowered strain of snapdragon *(Antirrhinum)* are crossed with plants from a white-flowered strain, the F₁ progeny all have pink flowers. Here Mendel's idea of dominance does not apply. But the F₁ plants still contain the information for making red and white flowers. When the F₁ snapdragons are selfed (Fig. 11.15, bottom), half the progeny form pink flowers, but the rest produce red or white flowers.

Although dominance does not apply, these results can be explained easily with the allele theory of inheritance. In Fig. 11.16, the red-flowered plants are assigned the genotype **RR;** white-flowered plants have the genotype **R′R′.** The F₁ progeny all have the genotype **RR′,** which is found by observation to give a pink-flowered phenotype. In the next cross, the pink-flowered parents (with genotype **RR′**) form equal numbers of **R** and **R′** gametes. These unite at random

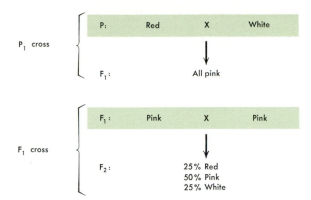

Figure 11.15. Results of crosses between red-flowered and white-flowered snapdragons. Mendel's dominance principle does not apply here.

so that four combinations result. One combination **(RR)** yields red-flowered plants; two combinations **(RR′)** give pink, and the fourth combination **(R′R′)** produces white-flowered plants. One would expect 25% red, 50% pink, and 25% white. This matches the observed progeny in the F_2 generation (Fig. 11.16).

Pairs of alleles such as this one, where the heterozygous condition gives a different phenotype than either of the two homozygous phenotypes, are said to show **codominance,** or **incomplete dominance.**

Dihybrid Crosses and Linkage

Many breeding programs are concerned with bringing two or more qualities together in the same plant. For example, a strain of corn with high resistance to fungus attack but with poor grain yield may be crossed with another strain that has low fungus resistance but good yield. Such a cross, involving two traits, is a **dihybrid cross.** The goal would be to produce plants that yield well and are disease-resistant. But the progeny might also include plants that combine poor yield with poor resistance. How many of the progeny can be expected to have the desirable combination of features? The answer will determine the effectiveness of the breeding program.

In a monohybrid cross, we were dealing with only *one* gene that had two alleles. In dihybrid crosses, *two* genes are involved, each with two alleles. It is important to distinguish clearly between "genes" and "alleles" because confusion between these terms will make it impossible to understand inheritance in dihybrid crosses.

The genes are carried on chromosomes, with many genes per chromosome governing many traits. Each kind of chromosome in the haploid set carries a particular group of genes, which are arranged in a definite sequence along the chromosome. Thus two genes on the same chromosome are physically coupled together, or **linked.** Two genes that are carried on different chromosomes are **unlinked.** Linkage limits recombination, which in turn limits the occurrence of new progeny types. Therefore, when we study two traits at once, we can predict the progeny only if we know whether the two genes are linked or not. To see the effect of linkage, let us compare an unlinked pair of genes with a linked pair in the garden pea.

In garden peas, the shape and color of the seeds are governed by a pair of unlinked genes. As to shape, round **(R)** is dominant over wrinkled **(r).** In seed color, yellow **(Y)** is dominant over green **(y).** Figure 11.17 shows how a plant that is heterozygous for both genes produces gametes. Remember that meiosis is preceded by a duplication of the chromosomes, so that each chromosome entering meiosis has two identical chromatids. Then in prophase of the first meiotic division, homologous chromosomes join together.

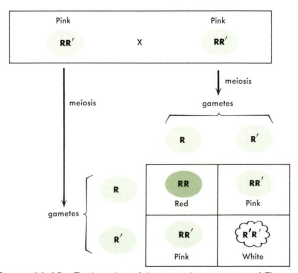

Figure 11.16. Explanation of the snapdragon cross of Figure 11.15. The two alleles are codominant.

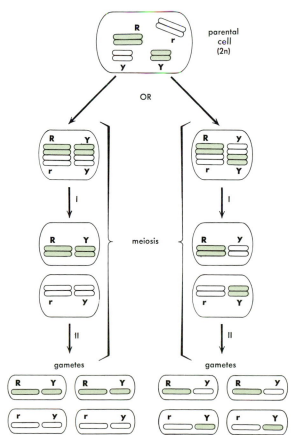

Figure 11.17. Gamete formation with two genes that are unlinked (on different pairs of chromosomes).

The resulting tetrad is a group of four chromatids. In Fig. 11.17 the long chromosomes form one tetrad; the short chromosomes form another. The tetrads then line up at the equator of the spindle. But tetrads move independently of one another. Therefore the two tetrads can assume two alternative positions relative to one another. In one arrangement (the left branch in Fig. 11.17), meiosis produces two gametes with genotype **RY** and two with genotype **ry.** The other arrangement (right branch, Fig. 11.17) results in two gametes with genotype **Ry** and two with genotype **rY.** Only one of the possible arrangements can occur in any one cell that undergoes meiosis. But meiosis occurs in many cells. Chance governs the positions that the tetrads take, so that the two arrangements will be equally common. Therefore, *if the two genes are unlinked, the four gamete types will be about equally numerous.*

Next, let us consider the formation of gametes when the two genes are linked. This can be illustrated with another pair of genes in the garden pea: flower color, with purple **(P)** dominant over red **(p);** and pollen shape, with long **(L)** dominant over round **(l).** Figure 11.18 shows the formation of gametes in a plant that is heterozygous for both genes. Here we assume that the parent has allele **P** on the same chromosome as allele **L,** and allele **p** is on the same chromosome as allele **l.**

Most cells that undergo meiosis will follow the sequence of events in the left branch of Fig. 11.18. Pairing brings the homologous chromosomes together in a tetrad. The coupling between **P** and **L** is maintained, as is the coupling between **p** and **l.** When meiosis is complete, half the gametes have the genotype **PL** and half have the genotype **pl.**

Some cells take the right branch of Fig. 11.18. Here, an event called a **crossover** occurs in the tetrad: two of the homologous chromatids break at the same point. Then the broken ends rejoin in such a way that the chromatids have traded parts. Tracing through the rest of meiosis, one can see that the crossover has introduced two new classes of gametes. The chromatids that took part in the crossover will result in gametes with the genotypes **Pl** and **pL.** The chromatids that were *not* involved in the crossover will form gametes with genotypes **PL** and **pl.** Meiosis with a crossover has formed four kinds of gametes, whereas meiosis without a crossover resulted in only two kinds of gametes. When a large number of cells undergo meiosis, only a minority will have crossovers between the two genes. Therefore, *the effect of linkage is to reduce the frequency of two classes of gametes out of the four.*

As Figs. 11.17 and 11.18 show, linkage affects the frequency of gamete types. However, we cannot directly see the genotypes of gametes. How, then, can we determine whether the genes in a pair are linked or unlinked? The easiest method is to subject the doubly heterozygous plant to a testcross.

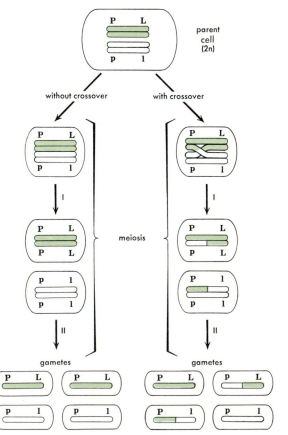

Figure 11.18. Formation of gametes with two genes that are linked (carried on the same pair of chromosomes).

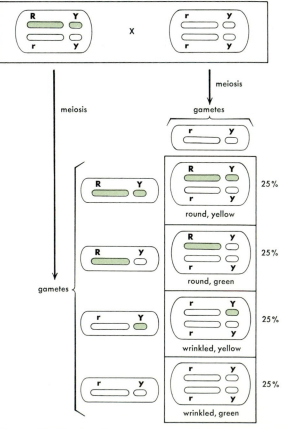

Figure 11.19. Results of a testcross with two genes that are unlinked. The four progeny classes are equally frequent.

Figure 11.19 shows what happens in the testcross when the genes are unlinked. As always in a testcross, one parent has only recessive alleles. This allows the other parent's alleles to be fully expressed in the progeny. The genes are unlinked, so the heterozygous parent has produced four equally common kinds of gametes. Therefore the progeny also include four equally common phenotypes.

Contrast this with the behavior of linked genes in a testcross (Fig. 11.20). Again there are four classes of progeny, but two classes are more common than the other two. The rare types are due to gametes that were formed by crossovers in meiosis.

One can see that the testcross of the double heterozygote can be useful in determining whether two genes are linked or not. But whatever the outcome, the testcross shows the relative frequencies of the various gamete types that a plant produces. This is useful information for breeding programs, because the gamete frequencies can be used to predict progeny genotype and phenotype frequencies. The mathematical methods for making these predictions cannot be described in the space we have here, but they are simple and can easily be learned from any elementary genetics text.

Though Fig. 11.18 shows only one crossover between two chromatids, it should be mentioned that more than one crossover can occur in a given tetrad during a given meiosis. Also, even though a single crossover involves only two chromatids, additional crossovers may involve other chromatids. Chance plays a large part in setting the locations where crossovers will occur along the chromosome. Thus crossovers occur less often between two closely spaced genes than between two distant genes. Two or more crossovers may occur between genes that occur at opposite ends of the chromosome, so that in testcrosses the genes may appear to be unlinked. Geneticists have developed mathematical methods for taking these factors into account in planning crosses and predicting results. Such methods have also been used to map the chromosomes, showing the locations where genes occur.

TRAITS THAT SHOW CONTINUOUS VARIATION

The preceding discussion has concentrated on traits that show discontinuous variation, but most economically important traits show *continuous* variation. For example, the seeds of corn *(Zea mays)* vary in protein content. In a study of this trait, plants that produced seeds high in protein were mated with plants that produce low-protein seeds. The F₁ contained progeny with a wide range of seed protein content (Fig. 11.21). This pattern suggests that protein content might be controlled by environmental factors rather than heredity. Undoubtedly the environment does play a part. But another kind of experiment shows that heredity is involved too. A field of corn was planted and seeds from the crop were used to replant the field again the next year—but only seeds from plants that gave the best protein yields were used. Every year, a similar selection and replanting was done. Figure 11.22 shows that the selection process gradually raised the average protein content of seeds produced in this field. In a parallel experiment, another field was replanted year after year, using only the seeds from plants that had the lowest protein yield in the preceding year's crop. In this experiment, the continued selection gradually decreased the average level of seed protein. These experiments show that protein content in the seeds depends on heredity: otherwise the protein content should not have been so affected by selection of the seeds.

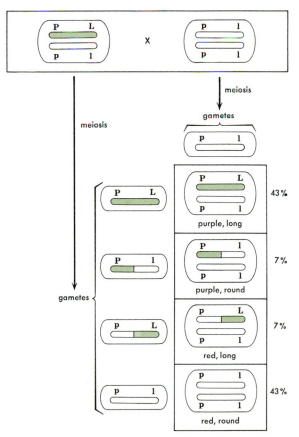

Figure 11.20. Results of a testcross when two genes are linked. The frequency of recombinant progeny depends on how closely the genes are linked (here they are 14% of the progeny).

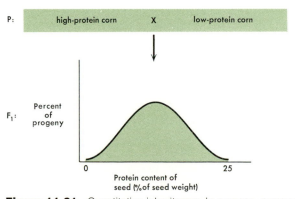

Figure 11.21. Quantitative inheritance. In crosses, progeny do not form distinct classes.

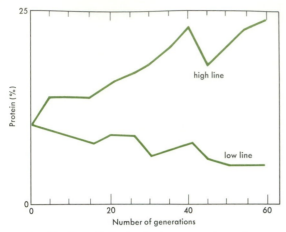

Figure 11.22. The effect of selection on seed protein content in *Zea mays*. Data from C. M. Woodworth, E. R. Leng, and R. W. Jugenheimer. "Fifty Generations of Selection for Protein and Oil in Corn," *Agron. J.* 44 (1952):60–65.

If quantitative traits such as seed protein content are under hereditary control, why don't the progeny fall into discrete classes? Aside from environmental effects, the central explanation seems to be that several genes contribute to the control of the trait. Two or more alleles might be present in the population for each of these genes. In matings, then, gametes and zygotes with many combinations of alleles could result. Each combination might give a somewhat different phenotype with respect to the trait that is being studied. If many genes are involved, the effect of changing any one gene's alleles might be too small to detect easily against the background of environmental effects. For plant improvement in such cases, selection programs like the one in Fig. 11.22 may be important, together with specific crosses between strains that have been selected for favorable features.

MATERNAL INHERITANCE

All the patterns of heredity mentioned so far have concerned genes that are carried in the nucleus, with a pollen grain and an egg cell making equal contributions. These include the great majority of cases. But anomalous patterns of heredity sometimes occur, in which the maternal (egg-producing) parent determines a trait in the progeny and the pollen seems to make no contribution. This can result from at least two causes. First, there are cases in which fertilization is bypassed and the flower forms seeds that start from a purely maternal cell (usually diploid) instead of a zygote. This process, called **apomixis,** occurs in the lawn dandelion. It is really a form of asexual reproduction and does not belong in a discussion of matings, but we mention it here because such cases can greatly confuse an attempt at controlled breeding. Second, *plastids* contain some genes that contribute to the development of the chloroplasts. These genes are passed from generation to generation by duplication and division of the plastids; their inheritance is distinct from that of nuclear genes. Since egg cells contribute plastids to the zygote while pollen cells usually do not, only the mater-

nal parent has a part to play in plastid inheritance. Effects of the chloroplast genes are especially noted in the inheritance of variegated color patterns in the leaves of some plants (e.g., variegated *Mirabilis*). Here the egg cell contains both normal plastids and plastids that cannot develop into green chloroplasts because of defective genes. Some cells in the embryo have only the defective plastids. These cells divide further to produce the distinctive white or yellow patches on the leaves and stem of the variegated plant.

KEEPING A USEFUL HYBRID

As mentioned before, plant breeders often try to combine valuable traits of two or more strains into a single line of progeny. For instance, it might be desirable to combine fungus resistance with a high crop yield. With woody plants that have a long life, the best method is often to breed until a desirable product has been obtained and then propagate the plants by means of cuttings from then on. This is regularly done with fruit trees and ornamentals such as roses. Likewise, hybrid potatoes are propagated by means of stem cuttings (segments of tubers). The vegetative propagation assures that the progeny will have traits nearly identical to the parents.

With most annual plants, including the cereal grains, vegetative propagation is not practical. Two alternatives exist, depending on whether the plant is normally self-pollinated (as in wheat) or cross-pollinated (as in corn). With a self-pollinating species, the approach is mainly one of selecting the favored progeny over many generations, as in Fig. 11.22. But in many species, such as corn (*Zea*), inbreeding leads to a gradual loss of vigor because damaging recessive alleles become homozygous. A single round of outcrossing immediately yields bigger, more productive plants. This phenomenon is called **hybrid vigor,** or **heterosis.**

How do we handle annual plants in which hybrid vigor is important? This is an important question because it applies to some of our most widely planted crops. The answer is that hybrid seeds must be prepared afresh by means of new matings every year. As an example, Fig. 11.23 shows the important **double-cross** method of breeding field corn (*Zea mays*). Commercial growers have developed many true-breeding strains of corn as a basic resource. They are weak and low in productivity for the reasons mentioned above, but they carry traits that are valuable in hybrids. To make the hybrids, the first step is to select four inbred strains that will give a good combination of traits; they are called A, B, C, and D in the figure. To make initial hybrids, A and B are planted together in a field. The plants of strain B are detasseled, so that they cannot make pollen[2]. Thus A is the **pollinating parent,** and B is the parent whose seeds will be collected (the **seed parent**). The B plants will form B × A hybrid seeds. Likewise, C and D are crossed so that the C plants will yield C × D hybrid seeds. The initial hybrid seeds

[2]To avoid the labor of detasseling, some breeding programs use strains in which the seed-producing parents are male-sterile (cannot produce pollen).

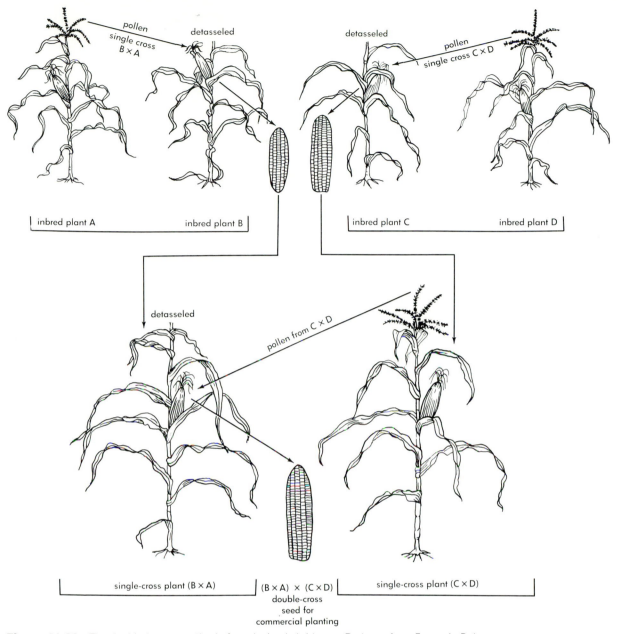

Figure 11.23. The doubled-cross method of producing hybrid corn. Redrawn from *Farmer's Bulletin* 1744, U.S. Dept. of Agriculture, Bureau of Plant Industry, Soils and Agricultural Engineering.

will grow into large, vigorous plants and could be sold directly. But there is a problem: the seed plants B and C, being inbred, are weak and do not produce many seeds. Therefore the B × A and C × D seeds are relatively few and expensive. To provide cheaper seeds, the B × A and C × D plants are crossed with one another as follows. Seeds from the B × A and the C × D crosses are planted in the same field, usually with one row of C × D for every three or four rows of B × A. The B × A plants are detasseled and will receive pollen from the fertile C × D plants. The B × A plants have hybrid vigor and produce many seeds at a low cost. Their seeds grow into large, productive plants with all the best traits of strains A, B, C, and D. Such seeds are used to sow the great grain fields of the American Midwest.

Since their introduction in 1940, the productivity of American cornfields has risen by over 250%.

GENETIC ENGINEERING

In recent years a new way of altering heredity has been drawing attention. Instead of crossing plants, the new method uses **recombinant DNA technology** (more on this later) to insert individual genes into cells. This method has been used to produce animal hormones and antibiotics, but so far it has not generated any improvements in plants. Nevertheless, it merits attention because we can expect much progress with it in the next few years. It is especially promising in cases in which we wish to transfer a gene between two species that are too distantly related to allow crossing. It also may be useful for improving the genes that already exist in a species.

For example, we might improve the food value of plant proteins. In corn, a family of proteins called the *zeins* make up most of the storage proteins. Unfortunately, the zeins are low in the amino acids lysine and tryptophan. If we could alter the genes for zein so that they incorporate these two amino acids, corn would become a perfect protein source. The method would be to isolate the genes for zein, alter them chemically, and insert them back into cells of the corn plant.

How can genes be isolated and inserted into a cell? This is the province of recombinant DNA technology. The experimental methods in use today involve enzymes that break and mend DNA, together with small units of DNA called **plasmids** (Fig. 11.24). Plasmids are ring-shaped DNA molecules that occur in bacteria and some (if not all) eukaryotic cells. Plasmids multiply within the cell, and sometimes they become incorporated into chromosomes. To make recombinant DNA, the plasmid ring is opened with enzymes, the desired gene is inserted, and the ring is resealed to make a hybrid plasmid. Then plant cells are induced to take up the plasmids (Fig. 11.24).

These methods are simple in principle but complex in practice. One problem is that cell walls present a barrier against the entry of foreign DNA. This can be overcome by working with cultured cells, from which the cell walls can be removed. First, blocks of tissue are placed in culture where they grow and divide into callus tissue (see Chapter 9 for details). When placed in a liquid medium, callus tissue is easily broken into separate cells that can be maintained as a cell culture. Enzymes can be added to digest the cell walls, leaving naked protoplasts. Genes can then be introduced by way of plasmids, or all the genes from two cells can be joined by fusing two protoplasts together.

The fusion of two whole cells has excited interest because it is a possible way to combine two quite unrelated crop plants into one species. However, such methods are still in the experimental stage. Cells as unrelated as plants and animals can be fused, but the hybrid cells usually do not survive for long. It is encouraging that in one case (the fusion of potato and tomato cells) a hybrid has been grown into a plant that formed flowers.

It is not enough just to inject genes into a cell: the genes must be supplied in such a way that the host cell reads and translates them. Nature has partly solved this problem for us; the natural host of a plasmid automatically reads the plasmid's genes. This is probably due to an evolved similarity between the DNA of the host and the plasmid. Thus, the quickest way to achieve an effective gene transfer is to use a plasmid that is already associated with the kind of plant we wish to modify. To this date, the most promising plasmid occurs in *Agrobacterium tumefaciens,* a bacterium that causes the cancerlike disease called *crown gall* in many dicot plants. When *Agrobacterium* enters a plant, it transmits

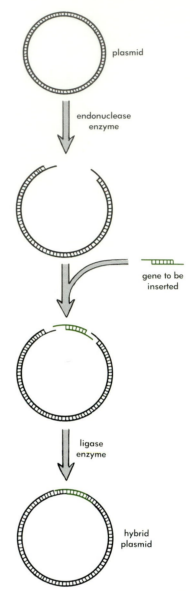

Figure 11.24. Inserting a gene into a plasmid. The hybrid plasmids can then be injected into cultured plant cells, which may be grown into plants.

a plasmid called *Ti* that remains in the cells even if the bacterium is later killed. The plasmid carries genes that are read by the plant cell and are reproduced along with the plant's own genes at the time of cell division. Some of the plasmid genes cause the cells to produce abnormal compounds, and they upset the hormone balance so that the infected cells multiply into a tumorlike growth. However, genetic engineers have developed harmless mutants of the *Ti* plasmid and have used them to transfer genes from one plant to another.

These methods use cultured cells, so the final step is to regenerate whole plants. The methods have been described in Chapter 9. Briefly, the cultured cells are allowed to settle on an agar surface, where they form a callus. Then the hormone concentrations are adjusted so that shoots and roots form **(Fig. 11.25, Color Plate 13).** With the callus as a stock of cells, many plants can be produced from a single culture. Unfortunately, at present few crop plants can be regenerated from cultured cells; further studies will be needed to expand our capabilities in this area.

.Perhaps the greatest problem in genetic engineering is that foreign genes can do more harm than good unless they are integrated into the plant's system of genetic control. At present, we know very little of the systems that control the genes in plants, but the potential offered by genetic engineering will be a great stimulus to research.

SUMMARY

1. Hereditary information is carried by genes, which are segments of DNA found mostly in the chromosomes.

2. Within a species, individual plants vary in ways that are passed on to the offspring. New genes arise through mutation, and new combinations of genes result from sexual reproduction.

3. Alleles are alternative forms of a gene that arise through mutation.

4. Asexual reproduction leads to a clone of offspring that are genetically identical to the parent.

5. The basic sexual life cycle includes (a) fusion of haploid gametes (syngamy) to produce a diploid zygote cell, (b) meiosis in diploid cells to produce new haploid cells, and (c) specialization of some haploid cells to form new gametes.

6. Within the sexual life cycle, mitotic divisions may occur in diploid cells, haploid cells, or both. Based on the phase in which mitosis occurs, sexual life cycles are classified as zygotic, gametic, or sporic. Flowering plants have a sporic life cycle; that is, mitotic divisions occur in both diploid and haploid cells, to produce distinct diploid and haploid bodies.

7. Meiosis divides diploid cells into haploid cells. At one step in meiosis, homologous chromosomes form pairs (synapsis) and exchange parts (crossing-over). Meiosis involves two successive divisions, which are accomplished by means of spindles.

8. Controlled matings (crosses) are the basis of plant breeding. In a single cross, attention is often focused on one trait (monohybrid crosses) or two traits (dihybrid crosses).

9. Flowering plants are diploid. If the two copies of a gene are different alleles (the heterozygous condition), one allele (the recessive) may not be expressed. In this case, the trait is said to show classical dominance. A testcross may be performed to determine whether an individual with the dominant phenotype is heterozygous.

10. In some cases, both alleles are expressed in the heterozygous individual, resulting in a phenotype that differs from the homozygous individual.

11. Two genes may be linked (carried on the same chromosome) or unlinked. Linkage can be detected by examining the ratios of progeny in a dihybrid testcross.

12. Variations in a trait may be discontinuous or continuous. The latter is common when several genes affect a trait. In this case, plant improvement requires elaborate programs of selection over many generations.

13. Some traits are controlled by cytoplasmic genes that are contributed to offspring chiefly by the maternal parent.

14. Hybrid plants may be superior as a result of hybrid vigor. A useful hybrid can sometimes be maintained by vegetative propagation. In other cases, fresh supplies of hybrid seed are prepared by crosses every year. High-yielding hybrid corn is produced by a double-cross method.

15. Genetic engineering involves the transfer of genes between organisms that may not be related closely enough for matings. Genes can be isolated and inserted into plasmids, which then carry the genes into cultured cells. Alternatively, cultured cells of different species can be fused together. The cultured cells may be grown into plants.

12 THE FLOWER

The flower initiates the sexual reproductive cycle in all Anthophyta (the flowering plants) and, in so doing, it usually terminates the growth of the shoot bearing the flower. With some plants—annuals, biennials, and some perennials—death follows flowering and seed set. With most perennials the capacity for vegetative growth is regained after sexual reproduction and the development of the next generation. In woody perennials, provision is always made for continued vegetative growth, through mixed buds or associated shoot buds following flower production. The function of the flower is to facilitate the important events of gamete (haploid reproductive cell) formation and fusion.

Of all the characteristics of flowering plants, the flower and fruit are the least affected by changes in the environment. For instance, we have seen that leaf shape is influenced by age, light, water, and nutrition. The basic morphology and anatomy of flowers and fruits are not affected quite so much and thus are important parts for angiosperm classification.

In addition, flowers have a high esthetic value, and the resulting fruits and seeds are of major importance in food production.

The essential steps of sexual reproduction, meiosis and fertilization, take place in the flower. The complete sexual cycle involves (a) the production of special reproductive cells following meiosis, (b) pollination, (c) fertilization, (d) fruit and seed development, (e) seed and fruit dissemination, and (f) seed germination. The seed completes the process of sexual reproduction in the angiosperms, and the embryo in the seed is the first stage in the life cycle of new individuals.

FLOWER STRUCTURE

A typical flower is composed of four whorls of modified leaves—(1) **sepals,** (2) **petals,** (3) **stamens,** and (4) a **carpel** or **carpels**—which are all attached to the modified stem end that supports these structures, the **receptacle** (Figs. 12.1 and **12.2, Color Plate 15**).

The sepals enclose the other flower parts in the bud and are generally green. All the sepals collectively constitute the **calyx.** The petals are usually the conspicuous, colored, attractive flower parts. Together, the petals constitute the **corolla.**

The stamens form a whorl, lying inside the corolla. Each stamen has a slender stalk, or **filament,** at the top of which is an **anther,** the pollen-bearing organ. The whorl or grouping of stamens is called the **androecium** (Fig. 12.3).

The carpel or carpels comprise the central whorl of modified floral leaves (Fig. 12.3). Collectively, the carpels are spoken of as the **gynoecium.** Each individual structure in the gynoecium is referred to as a **pistil.** A pistil may be composed of a single carpel, or of several united carpels in the center of the flower.

There are generally three distinct parts to each pistil: (1) an expanded basal portion, the **ovary,** in which are borne

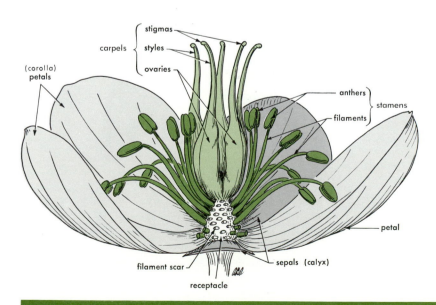

Figure 12.1. A diagram of a flower of Christmas rose (*Helleborus niger*). The perianth consists of two similar whorls; there are numerous stamens arranged in a spiral on a cone-shaped receptacle. Five separate carpels form the central whorl of floral parts.

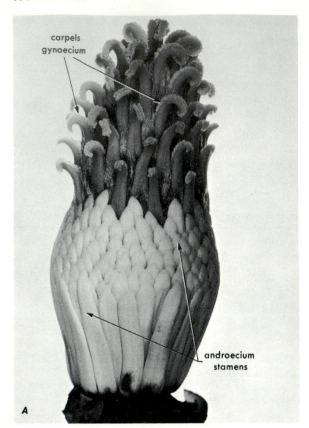

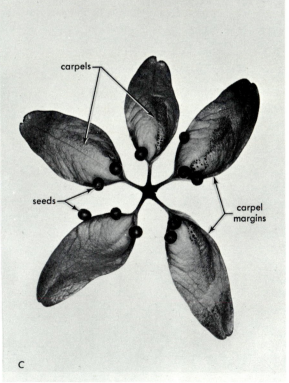

Figure 12.3. The essential floral organs. *A, Magnolia* sp., showing many separate stamens and separate carpels arranged in a spiral on the receptacle. *B,* regal lily (*Lilium regale*), showing six stamens with anthers and filaments, and three carpels in a single pistil with stigma, style, and ovary. *C, Sterculia plantanifolia:* dehiscence of matured ovary showing seeds attached to the margins of the five leaflike carpels.

the **ovules;** and (2) the **style,** a slender stalk supporting (3) the **stigma** (Fig. 12.1).

The term **perianth** is applied to the calyx and corolla collectively. It is frequently used to describe flowers, such as the tulip *(Tulipa),* in which the two outer whorls are similar in appearance. The individual parts of such a perianth are called **tepals.**

Sometimes individual flowers or compact clusters of flowers have a whorl of small leaves, or **bracts,** subtending them. A collection of bracts subtending flowers is called an **involucre.**

CARPELS AND STAMENS AS MODIFIED LEAVES

The question arises as to why carpels and stamens are considered modified leaves. One reason is because the early developmental stages of floral parts closely resemble those of leaves. The second reason, also circumstantial and speculative, concerns the shape of spore-bearing leaves in primitive angiosperms. If we were to examine the stamens of primitive flowers such as *Magnolia* or *Degeneria* (Fig. 12.4), we would see that the shape of the stamen is very leaflike. In addition, the folded open carpels of primitive angiosperms are leaflike in appearance. If a young pea pod, which is a single carpel, is opened carefully along its ventral suture (toward the axis), the margins can be folded back, showing the seeds. Each seed developed from an ovule, along the margin of a leaflike carpel (Fig. 12.17*D*).

The relationship between carpels and leaves is even more striking in *Sterculia platanifolia,* because in this plant there are five simple pistils united only by their stigmas (Fig. 12.3*C*). When these stigmas mature, they open to show five very leaflike carpels that bear seeds along their margins.

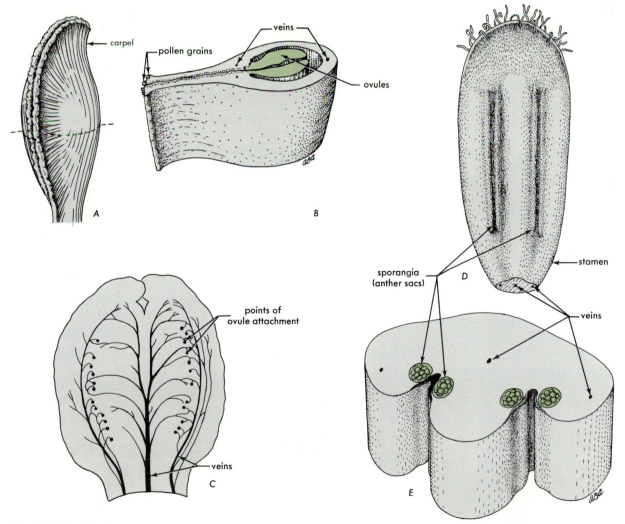

Figure 12.4. Primitive carpel and stamen. *A,* carpel of *Drimys piperita; B,* cross section of carpel (dotted line); *C,* carpel laid open; *D,* stamen of *Degeneria vitiensis; E,* cross section of stamen.

VARIATIONS IN FLORAL STRUCTURE

GENERAL DESCRIPTION

Complete and Incomplete Flowers

The parts of a typical flower—sepals, petals, stamens, and carpels—are all attached to the receptacle. A flower with all four sets of floral leaves is said to be a **complete flower.** An **incomplete flower** lacks one or more of these four sets. For example, there are (a) flowers that partially or completely lack a perianth (Fig. 12.5A and 12.5B), (b) flowers with carpels but no stamens, and (c) flowers with stamens but no carpels (Figs. 12.6 and 12.8).

Perfect and Imperfect Flowers

Unisexual flowers are either **staminate** (stamen-bearing) or **pistillate** (pistil-bearing) and are said to be **imperfect,** whereas bisexual flowers are **perfect.** When staminate and pistillate flowers occur on the same individual plant, as they do in corn *(Zea),* squash *(Cucurbita),* walnut *(Juglans)* (Figs. 12.6, 12.7, and 12.8), and many other species, the plant is said to be **monoecious.** In corn (Fig. 12.7), for example, the tassel (borne at the top of the stalk) consists of a group of staminate flowers, and the young ear is a group of pistillate flowers. When staminate and pistillate flowers are borne on separate individual plants, as in *Asparagus,* willow *(Salix),* and many other species, the species (or the plant) is said to be **dioecious.**

Floral Symmetry

In many flowers, such as those of columbine *(Aquilegia;* Fig. 12.9A) or cherry *(Prunus;* Fig. 12.10B), the corolla is made up of petals of similar shape that radiate from the center of the flower and are equidistant from each other. Such flowers are said to be **regular.** In these cases, even though there may be an uneven number of parts in the perianth, any line drawn through the center of the flower will divide the flower into two similar halves. They may be exact duplicates or mirror images of each other.

Irregular flowers, such as pea *(Pisum;* Fig. 12.9C) and mints *(Salvia;* Fig. 12.9B), have whorls with either (a) dissimilar flower parts, (b) parts that do not radiate from the center, or (c) parts that are not equidistant from one another. In most of these flowers, only one line divides the flower in equal halves; the halves are usually mirror images of each other. Some flowers, such as bleeding heart *(Dicentra),* though irregular, may be bisected by any number of lines into similar mirror images.

The irregular bean or pea flower has a corolla composed of the following (Fig. 12.9C): one broad conspicuous petal, the **standard** or **banner;** two narrower petals **(wings),** one on each side; and, opposite the banner, two smaller petals that are united along their edges to form the **keel.**

UNION OF FLOWER PARTS

In the flower of columbine illustrated in Fig. 12.9A, all parts

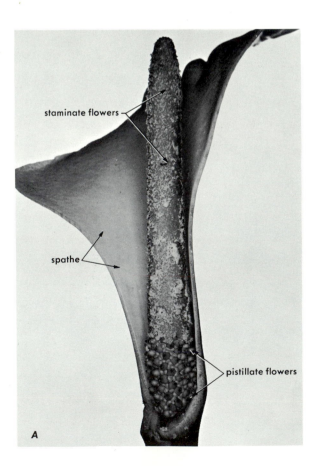

Figure 12.5. Flowers with perianth parts lacking. *A,* spathe of calla lily *(Zantedeschia aethiopica),* sepals and calyx absent; *B,* flower of *Clematis sp.,* petals absent and sepals prominent.

of the flower are separate and distinct; that is, each sepal, petal, stamen, and carpel is attached at its base to the receptacle. In many flowers, however, members of one or more whorls are to some degree united with one another, or are attached to members of other whorls. Union with other members of a given whorl is termed **connation.** Union of flower parts from two different whorls is known as **adnation.** A precise terminology has been developed to indicate these fusions. For example, syncarpy refers to the situation in which carpels of the gynoecium are connate.

Elevation of Flower Parts

In flowers of *Helleborus* (Fig. 12.1) and tulip (Fig. 12.10*A* and **12.11*A*, Color Plate 16**), the receptacle is convex or conical and the different flower parts are arranged one on top of another. The gynoecium is thus situated on the receptacle above the points of origin of the perianth parts and androecium. An ovary in this position is said to be **superior.** In the daffodil (Figs. 12.10*C* and **12.11*C***) the ovary appears to be below the apparent points of attachment of the perianth parts and the stamens. This is an **inferior** ovary. With an inferior ovary, the anatomy of the flower parts indicates that the lower portions of the three outer whorls—calyx, corolla, and androecium—have fused to form a tube, or **hypanthium.** The ovary in the daffodil is completely adnate with the hypanthium.

In a flower with a superior ovary, sepals, petals, and stamens arise from the outer, lower portion of the receptacle, below the point of origin of the carpels (Figs. 12.1 and 12.10*A*). The perianth and stamens are **hypogynous,** or with reference to the three outer whorls, a condition of **hypogyny** exists. In a flower with an inferior ovary, the perianth and stamens appear to arise from the top of the ovary, and they are **epigynous.**

In some flowers, such as the plum and apricot (*Prunus* sp.), the hypanthium does not become adnate to the ovary. The perianth parts and stamens arising from the rim of the cuplike hypanthium around the ovary are **perigynous** (Figs. 12.10*B* and **12.11*B***).

INFLORESCENCES

In most flowering plants, flowers are borne in clusters or groups. Morphologically, an **inflorescence** is a flower-bearing branch or system of branches. In manuals for the identification of flowering plants, many different terms are used to describe the various kinds of inflorescence. Only the most common ones are discussed and illustrated here (Fig. 12.12).

A very simple type of inflorescence called a **raceme** is found in such plants as currant *(Ribes)* and radish *(Raphanus)*. In this type, the main axis has short branches, each of which terminates in a flower. Each flower is on a short branch stem called the **pedicel.** The main axis of a raceme

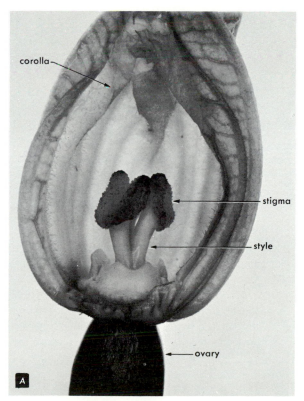
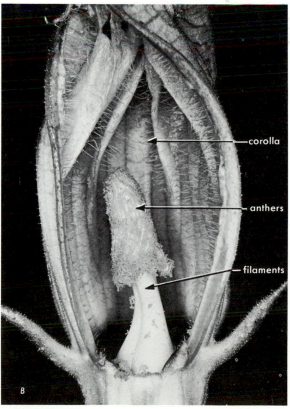

Figure 12.6. Imperfect flower. *A,* pistillate; and *B,* staminate flowers of squash (*Cucurbita Pepo*).

Figure 12.7. Flowers of corn (*Zea mays*). *A*, the tassel; *B*, exserted anthers of staminate flowers; *C*, pistillate flowers forming a very young ear; *D*, several pistillate flowers with attached styles.

continues to grow in length more or less indefinitely; the apical meristem persists. Primordia of leaves arise in the usual manner along the margin of this apical meristem, and in the axil of each leaf a flower is borne. The oldest flowers are at the base of the inflorescence and the youngest are at the apex.

In a simple raceme, the flowers are on pedicels that are about equal in length. In a **spike** (Fig. 12.12), the main axis of the inflorescence is elongated, but the flowers, each in the axil of a bract, are sessile (without a pedicel). The **catkin** is a spike that usually bears only pistillate or staminate flowers. The inflorescence as a whole is shed later. Examples

are willow, cottonwood *(Populus),* and walnut (Fig. 12.8).

In all the inflorescences just mentioned, the flowering axis is elongated. If it is short, flowers appear to be arising umbrellalike from approximately the same level. An inflorescence of this kind, in which pedicels are of nearly equal length, is called an **umbel** (Fig. 12.12). The onion *(Allium)* is a good example. The **head** is an inflorescence in which the flowers are sessile and crowded together on a very short axis. Members of the family Asteraceae (Compositae), including thistle *(Cirsium)* and sunflower, have this type of inflorescence (Fig. 12.12).

Inflorescences of the raceme type may be compound,

Figure 12.2. Flowers showing three different arrangements of flower parts. *A, Liriodendron tulipifera,* three sepals, six petals, numerous stamens, many carpels compressed in inner whorl. *B, Sedum* sp., five sepals, five petals, ten stamens, five carpels. *C, Lilium tigrinum,* perianth consists of two whorls, each with three similar sepals and petals, six stamens, and three carpels coalesced in one pistil.

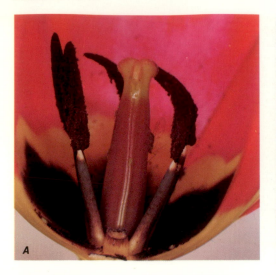

Figure 12.11. Elevation of floral parts. *A*, hypogyny in tulip flower *(Tulipa)*. *B*, perigyny in almond flower *(Prunus amygdalus)*. *C*, epigyny in daffodil flower *(Narcissus pseudonarcissus)*. *A* and *C*, about ×⅓; *B*, ×1.

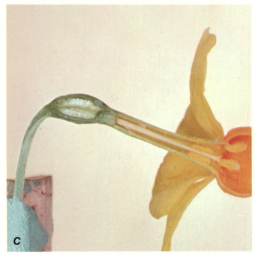

Figure 12.21. Floral guides for insects collecting nectar or pollen. *A, Viola,* splashes of color and guide lines leading to throat of corolla. *B*, section of nasturtium flower tubular spur for nectar and guide hairs leading to entrance to tube. *C*, mimicry—the flower of the orchid *Ophrys* resembles a female wasp. (*C*, courtesy of R. Norris.)

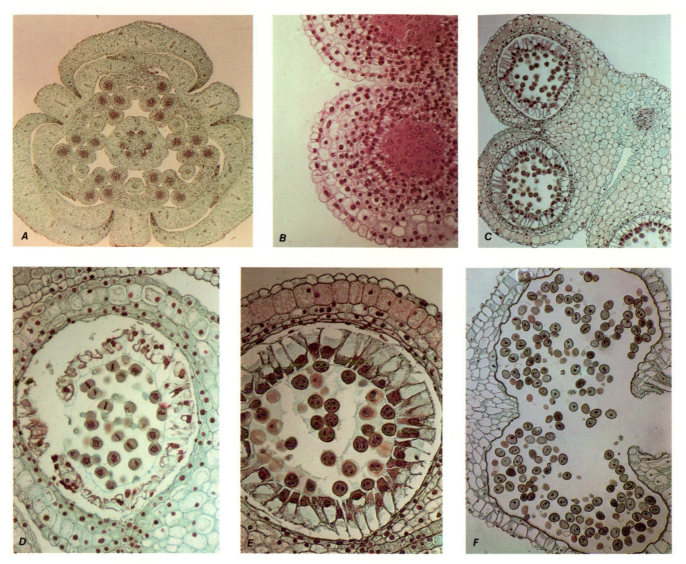

Figure 12.14. Pollen formation. *A,* cross section of lily (*Lilium* sp.) floral bud, three carpels in the center are surrounded by six anthers. *B,* cross section of two locules of an
anther containing microsporocytes. *C,* microsporocytes undergoing meiosis. *D,* anther with
closer view of same stage. *E,* microspore tetrad stage. *F,* microspore dividing mitotically to
form a pollen grain. Anther wall (right side) has split to release pollen grains.

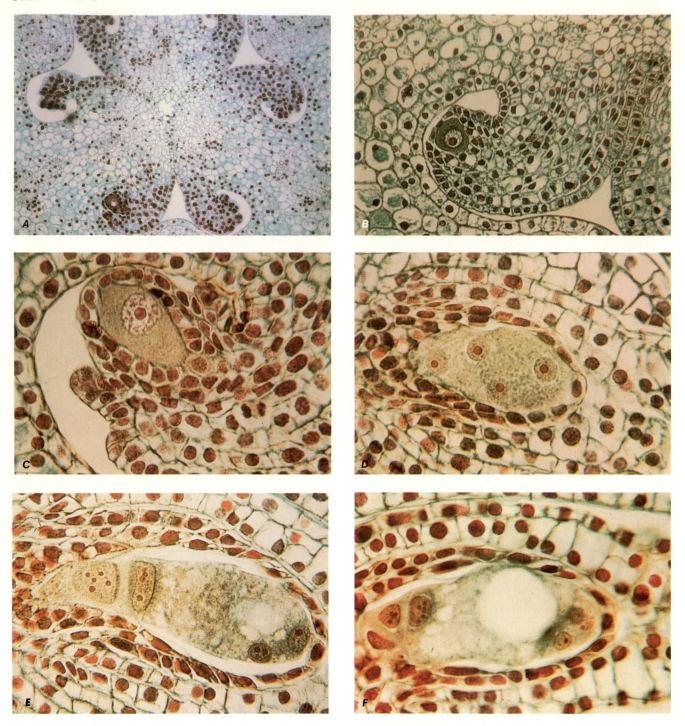

Figure 12.20. Embryo sac development. *A,* cross section of lily ovary showing location of ovules in three locules. *B,* ovule with a large megasporocyte. *C,* later stage, the integuments and single layered nucellus shown surrounding the megasporocyte. *D,* four megaspore stage. *E,* second four nucleate stage. *F,* mature embryo sac.

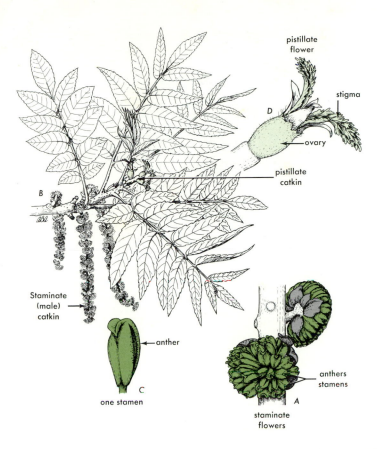

pistillate
flower

stigma

D

ovary

pistillate
catkin

Staminate
(male)
catkin

anther

C

one stamen

anthers
stamens

A

staminate
flowers

Figure 12.8. English walnut (*Juglans regia*) flowers. *A*, staminate flowers; *B*, pistillate flowers and staminate inflorescences (catkin); *C*, dehiscing anther; *D*, pistillate flower.

A

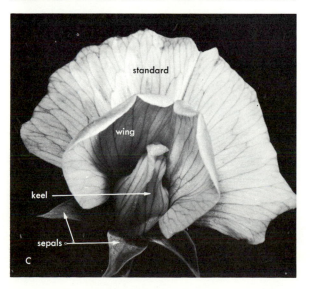

standard

wing

keel

sepals

C

B

Figure 12.9. Floral symmetry. *A*, regular flower of columbine (*Aquilegia* sp.); *B*, irregular flower of *Salvia* sp.; *C*, flowers and essential organs—flower of garden peas (*Pisum sativum*).

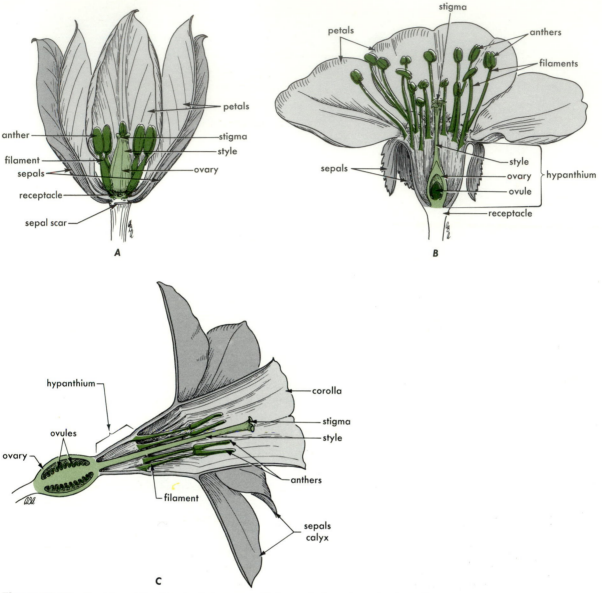

Figure 12.10. Elevation of floral parts. *A,* hypogyny (*Tulipa* sp.); *B,* perigyny in cherry flower (*Prunus avium*); *C,* epigyny in daffodil flower (*Narcissus pseudonarcissus*).

that is, branched. A branched raceme is called a **panicle** (Fig. 12.12). Compound spikes occur in wheat, rye, and certain other grasses.

In contrast to the raceme types of inflorescences, just described, is the **cyme** (Fig. 12.12). In the cyme, the apex of the main axis produces a flower that involves the entire apical meristem; hence, that particular axis ceases to elongate. Other flowers arise on lateral branches farther down the axis of the inflorescence and, thus, usually the youngest of the flowers in any cluster occurs farthest from the tip of the main stalk. The flower cluster of chickweed (*Cerastium*) is an example of the cyme.

ANGIOSPERM LIFE CYCLE

The floral organs necessary for sexual reproduction are the stamens (androecium) and carpels (gynoecium) (Fig. 12.1).

The perianth, composed of calyx and corolla, is a protective covering of the stamens and pistils, and when it is present, it also serves to attract and guide the movements of pollinators.

THE ANDROECIUM

Each stamen consists of an anther supported by a filament (Fig. 12.13A). The anther usually consists of four elongated and connected lobes called pollen sacs. Early in the development of the anther, the pollen sacs each contain a mass of dividing cells called **microsporocytes** (microspore mother cells; Figs. 12.13B and **12.14B, Color Plate 17**). Each microsporocyte divides by meiosis to form four haploid (1*n*) microspores (Figs. 12.13C and **12.14C**). The nucleus of each microspore then divides by mitosis to form a two-celled **pollen grain** that contains a **tube cell** and a smaller **generative cell** (Fig. 12.13E). The role of this two-celled,

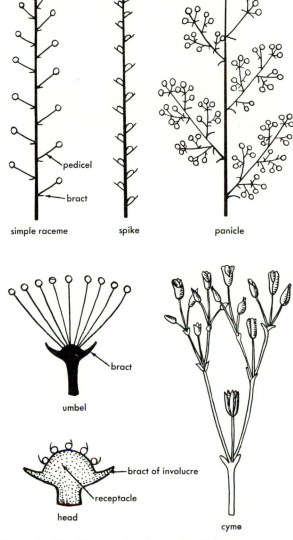

Figure 12.12. Diagram showing types of inflorescence.

labels: pedicel, bract, simple raceme, spike, panicle, bract, umbel, bract of involucre, receptacle, head, cyme

haploid, **male gametophyte** (gamete-producing plant) is to produce sperm which is necessary later for fertilization.

The pollen grain is surrounded by an elaborate, ornate cell wall that protects the gametophyte against dehydration. The pattern of the pollen wall is hereditary, and it varies widely among major groups of plants (Fig. 12.15). The walls contain a very hard material called sporopollenin that resists decay. As a result, pollen grains are among the most abundant fossils, and botanists have found them to be very useful in reasoning about the evolutionary history of plants.

After the pollen grains are mature, the anther wall splits open and the pollen are shed **(Fig. 12.14F).** In some manner (which will be discussed later), the pollen are transported to the stigma of adjacent or distant flowers. This process is called **pollination.** Once on the stigma, if the pollen and the stigma are genetically compatible, a pollen grain germinates to form an elongate **pollen tube** (Fig. 12.13F). The ornate pattern of the pollen grain wall serves as the means for receptive stigmas to recognize compatible pollen. The troughs

and ridges of the pollen wall (Fig. 12.15) apparently contain proteins in specific patterns and concentrations. If the patterns correlate with those of the stigma, metabolic events are triggered within the pollen grain to stimulate the pollen tube to grow into the pistil tissue. Some plants have developed different recognition systems that control pollen tube growth after pollen germination.

THE GYNOECIUM

The structure of the gynoecium depends on the number and arrangement of carpels comprising it. In the pea flower, for example, there is a single carpel forming the gynoecium (Fig. 12.17). It consists of three parts: (1) the **ovary,** an expanded basal portion, (2) the **style,** a slender stalk, and (3) a hairy irregular tip portion, the **stigma** (Fig. 12.17B). When the **pistil** is composed of one carpel, it is referred to as a simple pistil (Fig. 12.16A). In many other instances a compound pistil, consisting of two or more fused carpels, may occur (Fig. 12.16B).

The ovary is a hollow structure having from one to several chambers, or **locules** (Fig. 12.16). The number of carpels in a compound pistil is generally related to the number of stigmas (Fig. 12.1), the number of locules, and sometimes, the number of faces of the ovary (Fig. 12.16B). The pea has a single stigma, and its ovary has one locule (Fig. 12.16A). There are three stigmas in the tulip gynoecium; the ovary is three-sided (Fig. 12.16B), contains three locules, and is composed of three carpels.

Placentation

The tissues within the ovary to which the **ovules** are attached are called **placentae** (singular, **placenta;** Fig. 12.18). The manner in which the placentae are distributed in the ovary is termed **placentation.** When the placentae are on the ovary wall, as in the pea (Pisum) and bleeding heart (Dicentra; Fig. 12.18D), the placentation is **parietal.** When they arise on the axis of an ovary that has several locules, as in lilies, Fuchsia, and tulip, the placentation is **axile** (Fig. 12.18C). Less frequently, the ovules are on the axis of a one-loculed ovary, in which event the placentation is **central,** as in the primrose family (Figs. 12.18A and 12.18B).

Style and Stigma

The style is a slender stalk that terminates in the stigma. It is through stylar tissue that the pollen tube grows. In some flowers, the style is very short (Fig. 12.17B and 12.17C) or entirely lacking; in others, it is long. In Zea mays, the corn silks are the styles (Fig. 12.7C and 12.7D). In general, the style withers after pollination, but in some plants (e.g., Clematis) it persists and becomes a structure that aids in the dispersal of the fruit. The stigmatic surface often has short cellular outgrowths that aid in holding the pollen grains; and sometimes it secretes a sugary and sticky solution, the **stigmatic fluid.** In many wind-pollinated plants, such as the grasses, the stigma is much branched, or plumelike.

The Ovule

The ovule, the structure that will eventually become the seed, arises as a dome-shaped mass of cells on the surface of

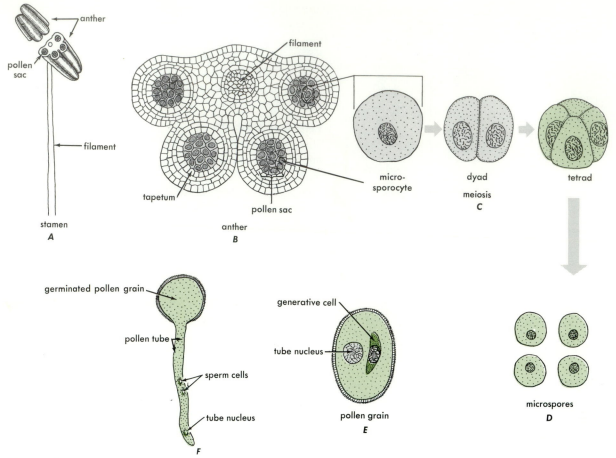

Figure 12.13. Development of pollen from a microsporocyte to the pollen grain. *A,* stamen; *B,* cross section of anther; *C,* development of tetrad of microspores from the microsporocyte by meiosis; *D,* four microspores; *E,* pollen grain; *F,* germination of pollen tube.

the placenta. The outer cells of the dome develop to form one or two protective layers, the **integuments** (Fig. 12.19*B* and 12.19*H*). The integuments do not fuse at the apex of the ovule, thus leaving a small opening, the **micropyle** (Fig. 12.19*B*). While this development is taking place, one of the internal dividing cells of the ovule, the **megasporocyte,** is enlarging in preparation for meiosis. The tissue within which the megasporocyte has differentiated is known as the **nucellus.** The ovule then is composed of one or two outer protecting integuments, along with the micropyle, nucellus, and megasporocyte (Fig. 12.19*B*).

EMBRYO SAC DEVELOPMENT

Meiotic division of the **megasporocyte** (megaspore mother cell) is considered the first step in the development of a seed. As a result of these divisions, a row of four cells called **megaspores** are produced in the nucellus. As a rule, three of the cells (the ones nearest the micropyle) disintegrate and disappear, whereas the one farthest from the micropyle enlarges greatly (Fig. 12.19*D* and 12.19*E*). This megaspore now develops into the mature **embryo sac.** The usual stages are as follows: (a) a series of three mitotic divisions, forming

an eight-nucleate embryo sac (Fig. 12.19*E,* 12.19*F,* and 12.19*G*); (b) migration of nuclei (Fig. 12.19*G*); (c) cell wall formation around nuclei (Fig. 12.19*H*).

At the micropylar end of the embryo sac there is one **egg cell** associated with two **synergid cells.** Since it is frequently difficult to differentiate the egg cell from the other two cells, all three cells are sometimes referred to as the egg apparatus. The two nuclei that migrated toward the center approach each other; these nuclei, known as **polar nuclei** (Fig. 12.19*G*), may fuse to form a single nucleus called a **secondary nucleus.** This center portion of the embryo sac is called the **central cell.** The three nuclei remaining at the end of the embryo sac opposite the micropyle form the antipodal cells. In cotton the antipodals disintegrate in the mature embryo sac (Fig. 12.19*H*). The embryo sac, which can be considered a seven-celled haploid plant (female gametophyte), is now mature and ready for fertilization.

Not every species of flowering plant produces a seven-celled embryo sac, nor do they all follow the same developmental sequence. However, the mature embryo sac must contain, at a minimum, an egg and polar nuclei.

In lily a slightly different sequence of embryo sac development occurs **(Fig. 12.20, Color Plate 18).** After the

Figure 12.15. Pollen grains as seen with the scanning electron microscope. *A, Iris sp.; B,* day lily (*Hemerocallis fulva*), *C,* cucumber (*Cucumis satius*); *D,* ragweed (*Ambrosia*).

meiotic division of the megasporocyte **(Fig. 12.20D),** four 1*n* nuclei are present in the immature embryo sac. The three nuclei at the chalazal end (opposite the micropylar end) of the embryo sac fuse together to form one 3*n* nucleus. At this stage, the embryo sac contains one 1*n* nucleus and one 3*n* nucleus. Both nuclei then divide two times by mitosis to form a total of eight nuclei—four 1*n* nuclei and four 3*n* nuclei **(Fig. 12.20E and 12.20F).** The nuclei will be dispersed so that the egg and synergids are each 1*n*, the three antipodals are 3*n*, one polar nucleus is 1*n*, and the other polar nucleus is 3*n*. When the two polar nuclei fuse with a sperm, the resulting endosperm will be 5*n*.

FERTILIZATION

Prior to fertilization, the pollen tube grows down through the stigma and style and enters the ovary (Figs. 12.13*F* and 12.19*I*). Many pollen grains may germinate, and their pollen tubes may grow through the pistil, but only one usually enters the embryo sac for fertilization. While the pollen tube is growing, the generative cell within it divides by mitosis to form two sperm cells. In some plants, like sunflower, the sperm cells form even before the pollen are shed from the anther.

Reaching the ovary, the pollen tube grows toward one of the ovules, usually enters the micropyle, penetrates one

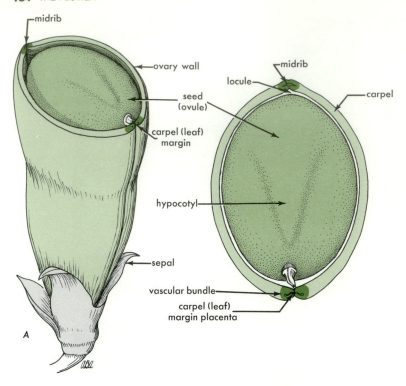

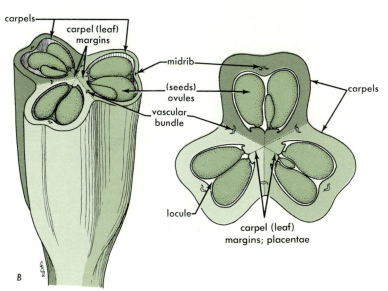

Figure 12.16. Comparison of gynoecia with one carpel and with three connate carpels. *A*, section of ovary of pea (*Pisum sativum*) consisting of a single carpel; *B*, section of ovary of *Tulipa* sp. showing three connate carpels.

or more layers of nucellar cells, enters the embryo sac (Fig. 12.19*I*), and approaches the egg cell and the central cell. These stages are not easy to study, but it appears that the tip of the tube ruptures, one sperm cell enters the egg, and the other sperm enters the central cell. Within the egg, sperm and egg nuclei fuse. Within the central cell, the other sperm nucleus and two polar nuclei fuse (Fig. 12.19*I*). This double fusion of egg with sperm and polar nuclei with sperm is called **double fertilization.** The **zygote** and **primary endosperm cell** result from these fusions, the antipodal cells and synergid cells degenerate, and conditions are set for further development of the seed and fruit.

One interesting question concerns how the pollen tube is directed into the embryo sac. The answer has been partially worked out for cotton (*Gossypium*). In the embryo sac of cotton, one or both of the synergid cells begin to shrivel and die before the pollen tube enters. It has been suggested that this activity results in the flow of some chemical substance from the synergid that possibly influences the direction in which the pollen tube grows. Two observations favor this hypothesis. First, at the base of both synergids is found a highly convoluted cell wall, known as the filiform apparatus (Fig. 12.19*G* and 12.19*I*). Cell walls in this configuration are commonly believed to be associated with active chemical

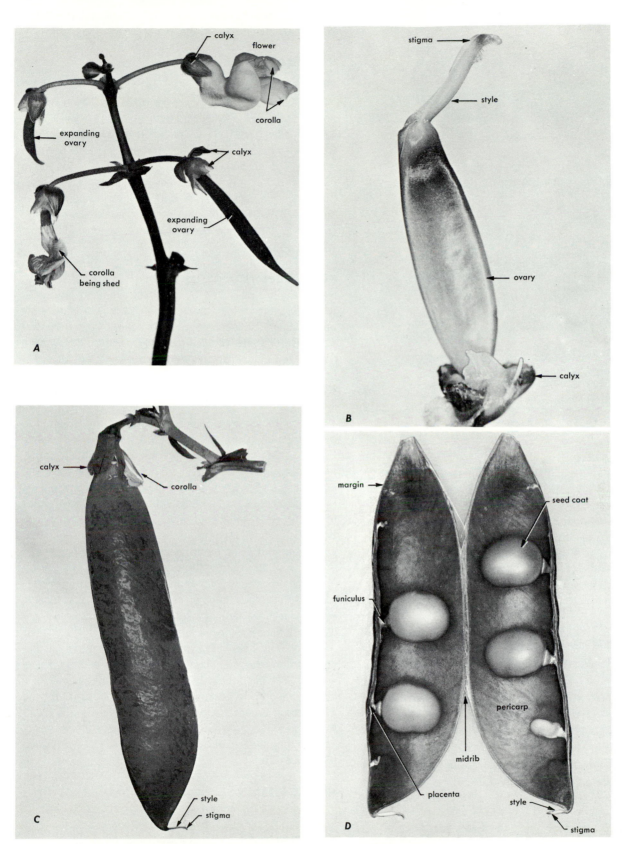

Figure 12.17. Development of legume fruits. *A*, bean (*Phaseolus vulgaris*) from flowers to young pods; *B*, pistil from a pea (*Pisum sativum*) flower; *C*, pea pod, unopened; *D*, opened pea pod showing developing seeds attached to carpel margins.

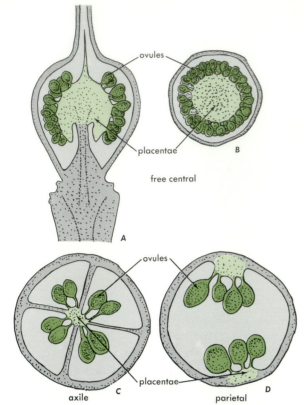

Figure 12.18. Three types of placentation. *A* and *B*, free central (*Primula* sp.); *C*, axile (*Fuchsia* sp.); *D*, parietal (*Dicentra* sp.).

secretion. The second bit of evidence is that the pollen tube actually penetrates one of the synergids and empties its contents into it. The two sperm then pass through the incomplete upper cell wall of the synergid; subsequently, one of them moves through the incomplete side wall of the egg to fertilize the egg cell, and the other moves deeper into the embryo sac to fuse with the central cell.

The fusions that occur in the embryo sac initiate the following changes in the entire ovary:

1. Development of the zygote to form the embryo plant.

2. Development of the primary endosperm cell to form the endosperm (the reserve food supply of the seed).

3. Development of integuments to form the seed coat.

4. Absorption or disintegration of nucellar tissue. In some plants, however, a portion of the nucellus, rather than the endosperm, may become the storage tissue of the seed. Such tissue of nucellar origin is called **perisperm.**

5. Development of ovary tissue forms the fruit.

6. Possible stimulation of accessory flower parts, such as the receptacle, sepals, or petals, to increased growth and incorporation into the fruit.

POLLINATION

Pollination is brought about either by wind or insects, and occasionally by water, birds, bats, and other small mammals. Flower structure and pollen type are generally adapted to one or the other of these pollinating vectors. In insect pollination, adaptation may be very complex, indicating a long association between the insect vector and plants.

POLLINATING VECTORS

Wind

Pollen is carried chiefly by wind and insects. Wind pollination is common in plants with inconspicuous flowers, such as in grasses (Fig. 12.7), walnuts (Fig. 12.8), oaks, and ragweeds (*Ambrosia*). Such plants usually produce pollen in enormous quantities. The flowers usually lack odor and/or nectar and hence are unattractive to insects. Furthermore, their pollen is light and dry and easily wind-borne. Their stigmas, in many cases, are feathery and expose a large surface to catch flying pollen. Hayfever victims suffer from windborne pollen.

Living Vectors

Plants that are pollinated by living vectors usually possess a colorful perianth that, together with other floral parts, may be arranged into a complex architecture (Figs. 12.9 and **12.11, Color Plate 16**). They also produce both a nutritive fluid, called nectar, and volatile compounds that have distinctive odors. Many investigators have shown that these characteristics are highly adapted for the attraction of pollinators.

Bees, for example, are able to detect and distinguish between many odors and colors and sugar concentrations. Bees can distinguish at least three colors in addition to ultraviolet, including the most common perianth shades (yellow and blue). Bees are directed to nectar-producing glands by splashes or lines of contrasting color **(Fig. 12.21A, Color Plate 16).** When the insect attempts to reach nectar or collect pollen, floral architecture ensures that the insect transfers pollen to the stigma. In nasturtium, for example, nectar is held in a long narrow tube or spur **(Fig. 12.21B).** A foraging bee directs its proboscis down the tube and usually rubs it against the stigma along the way. In this way, pollen, which may have been picked up in earlier visits to other flowers, is deposited on the stigma.

Orchid flowers have many very specialized mechanisms to ensure pollination. The petal shape and color of species of *Ophrys* closely resemble the appearance of a female wasp or fly. Male wasps or flies, emerging from the pupal stage before the females, mistake the *Ophrys* flower for females. The male lands on the flower and attempts copulation; repeated "pseudocopulation" results in pollination **(Fig. 12.21C).**

CROSS-POLLINATION AND SELF-POLLINATION

In a perfect flower, pollen grains can travel from the anthers to the stigmas within the same flower. This is called **self-pollination,** and it can lead to self-fertilization. Self-fertilization is of value when other plants of the same species are not available for mating, but it bypasses the genetic recombination that is the primary benefit of sexual reproduction. The alternative is **cross-pollination,** which occurs when pollen

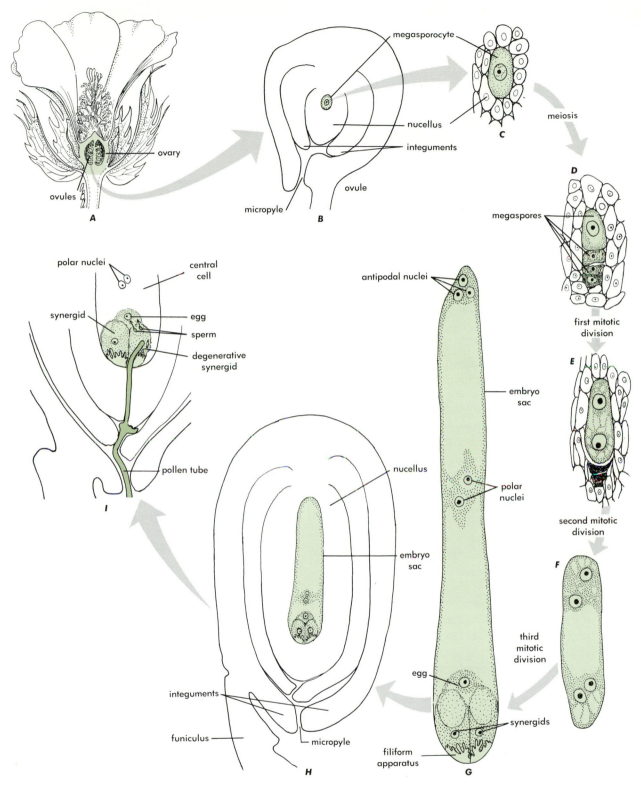

Figure 12.19. Embryo sac development in cotton (*Gossypium hirsutum*). *A,* mature flower. *B,* magnified longitudinal section of an ovule. *C,* megasporocyte before meiosis. *D,* four 1*n* megaspores after meiotic division of the megasporocyte; the three megaspores at the micropylar end disintegrate. *E,* first mitotic division of the megaspore. *F,* second mitotic division, *G,* third mitotic division and the resulting three antipodal nuclei, two polar nuclei, the egg, and two synergids. All have a 1*n* chromosome number; note the specialized cell wall in the synergids, and the filiform apparatus. *H,* mature embryo sac; the antipodals disintegrate at this stage in cotton. *I,* pollen tube penetration through the nucellus into a partially degenerated synergid. (*B, C, D, E, F, G,* and *H* redrawn from U. R. Gore, Amer. J. Bot. 19 (1932): 795–802; *I* modified and redrawn from W. A. Jensen, Amer. J. Bot 52 (1965): 781–797.

travels from one plant to another. The resulting cross-fertilization leads to offspring with new and possibly superior combinations of genes (see Chapter 11).

Plants vary widely in the extent to which they engage in self-fertilization. In some species, self-fertilization is the rule and cross-fertilization is rare. This is true of the garden pea, in which pollination is completed before the flower opens. Other species have features that limit or prevent self-fertilization. These features are quite varied. Some examples are:

1. *The dioecious condition:* Dioecious species form separate male and female plants, making self-pollination impossible.

2. *The monoecious condition:* With staminate and pistillate flowers on different parts of the plant, pollen must travel for a distance through the air to reach stigmas. This allows self-pollination but makes it less likely than in a plant with perfect flowers.

3. *Self-incompatibility:* In many plants, the stigmas have a chemical sense that detects pollen from the same flower and inhibits its germination or pollen tube growth.

4. *Timing of maturation:* The anthers may mature more slowly than the stigmas, so that cross-pollination has already taken place before the flower's own pollen grains are shed.

APOMIXIS

Apomixis refers to a type of reproduction in which no sexual fusion occurs but in which structures usually concerned in sexual reproduction are involved. Normally, egg nuclei will not start on the series of changes that result in embryo plants unless fertilization has occurred. Occasionally an embryo develops from an unfertilized egg; this type of apomixis is called **parthenogenesis.** Another type is that in which the embryo plant arises from tissue surrounding the embryo sac. These "adventitious" embryos (diploid) occur in *Citrus, Rubus,* and other plants.

Apomixis has variations other than those given above, but all forms of this phenomenon involve the origin of new individuals without nuclear or cellular fusion.

In some plants, deposition of pollen on the stigma is a prerequisite to apomictic embryo development, even though a pollen tube does not grow down the style and nuclear fusion does not take place. In such instances, there is evidence that hormones formed in the stigma, or furnished by the pollen, are transferred to the unfertilized egg cell, initiating there the changes that result in embryo development.

SUMMARY

1. The flower, the distinguishing structure of the Anthophyta, is formed of four whorls of parts that are specialized to carry out sexual reproduction, including pollination, fertilization, and seed production.

2. A floral apex has lost its ability to elongate and its potentiality for vegetative growth. The ability for vegetative growth is not regained until the completion of the two nuclear phenomena, meiosis and fertilization.

3. The four whorls of floral leaves are (1) the calyx, composed of sepals; (2) the corolla, composed of petals; (3) the androecium, composed of stamens; and (4) the gynoecium, composed of carpels.

4. There are two parts to a stamen, the pollen-producing anther and the filament.

5. A single unit of the gynoecium is frequently called a pistil. If it consists of one carpel, it is a simple pistil; if it consists of two or more fused carpels, it is a compound pistil.

6. A pistil consists of three parts: the stigma, the style, and the ovary. The stigma is receptive to pollen, and the ovary encloses the ovules. At the time of fertilization, the ovules consist of integuments, nucellus, and embryo sac.

7. Meiosis takes place both in the anthers and in the ovule.

8. Pollen can be considered a two-celled haploid plant protected by a cell wall, which is frequently elaborately sculptured. Two sperm are produced by the generative cell.

9. The embryo sac can be considered a seven-celled haploid plant. Two of its cells, the egg cell and the primary endosperm cell, are stimulated to develop by fusion with the sperm cells.

10. Pollination is the transfer of pollen from anthers to stigmas. A pollen tube growing down the style to the embryo sac serves as a pathway by which the two sperm reach the embryo sac.

11. Floral architecture facilitates pollination, which is most frequently carried out by wind or insect vectors.

12. Double fertilization, the union of one sperm with the egg cell and of a second sperm with the primary endosperm cell, occurs only in the Anthophyta.

13. The embryo plant develops from the zygote. The endosperm, developing from the fusion of sperm and primary endosperm nucleus, supplies some nourishment for embryo development and sometimes for seed development.

14. Variation in floral architecture arises (a) from variation in number of parts in a given whorl; (b) from symmetry in floral parts; (c) by connation of floral parts; (d) by adnation of floral parts; (e) by elevation of floral parts; and (f) by the presence or absence of certain whorls.

15. Flowers are either solitary on flower stalks or grouped in various inflorescences such as heads, spikes, catkins, umbels, panicles, racemes, or cymes.

16. There are essentially two types of pollination: cross-pollination, which results in much variation in the progeny, and self-pollination, which results in progeny that are quite similar.

17. Apomixis is a type of reproduction in which an unfertilized egg develops into an embryo without sexual fusion.

13 SEEDS AND FRUITS

PART ONE

EMBRYO AND SEED DEVELOPMENT

The seed completes the process of reproduction initiated in the flower. The embryo develops from the zygote, and the seed coat develops from the integuments of the ovule. As discussed in Chapter 12, at the time of fertilization the embryo sac in cotton consists of the egg cell, two synergids, and two polar nuclei (Fig. 12.19H), all with a 1n chromosome complement. In other plants the polar nuclei fuse to form the secondary nucleus, and in addition, three antipodal cells are present at the end of the embryo sac opposite the egg cell. The embryo sac is surrounded by nucellus, and all of these cells and tissues are enveloped by one or two integuments. This complete structure is the ovule (Figs. 12.19B and 13.1A). Double fertilization involves (a) fusion of egg (1n) and sperm (1n) nuclei to form a (2n) zygote, and (b) fusion of two polar nuclei with a second sperm to form the (3n) primary endosperm nucleus (Fig. 13.1B).

After fertilization, the zygote nucleus remains quiescent for a short time. The primary endosperm nucleus divides rapidly and soon forms the nutrient-rich storage tissue called **endosperm,** which may at first lack cell walls. In cotton, the first cellular stage of the endosperm consists of a single layer of cells on the periphery of the embryo sac (Fig. 13.1D). The zygote nucleus divides only after the endosperm has developed. The first divisions of the new generation result in a filament of two to four cells (Fig. 13.1C). The cell closest to the micropyle elongates and divides, becoming the **suspensor.** The suspensor is the structure that extends or supports the embryo in the endosperm (Fig. 13.1D). At about the same time, the cell furthest from the micropyle divides (Fig. 13.1D and 13.1E). This is the first in a series of divisions that will finally result in the embryo of the seed. Further divisions result in a globular stage and, finally, in a heart-shaped stage after the two cotyledons have developed (Fig. 13.1F). The major distinction between embryos of dicotyledonous and monocotyledonous plants is the number of cotyledons, two or one, respectively. In addition, the embryo always consists of an axis with a root apex at one end and a shoot apex at the other (Fig. 13.1G). The course of events just described is for cotton; there are many variations in the details of embryo development.

While development of the embryo is taking place, the nucellus, endosperm, and integuments are also undergoing changes that are characteristic of the group of plants to which the seeds belong. In the great majority of plants, nucellus and endosperm are required only for the initial stages of development. This is particularly true of the nucellus, which is generally used up as a nutritive source in early stages. It persists as a food storage tissue, the **perisperm,** in seeds of sugar beet *(Beta)* and a few other species. The endosperm persists as a food reserve in seeds of many monocot plants (Figs. 13.2 and 13.3). It is highly developed in grasses, some of which are of major economic importance, such as rice and corn. The endosperm persists as a food storage tissue in relatively few dicotyledonous seeds; the castor bean *(Ricinus communis)* is a good example.

In most seeds, integuments of the ovule become hard and horny seed coats in the mature seed. The scanning electron microscope shows the seed coats to be variously and sometimes beautifully sculptured (Fig. 13.4).

KINDS OF SEEDS

Seeds of angiosperms differ in two ways: (1) they have one or two cotyledons, and (2) they store food either in the embryo, in the endosperm, or more rarely, in nucellar tissue—the perisperm, which lies between embryo and seed coat.

When food storage occurs within the embryo, the normal vascular tissues of the embryo convey the solubilized food to meristematic regions, where it is required for growth. Food stored in the endosperm, outside the embryo, is in many cases absorbed through normal protodermal or epidermal cells. In other cases, such as onion, the cotyledons become specialized to act as absorbing organs (Fig. 13.10). In most grasses, the entire cotyledon has become a highly specialized absorbing organ known as the scutellum (Fig. 13.2).

COMMON BEAN

The fruit of the bean plant is a pod; within it are seeds. The external characters of the bean seed are more clearly seen after soaking it in water. Points of interest are the **hilum, micropyle,** and **raphe** (Fig. 13.5). The hilum is a large oval scar that is located near the middle of one edge, where the seed broke from the stalk, or **funiculus** (Fig. 12.19H), when the beans were harvested. The micropyle is a small opening in the seed coat (integument) at one side of the hilum, which was observed in the ovule as the opening through which the pollen tube entered. The raphe is a ridge at the side of the hilum opposite the micropyle. It represents the base of the funiculus, which is fused with the integuments. Conducting tissue present in the funiculus will be continued in

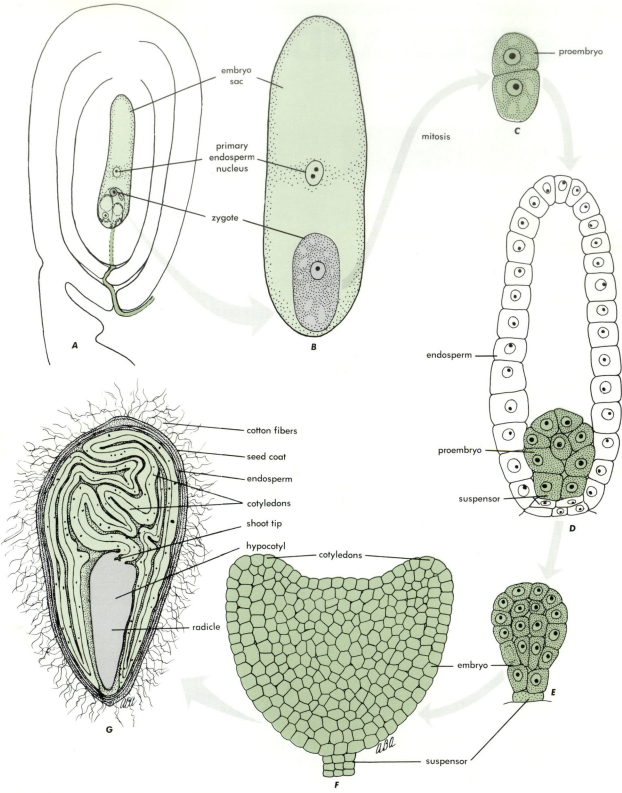

Figure 13.1. Embryo and seed development in cotton (*Gossypium hirsutum*). *A*, ovule after double fertilization containing a 2*n* zygote and a 3*n* primary endosperm nucleus. *B*, enlarged view of fertilized embryo sac. *C*, the zygote divides by mitosis to form a proembryo cell and a basal cell destined to become the suspensor. *D*, early stage of embryo (proembryo) and endosperm development. The early endosperm shown is a single layer of cells. *E*, the early embryo consists of a small globe of cells. *F*, heart-shaped stage of the embryo with newly formed cotyledons. *G*, mature embryo, with highly folded cotyledons, surrounded by seed coat to form a cotton seed. Note the seed coat fibers (really epidermal hairs) for which cotton is harvested. *A*, *C*, *D*, and *E*, redrawn after U. R. Gore, Amer J. Bot. 19 (1932): 795–807.

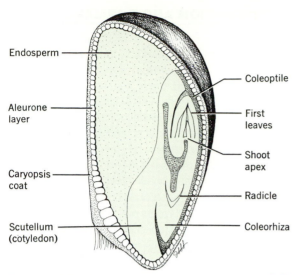

Endosperm

Aleurone layer

Caryopsis coat

Scutellum (cotyledon)

Coleoptile

First leaves

Shoot apex

Radicle

Coleorhiza

Figure 13.2. Median longitudinal section of caryopsis of yellow foxtail (*Setaria lutescens*).

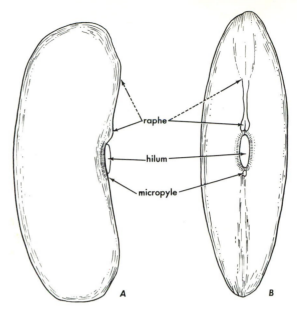

raphe

hilum

micropyle

A

B

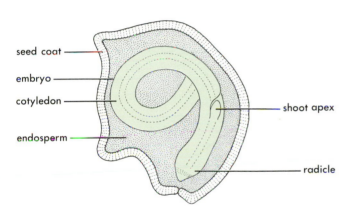

seed coat

embryo

cotyledon

endosperm

shoot apex

radicle

Figure 13.3. A longitudinal section through an onion seed (*Allium cepa*) showing the embryo coiled within the endosperm.

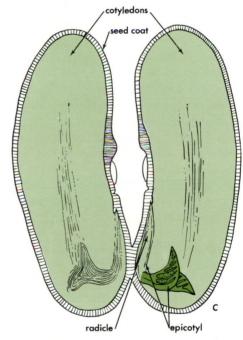

cotyledons

seed coat

radicle

epicotyl

C

Figure 13.5. Bean (*Paseolus vulgaris*) seed. A, external side view; B, external face or edge view; C, embryo opened.

Figure 13.4. The speed coats of many seeds are characteristically sculptured. A, morning glory (**Convolvulus arvensis**); B, California poppy (*Eschscholtzia californica*).

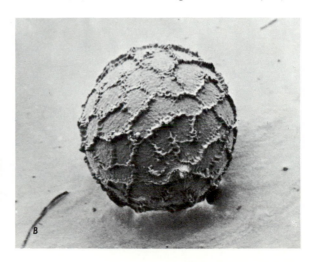

the raphe. An elevation or bulge of the seed coat adjacent to the micropyle marks the position of the radicle (embryo root) within the seed.

When the seed coat of a soaked bean is removed, the entire structure remaining is embryo; no endosperm is present. The following parts of the embryo can be observed: (a) a shoot consisting of two fleshy cotyledons, a short axis (the hypocotyl) below the cotyledons, and a short axis (the epicotyl) above the cotyledons, bearing several minute foliage leaves and terminating in a shoot tip; and (b) the root, or radicle (Figs. 13.5 and 13.8).

GRASSES

The so-called "seed" of corn, or of any other grass, is in reality a fruit—the caryopsis, or grain. It is a one-seeded, dry, indehiscent fruit with the pericarp (ovary wall) firmly attached to the seed. Pericarp and seed coats are so firmly attached to each other and to other tissues of the grain that it is impossible to peel them away.

A longitudinal section of a caryopsis of yellow foxtail (*Setaria lutescens*) is shown in Fig. 13.2. The endosperm constitutes the bulk of the grain and is composed of (a) an outermost layer (single row of cells) known as the **aleurone layer** and (b) a starchy endosperm. Cells of the aleurone layer contain proteins and fats but little or no starch.

Like all embryos of other seed plants, the grass embryo has an axis with a **shoot apex** and a **root apex** (Fig. 13.2). The shoot apex, together with several rudimentary leaves, are surrounded by a sheath—the **coleoptile.** The rudimentary root (radicle) is also surrounded by a sheath—the **coleorhiza.** At the juncture of shoot and root is a very short stemlike region. A relatively large part of the grass embryo is a single cotyledon called the **scutellum.** It is a shield-shaped structure that lies in contact with the endosperm. The outer cells of the scutellum secrete enzymes that digest the stored foods of the endosperm. These digested foods move from endosperm cells through the scutellum to the growing parts of the embryo. Unlike the bean, the cotyledon of grasses remains within the seed during germination and never develops into a green leaflike structure above ground.

The process of milling to make polished white rice removes the outside caryopsis coat, plus the entire protein-rich aleurone layer, and the outer layers of the endosperm. This means that most of the nutritious protein is removed before the rice is eaten. The removed material, composed of the outside layers and the embryo axis, is called bran.

ONION

The onion seed consists of a seed coat enclosing a small amount of endosperm. The embryo is a simple axis, the radicle and single cotyledon being quite prominent. The shoot apex is located at about the midpoint of the axis and appears as a simple notch. The embryo is coiled within the seed coat, with the radicle usually pointing toward the micropyle (Fig. 13.3).

SEED GERMINATION AND DORMANCY

GERMINATION

Seeds can remain viable for remarkably long periods. In one study, begun in 1878, Dr. W. J. Beal of Michigan State University buried jars containing seeds from several plant species. At 5-year and 10-year intervals a jar was opened and the seeds were tested for germination. Most species remained viable for at least 10 years, and one species, the moth mullein (*Verbascum blattaria*), still germinated after more than 90 years. However, this is not the record for seed longevity. Seeds from the Oriental lotus (*Nelumbo nucifera*) that were removed, still viable, from archeological diggings were thought to be more than 1000 years old. Recent radiocarbon dating of these seeds has established their age to be 400 years old, not 1000. Ancient seeds collected from Pharaoh's tomb reportedly have germinated; their age, however, has not been verified. Nevertheless, it is truly remarkable that seeds can continue to live over such long periods.

What induces germination? Many seeds germinate when provided with moisture, oxygen, and a favorable temperature. The water content of seeds is low, between 5% and 10%. The cytoplasm with contained organelles is scarcely recognizable because it is crowded between the large amount of reserved food materials (Fig. 13.6). At first, water is imbibed very rapidly, which results in the swelling of the pro-

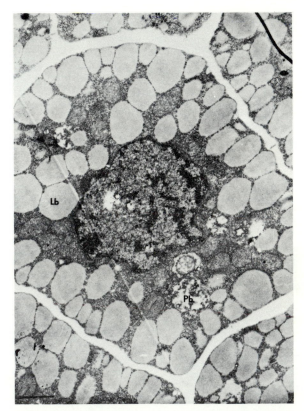

Figure 13.6. Scutellum cells in yellow foxtail (*Setaria lutescens*) in dry caryopsis. ×10,300.

toplasm with a reappearance of organelles (Fig. 13.7). Increase in water content makes possible an increased metabolic activity.

==The first indication that the processes of germination have begun is generally the swelling of the radicle.== In all cases, the radicle imbibes water rapidly and, bursting the seed coat and other coverings that may be present, starts to grow downward into the soil. This helps to assure that the young seedling has an adequate supply of water and nutrients when the shoot breaks through the surface of the soil. Although the succeeding steps of germination are essentially similar, there are variations. For instance, in germination of beans, peas, and onions, a structure with a sharp bend, or **hook,** is first forced through the soil (Figs. 13.8, 13.9, and 13.10), but the structure forming the hook is different in each case. Once above the ground, the hook straightens. In the case of bean, a straightening of the hypocotyl raises the cotyledons and shoot apex above ground (Fig. 13.8); a lengthening and straightening of the epicotyl pulls the shoot apex away from the cotyledons. The cotyledons, now exposed to sunlight, will conduct photosynthesis for a short period. Upon straightening of the pea epicotyl, however, the cotyledons remain in the ground, and only the apex and first leaves are raised upward (Fig. 13.9).

In onion, the primary root first penetrates the soil. Then a sharply bent cotyledon breaks the soil surface and slowly straightens out. The cotyledon of the onion is tubular, and

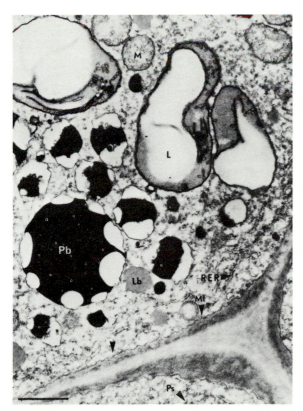

Figure 13.7. Radicle cells in yellow foxtail (*Setaria lutescens*) after 65 hours of germination. ×13,300.

its base encloses the shoot apex (Fig. 13.10). There is a small opening at the base of the cotyledon through which the first leaf finally emerges. In grasses, the situation is more complex. The shoot and root apices are enveloped by tubular sheaths known as the coleoptile and coleorhiza, respectively (Figs. 13.2 and 13.11). The primary root rapidly pushes through the coleorhiza. The root continues to grow but is eventually replaced by adventitious roots that arise from the lower nodes of the stem (Fig. 13.11). The coleoptile elongates and emerges above ground, becoming 2 to 4 cm long. At this time, the uppermost leaf pushes its way through the coleoptile and, growing rapidly, becomes part of the photosynthesizing shoot.

DORMANCY

==Many seeds will not germinate even when supplied with water, oxygen, and a favorable temperature; they are said to be== **dormant.** Various factors in different plants are associated with the breaking of dormancy. For instance, light is necessary for the germination of some grass seed and for some strains of lettuce *(Lactuca)* and perhaps other plants. Water and gases must pass through the seed coat. If the seed coat restricts the movement of water and gases or is impermeable to one or the other or to both, it must decay, be broken, or be scratched, allowing water and oxygen to reach the embryo before germination can start. In some seeds, the embryo itself fails to germinate when provided with favorable conditions. In orchids, the seeds are disseminated while embryos are rudimentary, and the embryos must develop further before germination.

Many kinds of seeds will not germinate unless the seeds, while on a moist substrate, are subjected for a time to temperatures close to freezing. At the other extreme, some seeds will not germinate unless they have been subjected to the high heat of a fire. Another type of dormancy is produced by the presence of natural chemical inhibitors. These occur in many fruits and serve to keep the seeds dormant while they are enclosed by the fruit. In other cases, inhibitors are present in the seeds themselves or may be produced by decaying leaves or forest litter. The influence of the plant growth substances on dormancy and seed germination is still unclear.

SUMMARY

1. Seeds store food within or outside the embryo. In most dicotyledons, such as cucumber and bean, food is stored in the two cotyledons. Cotyledons may also serve as absorbing and, later, as photosynthesizing organs. Food may be stored in an endosperm rather than in the cotyledons. In monocotyledonous seeds, food is stored in an endosperm.

2. In grasses, such as corn, a single cotyledon (scutellum) is a highly specialized absorption organ that does not emerge from the seed. In seeds like those of lilies and

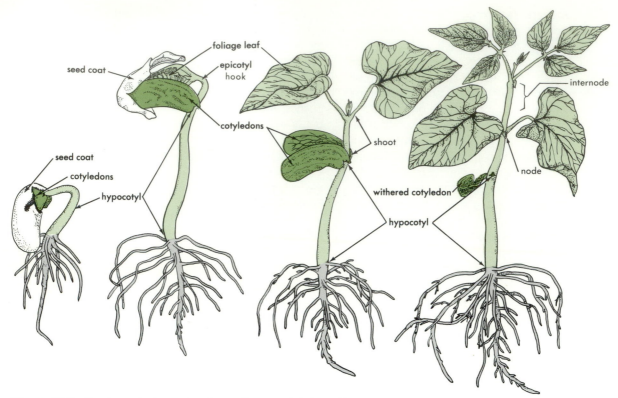

Figure 13.8. Seed germination and early seedling growth in bean (*Phaseolus vulgaris*).

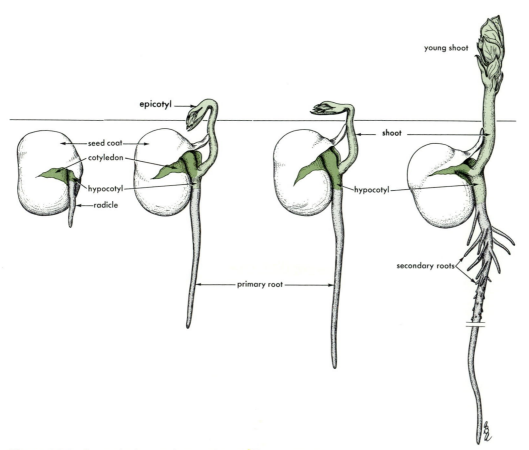

Figure 13.9. Stages in the germination of a pea (*Pisum sativum*) seed.

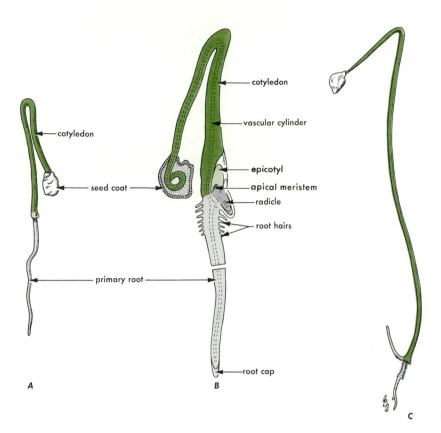

cotyledon

vascular cylinder

cotyledon

epicotyl

apical meristem

seed coat

radicle

root hairs

primary root

root cap

A

B

C

Figure 13.10. Germination of an onion (*Allium cepa*) seed.

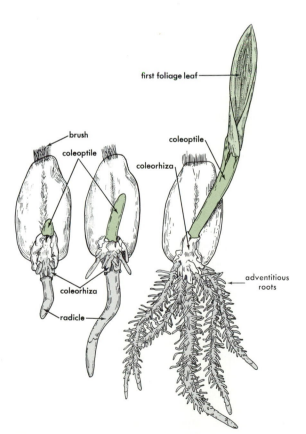

first foliage leaf

brush

coleoptile

coleoptile

coleorhiza

coleorhiza

adventitious roots

radicle

Figure 13.11. Germination wheat (*Triticum* sp.) caryopsis.

onion, the cotyledon emerges from the seed coat and becomes green, but its tip continues to absorb food from the endosperm.

3. Mature seeds are dormant, and depending on the species and the immediate environment of the seeds, they may remain viable and dormant from a few months to many years.

4. In germination, the cotyledons may be elevated above the ground, sometimes to become photosynthetically active, or in the case of some monocotyledons, to continue absorption of the endosperm. In other cases, the cotyledons remain below the ground.

5. Dormancy is usually broken by the provision of moisture, oxygen, and a favorable temperature. Other factors, such as light, the removal of chemical inhibitors, or the destruction of the seed coat, may be required in some instances.

PART TWO

THE FRUIT

A **fruit** is a ripened ovary of a flower; it may include other floral parts. In many cases, fruit development depends on pollination and the activity of indoleacetic acid, or other growth substances.

Fruits are auxiliary structures in the sexual life cycle; they occur only in the Anthophyta. Their development has unquestionably been an important factor in the successful evolution of land plants. Fruits protect seeds, aid in their dissemination, and may be a factor in timing their germination. Fruits are highly constant in structure and are thus very important plant parts in the classification of the Anthophyta. In everyday usage, the term *fruit* usually refers to a juicy and edible structure, such as an apple, cherry, or grape. Such structures as string beans, eggplant *(Solanum melongena),* and cucumbers, which are commonly called "vegetables," and "grains" of corn, wheat *(Triticum),* and other cereals are not popularly thought of as fruits. However, all the above are fruits in a botanical sense.

DEVELOPMENT OF THE FRUIT

Development of the fruit (pod) of pea is described as an example of the changes that may occur during transformation of the ovary into a fruit. Pistils of pea flowers are each composed of one carpel (Fig. 12.17); that is, one ovule-bearing leaf. Recall that the ovary of pea is formed by fusion of margins of the carpel and that ovules are attached to these margins (Fig. 12.16A). This cross section of the pea ovary shows the ovary wall, an ovule, and the locule. The ovary wall may be divided into three distinct layers: (1) an outer epidermis, (2) an inner epidermis, and (3) a middle zone consisting of several layers of cells. There are three carpellary bundles, one for each margin and one for the midrib opposite them.

Fertilization initiates a series of changes in the embryo sac and other tissues of the ovule that lead to the development of the seed. Tissues of the ovary wall undergo marked changes, developing into three layers. The fruit wall (developed from the ovary wall) is called the **pericarp;** the three more or less distinct parts, in order beginning with the outermost, are **exocarp, mesocarp,** and **endocarp** (Fig. 13.14, shown in almond). When the pod is mature, floral structures such as pedicel, calyx, withered stamens, style, and stigma, and even remnants of the corolla may also be present (Fig. 12.17C).

Mature pods usually show the presence of small, underdeveloped (abortive) ovules. It is probable that these ovules were not fertilized (Fig. 12.17D). In most plants, normal fruit development takes place only if pollination is followed by fertilization and if fertilization is followed by seed development. Some fruits, such as banana and navel oranges, develop without fertilization. These fruits, called **parthenocarpic fruits,** are seedless and are frequently different in shape and taste from normal seeded fruits.

KINDS OF FRUITS

In classifying the different kinds of fruits, the following criteria are taken into account: (a) the structure of the flower from which the fruit develops, (b) the number of ovaries involved in fruit formation, (c) the number of carpels in each ovary,

(d) the nature of the mature pericarp (dry or fleshy), (e) whether or not the pericarp splits **(dehisces)** at maturity, (f) if dehiscent, the manner of splitting, and (g) the possible role that sepals or receptacle may play in formation of the mature fruit.

Simple fruits are derived from a single ovary. They may be dry or fleshy; the ovary may be composed of one carpel or of two or more carpels; and the mature fruit may be dehiscent or indehiscent (Fig. 13.12). On the other hand, **aggregate fruits** and **multiple fruits** are formed by clusters of simple fruits. The difference between these types of fruits depends on the number of flowers involved in their formation. In strawberry *(Fragaria;* Fig. 13.17A) and blackberry *(Rubus;* Fig. 13.17B), the aggregate fruit is derived from the many ovaries of a single flower, all attached to a common receptacle. Mulberry *(Morus),* fig *(Ficus),* and pineapple *(Ananas comosus)* (Figs. 13.18 and 13.19) are multiple fruits consisting of the enlarged ovaries of several flowers, more or less grown together into one mass.

SIMPLE FRUITS
Pericarp Dry and Dehiscent

Legume or Pod.   This type of fruit, characteristic of nearly all members of the pea family, Fabaceae (Leguminosae), arises from a single carpel, which at maturity generally dehisces along both sutures (Fig. 12.17). Pods may be spirally twisted or curved, as in alfalfa.

Follicle.   An example of a follicle is the fruit of magnolia (Figs. 13.12A and 20.5). The follicle develops from a single carpel and opens along one suture, thus differing from the pod, which opens along both sutures.

Capsule.   Capsules are derived from compound ovaries, that is, an ovary composed of two or more united carpels. Each carpel produces a few to many seeds. Capsules dehisce in various ways (Fig. 13.12B, 13.12C, and 13.12D).

Silique.   This is the type of fruit (Fig. 13.12E) characteristic of members of the mustard family, Brassicaceae (Cruciferae). The silique is a dry fruit derived from a superior ovary consisting of two carpels. At maturity, the dry pericarp separates into three portions; the seeds are attached to the central, persistent portion.

Pericarp Dry and Indehiscent

Achene.   These fruits are commonly called "seeds," but as in sunflower (Fig. 13.12F), a carefully broken pericarp reveals that the seed within is attached to the placenta by its stalk. This pericarp can be separated easily from the seed coat, that is, from the layer of cells just beneath it. Achenes are indehiscent.

Grain or Caryopsis.   This is the fruit of the grass family, Poaceae (Gramineae), which includes such important plants as wheat, corn, and rice *(Oryza).* Like the achene, the grain

folicle

B

C

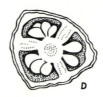

D

attachment

pericarp

seed

F

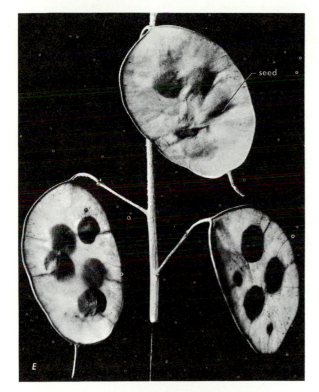

seed

E

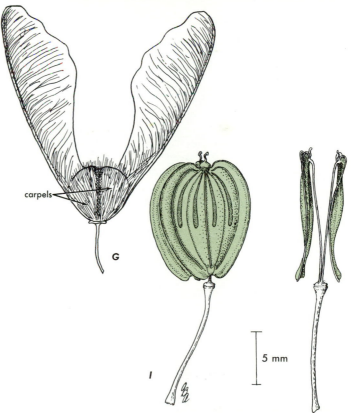

carpels

G

5 mm

I

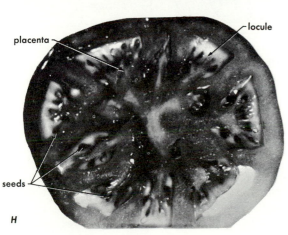

placenta

locule

seeds

H

Figure 13.12. Simple fruits. *A, Magnolia* sp., follicle; *B, Datura* sp. capsule; *C, D, Tulipa* sp. capsule; *E, Lunaria* sp. silique; *F, Helianthus annuus*, achene; *G, Acer* sp., samara. *H, Lycopersicon esculentum*, berry; *I, Heracleum hirsutum*, schizocarp.

is a dry, one-seeded, indehiscent fruit (Fig. 13.2). However, it differs from the achene in that the pericarp and seed coat are firmly united all the way around.

Samara. This is a dry, indehiscent fruit which may be one-seeded, as in the elm (*Ulmus*), or two-seeded, as in maple (*Acer;* Fig. 13.12G). These fruits are typified by an outgrowth of the ovary wall, which forms a winglike structure.

Schizocarp. This is the fruit characteristic of the carrot family, Apiaceae (Umbelliferae), which includes such common plants as carrot (*Daucus*), celery (*Apium*), and parsley (*Petroselinum*). The schizocarp is a dry fruit consisting of two carpels that split, when mature, along the midline into two one-seeded indehiscent halves (Fig. 13.12*I*).

Nut. The term *nut* is popularly applied to a number of hard-shelled fruits and seeds. A typical nut, botanically speaking, is a one-seeded, indehiscent dry fruit with a hard or stony pericarp (the shell). Examples are the chestnut (*Castanea*), walnut, and acorn (Fig. 13.13). An acorn, the fruit of the oak, is partially enclosed by a hardened involucral cup. The husk, or shuck, of the walnut, which has been removed before the product reaches the market, is composed of involucral bracts, perianth, and the outer layer of the pericarp. The hard shell is the remainder of the pericarp.

Pericarp Fleshy

Drupe. Examples of this type of fleshy fruit are cherry, almond, peach, and apricot, all of which are members of a genus in the rose family (Rosaceae). The olive (*Olea*) fruit is also a drupe. The drupe is derived from a single carpel and is usually one-seeded. However, if one examines young ovaries of flowers of almonds, cherries, plums, or other members of the group to which they belong, two ovules will be found, one of which usually aborts (fails to develop into a seed; Fig. 13.14). In olives, plums, and peaches, the drupe has a hard endocarp consisting of thick-walled stone cells. The exocarp is thin, forming the skin; and the mesocarp forms the edible flesh.

Berry. This fleshy type of fruit is derived from a compound ovary. Usually, many seeds are embedded in flesh that is both endocarp and mesocarp, although the line of demarcation may be difficult to see (Fig. 13.12H). The tomato is a common example of a berry.

The citrus fruit (lemon, orange, lime, and grapefruit) is a type of berry called a **hesperidium.** It has a thick, leathery rind (peel), with numerous oil glands, and a thick juicy portion composed of several wedge-shaped locules (Fig. 13.15). The peel of a citrus fruit is exocarp and mesocarp; the pulp segments are endocarp. The juice is in pulp sacs or vesi-

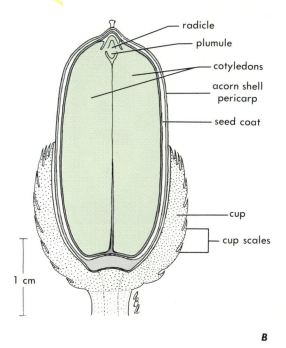

radicle
plumule
cotyledons
acorn shell pericarp
seed coat
cup
cup scales

1 cm

B

Figure 13.13. Acorns of oak (*Quercus* sp.). *A*, the fruits are a type of nut and the cups are fused involucral bracts. *B*, internal view of acorn of *Quercus Kellogii*.

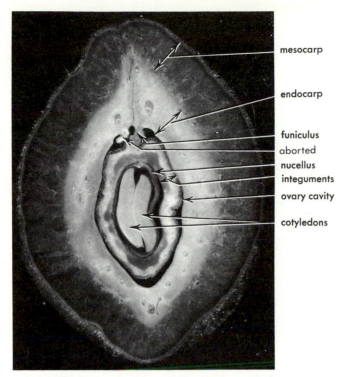

mesocarp

endocarp

funiculus
aborted
nucellus
integuments
ovary cavity
cotyledons

Figure 13.14. Drupe of almond (*Prunus amygdalus*).

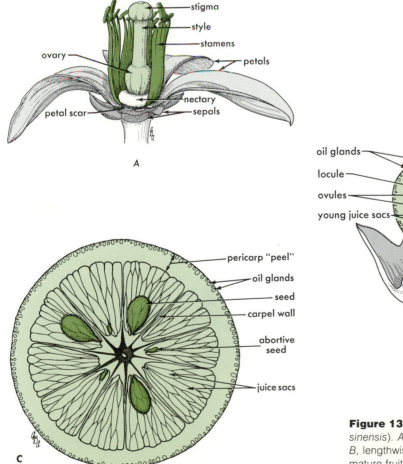

stigma
style
stamens
ovary
petals
nectary
petal scar
sepals

A

pericarp "peel"
oil glands
seed
carpel wall
abortive seed
juice sacs

C

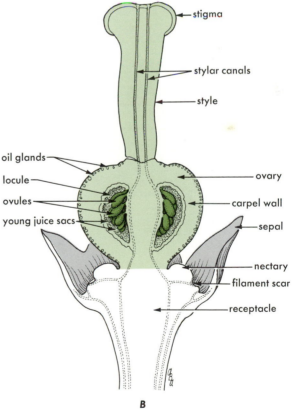

stigma
stylar canals
style
oil glands
locule
ovules
young juice sacs
ovary
carpel wall
sepal
nectary
filament scar
receptacle

B

Figure 13.15. Flower and fruit (hesperidium) of orange (*Citrus sinensis*). A, flower, dissected to show pistil and stamens, B, lengthwise section of maturing ovary; C, cross section of mature fruit.

cles, which are outgrowths of the endocarp walls (Fig. 13.15C); each mature vesicle is composed of many living cells filled with juice.

Another berrylike fruit is that of members of the family Cucurbitaceae, which includes watermelon, squash, pumpkin, and cucumber. This fruit is called a **pepo** (Fig. 20.14). The outer wall (rind) of the fruit consists of receptacle tissue that surrounds and is fused with the exocarp. The flesh of the fruit is principally mesocarp and endocarp.

Pome. This type of fruit is characteristic of a subfamily of Rosaceae, which includes apple (Fig. 13.16) and pear. The pome is derived from an inferior ovary (Fig. 13.16A). The flesh is enlarged hypanthium, and the core develops from the ovary (Fig. 13.16B).

COMPOUND FRUITS—
FORMED FROM SEVERAL OVARIES

Aggregate Fruits

An aggregate fruit is one formed from numerous carpels of one individual flower. These fruits, considered individually, can be classified as types of simple fruits. The strawberry flower has numerous separate carpels on a single receptacle. The ovary of each carpel has one ovule, and the ovary develops into a one-seeded dry fruit (achene). The receptacle to which these fruits are attached becomes fleshy; the whole structure, which we call a strawberry, is an aggregate of simple fruits, each an achene. The receptacle is stem tissue and consists of a fleshy pith and cortex with vascular bundles between them. The hull of the strawberry fruit is composed of persistent calyx and withered stamens. The achenes are usually spoken of as seeds.

Flowers of raspberry and blackberry and other species of *Rubus* have essentially the same structure as those of strawberry. In these flowers, many separate carpels attached to one receptacle develop into small drupes (Fig. 13.17A and 13.17B).

Multiple Fruits

A multiple fruit is formed from individual ovaries of several flowers. These fruits, considered individually, can be classified as types of simple fruits. The fig and pineapple are examples of multiple fruits; the individual fruits composing them are nutlets in fig and parthenocarpic berries in pineapple (Fig. 13.18).

The part of the fig that we eat is an enlarged fleshy receptacle (Fig. 13.19). Its flowers are very small and are attached to the inner wall of the receptacle. Both staminate and pistillate flowers occur and may be borne in the same or in different receptacles. Each ovary develops into a nutlet that is embedded in the wall of the receptacle. Thus, the fig is derived from many flowers, all attached to the same receptacle.

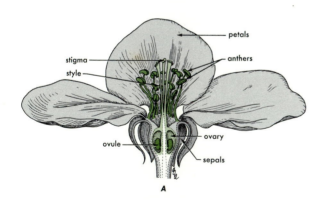

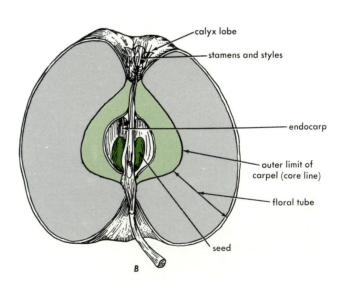

Figure 13.16. Flower and fruit (pome) of apple (*Malus sylvestris*). *A,* median longitudinal section of flower, showing inferior ovary and fused (adnate) petals and sepals; *B,* median section of mature fruit.

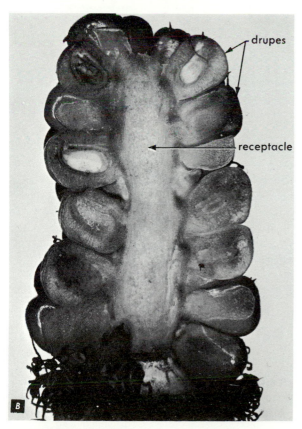

Figure 13.17. Aggregate fruits. *A*, strawberry (*Fragaria* sp.); *B*, blackberry (*Rubus* sp.).

Accessory Parts of Fruits

In the pineapple (Fig. 13.18), the edible portion is a thickened pulpy central stem in which berries are embedded. Tissues other than the ovary wall that form part of a fruit are referred to as **accessory.** Thus, much of the fleshy fruit of pineapple, apple, and strawberry is accessory.

KEY TO FRUITS

1. Fruit formed from a single ovary of one flower. *Simple fruits.*
   - A. Pericarp fleshy.
     1. The ovary wall fleshy and containing one or more carpels and seeds. *Berry.*
        - a. Ovary wall a hard rind. *Pepo.*
        - b. Ovary wall with a leathery rind. *Hesperidium.*
     2. Only a portion of the pericarp fleshy.
        - a. Exocarp thin; mesocarp fleshy; endocarp stony; single seed and carpel. *Drupe.*
        - b. Outer portion of pericarp fleshy, inner portion papery, floral tube fleshy; several seeds and carpels. *Pome.*
   - B. Pericarp dry.
     1. Dehiscent fruits.
        - a. Composed of one carpel.
          1. Splitting along two sutures. *Legume.*
          2. Splitting along one suture. *Follicle.*
        - b. Composed of two or more carpels.
          1. Two or more carpels dehiscing in one of four different ways. *Capsule.*
          2. Two carpels; separating at maturity, leaving a persistent partition wall. *Silique.*
     2. Indehiscent fruits.
        - a. Pericarp bearing a winglike growth. *Samara.*
        - b. Pericarp not bearing a winglike growth.
          1. Carpels, two to many, united when immature, splitting apart at maturity. *Schizocarp.*
          2. Carpels one, if more, not splitting apart at maturity; one-seeded fruits.
             - a. Seed united to the pericarp all around. *Caryopsis* or *grain.*
             - b. Seed not united to the pericarp all around.
               Fruit large, with thick, stony wall. *Nut.*
               Fruit small, with thin wall. *Achene.*

II. Fruits formed from several ovaries.
   - A. Fruits developing from one flower.
     Aggregate fruits (classify the individual fruits in key for simple fruits).
   - B. Fruits developing from several flowers.
     Multiple fruits (classify individual fruits in key for simple fruits).

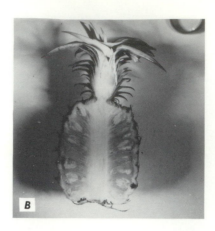

Figure 13.18. Pineapple (*Ananas comosus*), a multiple fruit. The flower is inferior, and the resulting fruits are berries, sterile in the cultivated species, which become embedded in the swollen edible rachis. A jagged-edged bract extends out over the fruit. The fruits are spirally arranged. *A*, exterior view, *B*, section through center showing central rachis surrounded by coalesced fruits.

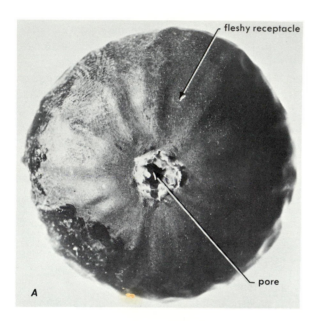

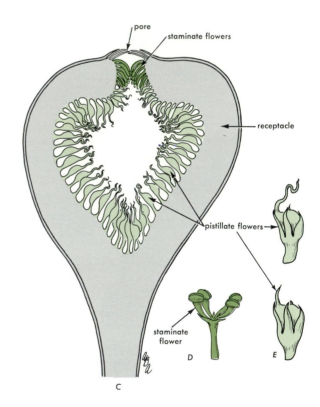

Figure 13.19. The fruit and flowers of fig (*Ficus carica*). *A*, end of fleshy receptacle, showing pore, which is lined with staminate flowers; *B*, receptacle turned inside out, showing pistillate flowers that have matured into small drupes. *C*, diagram of a fig, showing location of staminate and pistillate flowers. *D*, staminate flower; *E*, top: fertile long-styled pistillate flower of the edible Calimyrna fig; *E*, bottom: short-styled abortive pistillate flower.

DISSEMINATION OF SEEDS AND FRUITS

AGENTS IN SEED AND FRUIT DISPERSAL

The chief agents in seed and fruit dispersal are **wind, water,** and **animals.**

Wind

The structural modifications of seeds and fruits that aid in dissemination by wind are of several types. The most common of these are winged types (Figs. 13.12G and 13.20C) and plumed types (Fig. 13.20B).

Water

Fruits with a membranous envelope containing air, like those of sedges, or with a coarse, loose, fibrous outer coat, as in the coconut (Cocos), are well adapted for dispersal by water. A great variety of fruits and seeds float in water, even though they lack special adaptations to ensure buoyancy, and are readily transported long distances by moving water. In irrigated districts, irrigation water is a very important means by which weed seeds are distributed.

Animals

Many seeds and fruits are carried by animals, both wild and domesticated. Seeds with beards, spines, hooks, or barbs adhere to the hair of animals (Fig. 13.20E). An example is bur clover (Medicago denticulata).

Seeds of many plants pass through the digestive tract of animals without having their viability impaired. Fleshy, edible fruits are often eaten by birds (Fig. 13.20F) and carried by them long distances, and then the seeds are regurgitated or discharged with the excrement. Squirrels carry nuts, such as those of walnut and hickory (Carya), and the seeds of pines. Seeds of some aquatic and marsh plants and of mistletoe (Phoradendron), which are covered with a sticky material, are carried on the feet of birds.

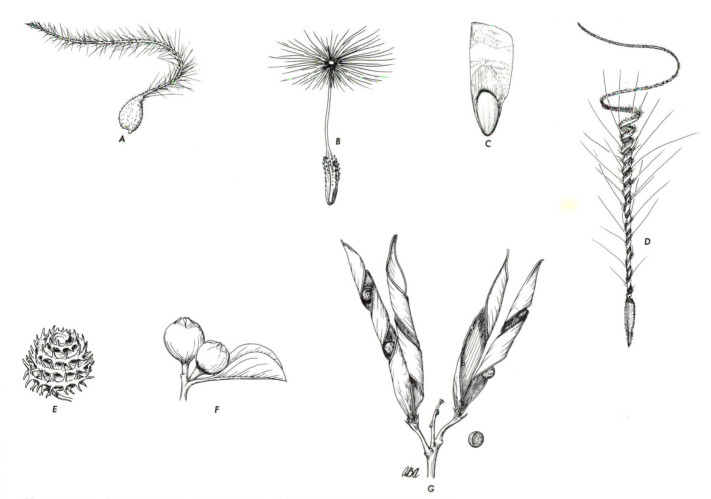

Figure 13.20. Fruits and seeds showing various devices that aid in dissemination. *Wind*: A, Clematis sp.; B, dandelion (Taraxacum vulgare). C, seed of Coulter's big-cone pine (Pinus coulteri). Attachment: D, cranesbill (Geranium sp.); E, bur clover (Medicago denticulata); F, fleshy edible fruit of Cotoneaster sp.; Violent dehiscence of pericarp; G, vetch (Vicia sativa).

SUMMARY OF FRUITS

1. A fruit is a ripened ovary, plus any other closely associated floral parts.

2. The initiation of fruit development generally depends upon pollination and fertilization.

3. Fruit development is accompanied by the development of the ovule into the seed.

4. There are *three different kinds of fruits*, classified on the basis of the number of ovaries and flowers involved in their formation:

   (a) *simple fruits*, derived from a single ovary

   (b) *aggregate fruits*, derived from a number of ovaries belonging to a single flower and on a single receptacle

   (c) *multiple fruits*, derived from a number of ovaries of several flowers more or less grown together into one mass

5. Simple fruits have a dry pericarp or a fleshy pericarp. If the pericarp is dry, it is dehiscent (splitting at maturity, allowing seeds to escape) or indehiscent (not splitting).

14 CLASSIFICATION AND EVOLUTION

The flowering plants, because they are so familiar and useful to humans, were the focus of the first 13 chapters. From this point on, we will include the many other life forms that botanists study.

The variety of life forms is tremendous, ranging from unicellular fungi and algae to mosses, ferns, and conifers. To explore them, we need a system of categories and names. This brings us to the science of classification, or **taxonomy,** which is perhaps the oldest branch of science (it dates back to the ancient Greeks, around 300 to 400 B.C.). In this chapter we will introduce a system of botanical names and will consider the problems that taxonomists encounter when they classify organisms. In so doing, we will also take up the study of evolution, which is important in defining groups of organisms.

THE SPECIES: UNIT OF LIFE IN TAXONOMY

In earlier chapters, we used the individual organism as the unit of life. But taxonomists are concerned with groups, in which the individual organism is just a short-lived member. Thus, in taxonomy a more useful unit of life is the **species.** All the organisms of a given kind make up a species; for example, all humans make up the species *Homo sapiens.*

The species concept is not always easy to apply. What do we mean when we say that two organisms are of the same kind? How much can they differ without being assigned to different species? Biologists often answer in terms of a **mating test,** which defines a unit called the **biological species:**

Two populations belong to the same biological species if their members interbreed (mate) and produce fertile offspring when they are brought together under natural conditions. If they coexist in the same area without interbreeding, they are of different biological species.

Unfortunately, the mating test has limitations. It cannot be applied to organisms that lack sexual reproduction, or to extinct organisms that are only known from fossil specimens. In these cases, species can only be distinguished by reference to physical or chemical traits. Furthermore, the mating test does not always give clear-cut results. Two populations may interbreed to a limited extent, producing offspring that are weakly fertile. How fertile must they be to pass the

mating test? The issue is debatable, and in such cases, taxonomists may disagree about the boundaries between species. Nevertheless, most species are distinct entities, and they make a sound starting point for the science of taxonomy.

EVOLUTION AND TAXONOMY

When taxonomists classify plants, they see evidence of an organizing principle at work in nature: species do not differ at random; they fall into groups. For instance, there are thousands of species that differ in detail and yet are clearly ferns; they are variations on a common theme. Likewise, there are thousands of flowering species, hundreds of conifer species, and so on. Through most of recorded history, the existence of groups was attributed to a divine plan of creation. The divinity was thought to have established an ideal pattern, something like a blueprint, for each species. Organisms assembled themselves to match the pattern as closely as possible, and differences between individuals were attributed to the imperfection of the physical world. With these ideas, the taxonomist's goal was to deduce the perfect pattern by examining the common features of individuals. Such thinking is said to be **typological** because it conceives of an ideal type of body for each species.

Typological thinking fell out of favor in the nineteenth century as scientists began to realize the immense age of earth and the great differences that exist between modern and ancient forms of life. The difference between dinosaurs and modern reptiles is a familiar example, but equally great changes have taken place in the botanical realm. For instance, consider the plants that are commonly called horsetails (Fig. 14.1A). You may have seen their characteristic jointed stems in marshy habitats. All the modern horsetails are herbs, usually a meter or less in height. But 300 million years ago, plants of the horsetail type grew to be tall trees (Fig. 14.1B); coal is partly made of their fossilized remains. Still farther back, 500 million years ago, there were no land plants at all. Plant life was limited to aquatic forms somewhat like simple modern seaweeds.

With such discoveries, by the 1850s many scientists were ready to believe that living species can **evolve,** or change their hereditary characteristics over the course of many generations. A convincing mechanism of evolution was proposed around the time of the American Civil War. On

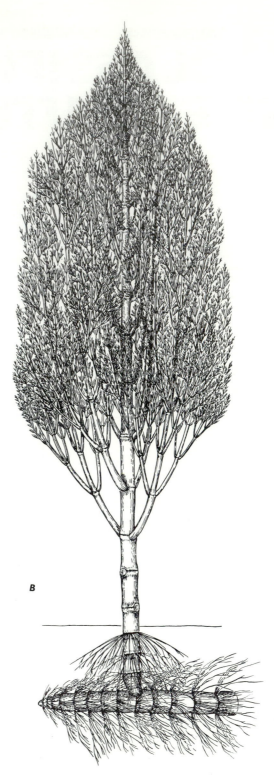

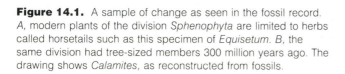

Figure 14.1. A sample of change as seen in the fossil record. *A,* modern plants of the division *Sphenophyta* are limited to herbs called horsetails such as this specimen of *Equisetum. B,* the same division had tree-sized members 300 million years ago. The drawing shows *Calamites,* as reconstructed from fossils.

opposite sides of the globe, two scientists independently worked out the same idea. Coincidentally, both were Englishmen. One was Alfred Wallace, who worked in the rain forests of Southeast Asia. The other, Charles Darwin, traveled around the world once but spent most of his life on a country estate in England. In 1859, Darwin published his ideas in a book called *The Origin of Species.* The book was immensely influential because it presented not just a hypothesis but a detailed, careful argument with much sup-

porting evidence. In a few decades, almost the whole scientific community adopted Darwin's view, and today no principle in biology is more important than evolution.

With the idea of evolution, the typological view of the species almost vanished. The concept of a perfect white pine or a perfect red oak is gone from science. A species is precisely what we see in nature, no more and no less: a set of organisms that resemble one another because they derive from common ancestors and that vary because heredity is subject to change. The quest for perfect types has been replaced with a search for the paths of change that led to modern forms.

NATURAL SELECTION

Darwin's theory gave the environment a guiding role in evolution. Darwin compared the effect of the environment to the artificial selection that a farmer practices when improving livestock or crop plants: individuals vary, and those with desirable traits are allowed to reproduce while others are not. (Artificial selection as it relates to plant breeding is described in Chapter 11.) Darwin imagined that the environment imposes a similar selective effect on wild species (Fig. 14.2). He called this effect **natural selection.**

It is important to realize that neither the environment nor the evolving species is aware that selection is taking place. A species does not "want" to evolve, nor does the environment "want" to shape the species. The environment simply provides an arena in which there are opportunities and hazards. Individuals vary in their ability to find food, evade predators, find mates, and resist disease. Those that fit the environment flourish and leave offspring; those that fit poorly leave few offspring. The survivors make up a population that is better attuned to the environment than before. In this way the organisms in each area become shaped to fit the local conditions. The process can lead to great changes when it operates for millions of years.

SOURCES OF VARIATION

Darwin's idea required spontaneous changes in heredity as a source of new forms that can be selected. He had plenty of examples: no two humans are alike, and the same is true of other organisms. Darwin guessed that changes in heredity occur at random; they do not arise to fit the needs of an organism. His theory also supposed that new traits are more or less faithfully passed on to succeeding generations. Darwin's ideas were bold speculations, for little was known of heredity in his day. But he was right: work in the twentieth century has shown that genes do change at random, and the changes are indeed passed on to the offspring. Three factors can introduce new genetic variation into a population:

1. *Mutations.* Hereditary information is carried in molecules of DNA, which are copied before cell division so that every cell gets a complete set. The copying process is remarkably accurate, but once in a while the enzymes make an error. Also, certain chemicals and radiations in the natural environment can alter DNA. Altered chromosomes may simply have their information modified, or they may break, and the fragments may rejoin in novel combinations. These changes in the DNA are called **mutations,** and the organisms that carry them are called **mutants** (Fig. 14.3). Mutant segments of DNA are copied and passed on as if they were normal, because in most cases the copying system cannot distinguish between normal and mutant segments.

2. *Migration.* When an organism travels to a new population, it may bring genes that the population previously did not have. Plant "migration" occurs when seeds, pollen, and other plant parts are carried by wind, water, or animals.

Figure 14.3. The effect of a mutation. *A,* common wild sunflower; *B,* mutant sunflower, known as sun gold. Most mutations produce less marked effects.

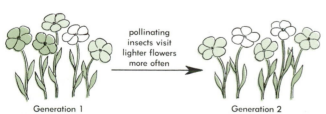

Figure 14.2. The principle of natural selection. In this hypothetical case, plants vary in flower color from leaf-green to white. Plants with lighter-colored flowers pass their genes on to the next generation more often because they are more visible to pollinating insects.

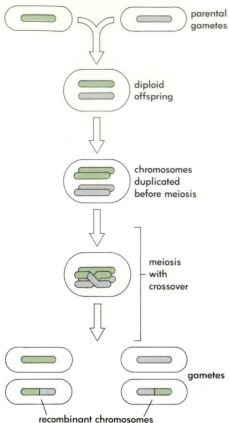

Figure 14.4. Recombination through sexual reproduction. Shading is used to distinguish chromosomes of the two parents. For simplicity, only one cell of the offspring is shown. Genes from separate individuals come together in the offspring. When the offspring makes gametes, meiosis gives each gamete a combination of genes from the two parents.

3. *Recombination.* Sexual reproduction usually results in offspring that have DNA from two parents (Fig. 14.4). Later, when the offspring produce their own gametes, the paired chromosomes are segregated by *meiosis* (see Chapter 11). During meiosis, the chromosomes exchange parts by the process of crossing-over. Further, during meiosis the pairs of chromosomes line up and separate without regard to parental origin. Therefore, each gamete combines genes and chromosomes from both of the organism's parents. This process of mixing DNAs is called **recombination.** It does not generate new genes, but it produces new patterns of heredity by combining traits that arose in separate individuals.

To summarize, mutation and recombination and migration work at random, and they provide the varied types of organisms that are needed before natural selection can produce evolution.

HOW NEW SPECIES ARISE: SPECIATION

Evolutionists believe that the millions of modern species arose from just one original form of life. For this to be true, one species must be able to split into two; and many splits must have occurred in the past. The division of one species into two is called **speciation** (Fig. 14.5). It requires two events:

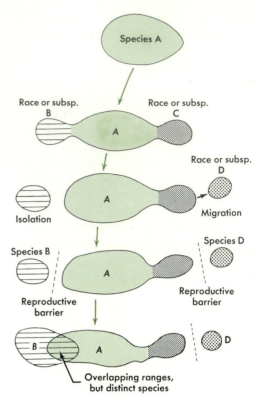

Figure 14.5. Diagram showing the sequence of events that leads to the formation of new races, subspecies, and species. One species (green, at top) extends into new habitats, and populations at the extremes become modified into races or subspecies. If the races or subspecies become isolated, they may evolve further and become unable to exchange genes with the original species; at this point they are recognized as distinct species. The reproductive barrier will keep the species distinct, even if later migration brings them back into contact (bottom).

(1) a population divides into two parts, and (2) the two populations accumulate different mutations and encounter different environmental conditions, so that they evolve into different forms (they **diverge**).

Sexual reproduction usually prevents speciation from occurring within a population. This is because speciation involves a gradual accumulation of genetic differences between two populations. Such differences cannot build up if members of the two populations mate and exchange genes; in this case, any gene that becomes common in one population will spread to the other population. Thus speciation does not occur unless the two populations are **reproductively isolated.** The isolation need not be complete, but the exchange of genes must be very infrequent. And the isolation must continue until the two populations have diverged so far that they can no longer mate successfully.

On the other hand, sexual reproduction can sometimes be an important source of new species. This can happen when plants of separate species mate and produce sterile offspring. (By "sterile" we mean that the offspring cannot produce functional gametes.) Though sterile, the hybrid offspring may be vigorous plants and may spread by asexual means. They have a mixture of traits not found in either parent alone.

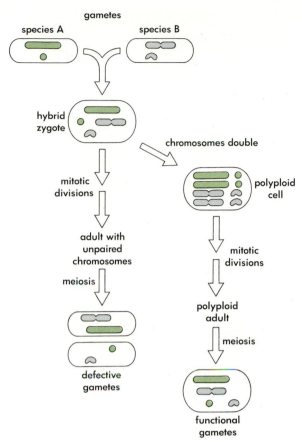

Figure 14.6. Why hybrids are sterile and how a doubling of chromosomes restores fertility. Left branch: the hybrid's chromosomes do not form matched pairs, hence they are distributed abnormally in meiosis, yielding nonfunctional gametes. Right branch: chromosome doubling permits normal pairing in meiosis, hence gametes receive one chromosome of each kind and are fully functional.

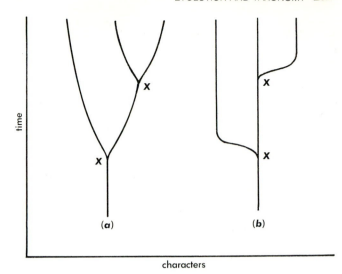

Figure 14.7. Two views of evolution. Each line represents a species passing through time. Horizontal movement indicates a change in characteristics, or evolution. An act of speciation, where a new species is formed, is indicated by x. A, the gradual model: each species is constantly changing. B, the punctuated equilibria model: change is confined to brief periods that are associated with speciation.

A sterile hybrid may become fertile later on by doubling its chromosome number. The plant is then called a **polyploid** because it has more than two sets of chromosomes. Figure 14.6 shows how polyploidy can restore fertility. The original sterility occurs because the chromosomes from the two parents differ too much to form pairs during meiosis. Therefore, meiosis miscarries and the hybrid forms abnormal gametes with extra or missing chromosomes. The problem can be corrected if an ordinary body cell doubles its chromosomes but then fails to divide for some reason (Fig. 14.6). A plant made of such cells can pair its chromosomes, perform meiosis, and make functional gametes. This is the start of a new species. Geneticists estimate that up to 40% of plant species may have arisen by hybridization and polyploidy.

CONSTANCY AND CHANGE IN EVOLUTION

At present, scientists are divided between two views on the way evolution progresses over long periods (Fig. 14.7). Darwin's view, called the **gradual** model of evolution, is illustrated in Fig. 14.7a. Here each species changes nearly all the time at a slow rate. In the more recent view, called the **punctuated equilibria** model, every species remains roughly

constant for millions of years (Fig. 14.7b). Changes are concentrated around the times when speciation is taking place, and the changes are relatively rapid (complete in thousands of years or less). In real species, the rates of change probably vary widely. Therefore both models are likely to apply at various times.

What keeps a species from changing for long periods? Natural selection. Most organisms at any moment are very well adapted to their environment. Random mutational changes are usually harmful, and the environment continually eliminates them. When natural selection works in this way to maintain the status quo, it is called **stabilizing selection.** By contrast, selection that leads to progressive change is called **directional selection.**

ADAPTIVE RADIATION

When operating for a long time, the processes described above can build a large set of related species from one ancestral population. Such an expansion can be very rapid when viewed on the geological time scale. For example, in a period of 20 to 40 million years, the flowering plants expanded from a small group to one of dominant status in most areas of the world, with levels of diversity approaching those present today. This is a brief period, when compared to the entire length of the fossil record (some 3500 million years).

The rapid multiplication of species is called **adaptive radiation.** It occurs when the ancestral species has the opportunity to expand into many ecological niches. An opportunity can arise in three ways: (1) through the extinction of competing groups; (2) through the evolution of advantageous new traits; or (3) through the origin of new environmental conditions.

The rise of the mammals illustrates radiation in response to the extinction of competitors. For 100 million years, reptiles occupied most of the ecological niches where mammals thrive today. Primitive mammals coexisted with the ruling reptiles, but they were limited to small, ratlike forms. Then, in a relatively brief time span, all the dominant reptiles died out. This happened about 65 million years ago. The mammals rapidly multiplied and diversified until they filled most of the vacated niches.

As mentioned above, the flowering plants also arose through a great adaptive radiation. They first appeared well before the ruling reptiles died out, around 120 million years ago. Thereafter, flowering plants exploded in numbers and diversity, taking over many niches that other plants had previously occupied. What caused the radiation of flowering plants? We can only guess. The best starting point is to look for traits that are advances over prior groups—advances so valuable that they persist in nearly all the modern flowering plants. Two sets of features come to mind: the structure of the xylem and the flower itself. Regarding xylem, the flowering plants have a division of labor between fiber cells that provide strength and vessels that conduct water. Earlier plants had both functions combined in tracheids. By specializing, fibers and vessels provide more strength and easier water flow. This may have allowed flowering plants to colonize environments that were too harsh for many earlier plants.

The flower offers still more advantages. Among other features, the carpels enclose and protect the ovules and seeds. Earlier plants had exposed ovules and seeds. The improved shelter made it safer to attract animals, which can transport pollen and seeds more effectively than wind and gravity. The benefits include more seeds produced with less expenditure of pollen, and wider dispersal. These features are discussed further in Chapter 20.

PRINCIPLES OF CLASSIFICATION

THE BINOMIAL SYSTEM

The study of plant taxonomy takes us back to the year 300 B.C., when Theophrastus (a student of Aristotle) set up the first formal classification of plants. He assigned about 500 plants to groups based on such features as growth habit (tree, shrub, herb), length of life (annual, biennial, perennial), and details of flower structure. For 2000 years there were few real advances over the work of Theophrastus.

The great sea voyages of the fifteenth through the eighteenth centuries were a powerful stimulus to taxonomy. The explorers brought back hundreds of novel organisms. Great botanical and zoological gardens were founded, some of which still flourish today (Kew Gardens in England, for example). Scientists were fascinated by the diversity of life, and thus some of the sharpest minds were turned to explaining it.

Modern taxonomy was established in that era, two centuries ago. Its principal founder was the Swedish botanist Carolus Linnaeus (Fig. 14.8). The basic need was for a simple, reliable way to name plants; a system that could grow

Figure 14.8. Linnaeus (1707–1778), the Swedish botanist who popularized the binomial system of nomenclature.

as new plants were discovered. Linnaeus provided the answer. Working in the great gardens, Linnaeus compared thousands of plant species. He saw that a system of names could be based on the structure of the flower, using traits such as those of Fig. 14.9. Flowers were especially useful because they have many well-defined features that vary from one species to another. Vegetative parts such as leaves offered fewer traits for comparison.

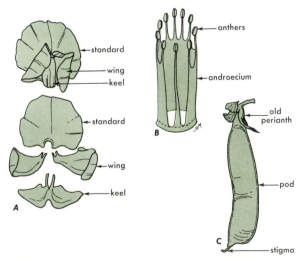

Figure 14.9. Diagram showing traits that are important in the classification of the sweet pea (*Lathyrus odoratus* L.). *A,* flower, exploded view; *B,* 10 stamens, 9 attached together and 1 free; *C,* pod with seeds.

Figure 14.10. *Tradescantia virginiana* (spiderwort). Note the jointed staminal hairs, from which the common name derives.

Table 14.1. Classification of the Sweet Pea

Classification Level	Name	Ending
Specific epithet	odoratus	—
Genus	Lathyrus	—
Family	Fabaceae	aceae
Order	Rosales	ales
Class	Dicotyledonae	ae
Division (phylum)	Anthophyta	phyta
Kingdom	Plantae	Not consistent

In 1753 Linnaeus published a book, *Species Plantarum,* in which he named about 6000 species and assigned them to 1000 groups called *genera*. The book was a foundation on which other taxonomists could build, and many of the groups that Linnaeus defined are still accepted today.

Linnaeus's book was valuable partly because it endorsed a convenient way of naming organisms. For each species, Linnaeus wrote a descriptive phrase of twelve words or less. He regarded this as the formal name, but for convenience he wrote a two-word abbreviation in the margin. For example, he described spiderwort (Fig. 14.10) with these words: *Tradescantia ephemerum phalangoides tripetalum non repens virginianum gramineum,* common name *Tradescantia virginiana*. Loosely translated, it means "the annual, upright Tradescantia from Virginia that has a grasslike habit, three petals, and stamens with hairs like spider legs, common name Tradescantia of Virginia."

Not surprisingly, taxonomists favored the two-word abbreviated name, or **binomial** (in this case it was *Tradescantia virginiana*). Today every species is given a binomial name, which is always underlined or printed in italics. The first word (always capitalized) is the **genus.** A genus (plural, **genera**) is a group of species that are similar enough to be obviously related. For example, the genus *Quercus* includes all the plants that produce acorns and have male flowers in catkins (i.e., the oaks). *Quercus alba* is the white oak. The second word in the name is the **specific epithet.** The combination of the generic name and the specific epithet is called the **species name.**[1]

HIGHER TAXONOMIC CATEGORIES

The binomial system of Linnaeus provides only two levels of classification. But when we survey the genera, we find that they, too, share important traits and can be grouped into sets. Genera with similar traits make up a **family.** For in-

[1]Rules for naming plants were set up at an International Botanical Congress in 1930 and are periodically revised. Each plant must have a binomial name, a descriptive paragraph must be published, and a preserved specimen must be presented for storage in a plant collection, or **herbarium.** The preserved **type specimen** is especially important because it connects the name with a real example of the plant and forms the basis for revising classifications in future years.

stance, all the plants that have roselike flowers comprise the family Rosaceae. They include not only the garden roses but also the trees that produce stone fruits (plums, peaches, and so on). In turn, families can be grouped into **orders,** orders into **classes,** classes into **divisions,** and divisions into **kingdoms.** Table 14.1 summarizes the main categories, using the sweet pea as an example. Note that taxonomists have agreed upon a standard set of letters to end the names at most levels.

A table of botanical divisions is printed inside the back cover of the book; you may wish to consult it as you study the various groups in later chapters. The table is a series of choices between alternative traits, and is called a **key.** Taxonomists use keys in their daily work to determine the names of plant specimens.

For want of space, our key mentions only a single trait at many decision points. This is misleading, for taxonomic groups are almost always defined by groups of traits rather than by a single trait. A comprehensive key would mention many traits at each level of classification, would cover all the levels from kingdom to subvariety, and would fill several large volumes.

In the coming discussion, we will often make statements that apply to taxonomic groups in general. Here the term **taxon** (plural, **taxa**) becomes useful. The word *taxon* can be used in place of any other group name when we wish to generalize; for instance, a species is a taxon, a kingdom is another taxon, and so on.

HOW TAXONOMIC GROUPS ARE DEFINED

In the course of history, taxonomists have approached the problem of defining groups from two different points of view (Fig. 14.11). The oldest approach, called **phenetic** classification, is to group organisms purely on the basis of appearance. Theophrastus used this method when he divided the plant kingdom into trees, shrubs, and herbs. As you can see in Fig. 14.11*A,* the phenetic approach can lead to groups in which the members are of mixed ancestry.

Since the acceptance of the concept of evolution, many taxonomists have attempted to define groups on the basis of common ancestry. This is called **phylogenetic** classification (*phylogeny* means "origin of groups"). In the strictest form of this approach (Fig. 14.11*B*), boundaries are drawn so that each group includes the common ancestor, all of its descendants, and nothing else. Such a group is said to be **monophyletic** ("of single origin").

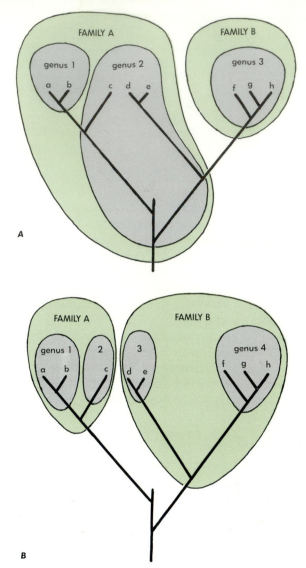

Figure 14.11. Two approaches to classification. *A,* phenetic classification, where groups are based entirely on similarity of the organisms. The resulting taxa may group unrelated organisms together. *B,* phylogenetic classification, where groups are based on common ancestry. In this case each group includes the common ancestor and all of its descendants. This method may place similar organisms, such as c and d, into separate groups.

In practice, most taxonomists compromise between the two approaches. Organisms are grouped phylogenetically when their evolutionary connections are well known; they are grouped phenetically when their ancestry is less clear, or when a strict phylogenetic approach would separate very similar organisms that still resemble the common ancestor.

How can we deduce the evolutionary connections between modern groups? Two approaches are available. First, the fossil remains of ancient plants provide ideas about ancestry. And second, we can reason about the course of evolution by comparing modern organisms with one another. On first thought, the approach via fossils seems most promising; but unfortunately, it provides only general hints about the paths of evolution. Millions of species existed at

each point in time, and only a small fraction of them (usually the most abundant) left fossil remains. The species that gave rise to new groups are therefore only occasionally found in fossil form. Furthermore, some plants and plant organs are not often preserved in the form of fossils: only plants that lived near swamps, lakes, or other bodies of water left abundant fossils, because the first step in forming fossils is usually burial of the organ in mud. And even in ideal habitats, most fossils are pollen grains, small fragments of wood, or imprints of leaves rather than whole plants. Delicate parts, such as flowers, are rarely preserved.

Since fossils are often inadequate, phylogenetic taxonomists must try to reconstruct the course of evolution by comparing modern organisms. They look for *primitive* and *advanced* traits, so as to define groups (wherever possible) on the basis of *shared advances.* Having said this, now let us see what it means.

A trait (such as the presence of tracheids in wood) is said to be **primitive** if it arose early in evolution, and **advanced** if it arose later. The word *advanced* does not necessarily mean "better" or "more sophisticated." It merely means "came later."

A trait is judged to be primitive if it is present in many taxa that are otherwise quite different. For example, consider the conducting cells found in the xylem (Table 14.2). Tracheids occur in all vascular plants, including ferns and conifers as well as flowering plants. The simplest explanation is that tracheids arose in an ancestral group that gave rise to all the modern vascular plants. By contrast, vessels occur almost exclusively in flowering plants. *Maybe* ancient ferns and conifers had vessels, and their descendants have simply lost them. However, a simpler idea is that vessels arose in the line that led to flowering plants after it had branched off from the lines that led to conifers and modern ferns. With this reasoning, tracheids must have come before vessels; therefore, tracheids are primitive and vessels are advanced.

Having defined primitive and advanced traits, the next step is to define taxa on the basis of shared advances. Figure 14.12 illustrates the reasoning. It is important to note that the evolutionary connections shown in the figure are *not* known in advance; we only see present-day organisms. The evolutionary connections are deduced as follows. Suppose that after studying a number of species we decide that traits A, B, C, and D are primitive while A′, B′, C′, and D′ are advanced. In this case, the common ancestor of the four species in Fig. 14.12 must have had traits ABCD. Next, we compare the living species with one another. Only two species have trait B′. The easiest explanation is that both species sprang from an ancestor that had trait B′. The lines

Table 14.2. Vascular Elements in Three Groups

Group	Tracheids?	Vessels?
Ferns	Yes	A few
Conifers	Yes	No
Flowering plants	Yes	Yes

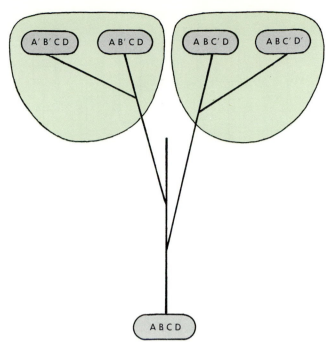

Figure 14.12. Forming taxonomic groups on the basis of shared advances. Each oval at the top is a living species; the oval at the bottom is the common ancestor. The letters within each oval represent traits that are present in the species. Traits A, B, C, and D are primitive; A', B', C', and D' are derived, or advanced, traits (for instance, A' is a modification of A). Lines that connect the species to one another were inferred from the traits of the modern species. The four species make up two genera (green shading).

Figure 14.13. Desert plants of the southwest United States (left) and Africa (right). The similarities are attributed to convergent evolution. Based on floral and anatomical features, these plants are placed in separate families: the Cactaceae (left) and the Euphorbiaceae (right).

connecting the species were drawn accordingly. The two species with trait B', and their common ancestor, share an advance over the ancestral condition. Therefore, they make up a probable genus. The rest of the connections and groupings in Fig. 14.12 were filled in by similar reasoning.

If you have a critical eye, you may object that other trees could have been drawn to produce the modern species of Fig. 14.12. If so, you have comprehended both the principles and the limitations of this method. Often there *are* several possible ways to reach the present from a given ancestral condition. Thus while most taxonomic groupings are stable and widely accepted, some are debatable, and new classifications are proposed from time to time.

Apart from the difficulty of deciding which traits are primitive, the greatest problem with phylogenetic classification comes from our assumption that shared advances imply common ancestry. The assumption is usually sound, but sometimes it fails and leads to wrong conclusions. Failures occur because similar traits sometimes arise independently in quite unrelated groups of plants. This is called **convergent evolution,** or **convergence.** Groups that are based on convergence are **polyphyletic.** Convergence often occurs when plants of different origins evolve under similar environmental pressures. For example, many plants in the deserts of Africa resemble the cactuses of American deserts (Fig. 14.13): they have similar fleshy green stems, spines,

and reduced leaves. But despite these similarities, a much larger set of traits having to do with anatomy and flower structure shows that the African and American species belong to quite different families, the Cactaceae and the Euphorbiaceae. Less specialized members of these families do not have the shared cactuslike features. Evidently the desert members of the two families evolved similar traits when they were subjected to similar selective pressures for a very long time. The situation is summarized in Fig. 14.14.

The case of the desert plants is relatively easy to settle because the Cactaceae and the Euphorbiaceae differ in a large cluster of floral traits. However, not all conflicts of evidence are so easily resolved. Sometimes the differences and similarities weigh so evenly that great arguments ensue about the evolutionary relationships. As a result, some taxonomists have abandoned the phylogenetic approach and are defining groups purely on the basis of phenetic similarities.

Whatever the taxonomist's philosophy, it can be difficult to define groups when several traits suggest rival groupings.

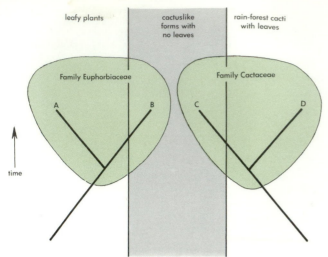

Figure 14.14. An example of convergent and divergent evolution. Convergent evolution has made the desert-dwelling forms B and C very similar in appearance. Divergent evolution has created differences between these forms and other members of their families. A, B, C, and D each represent one or more genera.

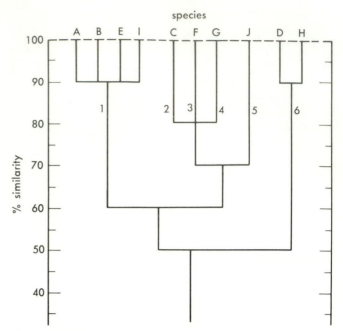

Figure 14.15. Dendrogram, showing percentage similarity between several plant species. The following interpretation illustrates how to read the graph: Species A, B, E, and I resemble one another at the 90% level; C, F, and G resemble one another at the 80% level. When these two groups of species are compared, they are similar at the 60% level.

In recent years a method called **numerical taxonomy** has been devised to provide more objective judgments. The numerical taxonomist records as many traits as possible without judging their relative importance: features such as the way leaf veins branch, the shapes of organs, the structure of hairs on the stem, and so on. Anything that differs between species can be listed; ideally the taxonomist stops only when time or money runs out. (However, at times a worker may be interested in a special part of the plant—perhaps the leaf—and the study will be limited accordingly.) Each trait is expressed as a number. The plant is given a score of 1 if the trait is present; 0 if absent. Then the numbers are put into a computer and a program is run to estimate the degree of similarity between pairs of species. The results are expressed in tables or in diagrams called **dendrograms** (Fig. 14.15). Such figures provide a more objective basis for reasoning about evolutionary connections between taxa, but they still leave room for subjective judgments on where to place boundaries.

EVOLUTION AND THE KINGDOMS OF LIFE

Since the time of Aristotle, humans have tried to divide the biological world into kingdoms. There is no agreement as to how it should be done, and yet any book that surveys a major fraction of the earth's biota must provide at least a hint of the way in which all the species fit together as a whole. Some ideas about the kingdoms of life will be discussed in the following pages.

The oldest idea distinguishes two kingdoms, the plants and the animals, on the basis of motility and nutrition: animals move about and ingest food; plants do not. When microbes were discovered, the swimming forms were ranked with the animals; the nonmotile ones with the plants.

During the nineteenth century, this simple scheme began to show weaknesses. Organisms such as slime molds were found to resemble plants in their reproduction and animals in their nutrition, and some plants were found to have swimming reproductive cells. The great German evolutionist Ernst Haeckel was one of several scientists who saw the need for a third kingdom to accommodate the unusual forms of life. For the third kingdom, he coined the term *Protista,* meaning "the first."

Haeckel was strongly influenced by Darwin's theory of evolution, which was still a novelty in the mid-nineteenth century. The term *Protista* shows that influence, for it implies that the simpler forms of life—most of them microscopic and aquatic—gave rise to the complex plants and animals.

This was also the time when biologists realized that plants and animals are made of cells. The step from unicellular to multicellular life was obviously a great event in the evolution of life, and some workers saw it as a logical boundary between kingdoms. In consequence, Haeckel's Protista came to be recognized as a kingdom of unicellular life.

In the present century, biochemistry and electron microscopy have distinguished prokaryotes from eukaryotes and have revealed great differences between the fungi and the green plants. To accommodate these findings, various workers have proposed additional kingdoms. A five-kingdom scheme proposed by the American ecologist Robert Whittaker has been especially popular. Whittaker placed the prokaryotes in the kingdom Monera, and he placed the unicellular eukaryotes in the kingdom Protista. Then he divided the multicellular eukaryotes on the basis of nutrition: photosynthetic organisms made up the kingdom Plantae, organisms that ingest solid food made up the kingdom Animalia, and organisms that absorb dissolved foods made up the kingdom Fungi. This scheme became popular because it emphasizes important ecological differences between groups, and because it emphasizes major steps in evolution: from prokaryotes to eukaryotes and from unicellular to multicellular life.

For all its strengths, the Whittaker scheme has a major shortcoming: it defines most kingdoms on the basis of traits that arose more than once in history. As a result, unrelated organisms are placed in the same kingdom, whereas close relatives are sometimes placed in different kingdoms. In other words, it reflects only the broadest aspects of evolution. This is especially true of the algae (see Chapter 16). For instance,

the green alga *Ulva* (sea lettuce) and the brown alga *Fucus* (a kelp) are placed together in the kingdom Plantae because both are multicellular and photosynthetic. But they have little else in common. With respect to chloroplast pigments, food reserves, flagellation, and other traits, sea lettuce is more like the unicellular green algae than it is like the kelps. Evidently, the green algae represent one line of evolution from unicellular to multicellular forms; the brown algae represent another line of evolution. From a phylogenetic point of view, these two groups should be kept intact and distinct.

To represent the paths of evolution more effectively, some taxonomists favor a modified version of the Whittaker scheme (Fig. 14.16). This scheme has been adopted in the present book. There are still five kingdoms, and they are broadly similar to those of Whittaker. However, this set of

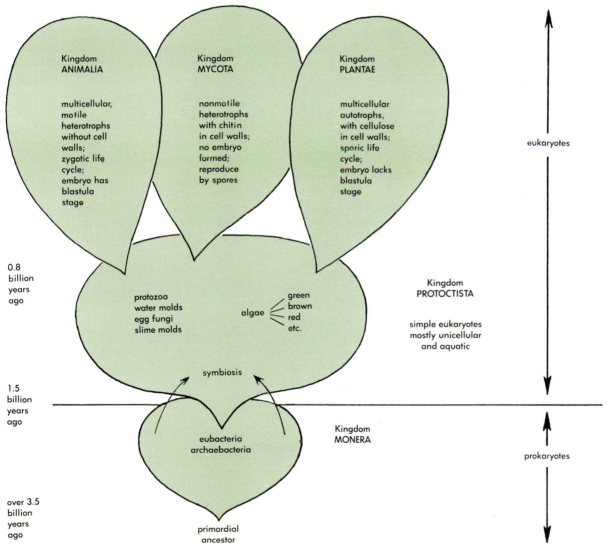

Figure 14.16. Five kingdoms of life and their origins. The times are very approximate, and the sizes of the kingdoms are not to scale. First came the prokaryotes, then unicellular eukaryotes, and lastly multicellular eukaryotes. The kingdoms Monera, Plantae, Mycota, and Animalia are so defined as to be (in principle) monophyletic. The kingdom Protoctista is an artificial group that combines all the eukaryotes not placed in other kingdoms. (Based on the classification system expressed in L. Margulis and E. Schwartz, *Five Kingdoms.* San Francisco: W. H. Freeman, 1981.)

kingdoms attempts to place the plant, animal, and fungal kingdoms on a monophyletic basis. This is done by basing the classification on combinations of traits that are likely to have arisen just once rather than many times in the course of evolution. The kingdom Animalia is limited to organisms that have a zygotic life cycle, lack cell walls, and form an embryo that goes through a hollow-ball (blastula) stage; the kingdom Plantae is limited to organisms (usually photosynthetic) that have cellulose walls and form an embryo; the kingdom Mycota (the "true" fungi) are organisms with chitin walls that lack a motile stage and reproduce by spores. The remaining eukaryotes make up a catch-all kingdom, the *Protoctista,* which includes all the eukaryotic algae and other organisms that do not fit into the preceding kingdoms (for the disposition of botanical groups, see the table inside the back cover). Unlike the old Protista, the kingdom Protoctista includes both unicellular and multicellular organisms. The name was proposed in the middle of the last century and has been sponsored in recent years by the American evolutionist Lynn Margulis (a leader in establishing the widely held view that plastids and mitochondria arose through symbiosis).

No system is perfect, and this one has its limitations. The kingdom Protoctista is a weak point: like the old Protista, it is artificial; its members make up several lines of descent that are not closely related to one another. To be consistent, the Protoctista should be divided into at least a dozen small kingdoms. Such a splitting has been avoided because the idea of a kingdom implies, to most people, a group of considerable size.

A greater problem concerns the prokaryotes and their relation to the other kingdoms of life, as will be discussed below.

RELATIONS BETWEEN PROKARYOTES AND EUKARYOTES

The oldest known fossils are of bacteria, found in rocks 3.5 billion years old. This gives the bacteria, or prokaryotes, a special importance in the classification of life. However, until a few years ago it was difficult to trace the evolution of these early forms. Their physical shapes are simple; modern and ancient specimens look alike; and the same shapes probably arose independently in many lines of descent. Their metabolism is quite varied and could provide clues to the past, but it is difficult to decide which traits are primitive and which are more recent adaptations to the varied habitats that bacteria occupy.

Recently, these problems were overcome by directly examining the nucleic acids in the cell. The base sequences in the nucleic acids are handed down from one generation to the next, faithfully for the most part, but with occasional mutational changes. When one species divides into two, the two species begin to accumulate mutational differences. Thus, by examining the base sequences found in different organisms, we can reason out the evolutionary relations between them.

Table 14.3. A Comparison of Archaebacteria, Eubacteria, and Eukaryotes[a]

Feature	Archaebacteria	Eubacteria	Eukaryotes
Membrane lipids	Branched *ethers*	Straight *esters*	Straight *esters*
Cell walls (when present)	Not based on muramic acid	Based on muramic acid	Not based on muramic acid
Nuclear membrane	Absent	Absent	Present
Ribosome size	30S, 50S	30S, 50S	40S, 60S
Sensitivity to anisomycin	Sensitive	Insensitive	Sensitive
Sensitivity to chloramphenicol	Insensitive	Sensitive	Insensitive

[a]Selected from a larger set of comparisons in C. R. Woese, "Archaebacteria," *Sci. Amer.* 244 (1981):98.

Karl Woese and his colleagues at the University of Illinois have applied this method to the bacteria. Studying ribosomal RNA, they discovered that the bacteria form two quite distinct groups, the *archaebacteria* and the *eubacteria.* All the common bacteria, including the Cyanobacteria, are eubacteria. The archaebacteria are less common because they are killed by oxygen; they occur in anaerobic habitats such as hot springs, salt marshes, and submerged decaying vegetation. The archaebacteria and eubacteria differ in many ways besides the ribosomal RNA: their walls are different, their membrane lipids differ, their transfer RNAs differ, and they are sensitive to different antibiotics (Table 14.3).

Most biologists assume that the eukaryotes arose from prokaryotic ancestors. But which group was the ancestor—the eubacteria or the archaebacteria? To find the answer, Woese and colleagues compared the eukaryotes and prokaryotes with respect to a variety of traits. They found that the eukaryotes resemble eubacteria in some ways, and they resemble archaebacteria in other ways (Table 14.3). This suggests that the three groups arose at about the same time from a common ancestor.

If these findings were honored in our set of kingdoms, then the prokaryotes should be divided into two groups, each with the same stature as the eukaryotes. This is not recognized in the five-kingdom system, but undoubtedly future schemes will find a way to express it.

SUMMARY

1. Taxonomy is the science of classification.

2. The species is the unit of life in taxonomy. A mating test is often used to define the limits of a biological species.

3. The theory of evolution is fundamental in taxonomy; all organisms are thought to be descended from a single original life form.

4. The theory of evolution proposes that organisms vary due to nondirected (random) changes in the genes. Mutation, recombination, and immigration bring novel

genes and combinations of genes into each population. Among the variant individuals, those that are well suited to the environment will reproduce more abundantly than others (the process of natural selection). As a result, each species becomes ever more perfectly adapted to the environment.

5. Speciation (the origin of new species) occurs when a population is split into two parts that diverge (evolve differently). To diverge, the two populations must be reproductively isolated. New plant species can also arise when a hybrid between two species becomes a polyploid.

6. Once adapted to the environment, a species may persist without change for millions of years due to stabilizing selection. Directional selection can cause rapid evolution when adaptation is incomplete due to new traits or new environmental conditions.

7. Evolutionists are divided as to whether evolution is slow and gradual or whether it consists of long periods of constancy punctuated with periods of rapid change at times when speciation is occurring (the theory of punctuated equilibria).

8. When an ancestral group encounters a new ecological opportunity, it may rapidly multiply and give rise to a large group of related species (the process of adaptive radiation). The flowering plants and many other groups arose in this way.

9. Modern taxonomy is based on the unifying work of Linnaeus, who popularized the binomial system of naming organisms. The two-word name indicates the genus and species.

10. Taxonomic categories form a heirarchy in which successive levels are the kingdom, division, class, order, family, genus, species, and variety. Subcategories are also recognized. At some levels the names are given standard endings.

11. In the phenetic approach, organisms are assigned to higher taxonomic groups on the basis of physical resemblances, regardless of evolutionary connections. This can result in quite unrelated species being assigned to the same group. In the phylogenetic approach, species are grouped on the basis of evolutionary relationships. This can result in placing very dissimilar organisms in the same group. Most of taxonomy is a compromise between these two approaches.

12. In the phylogenetic approach, evolutionary relationships are deduced by identifying primitive and advanced traits, so that groups are defined on the basis of shared advances over the primitive condition. Fossils are also useful in defining groups.

13. Convergent evolution can occur when species of different origin evolve in similar environments. Convergence can lead to errors in classifying plants.

14. Numerical taxonomy provides an objective measure of the similarity between species or other groups. Numbers are assigned to traits; as many traits as possible are considered; and the results are summarized in dendrograms by means of computer analysis.

15. The kingdoms recognized in this text are the Monera, Animalia, Plantae, Mycota, and Protoctista. The Monera are prokaryotic and were the first organisms, dating back to 3500 million years ago (mya). The other kingdoms are eukaryotic. Eukaryotes arose from prokaryotic ancestors about 1500 mya. Chloroplasts and mitochondria may have arisen by symbioses between early eukaryotes and prokaryotes. The first eukaryotes were unicellular (protista). Multicellular forms arose 1000 to 800 mya. The kingdoms Animalia, Plantae, and Mycota are monophyletic and are predominantly multicellular. All other eukaryotes are placed in the artificial kingdom Protoctista.

15 THE FUNGI

We commonly meet the fungi as mushrooms, mildews, molds, and the organisms that cause athlete's foot. A few fungi provide food for humans, but many more compete with us for food sources, and several of them use us as their own living food supply. The baking and brewing industries have an ancient partnership with the fungus known as yeast (Saccharomyces); and the medical profession has the makings of another long partnership with fungi, such as *Penicillium notatum,* which produce penicillin and other antibiotics. It is doubtful whether any other group of organisms is associated with human beings in as widely varied an array of relationships.

The great majority of fungi have a vegetative body composed of branching tubes (Fig. 15.1) called **hyphae** (singular, **hypha**). True cell walls, produced by the protoplast, surround the hyphae. The fungi are eukaryotes whose cell contents are similar to those of plants except that plastids and usually dictyosomes are missing. Most of the space within the hypha is occupied by large vacuoles.

Hyphae grow only at their tips, where the walls are soft and extensible under the turgor pressure from the protoplast. The region of wall that is left behind the growing tip becomes thicker and thus highly resistant to further stretching. Branches originate as bulges in the sides of the growing hyphal tips, where the walls are still soft. The branches behave exactly as their parent hyphae. Hence the system of hyphae progressively extends and ramifies as it absorbs nutrients and water from the substratum. The hyphae have a "chemical sense" and can orient their growth and branching toward sources of raw materials such as sugars, amino acids, water, and minerals. The orientation of growth by chemical signals is called **chemotropism.** It causes the fungus to progressively invade food sources such as moist slices of bread and decaying stored fruits.

The system of connected hyphae is the vegetative body of the fungus and is called a **mycelium** (Fig. 15.1). There is some cooperation between neighboring hyphae, but the degree of cooperation rapidly declines with distance. Thus, as the mycelium grows and spreads, its advancing hyphae tend to form local groups that function independently of one another, as if they belonged to entirely separate mycelia. Older parts of the mycelium may be abandoned, their usable contents withdrawn, and their empty walls sealed off by crosswalls that resemble abandoned tunnels. Such an event, or the accidental shearing of mycelia into parts by outside forces, readily converts a single mycelium into many mycelia. Thus, **fragmentation** is a mode of vegetative reproduction in nearly all fungi.

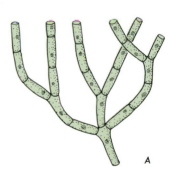

septate hyphae

A

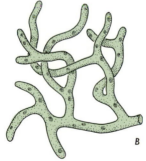

nonseptate (coenocytic) hyphae

B

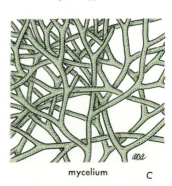

mycelium   C

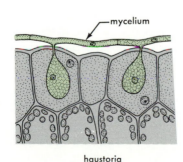

haustoria   D

Figure 15.1. Various aspects of the fungal vegetative body.

Mycelia often show sophisticated responses in which development is keyed to environmental limitations and opportunities. These responses use light, gravity, and chemical signals to direct growth and to select the best times for reproduction.

True fungi are solution feeders: they can take in only small molecules. In many cases their food requirements are very simple, consisting of an organic carbon compound such as a sugar, plus water, minerals, and perhaps a few vitamins (enzyme cofactors, required only in small amounts, which the fungus cannot build for itself). The vitamin most often needed by fungi is the same Vitamin B_1 that is essential in human nutrition. Despite the simplicity of their needs, mycelia exploit a wide range of organic materials as food sources. Solid materials can be attacked, but they must be rendered soluble before they can diffuse through the walls of the hyphae and reach the protoplast. Hyphae build and secrete enzymes that hydrolyze proteins to amino acids and polysaccharides to free sugars. Working outside the hyphae, these enzymes require water as a reactant and as a medium for support and transport. Thus mycelia tend to be active only in moist places.

CLASSIFICATION OF THE FUNGI

The informal term **fungus** refers to eukaryotic organisms that reproduce by means of spores and pursue a heterotrophic mode of life. Some 200,000 species answer this description. They vary widely and are probably derived from several different lines of evolution. In other words, the term *fungus* refers to a grade or level of life that has been reached through parallel evolution by several independent groups. From a phylogenetic point of view, the fungi should not all be placed in the same kingdom.

Despite the overall diversity, the great majority of fungi have many features in common and are likely to be derived from a common ancestor. In our classification scheme, these fungi make up the kingdom Mycota, the "true" fungi. They include all the familiar fungi, such as mushrooms and bread molds. Most are terrestrial organisms that form extensive mycelia, but a few (such as baker's yeast) are unicellular.

Members of the Mycota never have motile cells, and **chitin** is always the chief fibrous material in their cell walls. Chitin is a polysaccharide in which the sugar subunits have been modified by the addition of a nitrogenous group. The same substance is responsible for the strength of insect exoskeletons.

The rest of the fungi make up a very mixed array of forms, and their evolutionary relationships to the true fungi are very doubtful. In this book they are placed in the catch-all kingdom Protoctista.

This chapter will treat the groups listed below.

Kingdom Protoctista

Division Myxomycota (slime molds)

Division Oomycota (egg fungi or water molds)

Division Chytridiomycota (chytrids)

Kingdom Mycota

Division Zygomycota (zygote fungi)

Division Ascomycota (sac fungi)

Division Basidiomycota (club fungi)

Division Deuteromycota (imperfect fungi)

DIVISION MYXOMYCOTA

The **Myxomycota,** or *slime molds,* number about 300 species. They have a vegetative body that consists of a slimy mass of naked protoplasm without an external wall. The body has no definite shape; it creeps slowly by amoeboid movement on rotting tree trunks or dead leaves, engulfing bacteria and the spores of other fungi, or absorbing nutrients from dead organic matter. The absence of cell walls in most stages, the amoeboid movement, and the ability to take in solid food are characteristics that are usually associated with animals. Only in reproduction are the slime molds like other fungi: they produce spores with cell walls.

There are several groups of myxomycetes, two of which—the *acellular slime molds* and the *cellular slime molds*—are discussed below.

The **acellular slime molds** have a vegetative body called a **plasmodium** (Fig. 15.2). It is a large, easily visible mass of protoplasm with no internal division into cells—that is, it is **coenocytic.** The plasmodium starts as a microscopic, diploid cell (a *myxamoeba*) that creeps about by amoeboid motion. Feeding, growth, and nuclear division proceed without cell division, resulting in a large plasmodium with thousands of nuclei. If two plasmodia of the same species meet, they can join into a single unit.

Eventually the plasmodium moves to a drier location and produces spores. While the details are quite varied among species, the pattern shown in Fig. 15.2 is typical: a stalk forms, lifting a mass of protoplasm above the substratum; the surface of the mass develops a tough wall; and the contents divide into many spores. This is an example of a **sporangium** (literally, a "house of spores," an enclosure in which spores develop).

At some point in their development, the spores go through meiosis. They are dispersed by wind, and they germinate in the presence of water: the wall is ruptured, and the haploid cell escapes. Sometimes the cell has flagella and swims about for a time; sometimes it is a creeping amoeba. The amoebas feed and divide repeatedly. Eventually, pairs of amoebas or swimming cells fuse to form zygotes. If flagella were present, they are soon lost; and the diploid zygote forms a plasmodium to repeat the life cycle.

The **cellular slime molds** differ from the acellular types in that they are not coenocytic. Their haploid spores develop into amoebae that creep about, feeding and dividing but always remaining distinct cells with a single nucleus. When the population reaches a certain density, some of the amoebae stop moving and begin to emit a chemical signal, the compound cyclic AMP. Adjacent amoebae migrate toward

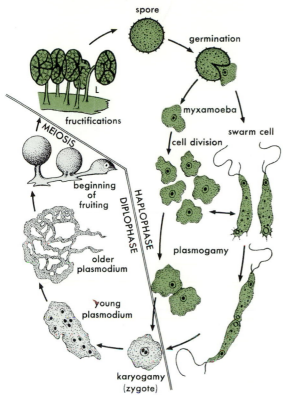

Figure 15.2. Life cycle of an acellular slime mold.

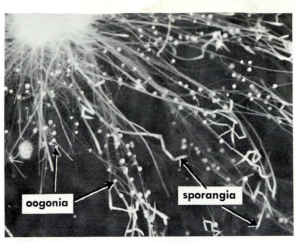

Figure 15.3. An oomycete growing on a hemp seed, ×3.

Most of the Oomycota form a mycelium that is *coeno-cytic*—that is, it consists of a single, much-branched cell with many nuclei (Fig. 15.1). Cross-walls, or **septa** (singular, **septum**) occur at special places: they cut off reproductive cells from the ends of hyphae, and sometimes they form at the bases of old hyphae.

Two important orders, the Saprolegniales and the Peronosporales, will be discussed below.

SAPROLEGNIALES

Fungi of this order usually live in fresh or salt water; hence their common name **water molds.** Most are saprophytic. Even the species that attack the gills of fish grow on tissues that have been weakened or have suffered injury. They may easily be cultivated by placing small pieces of meat, egg albumin, radish seeds, or dead flies in a dish of pond water (Fig. 15.3). After the mycelium has ramified through the food source, hyphae grow outward into the water. Reproductive cells are produced by these hyphae.

Asexual Reproduction

With ample food supply, there is little tendency to produce reproductive cells. If a well-developed mycelium is transferred to distilled water in which a food supply is lacking, asexual sporangia usually appear. Sporangia are formed by a cross-wall cutting off the tip of a hypha from the rest of the mycelium (Fig. 15.4A). The sporangium is multinucleate. The protoplasm of the sporangium divides into a large number of spores, each with one nucleus (Fig. 15.4B and 15.4C). Upon maturity, the spores are discharged from the sporangium. In most species these spores have flagella that permit swimming movements, and they are therefore known as **zoospores.** In *Saprolegnia,* each zoospore has two flagella attached to its anterior end that enable it to swim (Fig. 15.4D). After a time, the zoospores settle down, lose their flagella, develop a cellulose wall, and pass through a resting period. When they have resumed activity, they escape from the wall; the two newly developed flagella are now attached laterally, and the zoospores swim about for a period of time. Encountering a food object, the zoospore comes to rest and grows a hypha which will become a new mycelium.

the source of the chemical; where they stick together and produce still more cyclic AMP. In this way, thousands of amoebae form a mass called a **pseudoplasmodium** (*pseudo-* = false). The pseudoplasmodium moves about as a unit, but the amoebae remain as distinct cells within it. The pseudoplasmodium is dedicated to reproduction: some of its cells form a stalk, while other cells become spores in a naked mass at the top of the stalk. After dispersal by wind or water, the spores germinate to form a new set of amoebae, and the cycle is repeated.

DIVISION OOMYCOTA

In the years 1843 to 1847, a blight devastated the potato fields of Ireland. A quarter of a million persons died of starvation, and millions more emigrated to America. The blight was caused by *Phytophthora infestans,* a fungus whose genus name literally means "plant destroyer."

Phytophthora infestans is one of several major plant pests that belong to the Division **Oomycota.** The Oomycota also include many benign decomposers that dwell in soil and fresh waters. Apart from the effects of pathogenic members, the Oomycota are easily overlooked. They are small-bodied organisms; many are microscopic.

Three features distinguish the Oomycota from other fungi. First, they produce egg cells during sexual reproduction (*Oomycota* means "egg fungi"). Secondly, most members produce swimming spores with two anterior flagella. And thirdly, the vegetative body has walls of cellulose, rather than the chitin of the Mycota.

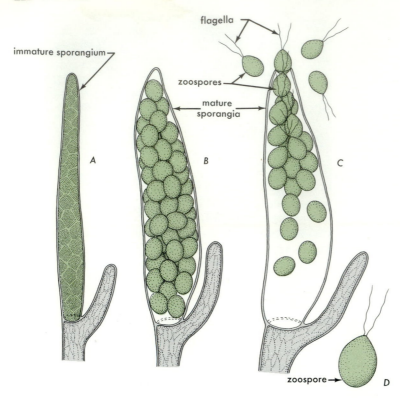

immature sporangium

flagella

zoospores

mature sporangia

zoospore

Figure 15.4. Zoosporangia of *Saprolegnia. A,* immature sporangium; *B,* mature sporangium; *C,* discharge of zoospores; *D,* a zoospore.

In a few species of water molds, one or sometimes both zoospore stages are suppressed.

Sexual Reproduction

When stimulated to reproduce sexually, the mycelium produces specialized hyphae called **oogonia** and **antheridia** in which gametes are formed. The nuclei in the vegetative mycelium are diploid; meiosis occurs in the sexual hyphae to produce the haploid nuclei needed for gametes.

Egg cells are formed inside the oogonia, which are spherical cells at the tips of short side hyphae (Fig. 15.5). When mature, each oogonium is three to four times the diameter of an ordinary hypha and may contain from one to twenty eggs. Each spherical egg contains one nucleus.

The antheridia are also formed at the tips of hyphae (Fig. 15.5). Each is separated from the rest of the hypha by a cross-wall. The antheridia are usually similar in diameter to ordinary hyphae.

The antheridium grows toward an oogonium; it is directed in its growth by a chemical signal (a *sex hormone*) released by the oogonium. When the antheridium touches an oogonium, it produces a short hypha called a *fertilization tube,* which punctures the oogonial wall and comes in contact with one or more eggs. If the oogonium contains several eggs, the fertilization tube usually branches, sending a branch to each egg. Nuclei (male gametes) from the antheridium migrate into each egg, and fertilization ensues. The zygote develops a heavy wall, becoming an **oospore,** and usually will not germinate for several months, even under favorable conditions. Upon germination, the oospore produces hyphae, which rapidly grow into a typical mycelium.

Some oomycete species are bisexual, both kinds of gametangia being borne on the same mycelium. But others have separate sexes. In these species, the antheridia and oogonia do not form unless a suitable partner is available. The system that signals the presence of a partner has attracted considerable scientific interest because it is a clearcut case in which environmental signals cause major developmental changes. By placing male and female mycelia in the same tank of water but separated by a membrane, it has been found that the female constantly emits very small quantities of a compound that functions as a sex hormone. The male detects this hormone and responds by producing antheridia. The antheridia in turn give off a second hormone that diffuses to the female mycelium and induces it to produce oogonia. Additional hormones induce the antheridium to grow toward the oogonium, cling to it, and produce the fertilization tubes. It has not yet been discovered how the hormones produce these responses, but special sets of genes are probably called into play.

PERONOSPORALES

Nearly all the Peronosporales are obligate parasites. In general, heavy-walled oospores persist through unfavorable periods, and various sorts of asexual spores bring about rapid multiplication under suitable conditions.

The most famous member of the Peronosporales is *Phytophthora infestans* (Fig. 15.6), the cause of the potato blight mentioned on page 221. In the 1840s Ireland had virtually a one-crop economy, featuring the potato. *Phytophthora* was present in the potato fields for many years prior to the blight years, gradually spreading, but conditions were never right

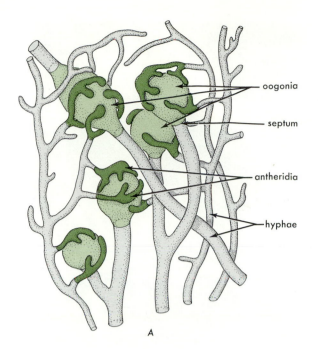

oogonia

septum

antheridia

hyphae

A

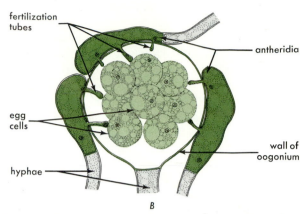

fertilization tubes

antheridia

egg cells

wall of oogonium

hyphae

B

Figure 15.5. Gametangia (gamete-forming structures) in *Saprolegnia*. *A*, oogonia and antheridia with associated hyphae; *B*, antheridia, oogonium with many eggs, fertilization tubes in place.

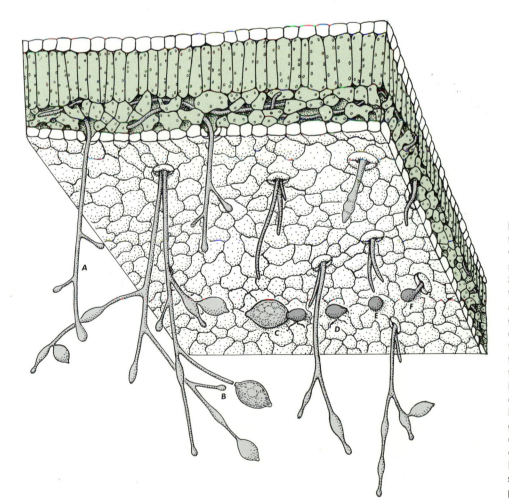

Figure 15.6. Asexual reproduction of *Phytophthora infestans*. The vegetative mycelium invades the spongy mesophyll of a potato leaf. Sporangiophores (*A* and *B*) grow through stomata. Multicellular sporangia (*B* and *C*) form. Each sporangium breaks off as a unit (*B*) and is carried by wind to another leaf surface (*C*). Zoospores emerge from the sporangium (*D*) and swim in the film of moisture on the leaf surface before encysting and becoming dormant (*E*). The dormant spore later germinates (*F*) and produces a germ tube that passes through epidermal cell walls or stoma, to produce a new mycelium within the leaf. By repeating these events the spores can quickly infect a whole potato field.

for maximum growth. Then, starting in 1843, there followed a period of unusually warm, humid summers during which the fungus nearly annihilated the potato crops. The disease kills foliage and rots the potato tubers. Farmers of the day had no idea of the cause and were skeptical when botanists discovered that a fungus was the culprit. Since then it has been found that *Phytophthora infestans* can be controlled (it still exists in potato fields) by spraying infected fields with fungicide, as well as by carefully disposing of infected tubers.

The mycelium of *Phytophthora* grows between the cells within a leaf or stem (Fig. 15.6). Parasitic hyphae **(haustoria)** penetrate the cells. Aerial sporangia arise on long aerial hyphae called **sporangiophores** that extend out through the stomata of the infected plant. The sporangia break loose and are disseminated by air currents.

Some of the sporangia eventually come to rest on the leaves of susceptible plants. When moisture on the leaf surface is sufficient, zoospores escape from them. They soon send out small hyphae called **germ tubes** that penetrate the host tissues and bring about new infections.

Downy mildews are a group of fungi in the Peronosporales that cause infections in many cultivated and wild plants (Fig. 15.7). They are easily identified by the sporangiophores that may, in severe cases, nearly cover the leaf. In the laboratory, the sporangia can usually be induced to germinate by floating them on cool water.

A small group of Peronosporales, the "white rusts," develop sporangia beneath the epidermis of such crop plants as mustards and spinach. We mention them because their sporangiophores do not grow out of the stomata as in the downy mildews. Instead, they collect in pustules under the

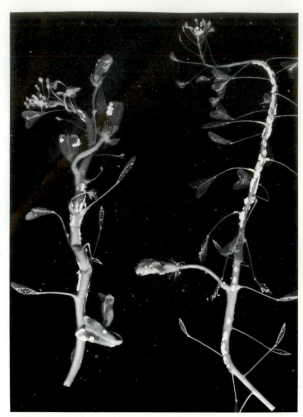

Figure 15.8. *Albugo* (a white rust) on shepherd's purse.

epidermis of the stem or leaf. Sporangia are cut off in chains from the tips of the sporangiophores. They accumulate in large numbers and finally rupture the epidermis, forming creamy-white pustules (Fig. 15.8).

Sexual reproduction in the Peronosporales is similar to that observed in the Saprolegniales.

DIVISION CHYTRIDIOMYCOTA

The true fungi are primarily terrestrial, but they are thought to be derived from aquatic ancestors. We may never be certain of their origin. But if we wish to see what the ancestors of true fungi were like, the present-day **Chytridiomycota,** or *chytrids,* may offer the closest resemblance. Most of the 450 chytrid species are aquatic, and they all have chitin as the fibrous component of the cell wall. They range from microscopic unicells without a mycelium to larger forms with a substantial mycelium. Chytrids have one unique feature not found in other fungi: they form zoospores with a single trailing flagellum. Thus, if the true fungi and the chytrids had a common ancestor, the ability to form flagella must have been lost in the transition to terrestrial life. Such a loss would be expected, for swimming cells are a liability in the terrestrial habitat.

The chytrids may provide a model for the evolution of the mycelium (Fig. 15.9). The simplest forms, such as *Olpidium,* are aquatic unicells that have no trace of a mycelium. They are parasites that attack algae and other fungi.

Figure 15.7. Downy mildew (*Bremia*) on a lettuce leaf.

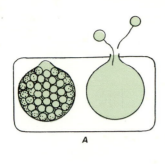

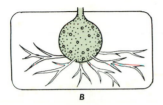

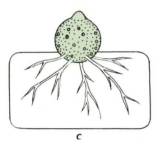

Figure 15.9. Variations in structure among the simple parasitic chytrids. In *A*, the entire organism becomes a simple saclike sporangium inside the host cell. In *B*, rhizoids increase the absorptive area. In *C*, the sporangium is outside the host cell; only rhizoids enter the host. (Redrawn from J. Webster, *Introduction to Fungi.* Cambridge: Cambridge University Press, 1970 p. 24.)

When the swimming spore finds a host, it makes a hole in the wall of a host cell and injects its own protoplast into the cell. Once inside, the chytrid absorbs food and divides into many zoospores. In effect, the whole chytrid becomes a sporangium, sheltered inside the host cell wall. The mature zoospores escape through a hole in the wall and swim away. Other chytrids, such as *Rhizophydium* (Fig. 15.9), behave similarly except that part of the chytrid cell remains outside the host and becomes a sporangium, while slender filaments called **rhizoids** grow into the host cell to absorb food. Rhizoids resemble ordinary hyphae but are too slender to admit nuclei. Chytrids of this type are not as well protected as those that are totally enclosed in the host cell, but the external sporangium may offer more space for spore formation. With the sporangium so exposed, one can imagine that chytrids with more rapid food absorption and better anchorage would have an advantage. The evolution of rhizoids may offer just such an advantage, since they provide both anchorage and a large surface for absorption. The most complex chytrids, such as *Allomyces*, have a fully developed mycelium with true hyphae, and they live as decomposers rather than parasites. Their mycelia are similar in complexity to those of the simpler true fungi (the Zygomycota).

Though chytrids are inconspicuous, many of them are important decomposers in lakes and streams. A few species parasitize crop plants, but none are major pests.

DIVISION ZYGOMYCOTA

The **Zygomycota,** for *zygomycetes,* are the simplest group of fungi that lack motile cells. They owe their name to the trait of converting the zygote into a resting spore called a **zygospore** (Fig. 15.11). They are chiefly terrestrial fungi that form an extensive coenocytic mycelium. In contrast to the Oomycota and the green plants, the walls of the zygomycete fungi have *chitin* as their fibrous framework.

The vegetative hyphae invade the food substrate and also grow on the surface, forming a cottony mycelium (Fig. 15.10). Soon after the mycelium is established, it commences asexual reproduction. This is the principal means of dispersal.

The zygomycetes reproduce asexually by means of sporangia, which form at the tips of aerial sporangiophores (Fig. 15.11). A cross-wall forms below the sporangium. The multinucleate protoplasm in the swollen sporangium then divides into many spores. Members of the class differ greatly in the details of asexual reproduction: for instance, the common bread mold, *Rhizopus nigricans,* forms special, thick hyphae **(stolons)** that arch through the air and are anchored to the substrate at intervals by finer branch hyphae, or **rhizoids** (Fig. 15.11). Clumps of sporangiophores arise at each anchor point. Other zygomycetes produce solitary sporangiophores, or sporangiophores that branch and bear several sporangia.

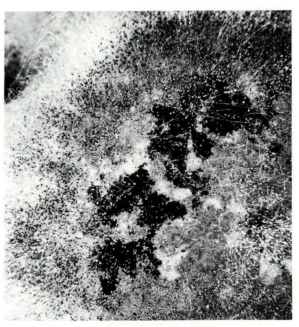

Figure 15.10. *Rhizopus* on peach, ×4.

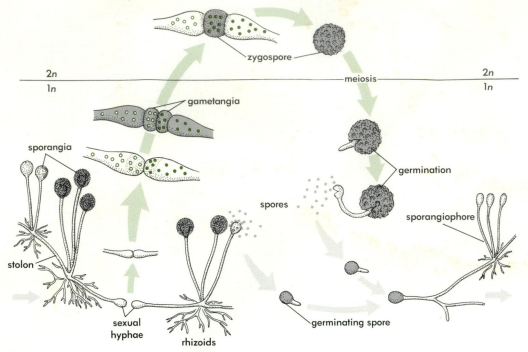

Figure 15.11. The life cycle of a zygomycete fungus (*Rhizopus stolonifer*). Green arrows trace the events in sexual reproduction; gray arrows trace the events in vegetative growth and asexual reproduction.

Many zygomycetes have sophisticated mechanisms that use light and gravity to direct sporangiophore growth so that the spores will be brought to locations suitable for dispersal. Other mechanisms may time the formation of sporangiophores on the basis of temperature, illumination, and nutritional conditions.

Spores are dispersed by various means. In some species the sporangial wall breaks down at maturity and the light, dry spores are spread by wind. In others the spores form a sticky mass that is carried by animals. Members of the genus *Pilobolus* build up pressure in the sporangiophore until the sporangium as a whole is discharged explosively and travels several meters through the air (*Pilobolus* literally means "hat thrower").

Sexual reproduction is much less common than asexual reproduction. It begins with the formation of specialized sexual hyphae (Fig. 15.11). These hyphae grow together, touch, and fuse at the tips. The two tips contain many nuclei and will serve as **gametangia,** or gamete-producing structures. Special enzymes dissolve the wall at the contact point so that the protoplasm of the two hyphae flows together in a single mass. At the same time, a cross-wall forms behind the contact point on each side. This walls off a single large cell with many nuclei from each parent. The nuclei fuse in pairs to form diploid zygote nuclei. The surrounding cell wall thickens, and the cell becomes a dormant spore (a zygospore). The diploid nuclei undergo meiosis to produce a new generation of haploid nuclei before the zygospore germinates. In all the life history of the zygomycete fungus, only the zygospore is diploid.

In some species of zygomycetes, a single mycelium can produce zygospores without a partner. But many species require a mating between sexual hyphae of two different mycelia of opposite "sexes." The partners look alike and contribute equally to mating, so mycologists designate them as "+" and "−" **mating types** rather than "male" and "female." The mycelia advertise their mating type to one another by emitting sex hormones, as in the oomycetes.

Most of the zygomycetes are saprophytes, although a few are parasitic on other members of the division. If left uncontrolled, some zygomycetes, such as *Rhizopus*, cause major losses of stored foods. A few of them cause human diseases.

DIVISION ASCOMYCOTA

The **Ascomycota,** or *ascomycetes,* are named for their trait of producing sexual spores inside a sac, or **ascus** (plural, **asci**), like marbles in a transparent bag (Fig. 15.14).

The life of an ascomycete typically begins with the germination of a haploid spore to produce a vegetative mycelium. As in the zygomycetes, the walls contain chitin instead of cellulose. But in the ascomycetes the mycelium is divided into cells by cross-walls. The cross-walls have a hole in the center (Fig. 15.12) that allows the passage of organelles from one cell to the next; however, until sexual reproduction occurs, each cell usually maintains just one haploid nucleus.

Sexual reproduction begins with the fusion of two hyphae (Fig. 15.13). Many ascomycete fungi are self-incompatible: fusion occurs only if the hyphae come from mycelia

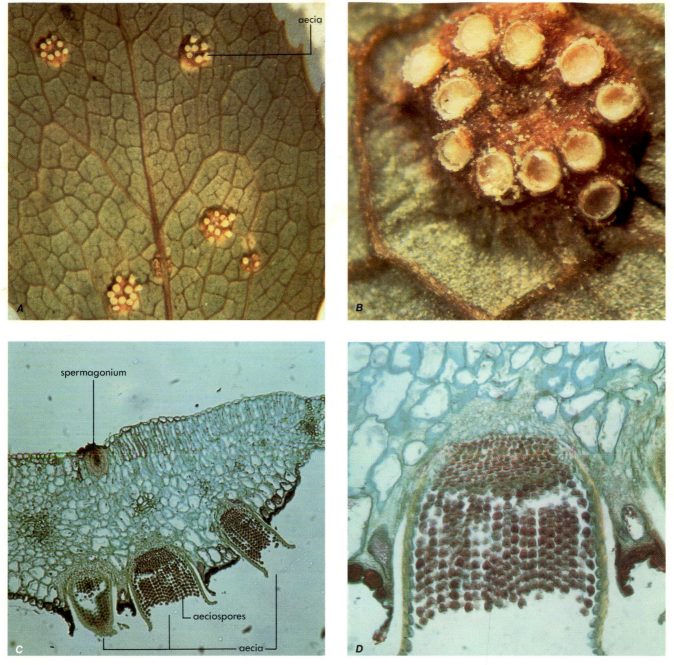

Figure 15.29. Stages in the life cycle of *Puccinia graminis* on barberry leaf. *A,* lower surface of barberry leaf *(Berberis vulgaris)* showing groups of aecia, ×1. *B,* enlarged view of groups of aecia, ×150. *C,* cross section of barberry leaf showing spermagonium on upper surface, aecia on lower surface, ×150. *D,* section of an aecium, ×1000.

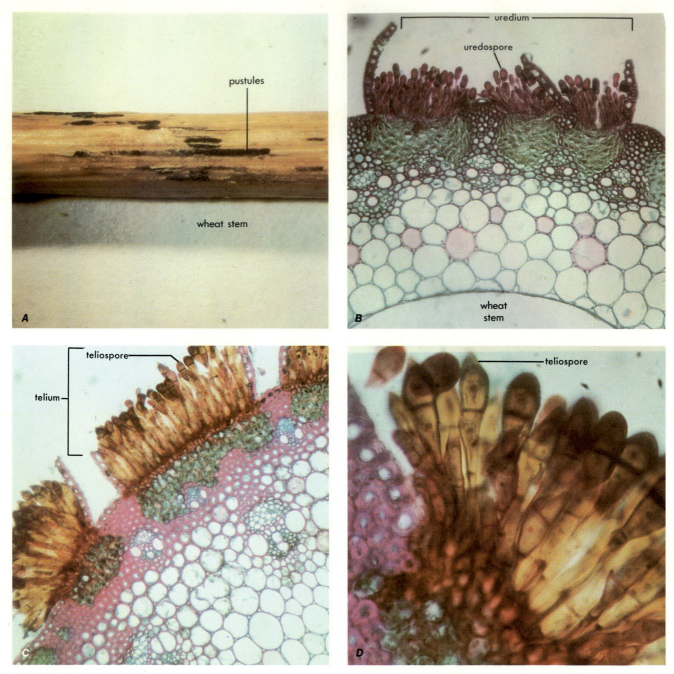

Figure 15.30. Stages in life cycle of *Puccinia graminis* on stem of wheat. *A,* sori on wheat stem, ×1. *B,* cross section of uredia, ×100. *C,* cross section of telia, ×100. *D,* enlarged view of telium showing teliospores, ×1000.

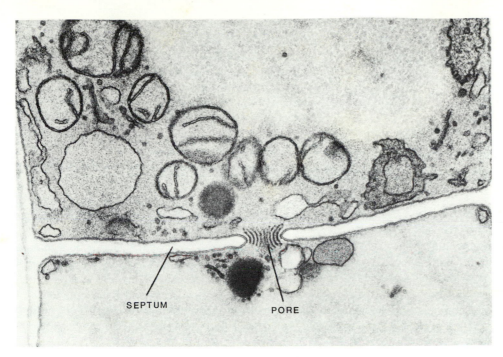

Figure 15.12. Cross-wall (septum) in a hypha of *Ascodesmis nigricans*. Note the central opening or pore. × 40,000. (Photo by C. E. Bracker, Purdue University; reproduced, with permission, from the *Annual Review of Phytopathology*, Volume 5, copyright © 1967 by Annual Reviews Inc.)

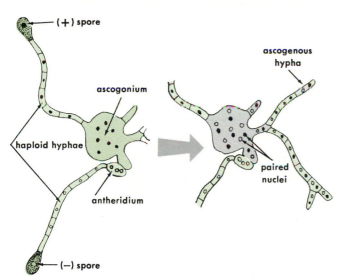

Figure 15.13. Diagram of conjugation and formation of $n + n$ ascogenous hyphae.

of opposite mating types. In some species, fusion can occur between any two hyphae, while in others, fusion involves specialized multinucleate hyphae called **ascogonia** and **antheridia,** and still others produce small free cells called **spermatia** instead of antheridia.

When contact is made, nuclei pass from the antheridium to the ascogonium. The (+) and (−) nuclei pair but do not fuse. The fertilized ascogonium produces hyphae into which the pairs of nuclei migrate. Hyphae with the paired nuclear arrangement are not truly diploid ($2n$), nor are they haploid ($1n$); rather, their nuclear condition is symbolized as $n + n$. A hypha that is $n + n$ is said to be *dikaryotic* (di = two; karyon = nucleus). Since these $n + n$ hyphae produce asci, they are called **ascogenous hyphae.**

The ascogenous hyphae grow and branch before they develop into asci. The formation of the ascus is outlined in Fig. 15.14.

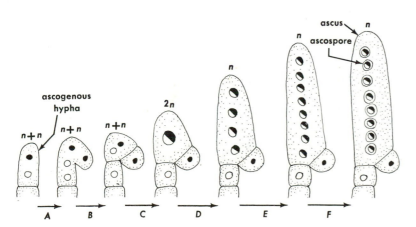

Figure 15.14. Ascus formation. *A*, paired nuclei divide by mitosis; hypha develops J shape. *B*, cross-walls form; penultimate cell is $n + n$. *C*, nuclear fusion (karyogamy) in penultimate cell. *D*, meiosis divides $2n$ zygote nucleus into four haploid nuclei. *E*, each nucleus undergoes mitosis. *F*, walls form around haploid nuclei, yielding eight ascospores inside ascus (old hyphal wall).

Most ascomycetes concentrate their asci within a complex structure called an **ascocarp,** which can be microscopic or as much as 15 cm in diameter. There are three types of ascocarps:

1. **Cleistothecium:** hollow, completely closed sphere (Fig. 15.15*A* and 15.15*B*).
2. **Perithecium:** hollow, flask-shaped body with narrow opening (Fig. 15.15*C* and 15.15*D*).
3. **Apothecium:** open, cup-shaped body (Fig. 15.15*E* and 15.15*F*).

The ascocarp contains both sterile (1*n*) and ascogenous (*n* + *n*) hyphae. Asci line the inner surface of the ascocarp, forming a fertile layer, or **hymenium** (Figs. 15.15*F* and 15.16). Sterile haploid cells, called **paraphyses,** also arise in the hymenium. The hymenial layer is supported and protected by haploid hyphae that form a layer called the **peridium** (Fig. 15.16). Altogether, the ascocarp is much more complex than any structure found in the oomycetes and zygomycetes. For this reason the ascomycetes are often called **higher fungi,** and the former groups, **lower fungi.**

In the ascomycetes, *n* + *n* hyphae are usually confined to the reproductive body and do not contribute to the vegetative mycelium. This contrasts with the basidiomycetes, to be discussed later.

Although sexual reproduction is important for producing new genetic combinations in the ascomycetes, asexual reproduction is more important as a means of multiplication and dispersal. Most species reproduce asexually by forming one spore after another at the tips of aerial hyphae (Fig. 15.17). Spores produced in this way are called **conidia,** and the hyphae that bear them are called **conidiophores.** Various species differ in the branching of the conidiophores. Conidia are usually produced in immense numbers and are light enough to travel for long distances on the wind. Fungi grow so rapidly that many generations of conidia and mycelia can be formed within a summer, and epidemics of fungal growth can occur. Conidia are usually killed by winter cold.

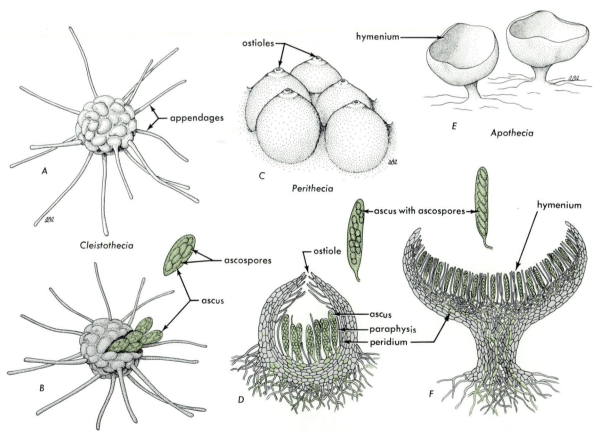

Figure 15.15. Diagram showing three types of ascocarps. *A,* and *B* cleistothecia, ×500; *C* and *D,* perithecia, ×150; *E* and *F,* apothecia, *E* ×¼, *F* ×10. (From *Comparative Morphology of Fungi,* by E. A. Gäuman; copyright 1928, McGraw-Hill Book Company. Used with permission of McGraw-Hill Book Company. *D,* redrawn from E. A. Gäuman. *The Fungi. A Description of Their Morphological and Evolutionary Development.* New York: Hafner Publishing Company, 1952.

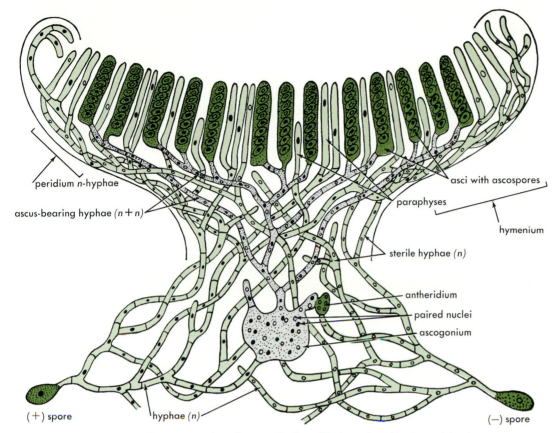

peridium n-hyphae

ascus-bearing hyphae (n + n)

asci with ascospores

paraphyses

hymenium

sterile hyphae (n)

antheridium

paired nuclei

ascogonium

(+) spore

hyphae (n)

(−) spore

Figure 15.16. Diagram of a cross-section of an apothecium. (Redrawn from *Fundamentals of Cytology*, by L. W. Sharp. Copyright 1943 by McGraw-Hill Book Company, New York. Used with permission of McGraw-Hill Book Company.)

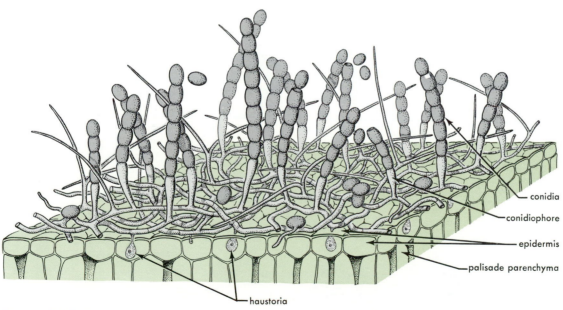

conidia

conidiophore

epidermis

palisade parenchyma

haustoria

Figure 15.17. The ascomycete *Erysiphe* on a leaf surface, producing conidia.

EXAMPLES OF ASCOMYCETES

There are at least 15,000 species of ascomycetes and so many variations that even an introductory survey of them would require many pages. Thus in the present book we will consider only a few illustrative examples.

Yeasts

The simplest ascomycetes are the yeasts. *Saccharomyces cerevisiae* is the yeast of baking and brewing. This fungus is important because it converts sugar (glucose) into alcohol and carbon dioxide. This trait is useful in gaining energy from glucose when molecular oxygen is not available (see Chapter 2). In the presence of oxygen, yeast (like other organisms) can completely oxidize glucose to carbon dioxide and water; although it gains much extra energy this way, it does not produce the economically valuable alcohol.

Yeasts are normally single-celled. After cell division the new cells separate; thus ordinary mitotic division is a form of asexual reproduction (Fig. 15.18). In older cultures the cells may hang together and form short chains. Some yeasts double in size and pinch in half. More often, though, division is a process of **budding:** new cells grow out from the mother cell much as a small bubble would form if a piece of thin rubber were made to expand through a small opening in some heavier material. The small "bubbles" are called **buds.** They enlarge and finally separate from the parent cell.

Ascus formation in yeasts is very simple. Two compatible yeast cells fuse, the nuclei fuse, and meiosis follows. The original wall forms the ascus. There is no ascocarp.

Most ascomycetes do nothing more with the diploid zygote nucleus than to divide it by meiosis, producing haploid nuclei. Some yeasts follow the same pattern. But in others, the diploid cell may reproduce asexually by mitosis and budding (Fig. 15.18). Later, meiosis finally intervenes to re-store the haploid condition. Yeasts vary greatly in the relative time spent in the diploid and haploid states. Some strains are diploid most of the time, a highly unusual trait among the sac fungi.

Powdery Mildews

These are parasites that attack plants and give a powdery appearance to the leaves. The genus *Erysiphe* is representative (Fig. 15.17).

The mycelium is generally confined to the surface of the leaves, flowers, or fruits. Haustoria penetrate epidermal and parenchyma cells, from which they secure nourishment. At first the mycelium appears like a delicate cobweb. Eventually, it assumes a white powdery appearance because of the development of numerous conidia that serve to spread the fungus from one plant to another during the growing season. (Sulfur dusted on the host plants at this stage prevents infections from the spores.)

For sexual reproduction, small black cleistothecia are formed, each enclosing one or several asci (Fig. 15.15*A* and 15.15*B*). The ascospores are usually discharged by force from the cleistothecium.

Although the powdery mildews may not kill their host, they weaken it and greatly reduce the crop yield. Powdery mildews cause diseases of apples, grasses, grains, grapes, cherries, and many other plants.

Monilinia

Brown rot of stone fruits, a very severe disease of peaches, cherries, plums, apricots, and nectarines, is caused by the disk fungus *Monilinia,* which infects mainly blossoms and fruits. The ascocarps of the order to which *Monilinia* belongs are cup-shaped or disk-shaped (apothecia) (Figs. 15.15*E*, 15.15*F*, and **15.19*A*, Color Plate 14**). They may be as much

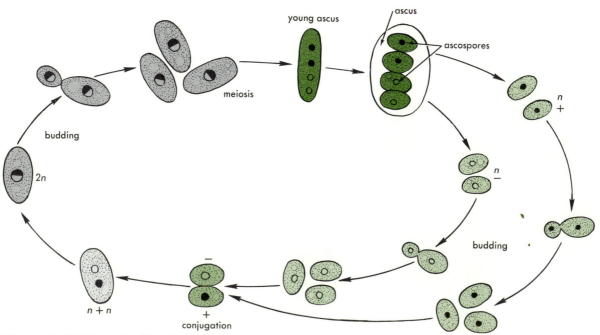

Figure 15.18. Life cycle of *Saccharomyces cerevisiae.*

as 15 cm in diameter, depending on the species, and are sometimes brilliantly colored. Members of a related genus, *Morchella* (the morels), are edible.

Claviceps

The genus *Claviceps* is parasitic on grasses, including grains. A dormant mycelium that replaces the mature grain is known as **ergot.** It possesses several alkaloids that have medicinal properties. Ergot constricts the blood vessels, particularly those that pass into the hands and feet, thus depriving the extremities of a normal blood supply. During humid summers in Central Europe, *Claviceps* may infect rye heavily. In centuries past, before the nature of the fungus was understood, ergot was milled along with the grains of rye. The contaminated flour, which perhaps contained as much as 10% of powdered mycelium, was baked in bread. A continued diet of bread from this flour resulted in much misery. Because of the contraction of the blood vessels and the limited supply of blood reaching feet or hands, gangrene sets in. Hands, arms, and legs die and finally drop off. The disease was known as "Holy Fire" because of a burning sensation in the extremities. Today, ergot is a valued drug used to control hemorrhage, particularly during childbirth, and to treat migraine.

With the knowledge of the poisonous nature of the ergot, diseased grain is no longer milled and *Claviceps* is not of concern in human diet. However, cattle may occasionally feed on grasses infected by ergot and thus be poisoned.

DIVISION BASIDIOMYCOTA

Based on the complexity and size of their reproductive structures, the most advanced of the higher fungi fall in the division **Basidiomycota.** They include all mushrooms, puffballs, and the bracket fungi that grow on the trunks of trees. These elaborate structures are called **basidiocarps.**

Not all the basidiomycetes have elaborate reproductive bodies. The feature that determines whether a fungus belongs to this class is the precise way in which sexual meiospores are formed. Rather than being enclosed in a common sac, the spores formed after meiosis are carried on separate stalks called **sterigmata** (singular, **sterigma**) (Fig. 15.22). The cell that undergoes meiosis and produces the spores is called a *basidium* (Latin for "pedestal"), and the spores are called **basidiospores.**

The basidiomycetes consist of two main groups or classes, as described below.

CLASS HOMOBASIDIOMYCETES

The Homobasidiomycetes comprise all the species that form elaborate fruiting bodies or basidiocarps. Some of these are parasitic but most of them are saprophytic. Some form mycorrhizae—symbiotic associations with the roots of higher plants, previously discussed in Chapter 6.

Others obtain their carbohydrates from wood and may even cause extensive decay of timbers in mines and buildings. Specialized tissues and organs are often formed: some

of the *n + n* hyphae become specialized to transport food and water, and others become tough and woody.

The vegetative body of such a fungus is a mycelium that starts from a haploid spore and invades its substratum, as in previous groups. The walls of the mycelium contain chitin rather than cellulose, and the hyphae are almost completely septate. There is a pore in the center of each cross-wall, but an elaborate caplike structure covers the pore so that organelles cannot easily pass between cells.

Sexual reproduction is highly developed in these fungi; they are not noted for their asexual reproduction.

Sexual reproduction begins with the conjugation or fusion of hyphae to produce a dikaryon with the paired, *n + n* nuclear condition. Some species are **homothallic,** conjugation taking place between any two cells of any two hyphae or even between two cells of the same hypha. Other species are **heterothallic,** and conjugation occurs only between hyphae from mycelia of different mating types.

Once formed, the dikaryon can grow and branch to form a long-lived *n + n* vegetative mycelium. The dikaryon may persist for many years, perhaps even for centuries. If conditions are not right, the production of sexual spores and basidiocarps may be postponed indefinitely.

In the *n + n* mycelium, division occurs in such a way that all the cells maintain the *n + n* condition. This requires a special set of events (Fig. 15.20). Growth of the tip cell of the hypha is accompanied by division of both nuclei. The tip cell develops a J-shape, and one of the nuclei migrates into the end of the J. Cross-walls are laid down as shown in Fig. 15.20. The lateral cell bends around, fuses with the trailing cell, and the *n + n* condition is restored in that cell. This happens each time a new cell is formed at the tip of the hypha. The curved and fused lateral cell is termed a **clamp connection.** Of all the fungi, only the basidiomycetes form a dikaryotic vegetative mycelium with clamp connections.

Although both haploid and *n + n* mycelia develop extensively, the *n + n* phase is of special interest because it gives rise to the spore-bearing body, or **basidiocarp,** in which the basidia are developed. The hyphae form a tangled mass in the substrate or host and emerge to form a basidiocarp—the mushroom, puffball, or bracket fungus. The basidiocarp of a typical mushroom such as *Agaricus brun-*

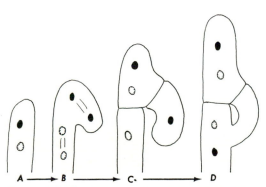

Figure 15.20. Formation of a clamp connection, which maintains the dikaryotic condition in the basidiomycete hypha.

Figure 15.21. *Agaricus brunnescens,* the common edible mushroom, ×1.

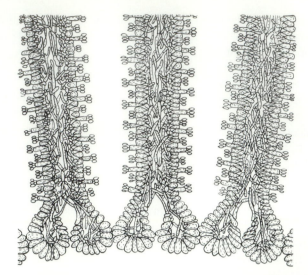

Figure 15.23. Section showing three gills of a mushroom.

nescens (Fig. 15.21) consists of a short upright stalk, or **stipe,** attached at its base to a mass of mycelium and expanding on top into a broad cap, or **pileus.** The underside of the pileus of *Agaricus* is formed by thin **gills** radiating outward from the stipe. These gills are lined with a hymenium, or spore-bearing layer.

Basidia develop from the tipmost cells of hyphae in the hymenium (Fig. 15.22). Here the paired nuclei finally fuse to give a diploid (2n) nucleus, which immediately undergoes meiosis. Sterigmata form at the tip of the basidium. The hap-

loid nuclei that result from meiosis are squeezed through the narrow necks of the sterigmata along with cytoplasm. The formation of a resistant wall completes each spore.

If the gill is very young, developing basidia on which basidiospores have not yet been formed can be observed; Fig. 15.23 shows a section through three gills. Note two stages of development of basidia and the loose tangle of hyphae that form the center of the gill. When a basidiospore is discharged, it is shot horizontally from its position on the basidium straight into the space between two gills. A basidiospore resembles a toy rubber balloon in that it has a large volume for its weight. It is shot from the basidium with a force sufficient to carry it about midway between the two gills. It then falls straight downward. It has been estimated that some basidiocarps may discharge as many as a million spores a minute for several days.

The Agaricales include not only edible fungi but also some of the most deadly of poisonous mushrooms. Those of the genus *Amanita* **(Color Plate 1)** are especially dangerous. It is said that the Eskimos used to pay high prices for dried specimens that, when taken in extremely small quantities, induce hallucinations. Experimentation is ill-advised, however, because even small dosages can be deadly. Although poisonous mushrooms are much less common than benign ones, mycologists advise enthusiasts to consult an expert before sampling mushrooms they have gathered from nature.

Not all Agaricales have gills. In some, the **pore fungi,** the lower surface of the cap, or bracket, has many small pores that extend upward to form small tubules (Fig. 15.24). The hymenium lines the tubules. Many bracket fungi found on the trunks of forest trees are pore fungi.

Some bracket fungi cause serious damage to forest trees and lumber in mines and wooden buildings. During World War II, much green lumber had to be used, and because fungi were able to feed on the improperly cured lumber, much loss resulted. One of the best-known fungi in this

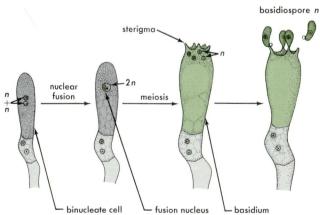

Figure 15.22. Basidiospore formation as it occurs in a mushroom (Homobasidiomycetes). (From *Cryptogamic Botany:* Vol. 1, *Algae and Fungi,* by G. M. Smith; copyright © 1955, McGraw-Hill Book Company; after H. Kniep, *Sexualität der Niederen Pflanzen,* copyright G. Fischer Verlag, Jena 1928; and after *Researches on Fungi,* Vol. VII *The Sexual Process in the Uredinales,* by A. H. R. Buller; University of Toronto Press 1922. Used with permission from McGraw-Hill Book Company, Gustav Fischer Verlag, the Royal Society of Canada, and the University of Toronto Press.)

Figure 15.24. The underside of the pore fungus *Polyporus brumalis,* × ⅓.

Figure 15.25. Basidiocarp of *Ganoderma applanatum,* × ⅓.

group is *Ganoderma applanatum* (Fig. 15.25). It causes a disease of forest trees known as *white-mottled rot,* to which many hardwoods and conifers are susceptible.

The basidiocarp of *Ganoderma applanatum* is a very elegant shelf-shaped bracket, or **conk.** The conks are perennial and grow to a large size. They are gray on top, and the lower surface has millions of pores lined with a creamy-white hymenium.

Other types of basidiocarps also occur, such as those in puffballs and bird's-nest fungi.

CLASS HETEROBASIDIOMYCETES

The fungi of this group are much less familiar to the average person than those of the group just described, because they do not produce elaborate basidiocarps. But economically they may be much more important. This class includes some of the most virulent crop destroyers: the **rusts** and **smuts,** which attack cereal grains such as wheat and corn and cost millions of dollars in lost harvests and preventive costs. Most members of this class are parasites on plants.

The hallmark of the smut fungi is the replacement of grains by black masses of spores that have a sooty look; hence, the name. Smuts vary in their method of attacking the host: for instance, some species attack only seedlings, and some attack flowers. Corn smut (*Ustilago maydis,* Fig. 15.26) causes local infections wherever the haploid spores

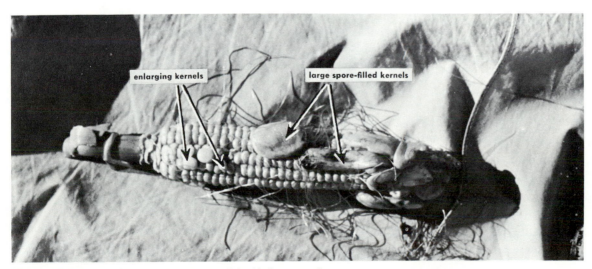

enlarging kernels   large spore-filled kernels

Figure 15.26. Smut disease of corn caused by *Ustilago maydis.*

land on the host plant. Conjugation occurs when two haploid mycelia grow together, and the resulting $n + n$ mycelium forms tumors in which many $n + n$ spores are formed (Fig. 15.26). After surviving through the winter, these spores germinate; the nuclei fuse and go through meiosis, and haploid basidiospores are formed to start another round of infections. Resistant strains of corn effectively limit this pest.

Some of the Heterobasidiomycetes have two hosts; that is, separate phases of their life history are passed on different plants. Such forms are **heteroecious.** If only one host is required to complete their life history, they are said to be **autoecious.**

The rust fungi owe their name to the rusty color of some of their spores, which form in abundance on the leaves and stems of infected grasses. An example of this group is *Puccinia graminis,* which causes common wheat rust and is a persistent and costly pest.

Puccinia graminis has two hosts: a grass, and a shrub known as barberry *(Berberis).* Various strains of the fungus attack different grasses. Variety *avenae* attacks oats and variety *tritici* attacks wheat. These host choices are hereditary with the fungus.

The two hosts have different roles in the life of the fungus (Fig. 15.27). Roughly speaking, the grass serves to multiply and spread the existing genetic types of the fungus population. This is achieved by asexual reproduction using food reserves that are drained from the leaves and stems of the grass plant. The grass is usually not killed, but its vitality is impaired and its seed production is reduced.

The barberry host is the sexual ground for the rust fungus. Here, matings between different mycelia produce new genetic combinations. These in turn move out to invade the grain fields. The barberry plant is not killed by the infection.

Our survey of the life cycle of wheat rust will begin with the basidiospores. These spores are carried by air currents in the spring, and they can only attack barberry shrubs. They germinate on barberry leaf surfaces and form a germ tube that penetrates the epidermis.

A mycelium develops in the tissues of the host. Pustules, called **spermagonia,** appear on the upper surface of leaves. A spermagonium is pear-shaped and has a small opening through the upper epidermis to the exterior of the leaf (Figs. 15.28 and **15.29C, Color Plate 19**). Hyphae in the spermagonium cut off large numbers of spermlike cells, the **spermatia.** These are forced out of the spermagonia through the pores.

Next, spermatia are carried by wind or insects to other spermagonia, where they contact and fuse with special **receptive hyphae.** An $n + n$ mycelium results. This is important because it brings the genes of two mycelia together, allowing the formation of superior rust strains. This benefit would be missed if spermatia were to fuse with receptive hyphae of the same mycelium. Appropriately, *Puccinia graminis* forms mycelia of distinct plus and minus mating types, and the spermatia and receptive hyphae must be of opposite types before they will fuse.

How do mycelia get from barberry to wheat? The newly formed $n + n$ mycelium on the barberry leaf invades the spongy mesophyll. Close to the lower epidermis the mycelium forms pustules called **aecia (Fig. 15.29, Color Plate 19).** Here $n + n$ spores **(aeciospores)** are formed. They are released when the pustule breaks through the leaf epidermis. Aeciospores, borne by the wind, attack only wheat plants.

Hyphae from the germinating aeciospores enter the wheat plant through its stomata. A delicate mycelium forms with haustoria that penetrate the cells of the leaf and extract

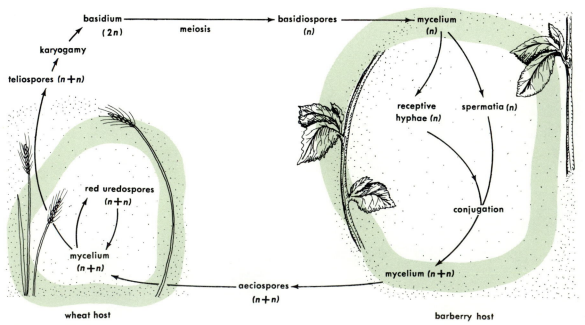

Figure 15.27. Summary of *Puccinia graminis* life cycle.

nutrients. About 10 days after infection, red *n* + *n* **uredospores** form. The pustules, or **uredia,** containing these spores open up **(Fig. 15.30A** and **15.30B, Color Plate 20).** Carried on wind currents, the uredospores attack other wheat plants at a rate that can attain epidemic proportions. Wheat rust gets its name because of these red lesions (uredia) in the stems.

As the wheat begins to mature, the production of uredospores gives way to the production of black teliospores by the same mycelium **(Fig. 15.30C** and **15.30D).** Whereas uredospores and aeciospores are killed by winter cold, teliospores are not; they are overwintering spores. Each cell of the teliospore begins with the dikaryotic (*n* + *n*) nuclear condition, but during winter, fusion occurs to form a zygote. Meiosis occurs in the warm, moist weather of spring. Then

each cell of the teliospore puts out a single short hypha, the basidium (Fig. 15.31), and four basidiospores are extruded. No host is involved at this stage; germination occurs at any moist surface where the spore happens to land. With this event the fungus has completed its cycle.

Wheat rust is a costly pest, and great efforts are being made to control it. Two approaches are involved. One approach is to breed new varieties of wheat that resist the common strains of rust. This is effective in the short term and has a side-benefit in that genes for better yields are also built into the new wheats. However, new strains of rust arise every year, due to sexual recombination on the barberry host. Among them are strains that can attack the new variety of wheat, and within a decade the once-resistant wheat cannot be planted without risking a rust epidemic.

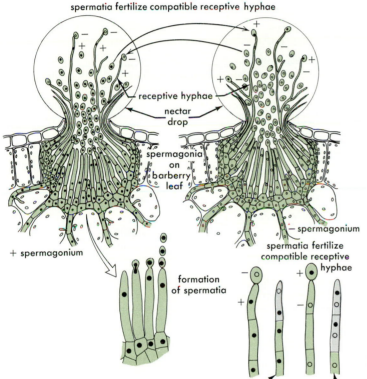

Figure 15.28. Reproduction in the wheat rust, *Puccinia graminis*. Formation of receptive hyphae and their conjugation on barberry.

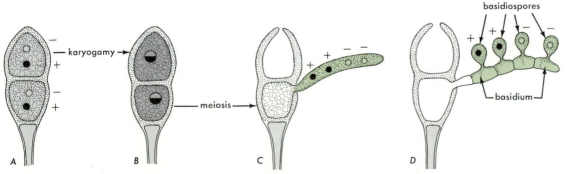

Figure 15.31. Germination of a teliospore, *Puccinia graminis*. *A*, teliospore with binucleate cells; *B*, 2*n* cells; *C*, meiosis and germination of a cell produces a hypha with four haploid nuclei; *D*, crosswalls form and basidiospores are extruded. In *C* and *D* the upper cell is left blank for clarity but would follow the same development as the lower cell.

Plant breeders must continually introduce new wheat varieties to keep ahead of the fungus.

The second method of controlling wheat rust is to eliminate the barberry plants. This approach dates back to the year 1660 in France. If successful, it would stop the formation of new rust strains. Unfortunately, barberry is widespread and inconspicuous, and it inhabits very rough mountainous areas as well as wheat-growing regions. This prevents a complete eradication, and so the battle between humans and wheat rust continues.

DIVISION DEUTEROMYCOTA (FUNGI IMPERFECTI)

About 24,000 species of fungi, in some 1200 genera, are known only by their vegetative and asexual reproductive structures. For convenience, they can be treated as a division (the **Deuteromycota**). However, it should be remembered that the grouping is highly artificial. They are often called the **Fungi Imperfecti** because their life cycles are incomplete (imperfect).

The structure of the hyphae, which are septate, and of the spores, suggests that many imperfect fungi may be ascomycetes; others may be basidiomycetes. The Fungi Imperfecti can be considered to be ascomycetes or basidiomycetes whose sexual stages have not been observed or no longer exist. Occasionally, sexual stages are discovered in species that have previously been classified as imperfects. When this happens, the fungus is renamed, but usually the old name for the imperfect stages is kept too. For instance, the imperfect fungus *Penicillium vermiculatum* is now known to form asci similar to those of the ascomycete genus *Talaromyces;* hence this fungus is also called *Talaromyces vermiculatum*.

Since the classification of imperfect fungi is very doubtful, the groupings are designated as **form genera** or **form families** to indicate that the members do not necessarily have a natural or family relationship. Some species have been named simply on the basis of the host upon which they were found.

Only a single form group, the form order Moniliales, will be discussed here; it is the largest in the class and has over 10,000 form species.

Penicillium and *Aspergillus* are probably the most widespread fungi. They are the common blue, green, and black molds that occur on citrus fruits, jellies, and preserves. Their conidia are in the air and soil everywhere.

In *Penicillium,* the spores are formed on profusely branched conidiophores (Fig. 15.32A). In *Aspergillus,* the tip of the conidiophore swells and conidia form in long chains radiating from this swollen tip (Fig. 15.32B and 15.32C).

Enzymes secreted by these fungi are particularly active in digesting starch and other carbohydrates. When purified, these enzymes are important industrial preparations. *Aspergillus oryzae* is used in the preparation of rice wine and soy sauce. Several species of *Aspergillus* are important in cheese manufacture. A disease resembling tuberculosis is caused by *Aspergillus fumigatus,* and several other *Aspergillus* species cause diseases in plants. Strains of *Aspergillus flavus* form compounds called aflatoxins that are toxic to animals.

Penicillium notatum has won deserved fame because **penicillin,** a drug derived from it, will inhibit the growth of bacteria without injuring human tissue. Substances such as penicillin, formed by one organism and inhibiting the growth of other organisms, are called **antibiotics.** Our knowledge of them dates back to 1870, but their significance was not fully appreciated until the discovery of penicillin.

In 1928, Alexander Fleming investigated extracts of *Penicillium;* after demonstrating their antibiotic properties,

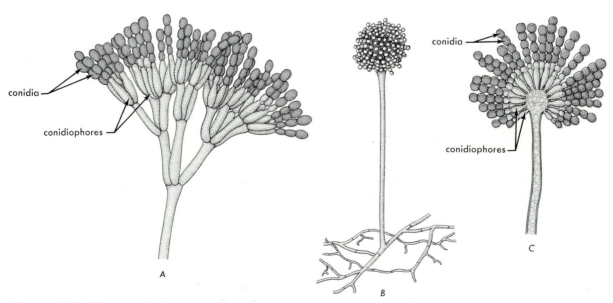

Figure 15.32. Diagrams of different forms of conidiophores. *A, Penicillium; B, C, Aspergillus; B,* three-dimensional appearance; *C,* optical section through *B.*

he gave the antibiotic the name *penicillin*. It was not until after 1940 that a concentrated effort was made to produce penicillin on a commercial scale, for use in World War II.

Many thousands of kilograms of mycelium of *Penicillium notatum* have been grown, and the species has been subjected to intensive study, yet no sexual stages have ever been observed. The absence of sexual reproduction presents the geneticists who are studying *Penicillium notatum* with a difficult breeding problem. Nevertheless, they have succeeded in growing strains that produce many times more penicillin per kilogram than the original strains.

Several forms in the Moniliales are adapted to capture and destroy microscopic animals. A few trap nematode worms by forming small rings that quickly constrict when stimulated by the contact of a nematode crawling through them. A nematode-trapping form is shown in Fig 15.33.

Since the sexual process has not been observed to occur in the Fungi Imperfecti and genetic variation is not thought to be a part of asexual reproduction, the question may be asked: Is genetic variation possible in the Fungi Imperfecti? The answer is, apparently, yes, at least in the laboratory. Careful genetic analysis has produced evidence that mycelia of imperfect fungi may sometimes fuse and bring together nuclei of different origin. By unknown mechanisms, nuclei are formed with genes from each of the original kinds of nuclei. Meiosis appears not to be involved; hence this way of forming recombinant nuclei is called the **parasexual cycle.** It is well-established in laboratory cultures where it may account for the great variability of some Fungi Imperfecti. It has not been established that the process is significant in wild populations.

THE LICHENS

The lichens are composite plants composed of algae and fungi. The algal components are generally single-celled forms belonging to either the Chlorophyta or the Cyanophyta. When free from the fungus, the algae can exist normally.

The fungal component is generally an ascomycete, but algal associations with basidiomycetes are also known. The fungal component of the lichen cannot live separated from the alga unless supplied with a special nutritive medium.

The association of a fungus and an alga in the lichen is generally believed to be **symbiotic;** that is, both the fungus and the alga derive benefit from the association. The alga furnishes food for the fungus, and the fungus supplies moisture, shelter, and minerals for the alga.

A section through a lichen thallus reveals four distinct layers (Fig. 15.34). The top and bottom layers consist of a compact mass of intertwining fungal filaments. The algal cells form a green layer beneath the top mass of fungal filaments, and a loose layer of hyphae lies directly below the algal cells.

Lichens are both widespread and varied (there are about 20,000 species). They form luxuriant growths on the frozen, northern tundras, where they supply feed ("reindeer moss") for animals. In the United States they frequently cover rocks, trees, and boards exposed to sun and wind **(Fig. 15.35, Color Plate 14).** They are slow but efficient soil formers; rocks are disintegrated slowly by their action. Lichens are pioneer plants, appearing before any other plants on recent lava flows. Lichens and mosses accumulate soil and organic matter sufficient to allow herbs and, later, shrubs and trees to become established.

For convenience in classifying, lichens can be divided into three groups simply on the basis of their vegetative body. This is a highly artificial grouping. Probably the most conspicuous lichens are members of the **fruticose** type **(Fig. 15.35C).** They consist of small tubules or branches. Fruticose lichens make extensive and attractive growths on oaks or the dead branches of certain pines. Some lichens have a leaflike plant body; they constitute the **foliose** type and

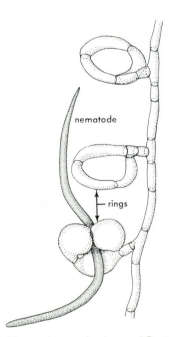

Figure 15.33. Nematode-trapping loops of *Dactylaria*. (Redrawn from E. A. Gäuman, *Fungi, A Description of Their Morphological Features and Morphological Development.* New York: Hafner Publishing Company, 1950.)

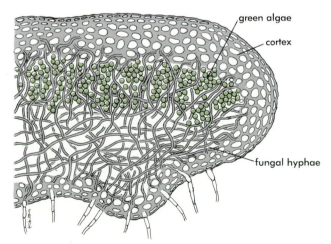

Figure 15.34. Cross section through a lichen.

are common in moister climates, where their growths on tree trunks or on fallen logs are extensive **(Fig. 15.35A)**. However, the **crustose** lichens are the most common type, as they grow extensively on rocks, bark, poles, boards, and other hard surfaces **(Fig. 15.35B)**. Indeed, any bright coloring on the dark rocks and gray granites of our northern mountains, on unpainted wood, or on bark of trees is quite likely to be caused by a growth of crustose lichens.

REPRODUCTION

Lichens often multiply by releasing pieces of the vegetative thallus. Many lichens produce special powdery bodies, composed of both a fungus and an alga, that serve as a means of vegetative reproduction. These bodies are called **soredia** (singular, **soredium**).

Sexual structures reflect the type of fungus present in the lichen. Since the fungal component is usually an ascomycete, ascocarps with asci and ascospores are usually formed.

Both the algal and the fungal components may produce spores. The algae can grow independently, but the young hyphae die if they do not find the required algal cells.

SUMMARY

1. The fungi are heterotrophic eukaryotes that lack photosynthesis, typically reproduce by spores, and have cell walls during at least part of the life cycle. About 200,000 species are known.

2. In this text, the fungi that lack flagella and have chitinous cell walls comprise the kingdom Mycota, which has four divisions: Zygomycota, Ascomycota, Basidiomycota, and the artificial division Deuteromycota. The fungi that have a motile stage are placed in the kingdom Protoctista. They include the divisions Myxomycota, Oomycota, and Chytridiomycota.

3. Most fungi form a vegetative body (a mycelium) composed of branching filaments (hyphae). They feed by absorbing dissolved solutes from solution.

4. Slime molds (division Myxomycota) lack cell walls except in spores. The vegetative body has amoeboid motion and engulfs solid food particles.

5. The chytrids (division Chytridiomycota) have chitinous walls, and they form zoospores with one trailing flagellum. Most are aquatic. The simpler chytrids illustrate possible steps in the evolution of the mycelium.

6. The egg fungi (division Oomycota) usually have reproductive cells with two flagella, and their cell walls contain cellulose. The mycelium may be diploid. Many species are aquatic. Some are important plant pathogens.

7. The zygote fungi (division Zygomycota) mate by fusing special hyphae. The zygote becomes a spore. Asexual spores are produced in sporangia. The mycelium is haploid, coenocytic, and is usually terrestrial. Most are saprophytic.

8. The sac fungi (division Ascomycota) form sexual spores inside a sac (the ascus). The vegetative body may be unicellular (yeasts), or it may be a multicellular mycelium that is divided by perforated cross-walls. The vegetative mycelium is haploid. Fusion of sexual hyphae produces dikaryotic (n + n) hyphae that produce asci. Most species form ascocarps composed of many haploid and dikaryotic hyphae. Asexual reproduction is usually by conidia. Most species are terrestrial saprophytes, but some are plant pathogens.

9. The club fungi (division Basidiomycota) produce meiospores on the surface of basidia. Mushrooms and brackets are the fruiting bodies (basidiocarps) of fungi in the Class Homobasidiomycetes. The basidiocarps consist of dikaryotic hyphae. The vegetative mycelium may be haploid or dikaryotic. Most members are terrestrial saprophytes. The rusts and smuts (Class Heterobasidiomycetes) lack basidiocarps and are plant parasites. Some of them use two plant species as alternate hosts.

10. The imperfect fungi (division Deuteromycota) have no known sexual reproduction. Reproduction is asexual, usually by conidia. Most form a septate terrestrial mycelium. This class includes the most common household molds and the fungi that produce penicillin and fine cheeses.

11. The lichens are symbiotic associations in which a fungus and an alga produce a common body. Widely varied forms occur. Reproductive structures include separate fungal and algal spores as well as bodies (soredia) that include both alga and fungus. Some lichens are important pioneers on bare rock.

16 ALGAE

The first eucaryotes were aquatic, and their descendants remained aquatic for at least a billion years before life spread to the land. Over that immense time, the aquatic plants evolved into many forms that are grouped under the informal heading **algae.** They number at least 25,000 living species and uncounted fossil forms.

The term *alga* represents a grade or level of life, which was reached by organisms from many lines of descent. It is not a formal taxonomic term. The only features found in nearly all algae are photosynthesis, an adaptation to aquatic life, and a lack of the traits that terrestrial plants evolved in response to their distinctive environment. Plants above the grade of algae have features that protect the gametes and offspring against dehydration: the gametes are formed inside a jacket of sterile cells, and the young plant grows to a multicellular embryo while still enclosed within parental tissues. Algae rarely form a sterile jacket and they never form embryos.

The algae include single cells, loosely organized colonies, flat, leaflike sheets, or intricately branched filaments (Fig. 16.1). In most algae, all the cells can carry on photosynthesis and obtain water and nutrients directly from their surroundings. Only rarely, in the kelps, do algal bodies become large and complex enough to differentiate their tissues into organs that resemble stems and leaves (Fig. 16.2). Algae lack true roots, and only the giant kelps form even a rudimentary vascular system. Their aquatic and semi-aquatic environments lack survival pressure for the evolution of vascular and supporting tissue. A body of such simplicity is called a **thallus,** and therefore the algae are called **thallophytes.**

Although algae are best adapted to wet environments, many species survive in seasonally dry habitats by going dormant between wet seasons or by forming symbiotic associations with fungi (see the discussion of the lichens in Chapter 15). Some algae drift or swim in open water; others are attached to the bottoms of streams, to soil particles, to tree trunks, or to rocky intertidal cliffs battered by surf. Still others form symbiotic associations with higher plants and with animals. Algae occur in the most severe habitats on earth. Some species grow on snow in perpetually freezing temperatures; others thrive in hot springs at temperatures of 70°C or more; a few live in extremely saline bodies of water, such as the Great Salt Lake in Utah; yet others survive the pressure and low light intensity conditions of 100 m or more below the surface of lakes or seas. They have even been detected 1 km from ground zero, six months after a 20-kiloton atomic bomb test in Nevada, and they were considered to be survivors of the explosion.

Probably the most commonly noticed natural urban habitats for algae are the sides of glass fish tanks. They can also generally be found around leaking faucets and in garden or park pools that are not kept "pure" with chemicals. Algae make up the "bloom" that occurs during the summer on many lakes or the scum found on ponds. Microscopic forms occur in most natural waters, including the top 75 m of the ocean. Here they constitute a primary food source for marine animals of all types and account for about 50% of the oxygen released into the atmosphere through photosynthesis.

ECONOMIC IMPORTANCE OF ALGAE

Algae are important in two basic but quite different ways. They are important to the entire biosphere because of the ecologically vital functions they perform. These functions include the production of carbohydrates, which places the algae at the base of food chains, the fixation of nitrogen, and reef building. They are also economically important to people because they serve as food, fodder, and fertilizer, and they also have many industrial and pharmaceutical uses.

ECOLOGICAL FUNCTIONS

Algae as Producers

In aquatic ecosystems, algae are the major producers. Shallow bodies of water may depend on productivity by higher plants such as rushes, water lilies, duckweed, or the like, or they may be fueled by a constant supply of decomposing plant material **(detritus)** washed down from the surrounding land. However, almost 70% of the globe is covered by deep water, and here algae are the only producers. They have been called the grasses of the oceans, converting solar energy into chemical energy that is passed up through the rich marine food chain, and releasing oxygen as an important by-product.

Algae occupy several distinct habitats in aquatic ecosystems. One habitat, along rocky oceanic coasts, is the **intertidal** zone (Fig. 16.3). This zone is especially severe for the growth of algae, because the plants are alternately exposed by low tide and inundated by high tide. When exposed, the plants are beset by desiccation, high temperatures, high light intensity, and increased salinity. When the tide comes in, all environmental factors change abruptly. In

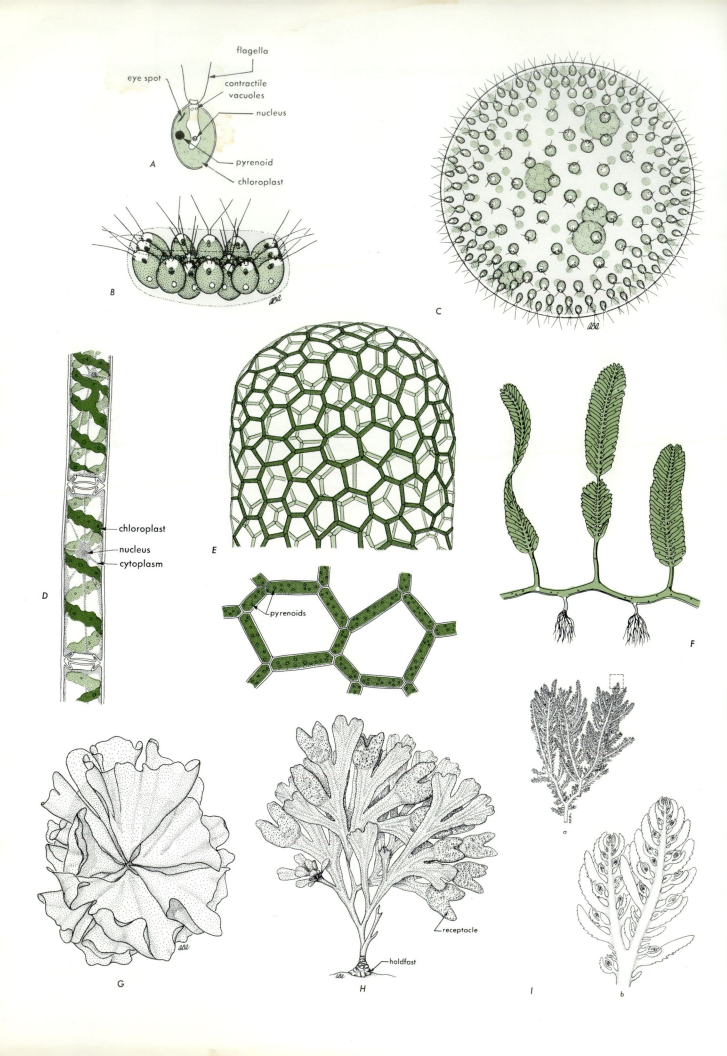

flagella

eye spot

contractile
vacuoles

nucleus

pyrenoid

chloroplast

A

B

C

chloroplast

nucleus

cytoplasm

D

E

pyrenoids

F

receptacle

holdfast

G

H

a

b

I

Figure 16.1. Algal diversity. Unicellular forms include (A) *Chlamydomonas* in the Chlorophyta. Colonial forms include (B) *Gonium* and (C) *Volvox*, both in the Chlorophyta. Filamentous forms can be separate and relatively simple in differentiation, as (D) *Spirogyra*, or (E) *Hydrodictyon*, or nonseptate and differentiated into prostrate, anchoring filaments and upright, photosynthetic filaments as in (F) *Caulerpa*, all in the Chlorophyta. Sheetlike forms include (G) *Porphyra* in the Rhodophyta. More differentiated, larger forms with leaflike fronds include seaweeds and kelps such as (H) *Fucus* and (I) *Ptilota filicina*.

Figure 16.3. The intertidal zone, exposed at low tide, along the central coast of California.

Laminaria dentigera

Figure 16.2. Example of a kelp, *Laminaria andersonii*, in the Phaeophyta.

winter, the plants may be locked in ice. The **neritic** zone is below the intertidal zone but is still relatively shallow and near shore. Species of attached algae in the intertidal and neritic zones have their own particular distribution: some occur at the upper end, others occur in the middle, and others in the lower region (Fig. 16.4). Their distribution probably reflects different degrees of adaptation to the stresses of exposure, for those species at the upper part of the intertidal zone are exposed more frequently and for longer periods of time than species farther down. Experiments have shown that algae of the upper part of the intertidal zone are actually more resistant to drying and to fluctuations in temperature, light intensity, and salinity.

Gradually, with increasing depth, the illumination becomes less and less favorable for photosynthesis. The **compensation depth,** or **compensation intensity,** is the point where positive growth is no longer possible (Fig. 16.5). In deep bodies of water, the habitat between the surface and the compensation depth is dominated by floating algae, the **phytoplankton.**

Phytoplankters are usually microscopic, but they include the large, floating *Sargassum* seaweed, namesake of the Sargasso Sea off Bermuda. The microscopic forms are typically unicellular and avoid sinking by swimming, by storing oil droplets, or by developing fine projections that extend out from the cell wall (Fig. 16.6). The average life span of a given cell is probably measured in hours or days; if it does not reproduce, the protoplast dies and the cell wall sinks to the bottom. Most phytoplankters are members of the divisions Bacillariophyta (diatoms) and Pyrrhophyta (dinoflagellates), but the Chlorophyta (greens), Cyanophyta or cyanobacteria (blue-greens), and Euglenophyta are also represented.

The quality of light in water changes with depth. Since red light is completely absorbed in the upper layers, a blue-green twilight prevails farther down. Experiments show that aquatic algae have adjusted their metabolism to the light at this depth. Figure 16.7 shows the action spectrum of photosynthesis for the green alga called sea lettuce (*Ulva taeniata*), which grows high up in the intertidal zone; superimposed on the same graph is the action spectrum for the red alga *Myriogramme spectabilis*, which grows at much greater depths. You can see that the peak has shifted and condensed to center around the blue-green region, 440 to 580 nm. Special accessory pigments (phycobilins) that are present in the red algae trap the light of blue-green wavelengths, eventually transferring this energy to chlorophyll.

Productivity and Pollution

The density of phytoplankton is not very high in the open ocean, perhaps a few thousand cells per liter, but the oceanic expanse is enormous and results in a high annual rate of gross productivity: 32.6×10^{16} kcal (3 quintillion, 260 quadrillion). This is about three times the annual productivity of

all the world's grassland and pasture, and four times that of all cropland.

As great as this productivity is, however, it is limited by factors such as light, temperature, and nutrients. If these limiting factors are ameliorated, productivity will increase manyfold. Such an increase slowly occurs in most bodies of water because silt accumulates, making the depth shallower and the water uniformly warmer, and vegetation encroaches along the edges and contributes increasing amounts of detritus. This process is called **eutrophication,** and it ordinarily takes centuries, millenia, or eons. However, eutrophication can be compressed into a matter of decades by human activities such as waste disposal. This is called **cultural eutrophication.**

As nutrients increase, algal and bacterial activity increase, resulting in more turbidity and lower light. Oxygen becomes limiting near the bottom; this limitation plus siltation prove fatal to fish eggs of some species. Algal species replace one another in an aquatic succession: generally greens are replaced by blue-greens, which may impart disagreeable odors or tastes to the water. As the base of the food chain changes, so does the entire chain. After 100 years of accelerating industrial and residential development along its shores, Lake Erie changed from a relatively clean lake with important fishing and recreational values to an aesthetic and economic loss. Plankton and other algae increased sixfold or sevenfold in some places, and the composition shifted toward the noxious blue-greens. Oxygen content of bottom waters declined 67%. Commercial catches of trout, whitefish, herring, sauger, walleye, and blue pike declined by 99.9%, while such fish as carp, shad, alewives, and smelt increased. Recently, the rate of pollution has declined and water quality has improved.

Red Tides and Blooms

One of the results of cultural eutrophication is the episodic appearance of **red tides,** or **blooms:** the rapid growth of phytoplankton populations until they become dense enough to color the water. This phenomenon can occur in bodies of water that range from small ponds to lakes to large coastal regions. Marine occurrences are often reported along the southwest coast of India, southwest Africa, southern California, Texas, Florida, Peru, and Japan. Factors associated with red tides or blooms are long hours of sunlight; shallow, warm, offshore water; and high levels of nitrogen and phosphorus.

Red tides are most often caused by dinoflagellate species of *Gymnodinium* or *Gonyaulax*. Both species produce a water-soluble toxin of high potency that affects animal nervous systems. It is related to curare poison, extracted from certain tropical flowering plants, and it is 10 times as effective as cyanide. Massive fish kills and the poisoning of shellfish result from red tides. A 1947 episode off Florida killed an estimated 500 million fish.

Other blooms are caused by the chrysophyte *Prymne-*

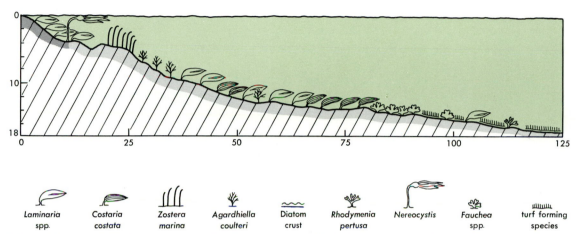

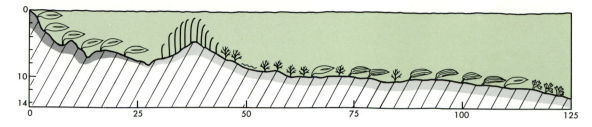

Figure 16.4. Representative transects of benthic (bottom attached) algae in the near-shore region of Puget Sound, Washington. Dark substrate is rocky. Elevation and length are in meters.

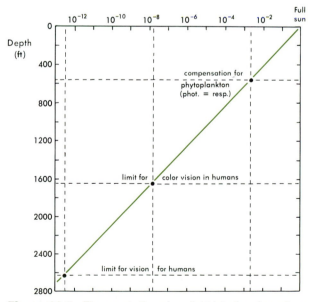

Figure 16.5. The penetration of sunlight into the clearest ocean water. The compensation intensity for most phytoplankton is 1/100 to 1/1000 the intensity of full sunlight at the ocean surface, compensation depth is about 550 ft (170 m).

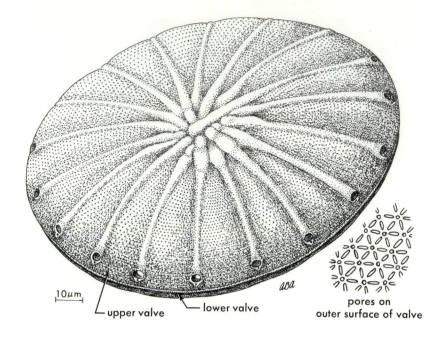

10μm

upper valve   lower valve

pores on
outer surface of valve

A

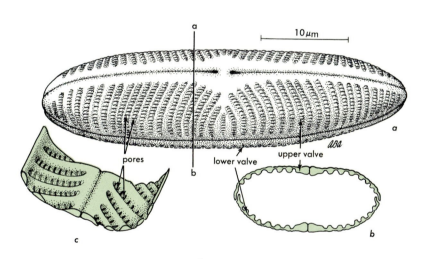

a

10μm

pores   lower valve   upper valve

a

b

c   b

B

Figure 16.6. Examples of phytoplankters. *A* to
D, diatoms *Asteromphalus elegans*, *Navicula
digitoradiata*, *Asterionella formosa*, and
Biddulphia biddulphia. *E* and *F*, dinoflagellates
Ceratocorys aultii and *Ceratium* sp.

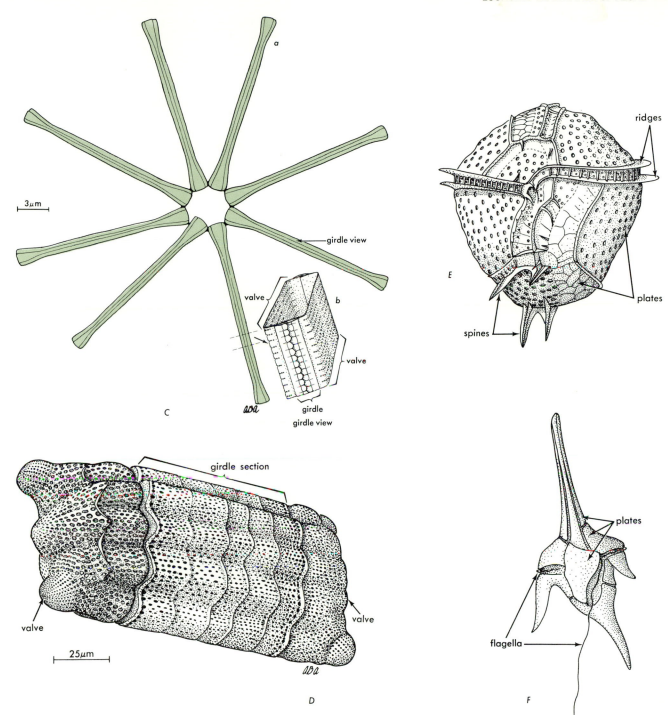

3μm

a

girdle view

valve

valve

b

girdle

girdle view

C

ridges

plates

spines

E

plates

flagella

F

girdle section

valve

25μm

valve

D

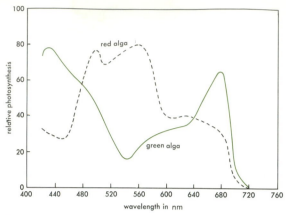

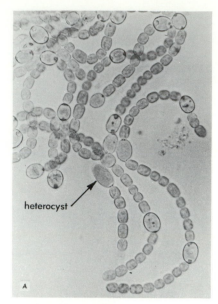

Figure 16.7. Action spectrum for photosynthesis in a green, surface-growing alga (solid line) and in a red alga growing at some depth (dashed line).

sium parvum and species of such cyanobacteria as *Microcystis, Anabaena, Nostoc, Aphanizomenon, Gloeotrichia,* and *Oscillatoria.* The blooms are not necessarily blue-green in color, however; a species of *Trichodesmium* causes red blooms, which give the Red Sea its name. Some of these organisms also produce toxins.

Nitrogen Fixation

Some bacteria, including about a fourth of all blue-green species, are able to assimilate, or **"fix,"** elementary nitrogen, N_2. The details of this metabolic pathway are still not clear, but the overall reaction can be written:

$$N_2 + 3H_2 \longrightarrow 2NH_3$$

The NH_3 can then be taken up by organisms, whereas N_2 cannot. All forms of biological nitrogen fixation contribute on the average 11 kg of nitrogen per hectare (ha) per year to terrestrial surfaces around the world. In exceptionally favorable areas, such as tropical grasslands, the amount added may be as high as 227 kg/ha. Cyanobacteria are cultivated in rice paddies and in some other kinds of agriculture as a form of free fertilizer. Living cells secrete free amino acids and peptides into the soil or water, and dried algal crusts can be plowed into the land. Specialized, thick-walled cells called **heterocysts** conduct most of the fixation; the resulting organic nitrogen then diffuses to other cells in the filament (Fig. 16.8) or out to the environment.

ECONOMIC USES

Algae as Food or Medicine

Seaweed—marine algae of moderate size, which usually grow in the intertidal or neritic zones—forms an important part of the human diet and medicine chest in several parts of the world. In the Orient, seaweed harvesting has been

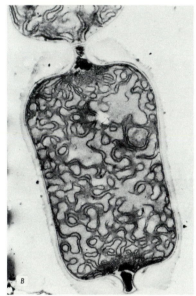

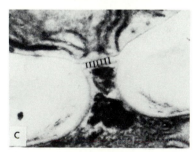

Figure 16.8. Ultrastructure of the heterocyst of the blue-green alga *Anabaena. A,* filaments of *A. azollae,* ×880, showing enlarged heterocysts (the large cells) in different stages of development. *B,* electron micrograph of a heterocyst of *A. cylindrica,* ×22,000, showing constrictions at heterocyst poles, three layers outside the cell wall, and contorted thylakoids (photosynthetic membranes) in the cytoplasm. *C,* an enlargement of *B,* showing microplasmodesmata connecting the cytoplasm of the heterocyst with that of the adjoining vegetative cell. The microplasmodesmata have been artificially darkened to enhance their visibility.

known for 5000 years. Shen Nong, the legendary Chinese "father of medicine," prescribed seaweed for certain ailments in 3600 B.C. Some 3000 years later, Confucius praised its curative value. For centuries, the Japanese have used algae as a healthful, tasty supplement to their rice diet. The demand for *nori* (the red alga *Porphyra*) has grown to such an extent in Japan that it is cultivated in shallow, intertidal bays (Fig. 16.9). The Polynesians in Hawaii were known to have utilized and named at least 75 species of *limu* (algae) as food sources. Some rare species were cultivated for the nobility in marine fish ponds. Dulce (the red seaweed *Rhodymenia palmata*) has been known as a food for 12 centuries in the British Isles. It was the Irish who discovered that small quantities of Irish moss (the red alga *Chondrus crispus*), when boiled with milk, would produce a jelly dessert that the French later called *blanc mange*. Brown algae off the coast of California (mainly *Macrocystis pyrifera;* Fig. 16.12) have been harvested for their content of iodine, which is added in trace amounts to the diet to prevent goiter, an enlargement of the thyroid gland in the neck.

It is not as food or medicine, however, that algae are most important today. With some exceptions, they do not have much nutritive value—in fact, their major constituents are largely indigestible. Algae are used more as condiments, garnishes, or desserts than as staple foods, much as we use lettuce, watercress, celery, or whipped cream. Iodine now is regularly obtained from other sources and added to table salt. Claims as to the health-giving value of seaweed do not have much foundation in fact.

Fodder and Fertilizer

Seaweeds not only contain such important trace elements as iodine, but they also contain large amounts of potash (potassium), nitrogen, phosphorus, and other characteristics of good fertilizer or cattle feed supplements. In historic times, the North American Indians and the Scotch-Irish used Irish moss as a fertilizer to build up poor soils for such diverse crops as corn and potatoes. As a fertilizer, seaweed compares favorably to barnyard manure: it enhances germination, increases the uptake of nutrients in plants, and

Figure 16.9. Nori (*Porphyra tenera,* Rhodophyta) culture, Sendai Prefecture, Honshu Island, Japan. *A,* distance view, showing the extent of the hibi nets in January, about the time of harvest. *B,* closer view, showing the hibi nets about 30 cm above mean low water in September, at the beginning of nori cultivation.

seems to impart a degree of frost, pathogen, or insect resistance. *Macrocystis* continues to be harvested off the California coast for use as a cattle feed supplement.

Cell Walls and Cell Wall Extracts

Peculiar characteristics of the cell walls of diatoms, brown algae, and red algae have led to many recent industrial, pharmaceutical, and dietary uses. It is in these areas that the algae have their greatest economic value. The characteristics result in products such as diatomite, agar, carrageenan, and algin, each of which we shall consider individually.

Diatomite. Diatomite, also called diatomaceous earth, is a sedimentary rock composed of the cell walls of unicellular algae called diatoms. As discussed, diatoms are important members of the phytoplankton (Fig. 16.6). One of the richest deposits of diatomite is located near Lompoc, California (Fig. 16.10). About 15 million years ago, this area was submerged beneath a warm, shallow sea. As diatoms flourished and died, their remains accumulated in bottom sediments. These particular remains persisted intact because the wall is composed of 95% silica, rather than carbohydrates such as cellulose. In more recent geologic time, the Lompoc area was uplifted above sea level and the diatom deposit was revealed by erosion. Several companies mine the diatomite for industrial and pharmaceutical use.

The walls of a diatom fit together like two halves of a Petri dish, and the glasslike case is perforated with hundreds of microscopic pores (Fig. 16.11). Each half is called a **frustule,** and the region of overlap is the **girdle.** Under the electron microscope, it is possible to see that even the pores have pores. The perforations form exquisite, symmetrical designs; each design is peculiar to a given species. Diatoms extract the opaline silica from the surrounding water by a process that is still not understood.

Diatomite makes a superior filter or clarifying material because the microscopic pores create a large surface area (0.5 lb—230 g—contain the area of a football field), and the rigid walls are incompressible. Diatomite is inert and can be added to many materials to provide bulk, improve flow, and increase stability. In these ways it is used in cement, stucco, plaster, grouting, dental impressions, paper, asphalt, paint, and pesticides. Diatomite is also used as an abrasive.

Figure 16.10. Aerial view of diatomite deposits near Lompoc, California.

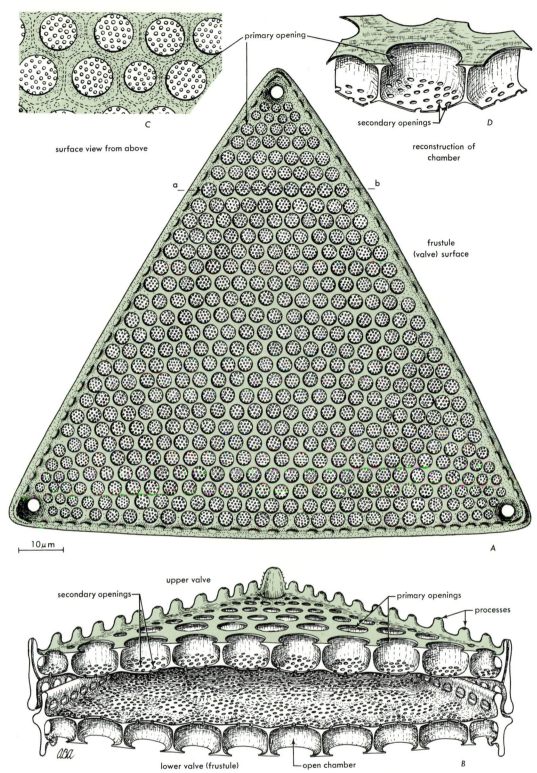

primary opening

secondary openings

D

reconstruction of chamber

C

surface view from above

frustule (valve) surface

a   b

10μm

A

upper valve

secondary openings

primary openings

processes

lower valve (frustule)

open chamber

B

Figure 16.11. Cell wall details of the triangular diatom *Triceratium favus*. *A*, face view; *B*, cross-sectional view taken at *A-B*; *C* and *D* details of the walls.

Agar. Less than 100 years ago, a physician's wife, Frau Fanny Eilshemius Hesse, discovered the use of agar as a bacterial culture medium. She passed the information on to her husband, who passed it on to Robert Koch, and Koch presented it to the world via his important scientific writings in the late nineteenth century.

Agar (or agar-agar) is a polysaccharide, analogous to starch or cellulose but chemically quite different. It is found in the walls of certain red algae, mainly *Gelidium* and *Gracilaria*. These seaweeds are commercially harvested by divers who descend 3 to 12 m in the warm near-shore waters off Japan, China, California, Mexico, South America, Australia, and the southeastern United States.

In addition to its bacteriological use, agar is important in the bakery trade. When added to icing, it retards drying in the open air or prevents running in cellophane packages. Because it is virtually indigestible, agar is also used medically as a bulk-type laxative.

Carrageenan. This is another polysaccharide from the walls of red algae, which is mainly harvested from Irish moss. The substance takes its name from the town of Carragheen, County Cork, along the south shore of Ireland, where its properties were first discovered.

Carrageenan reacts with the proteins in milk to make a stable, creamy, thick solution or gel. Consequently, it is used in ice cream, whipped cream, fruit syrups, chocolate milk, custard, evaporated milk, bread, and even macaroni. It is added to dietetic, low-calorie foods to bring back the appropriate mouth "feel" of nondietetic foods. Carrageenan is also used in toothpaste, pharmaceutical jellies, lotions of many sorts, and as a whiskey chaser for a cold "cure." Irish moss is commercially harvested in this country in Maine.

Algin. Since this long-chain polymer is made up of repeating organic acid units, it is also called alginic acid. It is the principal wall component of brown algae, constituting up to 40% of the middle lamella and primary wall by weight. Water is strongly adsorbed to algin, resulting in a thick solution. For example, one tablespoon of algin added to a quart (one liter) of pure water will increase the viscosity to approximately that of honey. In nature, algin may be valuable to intertidal brown algae during exposure by enabling water to be retained in and on the plant. Commercially, it has many uses (Table 16.1).

Hundreds of species contain algin, but only a few are commercially harvested: *Macrocystis pyrifera* along the California coast, and species of *Laminaria*, *Fucus*, and *Ascophyllum* off England. These large, brown marine algae are commonly called **kelps.** *Macrocystis* is the largest kelp known, and it has the fastest growth rate of any multicellular plant in the world. From a single-celled zygote, it grows into a young plant rooted on a rocky bottom 6 to 30 m below the surface, and then into a 60 m long mature giant—much of its shoot floating on the surface—in the course of a single growing season (Fig. 16.12). It has a rootlike region (the **holdfast** composed of many haptera) that anchors the plant to bottom substrate, a stemlike region (the **stipe**), and numerous leaflike **fronds** that arise all along the stipe and have gas-filled **bladders** at their base that increase buoyancy. Each stipe, growing as much as 30 cm a day, may live only a year or less, even though it has the potential for continuous tip growth. Other stipes continue to arise from the holdfast, and an average plant might live 4 to 10 years.

Macrocystis has been called the sequoia of the sea; indeed, it grows in such towering, dense stands, or "beds," that a scuba diver has the impression of swimming through a forest. The fast growth of *Macrocystis*, combined with its occurrence in dense beds, makes it commercially harvestable. Barges visit the beds every six weeks, cutting the tops one meter below the surface, then taking them to shore for processing.

Table 16.1. Some of the Products in Which Algin Is Used

Food Products	Industrial
Bakery icings and meringues	Water base paints
Salad dressings	Wall joint cements
French dressings	Welding rod coatings
Dietetic dressings	Textile print paste
Pickle relish	Textile sizing
Meat sauces and pepper sauces	Latex creaming and thickening
Orange concentrates	Adhesives
Fruit drinks	Paper coatings
Dietetic beverages and drinks	Corrugated paper
Beer	Paper sizing
Ice cream and related frozen desserts	Paper cartons for foods, soaps, and detergents
Egg nog	Food wrappers
Creamed cottage cheese	Waxed cardboard cartons
Cream cheese	Boiler compounds
Pasteurized cheese spreads	Can sealing compounds
Canned buttered vegetables	Battery plate separators
Canned chow mein	Mold release coatings
Canned meat stews	Wax cleaner polishes
Cake mixes	Ceramic glaze
Puddings, pie and cake fillings	Sugar beet clarification
Fountain syrups	Finger paints
Buttered pancake syrups	
Berry syrups	*Pharmaceutical*
Chip dips and mixes	Antibiotic tablets and suspensions
Instant dessert mixes	Dental impression compounds
Chocolate drink	Toothpaste
Delicatessen salads	Surgical jellies
Candy	Mineral oil emulsions
Dessert and salad gels	Medicated rubbing ointments
Breading batters	Tranquilizing tablets
	Hand lotion
	Facial beauty masque

young lateral blades

terminal blade

D

C

mature blade

5 cm

haptera

B

A

holdfast

Figure 16.12. The kelp *Macrocystis pyrifera*, including details of the blades (*C* and *D*) and holdfast (*B*).

Table 16.2. Major Characteristics of 10 Algal Divisions[a]

Division	Chlorophylls and Accessory Pigments	Cell Wall	Storage Product	Flagella	Habitats	Number of Species	Notes
Cyanobacteria (blue-greens)	a + phycobilins (phycocyanin and phycoerythrin)	Mureins and sugars + sheath	Cyanophyte starch (= amylopectin)	none	All, but often polluted water; 75% marine	1500	Asexual reproduction only; prokaryotic; prominent in blooms; some fix N
Rhodophyta (reds)	a (+ d in some) + phycobilins	Cellulose + sometimes agar or carrageenan	Floridean starch (= amylopectin)	none	98% marine; often warm, deep water	4000	Mostly seaweeds; complex life cycles
Chrysophyta (golden algae, class Chrysophyceae only)	a + c + fucoxanthin	Cellulose	Fats, oils, chrysolaminaran	2, unequal, anterior	Mainly fresh water	300	
Xanthophyta (yellow-greens)	a (+ c in some)	Cellulose or none; sometimes with Si and diatomlike	Oil or chrysolaminaran	2, various	Mainly fresh water	400	Sometimes placed in Chrysophyta
Euglenophyta	a + b	None	Paramylon (paramylum)	1–3, equal, anterior	Mainly polluted fresh water	450	Many animal-like properties
Chlorophyta (greens)	a + b	Cellulose	Starch	2, equal, anterior	Widest distribution of any division	7000	Possible precursor of higher plants
Charophyta (stoneworts)	a + b	Cellulose	Starch	2, equal, anterior	Often bottom dwellers in fresh water	250	Differentiated into nodes and internodes; gametangia with sterile jacket, sometimes put in Chlorophyta
Bacillariophyta (diatoms)	a + c + fucoxanthin	Silica + pectin	Fats, oils, chrysolaminaran	generally none	Mainly aquatic	8000+	Prominent in phytoplankton; sometimes put in Chrysophyta
Pyrrhophyta (dinoflagellates, class Dinophyceae only)	a + c + peridinin	None or internal plates of cellulose	Starch(?), oil in some	2, unequal, lateral	93% marine	1000	Prominent in phytoplankton
Phaeophyta (browns)	a + c + fucoxanthin	Cellulose + algin	Laminaran, mannitol	2, unequal, lateral	99.7% marine; often, shallow water	1500	No unicellular forms; includes the kelps (another name for brown seaweeds)

[a]A few small groups have not been included.

ALGAL CLASSIFICATION

As shown in Table 16.2, there are many algal divisions that differ in biochemistry, cytology, and habitat. Differences in cell wall construction, number and placement of flagella, and type of chlorophyll are quite fundamental and significant. We shall briefly survey seven of the divisions.

CYANOBACTERIA: THE BLUE-GREEN ALGAE

Unlike true algae, the blue-greens are prokaryotic and are closely related to bacteria. In fact, many taxonomists call them **Cyanobacteria** and place them with other bacteria in the kingdom Monera. They lack an organized nucleus, membrane-bound organelles, and endoplasmic reticulum. They contain gas vacuoles, photosynthetic membranes, and storage granules of material that has been likened to lipid, glycogen, and protein (Fig. 16.13). The wall resembles that of certain bacteria, being composed of muramic acid, glucosamine, glutamic acid, glucose, xylose, mannose, and several other mureins and sugars. Cellulose is not present. Virtually all free-living forms are surrounded by a mucilaginous sheath (Fig. 16.14).

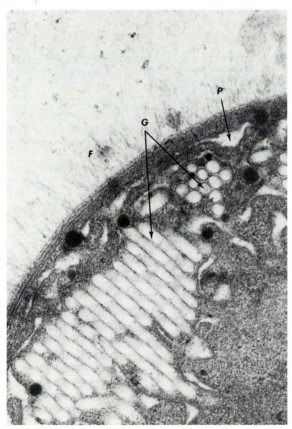

Figure 16.13. Ultrastructure of the blue-green alga *Anabaena flos-aquae*, ×73,500, showing microfibrils (*F*) in the outer sheath, a four-layered cell wall, gas vesicles (*G*) in longitudinal section and cross section, photosynthetic membranes, dark-staining lipid bodies, and many granular ribosomes.

Morphological diversity is limited to unicellular, colonial, and simple filamentous forms (Fig. 16.14). There are no flagellated cells, and sexual reproduction has never been documented. Instead, they reproduce asexually by fragmentation of filaments, the formation of resting cells, or simple fission. Movement is passive except in some filamentous forms, which rotate and glide by an unexplained mechanism. Some forms, such as *Nostoc* and *Anabaena,* possess scattered specialized cells called heterocysts, which are the site of nitrogen fixation, as mentioned earlier (Fig. 16.8).

RHODOPHYTA: THE RED ALGAE

Cyanobacteria and Rhodophyta have a few characteristics in common. Both possess only chlorophyll *a* and identical accessory pigments called phycobilins, neither possesses flagellated cells, and both form a nonstarch storage product related to amylopectin. The differences, however, are much more striking. The Rhodophyta are eukaryotic, possess cellulose walls, and exhibit some of the most complex life cycles—incorporating both sexual and asexual reproduction—of any group of plants. Their range of morphological diversity is also greater than that of the blue-greens, including sheetlike forms and complex, finely branched filamentous forms with holdfast and stipe (Fig. 16.1). In some species, pits connect the cytoplasm of adjacent cells.

The red algae are almost exclusively marine and are most abundant in warm water, often extending to some depth. Again like the Cyanobacteria, some forms are lime-encrusted and participate in tropical reef building.

EUGLENOPHYTA

This division is a small one, consisting mainly of unicellular, motile species. Members of this division have several protozoan features, such as the lack of a cell wall (although the outer cytoplasm may be modified for rigidity), rapid movement, and the ability to produce a nonstarch food reserve called paramylon. Many Euglenophyta lack chloroplasts and either engulf or absorb food. Exposure to ultraviolet radiation can halt chloroplast division but not cell division; the result is that within a few generations some progeny cells lack chloroplasts. These cells continue to live as heterotrophs and produce progeny just like themselves.

CHLOROPHYTA: THE GREEN ALGAE

The green algae are predominantly freshwater forms, but they also exist in salt water, on snow, in hot springs, on soil, on branches, and on the leaves of terrestrial plants. They form the second-largest division within the algae. (Diatoms are the largest.)

Cytologically, this group shares several important traits with higher plants: chlorophyll *a* and *b,* similar accessory pigments, starch as a storage product, and cellulose cell walls. For this reason, many taxonomists hypothesize that ancient Chlorophyta served as the ancestors of all higher plants. The most complex greens have well-developed

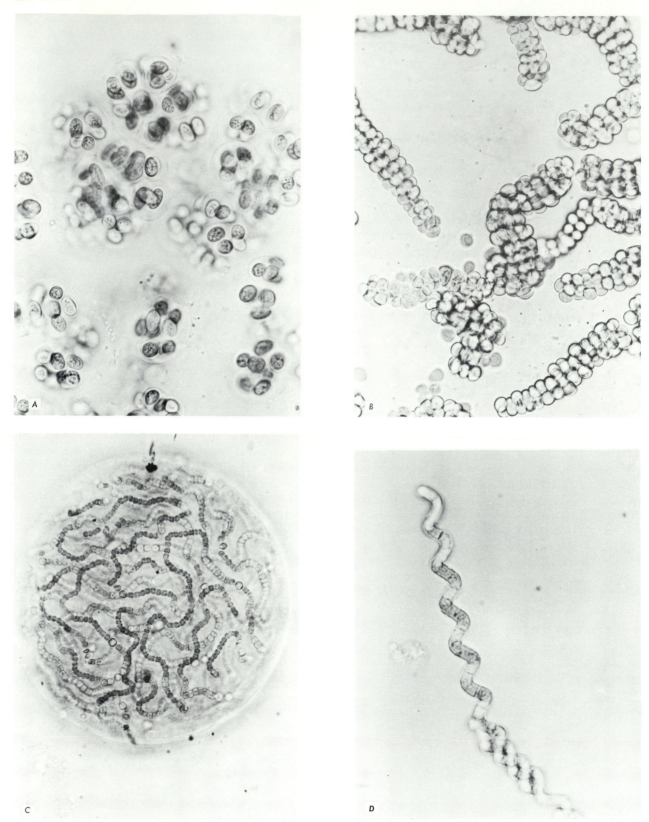

Figure 16.14. Diversity of form in the Cyanobacteria. *A* and *B*, loosely organized colonies of *Gloeocapsa* sp. and *Fisherella musicola,* both ×800; note the pronounced gelatinous sheath around *Gloeocapsa*. *C*, a colony of the filamentous *Nostoc* sp., ×880, with enlarged heterocysts visible, all embedded in a gelatinous matrix. *D*, the spirally filamentous *Arthrospira* sp.; two filaments are interwined along one portion of the figure, ×880.

prostrate portions and upright portions, indicating a high degree of differentiation. This type of habit is called **hetero-trichy.** Other forms (as shown in Fig. 16.1) are unicellular, filamentous (with or without cross-walls), colonial, and sheetlike.

BACILLARIOPHYTA: THE DIATOMS

We have already discussed these organisms earlier in the chapter as one of the major components of phytoplankton. They exist as single cells (Figs. 16.6 and 16.11)—sometimes stalked and sedentary, sometimes floating free—or as filaments. As seen in a face view, their silica walls are either circular or elongate, giving rise to two main orders, Centrales and Pennales. Some diatoms are capable of a gliding motion even though cilia and flagella are absent.

Diatoms contain chlorophyll *a* and *c*, store carbohydrates as well as oil, and their unusual silicate wall is embedded in a pectin matrix. They are mainly aquatic but are also found in soil.

PYRRHOPHYTA: THE DINOFLAGELLATES

Together with the diatoms, this group dominates the phytoplankton. Most of its species are unicellular, flagellated, and marine (Fig. 16.6), but some live in fresh water and a few are colonial or filamentous. They cause "red tides" when they multiply in great numbers. The cells are either naked, that is, with only a thickened membrane around them as in Euglenophyta, or they have a unique armor made of cellulosic plates beneath the cell membrane. Usually two flagella are present; both emerge from the same pore but are otherwise different. One flagellum is flat and ribbonlike and encircles the cell in a transverse groove; it causes rotation and some forward movement. A second flagellum trails behind and provides forward movement while at the same time acting as a rudder.

Dinoflagellates contain chlorophyll *a* and *c* and a brown pigment, peridinin, which gives the cell a green-brown or orange-brown color. True starch appears to be stored. Some forms are luminescent and contribute to the glow of water when it is disturbed at night, as in the wake of a ship.

PHAEOPHYTA: THE BROWN ALGAE

The brown algae include filamentous forms, sheetlike forms, and large kelps with more complex differentiation in anatomy and morphology than any other algae (Figs. 16.1, 16.2, and 16.12). Unicellular forms are unknown. Chlorophyll *a* and *c* are present, plus the accessory pigment fucoxanthin, which gives these plants their characteristic brownish color (though color can range from olive-brown to golden-brown to practically black). Carbohydrate is stored as mannitol or laminaran, not as starch. The cell walls are composed of cellulose and often a great deal of algin as well.

Kelps such as *Macrocystis* (Fig. 16.12) are complex anatomically as well as morphologically. If the stipe is sectioned and examined under the microscope, several regions are apparent (Fig. 16.15). Cells in the outermost layer not only are protective, but they remain meristematic and contain chloroplasts as well. To distinguish this unique tissue from epidermis, it is given the name **meristoderm.**

A broad cortical region is composed of parenchyma-like cells. Mucilage-secreting cells form definite canals through the cortex. Loosely packed filaments of cells fill the central region, the **medulla.** Some inner cortex cells next to the medulla appear to function as sieve-tube members: they have sieve plates, form callose, adjoin one another to form continuous tubes, and are known to translocate mannitol by a mechanism resembling that found in vascular plants (Fig. 16.16). However, the kelps contain no tissue that resembles xylem. The value of a photosynthate-conducting system in these large plants is easy to perceive. The mass of floating fronds considerably reduces the penetration of light into the water below; consequently, the lower part of the stipe and the holdfast may be below the compensation depth. They must be nourished by food translocated from above.

Unlike many of the other algae, the dominant generation of the kelps is diploid. The dominant generation is also larger and more complex than any other alga.

REPRODUCTION

Knowledge of reproductive processes in the algae is incomplete and largely confined to the Rhodophyta, Phaeophyta, and Chlorophyta. In some other divisions, we have detailed knowledge for a few species, genera, or families. There is no typical life cycle for the algae as a group.

ASEXUAL REPRODUCTION: VEGETATIVE

Reproduction is said to be **vegetative** if it involves no specialized reproductive cells. Among algae, vegetative reproduction is widespread and varied.

One form of vegetative reproduction is the longitudinal splitting of a biflagellate motile *Euglena* into two uniflagellate daughter cells. By another mechanism, diatoms may divide vegetatively for up to five years before reproducing sexually. Recall that the cells have two sections that fit into each other like the top and bottom of a Petri dish. After division, a new wall forms within the old wall (Fig. 16.17). This means that the new cell receiving the lower and smaller wall will be smaller than the parent cell. Eventually, a minimum size is reached and division stops. Full size is regained at the time of sexual reproduction.

Vegetative reproduction in filamentous forms occurs by a simple fragmentation of the filament. In some species, specialized cells seem to be associated with the breaking of the thallus. In species of the Cyanobacteria, notably *Oscillatoria*, the point at which fragmentation occurs is marked by a dead cell. Separation at these dead cells results in short filaments that have a gliding motion. This enables them to change their position in the mucilaginous mass in which the filaments grow, and gradually to enlarge the mass. Other

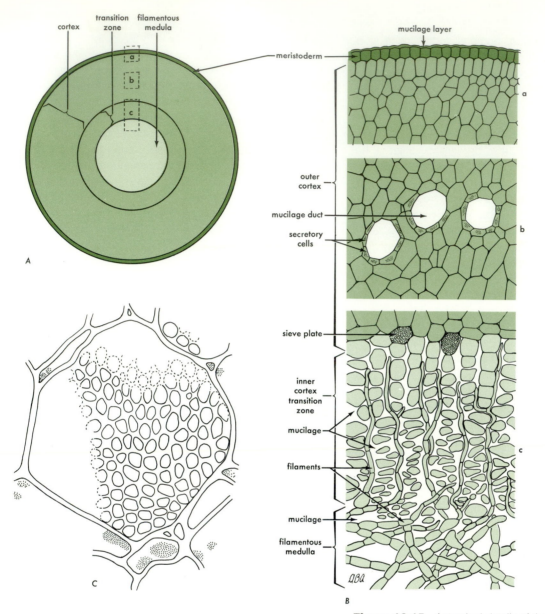

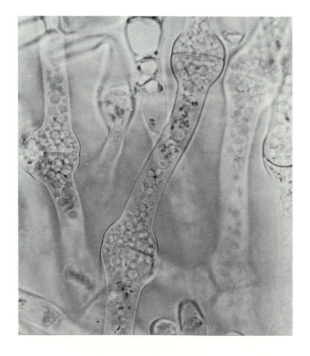

Figure 16.15. Antomical details of the stripe of *Postelsia*. *A,* diagrammatic cross section, showing the major regions and the location of detail drawings *B* and *C. B,* details of the meristoderm region, cortex with mucilage canals, and inner cortex with sieve-tubelike cells in a transition region near the central medulla. *C,* detail of the end wall of a sieve-tubelike cell, showing pores through which mannitol and other substances can flow.

Figure 16.16. Photograph of sieve-tubelike cells, with swollen junctures, from the kelp *Laminaria groenlandica,* about ×900.

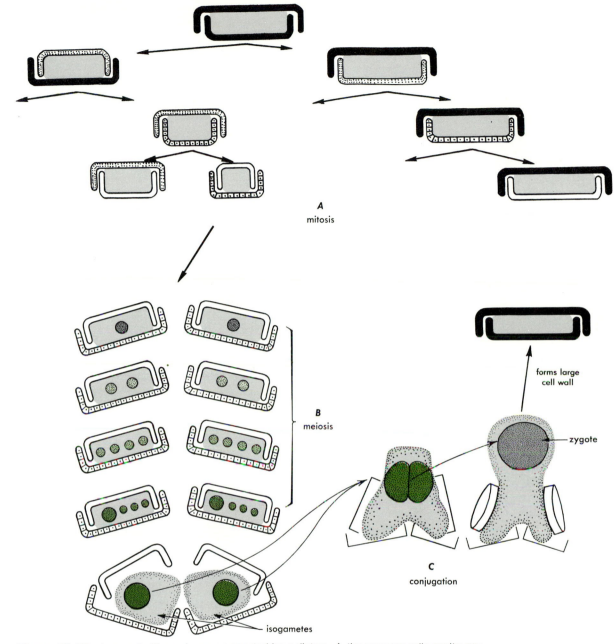

Figure 16.17. A gametic life cycle as represented by a diatom. *A,* the progeny cells continually become smaller when, after cell division, the new cell forms within the smaller valve of the old cell wall. *B,* meiosis occurs when cells become smaller than a critical size; three of the nuclei with the *n* number of chromosomes degenerate and the single remaining haploid cell becomes a gamete. *C,* fusion of the two gametes to form a zygote, which forms a new, large wall.

species modify somatic cells by adding a thick wall, and these cells function as **resting spores,** living through a hostile period of time in dormancy when all other somatic cells die.

ASEXUAL REPRODUCTION: MITOSPORES

In addition to vegetative and sexual modes of reproduction, many algae reproduce by means of asexual spores. A **spore** is a simple, usually unicellular, unit, capable of producing a new organism by itself. Asexual spores preceded by mitosis can be called **mitospores** to distinguish them from spores preceded by meiosis, which are called **meiospores.** Since

only mitotic divisions are involved, the genotypes of all asexual spores arising from one and the same parent are identical with each other and with the parent plant. They produce a population of genetically identical individuals (a **clone**). When a population has the genotype best fitted to a given growing condition, or locality, it multiplies by asexual reproduction and populates that locality.

Mitospores are either motile or nonmotile. Motile spores are called **zoospores,** and nonmotile spores are frequently called **aplanospores** (or, in *Polysiphonia,* **carpospores**). Mitospores are produced in specialized cells called **sporangia.** In many algae, the sporangia show little, if any, dif-

Labels in figure:
- A mitosis
- B meiosis
- C conjugation
- forms large cell wall
- zygote
- isogametes

ference in appearance from ordinary vegetative cells. The number of spores produced depends on the species, but it is typically 16 to 64.

Most zoospores are pear-shaped or spherical. Depending on the species, they have two, four, or many flagella. After liberation from the sporangium, zoospores are motile for as short a time as a few minutes or for as long as three days. Their movement and periods of activity are frequently affected by light. After their period of activity, they settle to the bottom of a pond or culture tank, lose their flagella, secrete a cell wall, and begin to develop a new thallus by cell division.

SEXUAL REPRODUCTION

Sexual reproduction is responsible for variation within a population. Since sexual reproduction is frequently associated with resistant cells that carry the plants over seasons unfavorable to growth, all of the new plants in a given population formed by sexual reproduction will have different genotypes when growth is resumed; those plants best adapted to the new growing conditions will become established.

As described in Chapter 11, sexual reproduction begins with the fusion of haploid (1*n*) gametes. By cell division, the resulting zygote may produce a diploid plant (the **sporophyte** generation). Later, meiosis restores the haploid chromosome number. The cells that undergo meiosis are called **meiocytes,** and reproductive cells that result from

meiosis are called **meiospores.** Meiospores can be motile or nonmotile. If they lodge in an appropriate environment, they germinate and, by mitotic divisions, may produce a haploid plant (the **gametophyte** generation). The gametophyte generation produces gametes.

Gametes are produced in cells called **gametangia.** A gametangium, in the most unspecialized case, may be indistinguishable from an ordinary vegetative cell, except that several additional mitoses occur within it, giving rise to 16 to 32 cells. In the unicellular, motile green alga *Chlamydomonas*, these cells strongly resemble the vegetative cells, but they may be smaller (Figs. 16.18 and 16.21). In the simple filamentous species of the green alga *Ulothrix* (Fig. 16.19), an ordinary vegetative cell gives rise to 8 to 64 flagellated cells. When only 8 motile cells are formed, they germinate directly, acting simply as zoospores (mitospores), and the cell in which they are formed is called a sporangium. However, when a greater number of cells is formed, the cells are smaller and they behave as gametes that fuse in pairs. When gametes resemble each other, they are called **isogametes,** and the species that produces them is **isogamous**. Thus, *Chlamydomonas* and *Ulothrix* are isogamous.

In some species, the gametes differ in size, one being slightly smaller than the other. Gametes with only a slight difference in size are said to be **anisogametes** (heterogametes), and species expressing this state are **anisogamous species** or exhibit **anisogamy** (heterogamy) (Fig. 16.20).

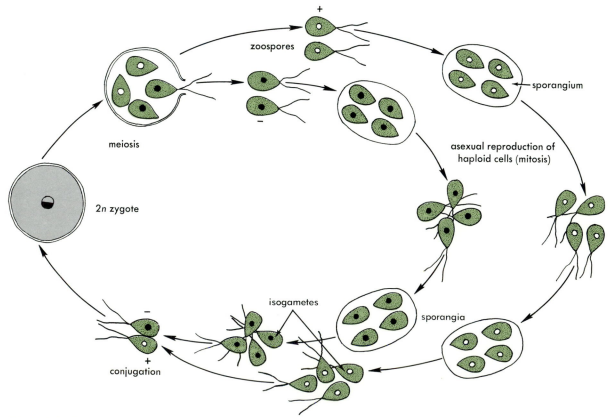

Figure 16.18. A zygotic life circle is represented by the green alga *Chlamydomonas*. Meiosis occurs during the germination of the zygote, thus the zygote is the only diploid (sporophytic) cell.

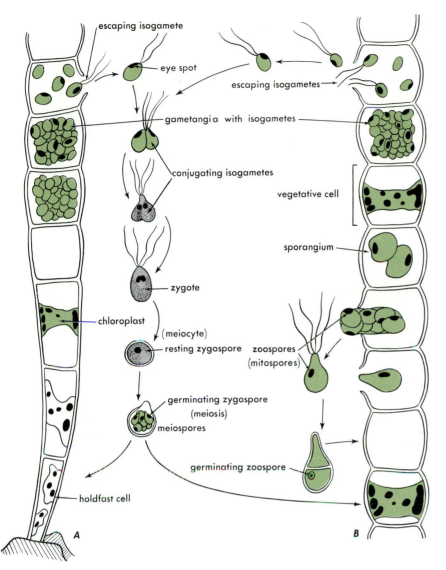

Figure 16.19. A zygotic life cycle as represented by the filamentous green alga *Ulothrix*. As with *Chlamydomonas*, only the zygote is diploid. Asexual reproduction is also shown: sporangia produce zoospores that develop into new filaments.

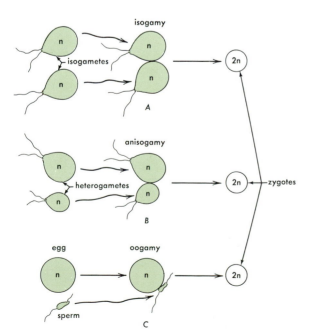

Figure 16.20. Differences in gametes. *A*, in isogamy the gametes are identical. *B*, in anisogamy the gametes are slightly different in size or activity. *C*, in oogamy, the gametes are very different in size and activity and are called sperm and eggs.

In other species, the gametes differ greatly in size and motility. One type of gametangium produces many small, motile gametes called **sperm.** A sperm-producing gametangium is known as an **antheridium.** Another vegetative cell may become quite enlarged, sometimes becoming flask-shaped. The protoplast retracts slightly from the cell wall. It has become an **egg** cell, and the cell in which it is produced is known as an **oogonium.** The sperm are liberated from the antheridium and swim toward the oogonium. One sperm enters the oogonium, either through the funnel end or by a breaking down of the oogonial cell wall. Union of egg and sperm now ensues; this process is known as **fertilization.** Plants that form distinct eggs and sperm are said to exhibit **oogamy.** This is true of *Laminaria, Fucus,* and *Polysiphonia* (Fig. 16.20).

How does a sperm cell find an egg cell? It could happen by chance, much as wind-blown pollen grains may land on flowers of their own species. However, there is considerable evidence that the female gametangia or gametes in plants produce chemical agents that attract the sperm to the close proximity of the egg cells. For instance, a solution prepared from the mature female filaments of the alga *Oedogonium* may be drawn up into a capillary tube. When the tube is placed in a suspension of sperm, the sperm collect at the tip of the capillary tube and some of them eventually enter the tube. This attraction seems to be species-specific for *Oedogonium.* Once the gametes fuse, a diploid zygote cell is formed. Further cell division produces the sporophyte generation.

Alternation of Generations

Three types of life cycles can be defined on the basis of where mitotic divisions occur (see Chapter 11). All three types are found among the algae, and examples are shown below.

Gametic Life Cycle. The diatoms are unicellular forms that multiply asexually by cell division for periods as long as five years. With each division, one of the cells is smaller than the parent cell (Fig. 16.17). At the end of a period of mitotic divisions, meiosis occurs and four gametes are produced, not all of which may function. Depending on the species, these most generally are isogametes, but they may be anisogametes or even eggs and sperm. After fusion, the zygote increases greatly in size and, again depending on the species, may begin a rest period. The germination of the zygote is by mitotic cell division. Note that in diatoms the products of meiosis are gametes, not meiospores. The gametes are the only haploid cells in the life cycle; this is the characteristic feature of a gametic life cycle.

Zygotic Life Cycle. *Chlamydomonas* and *Ulothrix* are examples of a zygotic life cycle, in which the zygote is the only diploid cell (Figs. 16.18 and 16.19). In each of them, two haploid gametes fuse to produce a diploid zygote. Meiosis occurs in the germination of the zygote, which is the only diploid cell in the life cycle.

Sporic Life Cycle: Alternation of Isomorphic Generations. In the sporic life cycle, both haploid and diploid cells undergo mitotic divisions. The result is an alternation of haploid and diploid generations. The generations are said to be **isomorphic** if the haploid and diploid bodies are alike; they are said to be **heteromorphic** if they differ. *Ectocarpus* (Phaeophyta) and *Polysiphonia* (Rhodophyta) will serve to illustrate alternation of isomorphic generations. In *Ectocarpus* (Fig. 16.21), specialized reproductive cells are formed. The gametangia are grouped together, forming elongated, slightly curved structures. Each cell in this structure is a gametangium that has resulted from mitotic divisions and will produce one motile gamete with the haploid chromosome number. They conjugate to form the usual diploid zygote. The zygote, upon germination, grows into a diploid sporophyte (Fig. 16.21C) whose thallus is identical in appearance to that of the gametophyte thallus. Two different kinds of sporangia are formed on this sporophyte plant. One of them looks exactly like the group of gametangia found on the gametophyte plant. Like the gametangia, all sporangium cells are formed through mitotic divisions, and each cell produces a motile diploid zoospore that can germinate directly into a new diploid plant (Fig. 16.21B). The other type of sporangium found on the sporophyte plant is spherical; after meiosis, it produces numerous motile zoospores or meiospores that have the haploid chromosome number. These meiospores grow into new gametophyte plants (Fig. 16.21A). Haploid gametes, haploid meiospores, and diploid zoospores are all identical in appearance.

Polysiphonia. In the Rhodophyta, an alternation of isomorphic generations is complicated by a second, distinctive sporophyte phase. The delicate feathery thalli of male and female gametophytes and one sporophyte of *Polysiphonia* are identical in appearance and growth (Fig. 16.22). No motile cells are produced. The nonmotile sperm **(spermatia)** are produced in great profusion near the tips of male gametophyte filaments (Fig. 16.22C). The oogonia **(carpogonia)** form near the tips of the female gametophyte. The oogonia are flask-shaped cells with long slender necks **(trichogynes)** that protrude from a protecting envelope formed from the cells at the bases of the oogonia (Fig. 16.22D). Spermatia are carried by water currents to make contact with the necks of the oogonia. Nuclei enter the necks and pass downward to unite with the female nuclei.

Fertilization stimulates the development of an array of short diploid filaments (the **carposporophyte**), which eventually produces diploid mitospores (**carpospores**). The original protective envelope of haploid filaments is also stimulated to grow. It enlarges and gives rise to a surrounding protective case formed of haploid filaments (**cystocarp;** Fig. 16.22E). Thus, the first sporophyte generation (carposporophyte) in *Polysiphonia* and many other Rhodophyta is composed of short diploid filaments that produce diploid spores. This sporophyte is protected by an envelope of gametophyte cells. When discharged, the diploid carpospores give rise to a second sporophyte (**tetrasporophyte**), iden-

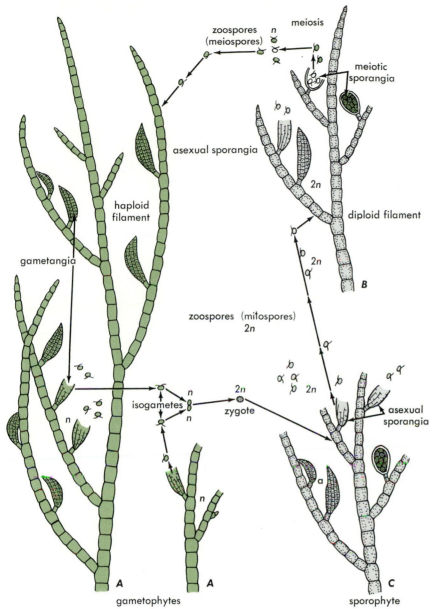

Figure 16.21. Sporic life cycle, alternation of isomorphic generations differing only in having diploid and haploid chromosome numbers, as represented by the filamentous brown alga *Ectocarpus*. Asexual reproduction by means of zoospores also occurs.

tical in appearance to the two gametophyte plants. Meiosis occurs in sporangia in the sporophyte plant in cells formed between rows of axial and pericentral cells (Fig. 16.22G). Each sporangium produces four meiospores **(tetraspores)** that upon germination give rise to male or female gametophyte plants similar in appearance to the sporophyte plants.

Sporic Life Cycle: Alternation of Heteromorphic Generations

Kelp. Alternation of heteromorphic generations occurs in both the Chlorophyta and Phaeophyta but is most highly developed in the Phaeophyta, particularly in the kelps of the order Laminariales. The sporophyte generation of the kelps is large and well known; its vegetative characteristics

have been discussed in some detail earlier in the chapter. The sporangia arise from meristematic cells, the meristoderm, that form the outermost cells of the fronds (Fig. 16.23B). The sporangia usually arise in groups **(sori)**, and adjacent cells elongate to protect the developing sporangia. Depending on the species, each sporangium produces 8 to 64 motile meiospores.

The meiospores germinate into male or female gametophyte plants, both of which consist of a small branched filament (Fig. 16.23C and 16.23D). The antheridia are produced in large numbers, either singly or in groups, at the ends of short branches. There appear to be no specialized oogonia, as eggs may be produced in any of the cells of the female gametophyte (Fig. 16.23E). The eggs are usually

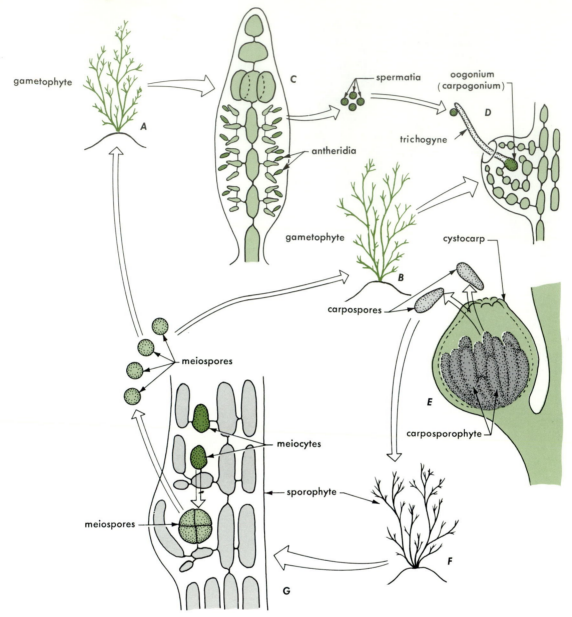

Figure 16.22. Sporic life cycle, alternation of isomorphic generations, as represented by the red alga *Polysiphonia*. *A*, male gametophyte plant. *B*, female gametophyte plant. *C*, production of spermatia at the tip of male gametophyte. *D*, the carpogonium (oogonium) is protected by filaments and spermatia become attached to the protruding trichogyne. *E*, the zygote produces a body of 2*n* filaments, which form 2*n* carnospores; and the gametophytic cystocarp enlarges and surrounds the carposporophyte. *F*, a sporophyte, identical in appearance to the gametophyte, arises from the carpospores. *G*, some cells of the sporophyte give rise to meiocytes which, through meiosis, produce four meiospores. The vegetative structure of both gametophytes and the sporophyte are identical.

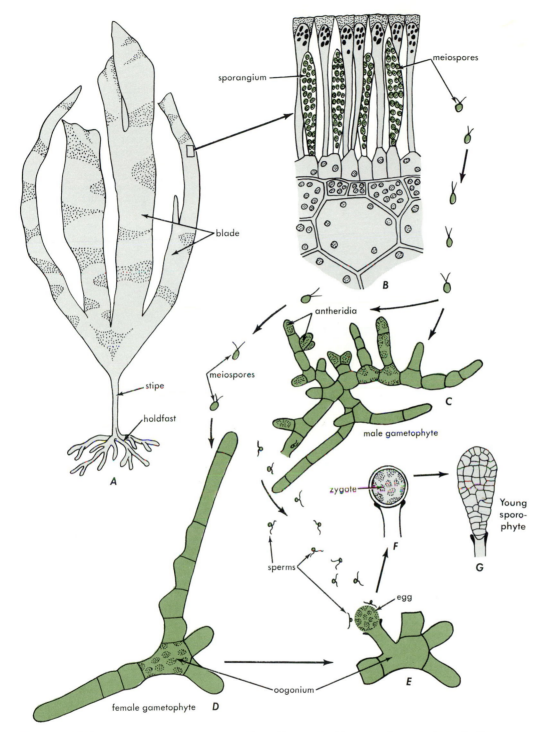

Figure 16.23. A sporic life cycle, with alternation of heteromorphic generations, as shown by the brown alga *Laminaria*.

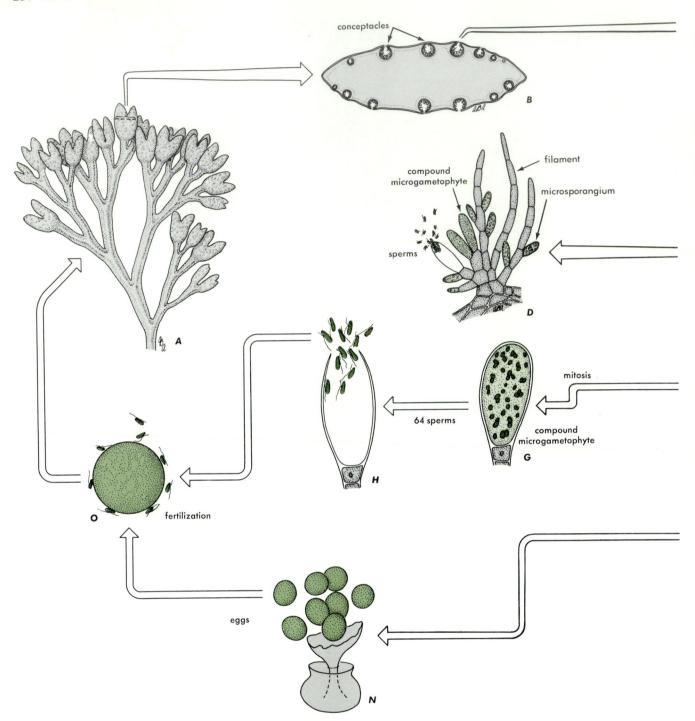

Figure 16.24. A sporic life cycle, as shown by *Fucus*. *A*, dichotomously branching blade with swollen tips. *B*, cross section of tip showing cavities (conceptacles). *C*, the conceptacles contain sterile filaments (paraphyses), branching filaments bearing microsporangia and microgametophytes, megasporangia, and large megagametophytes. *D*, enlarged view of branching filaments with microsporangium and developing microgametophytes. *E* and *F*, formation of microspores by meiosis in the microsporangium. *G*, the microsporangium wall and the enclosed gametophytes, making up a compound microgametophyte with 64 cells. *H*, sperm cells escaping from the sporangial wall. *I*, *J*, and *K*, formation of megaspores by meiosis in the megasporangium. *L*, mitotic division of the megaspores. *M*, the megasporangial wall and the four enclosed gametophytes, making up a compound megagametophyte with eight cells. *N*, eggs escaping from the megasporangial wall. *O*, fertilization occurs in the open water.

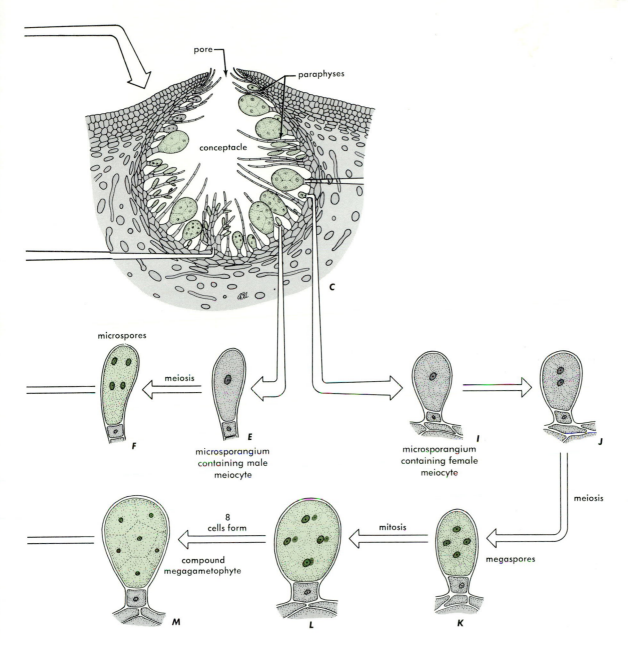

discharged from the oogonium before fertilization but remain attached to it for some time. The egg is thus fertilized outside of the oogonium in open sea water.

Fucus. Alternation of heteromorphic generations has reached an extreme in *Fucus* (Fig. 16.24). The diploid generation is conspicuous, and the haploid phase has been reduced to a few cells. These do not develop into free-living vegetative gametophytes but become sperm and eggs, and fuse to produce a new diploid generation.

The thallus of *Fucus* is shown in Fig. 16.24A. Reproductive structures occur on the flattened ends of the branches, where they appear to the eye as small raised areas. In section view, these areas prove to be cavities, or **conceptacles** (Fig. 16.24B and 16.24C). Sporangia occur in the cavities, along with many sterile hairs. Two kinds of sporangia occur:

microsporangia, which give rise to sperm-producing cells; and *megasporangia,* which give rise to egg-forming cells. In some species, the two kinds of sporangia occur on separate male and female plants; in others, both kinds of sporangia occur in the same conceptacle.

Microsporangia are borne at the ends of short branching filaments (Fig. 16.24C, 16.24D, and 16.24E). The microsporangium contains a single large cell, which is called a **male meiocyte** since it goes through meiosis and is the ultimate source of sperm. Meiosis divides the meiocyte into 4 cells, the **microspores** (Fig. 16.24F). Each microspore can be viewed as a distinct haploid plant, or gametophyte, though it remains within the original sporangium wall. Four rounds of mitosis follow, so that the sporangial wall now encloses four gametophytes with 16 cells each. The whole set of 64 cells is called a **compound microgametophyte;** the word *compound* indicates that it includes four individual gametophytes (Fig. 16.24G). Finally, each cell develops into a sperm; and the sperm are shed into the cavity.

The **megasporangium** also contains a single large cell, the **female meiocyte** (Fig. 16.24C and 16.24I). Meiosis converts this cell into four haploid **megaspores,** which remain within the sporangium wall (Fig. 16.24K). The haploid cells can be viewed as four distinct female gametophytes. Each gametophyte divides once by mitosis, so that the sporangial wall now encloses four gametophytes with two cells each (Fig. 15.24M; not all cells are visible). Collectively, they represent a **compound megagametophyte.** Each cell develops into an egg that is shed into the cavity. Eggs and sperm escape from the conceptacle to the open sea, where fertilization takes place.

SUMMARY

1. Algae are photosynthetic, nonvascular, relatively undifferentiated organisms. Their gametangia are unicellular—that is, they lack a jacket of sterile protecting cells around the developing gametes—and they do not form embryos.

2. The algae consist of more than 25,000 species belonging to 10 divisions. Although most species are aquatic, the algae as a group are found in a great diversity of habitats including some extreme environments.

3. Ecological importance of the algae centers around their role as phytoplankton at the base of aquatic food chains. Water pollution can lead to cultural eutrophication, red tides, and blooms. Some blue-green algae are also important because they fix atmospheric nitrogen into forms that can be utilized by other organisms.

4. The algae are economically important mainly because of their cell wall materials (agar, algin, carrageenan), but they also have some medical, food, fodder, and fertilizer uses.

5. Algal classification is based on biochemical and cytological traits, as well as gross morphology. The Cyanobacteria are prokaryotic forms that can be placed, with other bacteria, in the kingdom Monera. The rest of the algae are eukaryotic. The Rhodophyta share some biochemical traits with the Cyanobacteria but, in contrast, exhibit complex life cycles and a greater range of morphological diversity. The Euglenophyta are mainly unicellular and have some protozoan traits, such as the absence of a cell wall. The Chlorophyta are thought to be precursors of higher plants because they share the same chlorophylls and storage products as higher plants.

Their species exhibit a great range of morphological diversity. The Pyrrhophyta include the organisms that cause certain kinds of red tides. The Phaeophyta include the largest, most anatomically complex algae, called kelps. Photosynthate is translocated in some kelps through cells that closely resemble sieve cells. Tissue resembling xylem, however, is absent.

6. Reproduction in the algae includes asexual and sexual types. There is no typical life cycle for the algae as a group; various algae show sporic, zygotic, and gametic life cycles. Asexual reproduction may involve cell division in unicellular forms, fragmentation of filaments, or the production of mitospores. The Cyanobacteria only reproduce asexually. Sexual reproduction may involve isogamy, anisogamy, or oogamy, and the alternation of generations that results may be of identical-looking generations (isomorphic generations) or of different-looking generations (heteromorphic generations).

7. A generalized life cycle, summarizing many of the reproductive terms used in this chapter, is shown in Fig. 16.25.

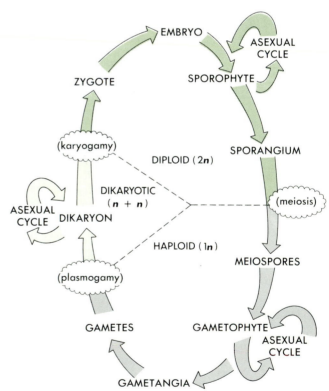

Figure 16.25. Imaginary, complete life cycle. None of the plants we examine will possess all these phases, but collectively all the plants will exhibit all these phases.

17 BRYOPHYTES

GENERAL CHARACTERISTICS

On the banks of streams and on the moist trunks of forest trees grow multitudes of small, delicate plants called **bryophytes.** The most familiar forms are the mosses, which include at least 14,000 species. Less familiar are the liverworts and the hornworts.

Bryophytes thrive in habitats that are intermediate between aquatic and terrestrial. Correspondingly, they resemble the aquatic algae in some ways, and in other ways they resemble fully terrestrial plants. Bryophytes are alga-like in that:

1. They produce free-swimming sperm cells that must travel through open water to reach the eggs.

2. They have no vascular system, but some members do have a primitive type of conducting tissue.

3. They have no lignified tissues (with the consequence that, on land, they cannot grow tall).

4. They lack roots and true leaves.

Because of the preceding traits, the bryophytes are especially successful in habitats that are perpetually wet, such as the hillside beside a waterfall. Nevertheless, many bryophytes can survive long periods of dehydration and are found on tree trunks and stones that are dry for the greater

Figure 17.1 Habit of *Riccia*.

part of the year. They are shriveled and dormant during the dry spell, but they quickly revive with the first rain.

Bryophytes resemble land plants in a variety of ways. Some have a cuticle and stomata, which help to restrict dehydration. However, the chief adaptations to life on land are associated with reproduction and dispersal:

1. Bryophytes produce wind-borne spores that are protected from dehydration by a wax-impregnated wall.

2. In common with all land plants, the bryophytes shelter their gamete-producing cells inside a jacket of protective cells (a "sterile jacket"). The protective jacket is probably an adaptation to land life, serving to protect the gametes against dehydration and to assure that sperm will only be released when water is present. This contrasts with the algae, in which the gamete-producing cells rarely have any protection beyond the wall of the parent cell. Eggs are formed in flask-shaped **archegonia** (Fig. 17.2), and sperm are formed in rounded **antheridia.**

3. In common with all land plants, the bryophytes shelter the diploid offspring inside the parental body until it has grown into a multicellular **embryo.** This protects the offspring against dehydration during the critical early stages of development. Algae do not shelter their offspring; either the gametes are shed before fertilization, or the offspring are released while they are still zygotes.

Like all land plants and some algae, the bryophytes have a sporic life cycle; that is, they produce both gametophyte (haploid) and sporophyte (diploid) bodies. But the bryophytes differ from all other land plants in that their chief photosynthetic body (the leafy moss plant, for instance) is the gametophyte. The sporophyte is smaller and usually depends on the gametophyte for nutrition. The opposite is true of all other land plants; their conspicuous body is the diploid sporophyte.

BRYOPHYTES AND THE EVOLUTION OF HIGHER PLANTS

How do the bryophytes fit into the evolutionary process that led from algae to higher plants? At present, we are not at all certain of the answer. Some taxonomists think the bryophytes are a completely separate line of evolution. Others think bryophytes are degenerate forms that arose when vascular plants returned to a semiaquatic habitat. The fossil record is of little help in resolving the issue, because the early record of the bryophytes is very poor. The first fossil bryophytes occur in rocks that were formed about 350 million years ago, when more advanced plants were already plentiful.

ECONOMIC IMPORTANCE

Bryophytes occur on soil, trunks of trees, and rocks; many species are truly aquatic. They are not extremely important economically, but they do have an interesting, if short, ethnohistory. For example, native American Indians in Utah ground up mosses, such as *Mnium* and *Bryum,* into a paste and applied it as a poultice to treat burns and bruises. In Europe, the growth of mosses such as *Dicranoweisia* was encouraged on shingle roofs to make them watertight.

Sphagnum is by far the most important moss economically. The surface layers of stems and "leaves" of *Sphagnum* are composed of alternating files of living green cells and large empty hyaline cells (Fig. 17.12). The hyaline cells are capable of absorbing water or other fluids in great quantities as high as 20 times their own weight. In the 1880s in Germany it was discovered, quite by accident, that *Sphagnum* moss pads make excellent absorbent bandages. A workman in Kiel lacerated his arm while working in a peat moor. Since he had no other bandage material, he packed his wound with dry peat *(Sphagnum).* Ten days later, when he was able to obtain treatment, the wound had healed. After that, disinfected bags filled with dry *Sphagnum* became a common wound dressing in Germany. The use of *Sphagnum* as wound dressings was not popular outside Germany until World War I. At that time they were used on a large scale by the Allied Armies as well. The British, for example, used 1,000,000 *Sphagnum* dressings each month. *Sphagnum* dressings were used very little in the United States, but the American Red Cross did publish inquiries and instructions on finding useful *Sphagnum.* More recently, *Sphagnum* is still used as packing for vegetables, in potting mix to increase the water-holding capacity of the soil; and old compressed *Sphagnum* is used as a fuel in the form of a low-grade coal known as peat.

CLASSIFICATION

The Bryophyta are divided into three classes: **Hepaticae** (liverworts), **Anthocerotae** (hornworts), and **Musci** (mosses). The Hepaticae are the simplest and perhaps the most primitive bryophytes. They consist of a flat, ribbonlike, green thallus that produces gametes and a meiospore-producing capsule or sporangium usually embedded within the gametophyte thallus. The name *liverwort* is very old, having been used in the ninth century. It was probably applied to these plants because of their fancied resemblance to the liver and the belief (the "Doctrine of Signatures") that plants resembling human organs would cure diseases of the organs they resembled. At any rate, a prescription for a liver complaint in the 1500s called for "liverworts soaked in wine."

Mosses are the best-known class of Bryophyta. They have simple "stems" and "leaves," and sporophytes are borne on green gametophytes. They have either an upright or prostrate leafy gametophyte and sometimes form extensive mats on moist, shaded soil. The sporophyte is raised above the leafy gametophyte.

CLASS HEPATICAE

GENERAL CHARACTERISTICS AND DISTRIBUTION

Hepaticae are commonly known as liverworts. There are some 9,000 species, the great majority growing in moist, shady localities. The gametophyte is the prominent plant, and the sporophyte is partially or wholly dependent on the gametophyte. The gametophyte is green and grows either as a flat ribbon or as a leafy shoot. In either event, the plant body is frequently called a thallus, even though its internal structure does not correspond to that of the thallophytes. Of the four orders in the Hepaticae, we shall consider briefly the characteristics of three genera: *Riccia* and *Marchantia* in the order Marchantiales, and *Porella* in the order Jungermanniales.

Gametophytes of the Marchantiales are small, green, ribbon-shaped plants (Figs. 17.1 and 17.2A). They branch regularly by a simple forking at the growing tip, resulting in a number of Y-shaped branches. In some species, a rosette may be formed. The upper surface of the thallus is composed of cells adapted for photosynthesis and arranged to form air chambers that open to the surface by a pore (Figs. 17.2A and 17.6A). Several types of storage cells generally make up the lower surface. **Rhizoids,** specialized elongated cells, extend downward from the lowermost layer of cells and anchor the gametophyte to the substrate. The thallus is thus differentiated into distinct upper and lower portions. Scales, frequently brown or red, are also formed on the lower surface (Figs. 17.2A).

RICCIA

Riccia is a widely distributed genus, and although it requires water for active growth, most species can tolerate considerable drought. Several species are aquatic, growing either on mud or on the surface of small ponds.

Gametophyte

The gametophyte is a small green thallus, frequently forming a rosette (Fig. 17.1). Tissue on the lower side is composed of colorless cells that sometimes contain starch. Tissue on the upper side consists of vertical rows of chlorophyll-bearing cells, between which are air chambers (Fig. 17.2A). The gametangia are embedded in deep, lengthwise depressions or furrows on the upper surface of the thallus. Antheridia and archegonia are usually found on the same gametophyte (homothallic).

Antheridia. Antheridia of different representatives of the Bryophyta are similar, though they may vary somewhat in shape. In *Riccia,* they are pear-shaped and composed of two types of cells: fertile and sterile (Fig. 17.2B). Fertile cells give rise to sperm, which are relatively numerous, small, and dense with protoplasm. Sterile cells form a protective jacket, one cell in thickness, around the fertile cells.

Mature sperm consist mainly of an elongated nucleus

with two long flagella. They may be shot with considerable force from the mature antheridium or they may be extruded slowly in a single mucilagenous mass. In any event, they do not leave the antheridium until enough moisture is present to allow them to swim about.

Archegonia. The archegonium of *Riccia* is a flask-shaped structure consisting of two parts: (1) an expanded basal portion, the **venter,** and (2) an elongated **neck** (Fig. 17.2C and 17.2D). Four **cover cells** are located at the top of the neck. Each archegonium contains a single egg cell, which is located in the venter. A short stalk attaches the archegonium to the gametophyte. Archegonia of a similar structure are found in other Bryophyta.

Shortly before the egg cell is mature, the cover cells separate. At the same time, the cells in the center of the neck dissolve, so that an open canal connects the venter with moisture outside the archegonium. Free-swimming sperm move toward certain chemical substances formed by the archegonium. Several sperm may enter the archegonium, but only one sperm fertilizes the egg in the venter.

Sporophyte
As a result of fusion of sperm and egg nuclei, the zygote contains the diploid, or 2n, set of chromosomes. Mitosis now proceeds in a more or less orderly manner until a spherical mass of some 30 or more similar cells is formed (Fig. 17.2D). These cells are partially dependent on the gametophyte. This mass of undifferentiated cells comprising the young sporophyte is an **embryo.** It is located within the venter of the archegonium.

Further development of the embryo involves differentiation of an outer layer of cells to form a protective jacket of sterile tissue surrounding a mass of cells that are capable of forming spores. This spore-forming tissue is called **sporogenous tissue.** The sporogenous cells continue to divide by mitosis until many have formed. Each of these cells, now known as sporocytes, will divide by meiosis to produce four spores with the haploid chromosome number. Remember that meiosis involves two cell divisions and results in a reduction in the chromosomes by one-half (see Chapter 11).

The mature sporophyte of *Riccia* (Fig. 17.2E) is called a **capsule.** It is composed of a jacket of sterile cells enclosing a mass of spores. The sporophyte wall breaks down when the spores are mature. These spores remain embedded in the gametophyte thallus and are not released until after the death and decay of the gametophyte. As each spore germinates, it grows into a typical gametophyte plant.

Significant Steps in the Life History of *Riccia*
1. The gametophyte, a small, green, flat plant, absorbs water and mineral salts from the soil and carbon dioxide from the air. It contains chlorophyll and can synthesize food. All gametophytic nuclei are haploid (contain 1n set of chromosomes).

2. Gametangia (antheridia and archegonia) develop in a furrow on the upper surface of the gametophyte.

3. Each antheridium produces thousands of sperm.

4. A single egg is formed in the venter of each archegonium.

5. Gametes (eggs and sperm) contain 1n set of chromosomes.

6. When sufficient moisture is present, mature sperm are extruded from the antheridia. The neck of the mature archegonium is opened, and sperm swim down to the egg in the venter.

7. Each egg is fertilized by a single sperm.

8. The zygote, as a result of the union of two haploid gametic nuclei, is diploid.

9. An embryo sporophyte, consisting of a spherical mass of undifferentiated cells, partially dependent on the gametophyte, develops from the zygote. Each nucleus of the embryo sporophyte is diploid (2n).

10. The cells of the embryo differentiate into a sporangium that consists of a jacket of sterile cells enclosing a mass of fertile cells.

11. Sporocytes, each with 2n nuclei, form in sporangia.

12. Sporocytes undergo meiosis. A quartet of spores forms from each sporocyte. All sporocytes are 2n and spores are 1n.

13. Upon death and decay of old gametophytes, spores are liberated from the capsules and new gametophytes from them.

MARCHANTIA
Marchantia may grow in large mats on moist rocks and soil in shady locations **(Fig. 17.3A, Color Plate 21).** It is widely distributed and fairly common on banks of cool streams. It is somewhat better adapted to growing on land than is *Riccia,* but considerable moisture is still required for active growth and for fertilization.

Gametophyte
The gametophyte of *Marchantia* differs from that of *Riccia* in that gametangia on special disks are raised some distance above the vegetative thallus **(Figs. 17.3C and 17.3D).** The thallus of *Marchantia* is similar to that of *Riccia* but is much coarser. It is strap-shaped, with a prominent midrib and shows dichotomous branching **(Fig. 17.3B).** The tips of the branches are notched. An individual thallus is generally 1 to 1.5 cm broad. Their length (1 to 1.5 cm) and the degree of branching depend on growth conditions. It is usual for new growth to overlap the older decaying ends of adjacent thalli. On its upper surface are polygonal areas, with a small but conspicuous pore in the center (Fig. 17.6). These areas, with their air pores, mark the outlines of air chambers, each of which is filled with short filaments of cells containing chloroplasts. As in *Riccia,* the lower surface of the thallus is

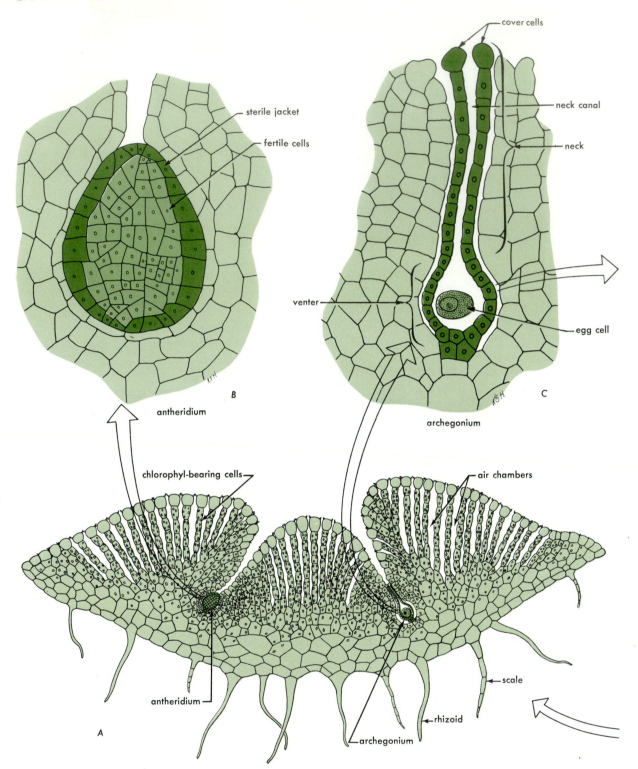

Figure 17.2 *Riccia. A,* cross section through a thallus: gametangia are on the upper surface of the thallus between the photosynthetic filaments or in a notch in the thallus. *B,* an antheridium; note jacket of sterile cells surrounding the developing sperm cells. *C,* archegonium on thallus, neck canal open, egg cell in venter. *D,* young sporophyte developing in venter of old archegonium. *E,* sporophyte with spore mother cells (sporocytes) still enclosed in venter of old archegonium. *F,* meiosis occurs within the sporophyte. *G,* the venter now contains fragments of the old spore case (sterile cells of the sporophyte) and spores. *H,* a spore.

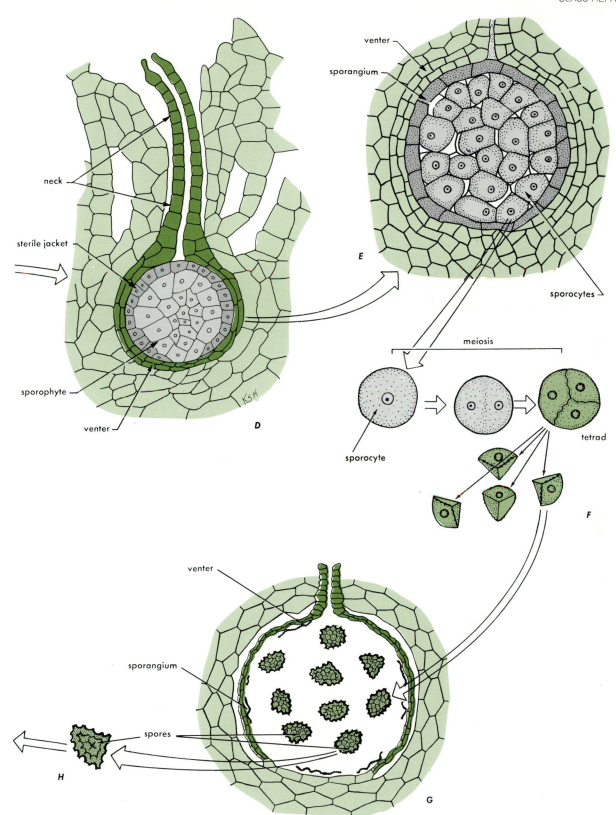

Fig. 17.2 continued

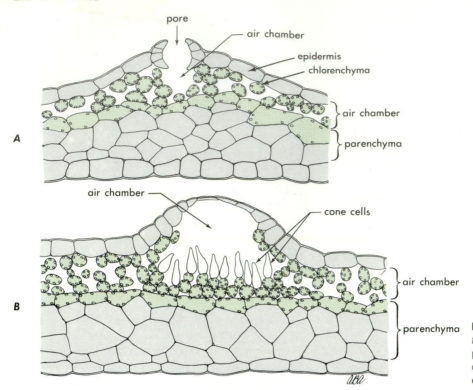

Figure 17.6 Cross section of *Marchantia. A* and *B. Concephalum* sp. gametophyte thallus showing pore, chlorenchyma, parenchyma storage tissue, and epidermis.

composed of colorless cells, some of which are modified for storage. Rhizoids, which anchor the thallus, grow from the cells covering the lower surface. Several rows of scales are also attached to this lower surface.

Asexual Reproduction. *Marchantia* reproduces asexually in two ways: (1) older parts of the thallus die and younger portions, no longer attached, develop into new individual plants; and (2) small cups, known as **gemmae cups,** form on the upper surface, and small disks of green tissue, called gemmae, grow from the bottom of these cups **(Fig. 17.3***B***).** Gemmae, when mature, break off from the thallus and are distributed into nearby areas. Raindrops are often the agents that break off gemmae and scatter them away from the thallus. New gametophyte plants grow from gemmae.

Sexual Reproduction. *Marchantia* is heterothallic; that is, gametophytes have either antheridia or archegonia. The gametangia, however, are very similar in structure to those of *Riccia.* Antheridia are pear-shaped bodies composed of a jacket of sterile tissue surrounding sperm **(Fig. 17.4***A***, Color Plate 22).** They are borne on disks raised above the thallus on slender stalks, called **antheridiophores** **(Fig. 17.3***C***).** Antheridiophores are modified portions of the thallus having furrows, rhizoids, and air chambers. Antheridia develop in cavities on the upper surface of the disk, the youngest antheridia being close to the outer margin of the disk. Mature sperm are extruded in a mucilagenous mass.

Archegonia are flask-shaped and have the same structures as those of *Riccia*—venter, neck, and cover cells **(Figs. 17.3***E***, 17.4,** and **17.5).** They are borne on specialized branches called **archegoniophores.** The disk at the top of the archegoniophore typically has eight or more lobes. In development, the lobes bend downward and then grow inward toward the stalk. Thus, the youngest portion of an older disk bearing archegonia is on the underside and close to the stalk **(Fig. 17.4***B***).** Archegonia develop in the lower surface of the disk in tissue similar to that found on the upper surface of the thallus. Fingerlike processes grow out from between the lobes. The archegonia mature, and fertilization takes place when the disk of the archegoniophore is but slightly elevated above the thallus.

Sporophyte

Zygotes develop, as in *Riccia,* into diploid embryos, which are spherical masses of undifferentiated, dependent tissue **(Fig. 17.5***A,* **Color Plate 22).** Subsequent growth, however, is more complicated than in *Riccia.* Some cells of the embryo divide to form a mushroomlike growth that becomes embedded in gametophyte tissue. This growth is called the **foot.** Cells in the central portion of the embryo form a seta. The remaining lower cells develop into a **sporangium (Fig. 17.5***B* **and 17.5***C***).** Lengthening of the seta suspends the sporangium below the disk. While this development is taking place, the stalk of the archegoniophore elongates, lifting the disk with its archegonia well above the thallus.

Fertilization stimulates enlargement of the old archegonium, which keeps pace with the enlargement of the developing sporophyte. As a result, the sporophyte is continually enclosed within the archegonium, which, because of its increase in size and change in function and shape, is now known as the **calyptra.** In addition, the surrounding gametophyte tissue produces two other envelopes that protect the sporophyte.

The mature sporophyte **(Fig. 17.5C)** is composed of a **foot,** a stalk or **seta,** and a **sporangium.** The foot is an absorbing organ. The seta serves to lower the sporangium away from the archegonial disk and thus to facilitate distribution of spores. Before meiosis, the sporangium or capsule is composed of a jacket of sterile cells surrounding a mass of sporocytes, among which are a number of sterile elongated cells. Four spores, each containing $1n$ chromosomes, develop from each sporocyte. The elongated sterile cells are transformed into spiral elements, called **elaters,** that change shape under the influence of varying moisture conditions. Their twisting motion as they imbibe and lose water aids in dispersal of spores **(Fig. 17.5D).**

Dissemination of spores is aided by two structural features not found in *Riccia:* (1) the sporangium of the sporophyte hangs from the lower side of the raised archegonial disk, and (2) elaters help to empty the spore case. Spores develop immediately into new gametophytes if they land in an appropriate habitat.

Significant Steps in the Life History of *Marchantia*

1. The green gametophyte thallus absorbs water and mineral salts from soil and carbon dioxide from air.

2. Gametangia (antheridia and archegonia) are borne on upright branches called antheridiophores and archegoniophores.

3. The zygote is diploid ($2n$).

4. The embryo sporophyte, consisting of a spherical mass of undifferentiated cells, is formed from the zygote.

5. Cells of the embryo sporophyte differentiate into foot, seta, and sporangium.

6. The foot is an absorbing organ. It is embedded in the gametophyte, from which it receives nourishment.

7. The sporangium consists of a jacket of sterile cells surrounding a mass of sporogenous tissue interspersed with elongated sterile cells.

8. The seta is a stalk that, in lengthening, lowers the sporangium below the surface of the archegonial disk.

9. Repeated mitotic cell divisions in sporogenous tissue result in sporocytes each one of which contains $2n$ chromosomes.

10. Sporocytes divide by meiosis into tetrads of spores; each spore contains $1n$ chromosomes.

11. The elongated sterile cells are transformed into spiral elaters.

12. The presence of elaters and the hanging position of the sporophyte aid in disseminating spores.

13. Spores germinate immediately into new gametophyte plants.

JUNGERMANNIALES

This is the largest order of liverworts containing two-thirds of all species in the Hepaticae. The gametophytes of these liverworts are leafy, in contrast to thallus plants; that is, the gametophytes have stemlike and leaflike organs that may be held somewhat erect, off the ground. Rhizoids anchor the aboveground portion. A very common, widespread genus in this order is *Porella* (Fig. 17.7).

Porella plants occur in dense mats. Shaded, older portions of the stems lack leaves, and rhizoids arise in those regions. Some species, such as the one illustrated, are nearly devoid of rhizoids. Younger portions of stems are densely

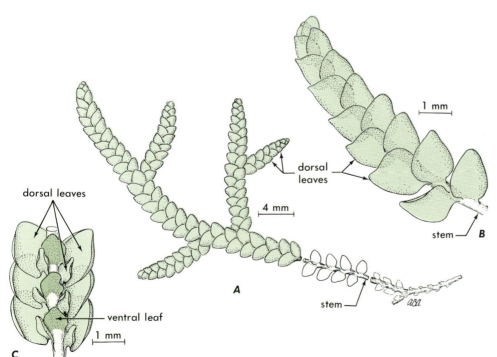

Figure 17.7 *Porella* sp. a leafy liverwort. *A,* habit sketch. *B,* detail of two dorsal ranks of leaves. *C,* detail of smaller, ventral leaves.

clothed with leaves that arise in three ranks. The lower (ventral) rank of leaves is much smaller than the upper two (Fig. 17.7C). The leaves are one cell thick and lack a cuticle.

Antheridia occur in the axils of leaf and stem, and archegonia occur on short, leafy branches. Sporophytes resemble those of *Marchantia*.

CLASS ANTHOCEROTAE

The Anthocerotae have the simplest gametophytes of the Bryophyta. They are small, green thallus plants with little

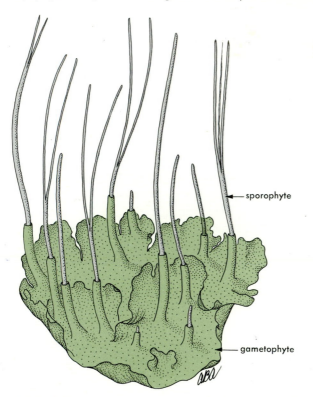

Figure 17.8 *Anthoceros* sp., a hornwort. Gametophyte with upright, dependent sporophytes.

internal differentiation of vegetative tissues. They are slightly lobed, with numerous rhizoids growing from the lower surface (Fig. 17.8). The antheridia are similar in structure to those encountered among the Hepaticae (Fig. 17.9A). They are located in roofed chambers in the upper portion of the thallus. The archegonia are embedded within the thallus and are in direct contact with the vegetative cells surrounding them (Fig. 17.9B).

The sporophyte (Figs. 17.8 and 17.10) of the Anthocerotae is in striking contrast to those of the Hepaticae. The subepidermal cells contain chloroplasts, and typical stomata are found in the epidermis. A foot embedded in the thallus serves as an absorbing organ. The sporangium is an upright elongated structure. Sporogenous tissue forms a cylinder parallel with the elongated axis of the sporangium. Spores mature in progression from the top down. A meristematic region (i.e., region of continuous cell division) lies just above the foot and continually adds new cells to the base of the sporangium. Its presence means that spores are produced over long periods.

Under exceptionally favorable growing conditions, the sporophyte lengthens greatly. Some sporogenous tissue at the base of the sporangium is sometimes replaced by a conspicuous conducting strand. The foot enlarges and, through decay of the gametophyte, comes into more or less direct contact with the soil. Such sporophytes are capable of surviving independently for some time.

In one sense, the hornworts differ markedly from other bryophytes. They have just one large chloroplast per cell, and the chloroplasts have pyrenoids. A similar situation occurs in many green algae (see Chapter 16). By contrast, the mosses and liverworts lack pyrenoids and have many chloroplasts per cell. The difference suggests that hornworts may derive from a different line of evolution than the other bryophytes.

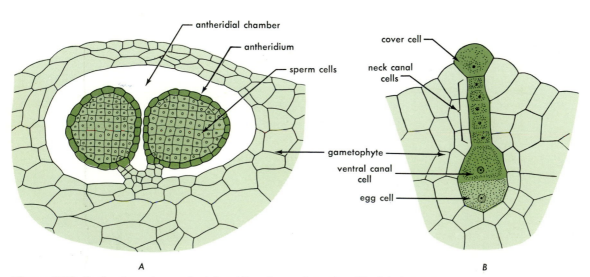

Figure 17.9 Section through an antheridium (A) and an archegonium (B) of *Anthoceros* sp.

Figure 17.3. *Marchantia. A,* habit, shows dichotomously branching thallus with prominent midrib, gemmae cups, antheridial heads, and archegonial heads with mature (yellow) sporophytes, ×1/6. *B,* gemmae cups, ×1. *C,* antheridial heads, ×2. *D,* archegonial heads, ×1.5. *E,* undersurface of an archegonial head: the sporophytes are enveloped by a protective perianth. Immature sporophytes are green; ripening sporophytes show yellow, ×3. *F, G,* and *H,* time sequence showing opening of sporophyte capsule (arrows), ×3. *I,* mass of spores being extruded from sporophyte capsule, ×10.

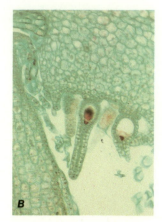

Figure 17.4. Light micrographs of gametangia of *Marchantia*. *A*, antheridium in antheridial head. *B*, archegonium suspended from lower surface of archegonial head. *C*, details of archegonia.

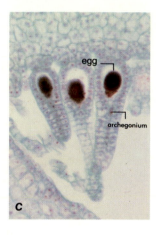

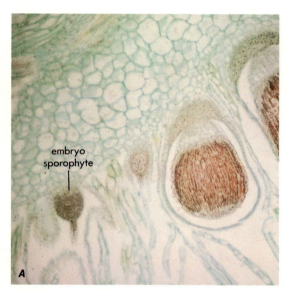

Figure 17.5. Light micrographs of developing sporophyte of *Marchantia*. *A*, embryo sporophyte. *B*, elongating sporophyte showing three regions. *C*, mature sporophyte. *D*, higher magnification of mature sporophyte showing spores and elaters.

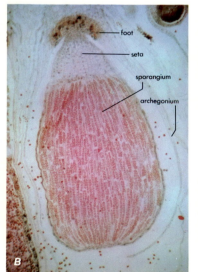

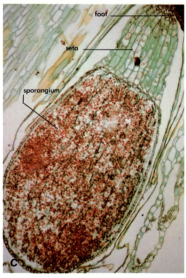

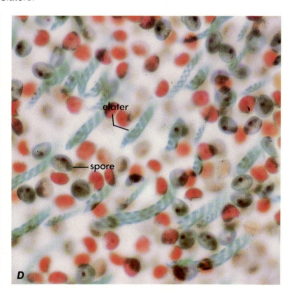

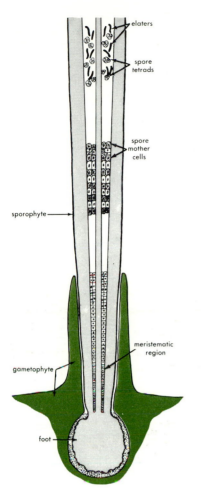

Figure 17.10 Longitudinal medium section through sporophyte of *Anthroceros* sp.

CLASS MUSCI

GENERAL CHARACTERISTICS

Although the mosses are small plants, they are nevertheless conspicuous. They frequently cover rather large areas of stream banks. They grow on rocks and trees and are sometimes submerged in streams. Although some mosses are able to resist considerable drought, all require moisture for growth and reproduction.

Mosses as a class show great structural uniformity. *Funaria*, *Sphagnum*, and *Mnium* are examples (Figs. 17.11, 17.12, and 17.15). Gametophytes of nearly all species have two growth stages: (1) a creeping, filamentous stage (the **protonema**) (Fig. 17.11), from which is developed (2) the moss plant with an upright or horizontal stem bearing small, spirally arranged green leaves (Fig. 17.15).

GAMETOPHYTE

Germinating spores of mosses do not develop directly into a leafy gametophyte but first become a filamentous structure. This early stage of the gametophyte is called the protonema (Fig. 17.11). The protonema is not a permanent structure, although it may branch considerably under favorable conditions and cover a rather large area of soil, sometimes forming a green coating resembling algal growth. Cells composing the protonema contain numerous chloroplasts. At various locations, cell divisions along the protonema produce swollen regions called **buds.** From them develop the upright gametophyte stage.

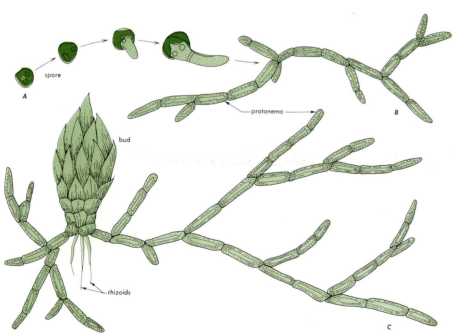

Figure 17.11 *Funaria* sp., a moss. *A*, germination of spores; *B*, protonema; *C*, protonema with bud.

The mature stem of the leafy gametophyte shows a wide range of differentiation, depending on the species, age, and environment. Elongate cells called **rhizoids** anchor the gametophyte to the substrate (Figs. 17.11, 17.15, and 17.17). These structures are not roots and apparently are not involved in absorption. The simpler mosses (Fig. 17.13B) have an outer layer of thick-walled epidermal cells (stereids), surrounding an undifferentiated cortex made up of parenchymalike cells. Epidermal cells of the peat moss *Sphagnum* (Fig. 17.12B, 17.12C, and 17.12D) are large and empty, with pores that open to the outside.

Some mosses, such as *Mnium* (Figs. 17.13A, 17.13B, and 17.14) have a central conducting strand in their stems. It is made up of elongated, thin-walled **hydroids,** which are dead, empty cells that conduct water (Fig. 17.14). Their end walls are oblique and sometimes very thin, perforated with pores, or partly dissolved. Experiments with dyes show that translocation of water can occur in these cells. Some moss "leaf" midribs also contain hydroids, but only in one genus are these known to connect with the hydroids of the stem. Hydroids show some resemblances to tracheids, but lack specialized pitting and lignified walls. No lignin has been detected in bryophytes.

A few of the most specialized mosses also contain cells resembling the sieve cells of vascular plants. Between the epidermis and central strand, elongated cells with oblique end walls **(leptoids)** may occur (Fig. 17.14). These cells are alive, but their nuclei are degenerate and inactive. They have many plasmodesmata in their end walls, and callose may be present. Studies show that sugars may be translocated through these cells.

It is important to emphasize that: (1) most mosses do not contain leptoids; (2) there are important differences between hydroids and tracheids, and between leptoids and sieve cells; (3) there is as yet no firm evidence to show that

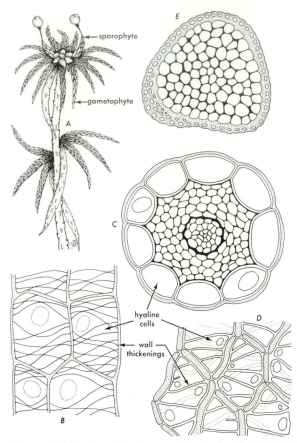

Figure 17.12 *A,* habit of sporophyte plant of *Sphagnum papillosum. B,* external view of stem showing large hyaline cells. *C,* cross section of branch stem showing external hyaline cells and thick walled cells of central axis. *D,* leaf cells; note the narrow chlorophyll-containing cells and the larger hyaline cells with peculiar bandlike thickenings. *E,* cross section of *Tetraphis pellucida* stem to show its simple structure.

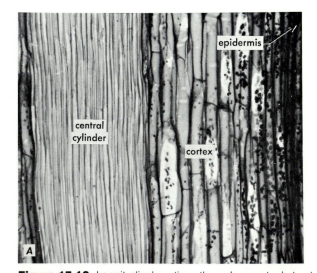

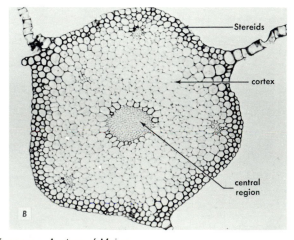

Figure 17.13 Longitudinal sections through gametophyte stems of mosses. *A,* stem of *Mnium* sp. showing central cylinder of sclerenchymalike cells, × 100. *B,* cross section of *Mnium* sp. seta showing thick-walled stereids, thin-walled cortex, and central region of partially collapsed hydroids, × 72.

these cells are phylogenetically related to conducting cells of vascular plants.

Sexual Reproduction

The gametophytes of many mosses are **monoecious** (homothallic); that is, both antheridia (sperm-producing) and archegonia (egg-producing) are produced by the same gametophyte (Fig. 17.15). Some mosses are **dioecious** (heterothallic); in other words, antheridia and archegonia are produced on separate gametophytes. In *Mnium,* shoots bearing antheridia are easily recognized: the leaves that surround these shoots are spread around them somewhat like petals of a flower. The group of antheridia appears as an orange spot in the center of the terminal cluster of leaves (Figs. 17.15 and 17.16*A*).

Antheridia of most true mosses are enclosed by cells forming a **sterile jacket.** These cells contain chloroplasts that become orange-red when the antheridium ripens. The sperm consists mainly of an elongated nucleus and each

has two flagella. The antheridia are surrounded by club-shaped, multicellular, sterile hairs **(paraphyses)** (Fig. 17.16*B*) with conspicuous chloroplasts.

The archegonia have a long **neck,** a thickened middle portion called the **venter,** which contains the egg, and a long **stalk** (Fig. 17.16*C*). When sufficient moisture is present, sperm are extruded from the antheridium. The neck cells of the archegonium separate, opening a passage to the egg. Sperm swim down, possibly responding to a chemical signal, and one fuses with the egg. Mosses are dependent on free water for fertilization.

SPOROPHYTE

Soon after fertilization, the zygote begins to develop into a spindle-shaped embryo that differentiates into a sporophyte consisting of a **foot, seta,** and **sporangium** or **capsule.** The foot penetrates the base of the venter and grows into the apex of the leafy shoot. It absorbs water and nourishment from the gametophyte for its growth and development. The

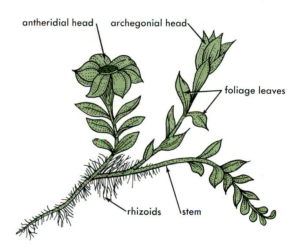

Figure 17.15 *Mnium* sp., showing location of gametangia.

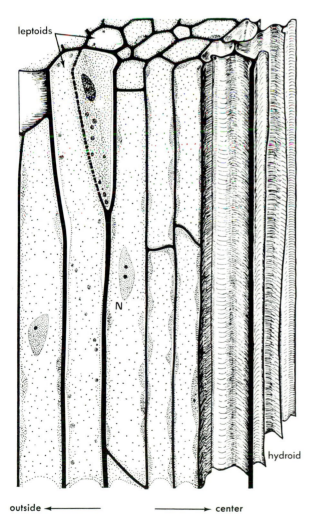

outside ← ——— ——— → center

Figure 17.14 Partial stem portion showing "phloemlike" leptoids and "xylemlike" hydroids.

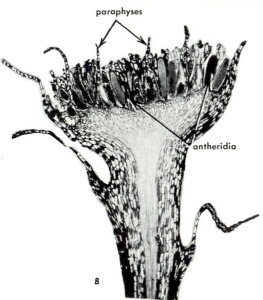

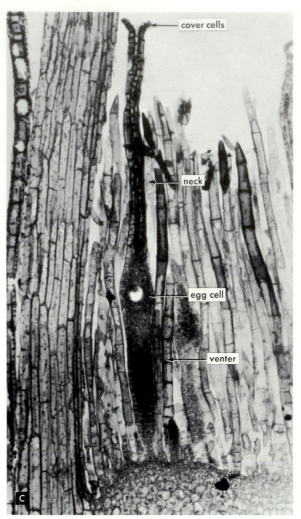

Figure 17.16 Microscopic views of moss gametangia. *A,* antheridial heads of *Polytrichum sp.; B,* antheridial head of *Mnium* sp., × 35. *C,* archegonium surrounded by paraphyses (*Mnium* sp.), × 50.

seta elongates rapidly, raising the sporangium 1 cm or more above the top of the leafy gametophyte (Fig. 17.17*A*). The old archegonium increases in size as the sporophyte enlarges. When the seta elongates, the top of the expanded archegonium is torn from its point of attachment to the gametophyte and elevated with the sporangium. The upper end of the old archegonium, now known as the **calyptra,** remains for a time as a covering for the sporangium (Fig. 17.17*A*).

The sporangium of mosses (Figs. 17.17 and 17.18) generally measures 1 to 3 mm in diameter and 2 to 6 mm long. It is surrounded by an epidermal layer composed of cells similar to those in the epidermis of higher plants. Stomata occur in the epidermis covering the lower half of the sporangium. The sterile tissue forming the inner portion of the sporangium can be conveniently divided into three regions, each of which can be recognized by the type of cells comprising it. The fourth region, composed of sporogenous cells, forms a layer around the columella (Fig. 17.17*B*). The cells of the sporogenous tissue (sporocytes) may increase in number by mitosis; each cell contains the diploid number of chromosomes. Meiosis takes place next, and haploid spores containing 1*n* chromosomes result. They are the first cells of the gametophyte generation.

The columella projects upward, forming a small dome above the main mass of the sporangium. The four or five

A

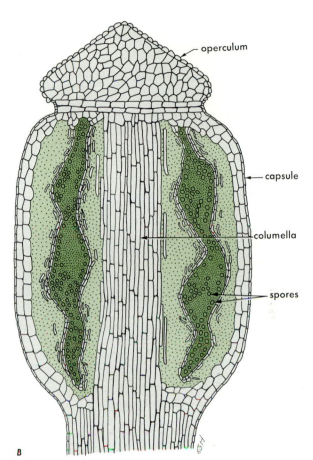

B

Figure 17.17 *A, Mnium* sp., showing gametophytes with attached sporophytes. *B,* median section through a mature but unopened capsule of *Mnium* sp.

outer layers of cells of this dome differentiate into a dry, brittle cap called the **operculum** (Figs. 17.17*B* and 17.18). The cells immediately beneath the operculum form a double row of triangular **peristome teeth** (Fig. 17.18*B* and 17.18*C*). The broad bases of the teeth are attached to the thick-walled deciduous cells that form the **annulus** around the upper end of the sporangium. When the sporangium matures and becomes dry, thin-walled cells of the annulus break down, allowing the operculum to fall away and expose the peristome teeth. By this time, most of the thin-walled cells within the columella have collapsed and the cavity thus formed is filled with a loose mass of spores. Peristome teeth are rough and are very sensitive to the amount of moisture in the air (Fig. 17.18*B* and 17.18*C*). When they are wet or the atmospheric humidity is very high, they bend into the cavity of the sporangium; when dry, they straighten and lift out some of the spores, which are then disseminated by air movements. If a spore comes to rest on moist soil and if illumination and temperature are favorable, it germinates and grows into a protonema.

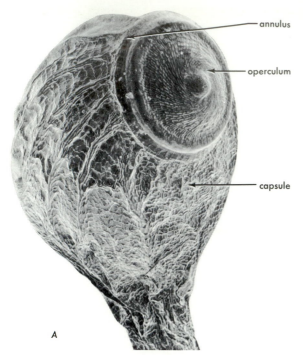

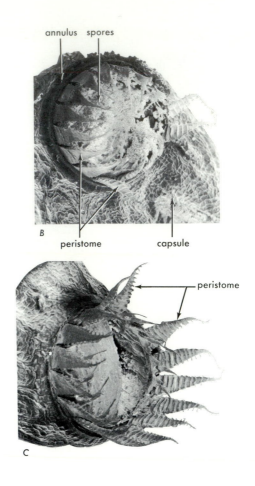

Figure 17.18 Scanning micrograph of a capsule of *Funaria* sp. *A*, capsule with operculum in place; *B*, operculum has been shed, peristome teeth partially open, spores attached to them; *C*, peristome teeth fully open, capsule empty, × 300.

SUMMARY OF MOSS LIFE HISTORY AND CHARACTERISTICS

1. The gametophyte consists of (a) a filamentous, branched, algal-like structure called a protonema, and (b) leafy shoots that develop from buds on the protonema. The shoots consist of a stalk bearing rhizoids at its lower end and leaves throughout its length. The gametophyte is green and able to synthesize food. All gametophyte nuclei are haploid.

2. Antheridia and archegonia develop at the apex of the leafy gametophyte.

3. Each antheridium produces hundreds of sperm.

4. A single egg is formed in the venter of each archegonium.

5. Gametes (eggs and sperm) each contain $1n$ chromosomes.

6. When sufficient moisture is present, mature sperm are extruded from the antheridium. The neck of the mature archegonium is open and sperm swim down it to the egg in the venter of the archegonium.

7. The egg is fertilized by one sperm.

8. The zygote, as a result of the union of gametic nuclei, is diploid.

9. An embryo sporophyte, consisting of a spindle-shaped mass of undifferentiated cells, partially dependent on the gametophyte, develops from the zygote.

10. The embryo develops into a sporophyte consisting of a foot, seta, and sporangium. The sporangium contains sporogenous tissue and several types of sterile tissues, among which are the operculum, annulus, and peristome.

11. Sporocytes, each containing $2n$ chromosomes, form from sporogenous cells.

12. Sporocytes undergo meiosis. Four spores result from each sporocyte.

13. When spores germinate, they form the protonema.

14. In the central strand of some mosses, such as *Sphagnum* and *Mnium,* there are specialized cells called hydroids that may be involved in water conduction.

15. Cells specialized for sugar transport, called leptoids, are also found in some mosses.

18 LOWER VASCULAR PLANTS

Some 400 million years ago, plants spread from aquatic to terrestrial habitats. The early forms probably were low-growing plants that hugged stream banks and the shores of lakes and seas, something like modern bryophytes. In these low-growing forms, turgor pressure is enough to hold the body erect, and water can move throughout the body by diffusion and protoplasmic streaming. But taller plants require stronger supporting tissues and more efficient conducting systems. The origin of the *vascular system* met the need for both support and conduction, and allowed land plants to reach the size of trees. The addition of roots made it possible for some of these plants to spread into habitats where the only available water was below the ground surface. A great adaptive radiation took place, and today a large majority of plant species have vascular systems. They are informally called **vascular plants,** and they include all the seed-bearing species (six divisions) as well as four divisions of plants that reproduce by spores rather than seeds.

When we try to reconstruct the history of plant life, the spore-bearing vascular plants become especially important. Spores are regarded as a primitive character because they occur in algae and bryophytes as well as vascular plants. Thus the term **lower vascular plants** is often applied to the forms that produce spores rather than seeds.

The four divisions of lower vascular plants are the **Psilophyta,** the **Lycophyta,** the **Sphenophyta,**[1] and the **Pterophyta.** The latter are the ferns, and they include some 10,000 species that dwell in a wide range of habitats. Some ferns are small woody trees. The other divisions have many fewer species and are mostly small herbaceous forms.

The modern lower vascular plants are remnants of much larger groups that flourished 395 to 260 million years ago (Table 18.1 places this time into a larger perspective of geological time). Extremely lush swamp forests were common at this time (Fig. 18.1). The forests were dominated by Lycophyta, which included both herbs and trees. Second in abundance were giant Sphenophyta such as *Calamites,* shown earlier in Fig. 14.1. Ferns were third in abundance, and they too included tree-sized forms. Because the land was low, the seas alternately inundated the forests and then retreated, so that one forest after another was buried and later regrown. The buried organic remains have been compressed and changed through time. Today, they form the coal reserves of the world. Burial of plants and transformation into fossil fuels have continuously taken place throughout time, but not at the rate it did during the Coal Age. During the Permian period, starting about 260 million years ago, the climate seems to have cooled and the land was uplifted, ending the dominance of the lower vascular plants; at that time most of the larger forms became extinct.

While each division of lower vascular plants has unique features, the four divisions have several points in common. By definition, all have vascular systems and lack seeds. In

Figure 18.1 Reconstruction of a Coal Age forest, courtesy of the Field Museum of Natural History.

---

[1] We choose to use Sphenophyta for the division to which the horsetails belong, because the alternative designation, Arthrophyta, might be confused with Anthophyta or Arthropoda.

Table 18.1. Geologic Time and the Dominance of Different Plant Groups through Time

Era	Period	Epoch or part	Began (millions of years ago)	Dominants
Cenozoic	Quaternary	Recent	Last 12,000 years	Flowering plants
		Pleistocene	2.5	
	Tertiary	Pliocene	7	
		Miocene	26	
		Oligocene	38	
		Eocene	54	
		Paleocene	65	
Mesozoic	Cretaceous	Upper	90	
		Lower	136	Gymnosperms
	Jurassic	Upper	166	
		Lower	190	
	Triassic	Upper	200	
		Lower	225	
Paleozoic	Permian	Upper	260	
		Lower	280	Lower vascular plants
	Carboniferous (Coal Age)	Pennsylvanian	325	
		Mississippian	345	
	Devonian	Upper	360	
		Middle	370	
		Lower	395	Algae
	Silurian		430	
	Ordovician		500	
	Cambrian		570	
Proterozoic	Precambrian		4,500–5,000	

addition, they all have swimming sperm cells and require free water for fertilization. And finally, like other vascular plants, the sporophyte is the dominant phase of the life cycle.

DIVISION PSILOPHYTA

This division is represented today by a single family, the Psilotaceae, with two genera, *Psilotum* and *Tmesipteris*. They are rare plants found mainly in the tropics, one form of *Psilotum* growing as far north as Florida. *Psilotum* is particularly interesting because it has many features in common with the oldest known fossilized vascular plant: *Cooksonia*, which is found in 395 million-year-old rocks of Czechoslovakia and Wales (Fig. 18.2). *Cooksonia* was probably less than 10 cm tall, made up of dichotomizing branches less than 4 mm in diameter that terminated with sporangia. The spores had a waxy cuticle, indicating that they were adapted for dissemination on land. The xylem in the stem was a solid, central

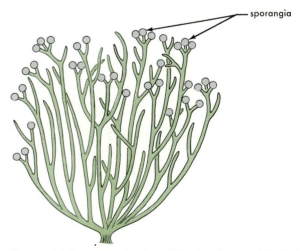

sporangia

Figure 18.2 A reconstruction of *Cooksonia*, one of the first vascular plants. The aboveground portion, shown here, was less than 10 cm tall.

core; there was no pith. We know nothing of the root system.

The living species *Psilotum nudum* is shown in Fig. 18.3. Upright green stems arise from a branched rhizome that has many rhizoids and is always associated with a fungus. There are no roots. The leaves are small and scalelike, and they appear to be outgrowths of the epidermis. Each leaf is supplied by a vein that departs from the vascular cylinder of the stem without creating a gap. Such leaves are called

microphylls, whereas the more familiar leaves of higher plants are called **megaphylls.** In plants that have megaphylls, a gap occurs in the vascular cylinder above the point where a trace departs to supply the leaf.

The vascular system of *Psilotum* is a cylinder of xylem (composed of tracheids) surrounded by phloem (Fig. 18.4). An endodermis with a Casparian strip separates the cylinder from the surrounding cortex. This is reminiscent of the root in higher plants and raises questions concerning the evolution of roots and the original function of the endodermis.

Sporangia are borne in axils of some of the leaves (Fig. 18.3). Meiosis occurs in the sporangia, forming tetrads of meiospores. When the meiospores are ultimately released from the sporangia, they may land in a suitable habitat, germinate, and divide by mitosis repeatedly to form a gametophyte.

Psilophyta gametophytes are little more than a nonpho-

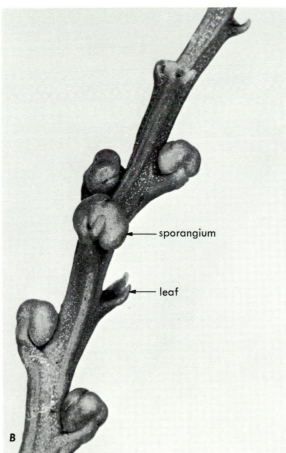

Figure 18.3 *Psilotum nudum. A,* plant; note sporangia on branches on left. *B,* end of a branch, showing scale leaves and sporangia at nodelike swellings.

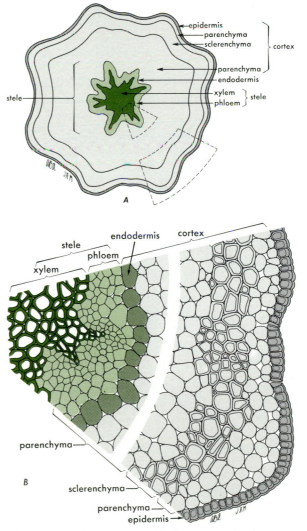

Figure 18.4 Cross section of *Psilotum* stem. *A,* diagram to show the arrangement of tissues. *B,* enlarged sector showing cellular detail.

tosynthetic axis of parenchymatous tissue several millimeters long (Fig. 18.5). Xylem and phloem are rarely present. Gametophytes grow on the surface of soil, rock, or bark, or within the soil. Nutrients are provided via a symbiotic or parasitic relationship with a fungus, which ramifies throughout the tissue.

Gametangia are scattered over the surface of the gametophyte: flask-shaped archegonia with eggs, and spherical antheridia with spirally coiled, multiflagellate sperm.

After a sperm swims through free water and reaches an egg, fertilization occurs and a diploid zygote cell begins dividing. The young sporophyte (embryo) consists of a shoot-root axis attached to the gametophyte by a **foot.** While the shoot-root portion is forming, the foot enlarges by repeated cell divisions, sending haustorial outgrowths into the gametophytic tissue. The foot anchors the sporophyte and absorbs nutrients until the shoot-root axis becomes physiologically independent and a mature sporophyte results.

DIVISION LYCOPHYTA

Fossil representatives of the Lycophyta far outnumber the living forms. There are several extinct orders, Lepidodendrales being one of the best known. Extinct members of this order include large trees that dominated the Coal Age forests. *Lepidodendron* (Fig. 18.6) is an example. It grew to at least 35 m in height and had straplike leaves with sporangia near the ends. A cross section of the trunk reveals pith, primary and secondary xylem, cambium, and an enormous amount of cortex and cork. Though the trunk reached 1 m in diameter, very little of its area served for conduction or woody supporting tissue. The roots were dichotomously branched.

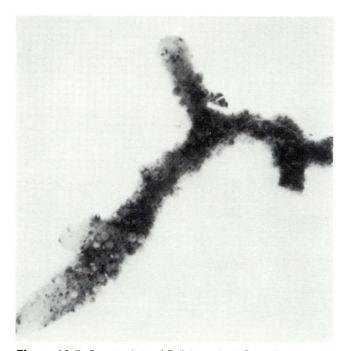

Figure 18.5 Gametophyte of *Psilotum,* about 3 mm long.

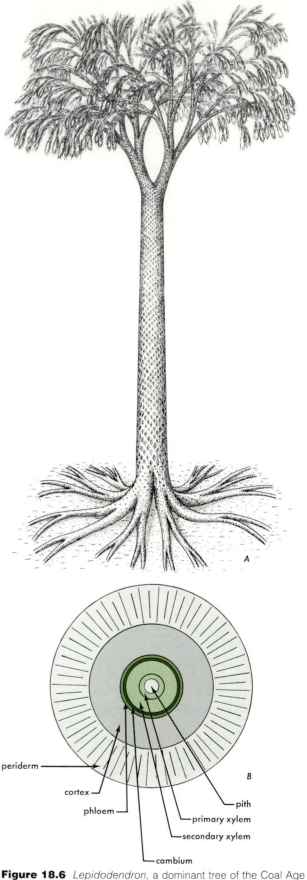

Figure 18.6 *Lepidodendron,* a dominant tree of the Coal Age forest. *A,* reconstruction of the entire plant, about 30 m tall; *B,* cross section of the trunk.

Living Lycophyta are herbaceous. Their stems have a solid core of xylem, and their leaves are microphylls arranged in spirals along the stem. As in the fossil members, sporangia occur along the branches.

Three genera of Lycophyta are common in the Northern Hemisphere: *Lycopodium, Selaginella,* and *Isoetes.* We will consider mainly the first two genera.

LYCOPODIUM

Approximately 400 species are in this genus. Most are trailing plants, many forming short, upright branches that loosely resemble pine seedlings (Fig. 18.7). They are frequently called "ground pine" or "club moss." Although widely distributed, they are most abundant in subtropical and tropical forests. They cannot grow in arid habitats. Several species grow in the eastern and northwestern United States, but none occur in the more arid states of the Southwest. Some eastern species are in danger of extinction because they are popular as Christmas decorations.

Mature Sporophyte

The main stem branches freely and is prostrate. Upright stems, approximately 20 cm in height, grow from the horizontal stem. Both types of stems are sheathed with small green leaves. Small but well-developed adventitious roots

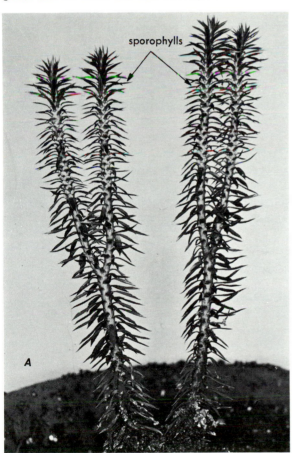

Figure 18.7 Sporophylls in *Lycopodium. A* and *B,* sporophylls similar to vegetative leaves in *L. selago. C,* sporophylls different from vegetative leaves and grouped in terminal strobili in *L. obscurum.*

arise irregularly from the underside of the horizontal stem. As in many higher plants, the primary root, which grows from the embryo, is not long-lived.

The xylem in a *Lycopodium* stem is a complex system of anastomosing strands, so that its distribution in the stem varies greatly in different cross sections of the stem. The phloem is present between the strands of xylem and thus, too, forms a complex system of anastomosing strands. The xylem is composed of tracheids, whereas the phloem contains sieve cells and some parenchyma cells. An endodermis encircles the vascular cylinder. *Lycopodium* stems lack pith, as do roots of most higher vascular plants.

Leaves of *Lycopodium* show a well-developed epidermis with stomata, a mesophyll with many air spaces, and a midvein (Fig. 18.8).

Reproduction

Lycopodium reproduces asexually or through an alternation of generations involving distinctive gametophytic and sporophytic generations (Fig. 18.9). Both generations are independent, and the sporophyte is the dominant generation. Meiospores are borne in sporangia in or near the leaf axils (Figs. 18.7 and 18.9*B*), sheathing the stem. Spores (Fig. 18.9*C*) and sporangia of a given species are all alike. Leaves bearing sporangia are called spore-bearing leaves, or **sporophylls.** In some species, they closely resemble ordinary nonspore-bearing leaves in their structure and appearance. In other species, they differ from sterile leaves in size, shape, position, and color. Such modified sporophylls are grouped together closely at the ends of stems, forming a **cone** or **strobilus** (Figs. 18.7*C* and 18.9*A*).

Gametophyte.   The meiospores germinate and develop into gametophytes (Fig. 18.9*D*). Gametophytes in some species grow above ground and are green; in other species, gametophytes are subterranean and lack chlorophyll. The gametophytes are always associated with a fungus. Both male

and female gametangia are found on the same gametophyte. Gametangia are borne on the upper portion of the gametophyte (Fig. 18.9*E*). Fertilization occurs when sufficient free water is present to allow sperm to swim to mature archegonia.

Sporophyte Embryo.   The embryo sporophyte possesses (a) a well-developed foot, (b) rudiments of a short primary root, (c) leaf primordia, and (d) a short shoot apex (Fig. 18.9*I*). The embryo grows directly into the mature sporophyte plant.

SELAGINELLA

Selaginella species resemble those of *Lycopodium* in their general appearance but are smaller. They number more than 700. Although they are widely distributed, most of them are tropical; a few grow in temperate zones. Some species are adapted to withstand periods of drought and thus may grow in relatively dry localities.

Mature Sporophyte

As in *Lycopodium,* the sporophyte of *Selaginella* generally consists of a branched, prostrate stem with short, upright branches, usually only 10 cm high. In some species, the stem is upright and slender and taller. Both horizontal and upright stems are sheathed with small leaves in four longitudinal rows or ranks.

Two species of *Selaginella* are shown in Fig. 18.10. One of them, *Selaginella watsonii,* grows in exposed rock crevices in the high Sierra Nevada and is able to withstand periods of drought. The more delicate *Selaginella emmeliana* grows best in a humid environment; it is frequently grown as an ornamental plant.

The vascular system in stems of *Selaginella* consists of one to several branching strands, each having a central core of xylem surrounded by phloem. The cross sections in Fig. 18.11 show a single strand in a stem of *Selaginella*. It occurs in a large air space and is supported by strands of radially arranged endodermal cells. Vessels are present in the xylem of several species of *Selaginella*. Parenchyma and sclerenchyma tissue form a cortex, which is bounded externally by an epidermis.

Reproduction

As in *Lycopodium,* spores are borne in sporangia, which grow in or near axils of sporophylls. Although sporophylls do not differ greatly in appearance from sterile leaves, they are always grouped to form cones, or strobili, at the ends of upright branches.

Two types of sporangia are formed: **megasporangia** (Greek *mega,* "large") and **microsporangia** (Greek *micro,* "small"). As the names indicate, megasporangia produce larger spores. A single strobilus usually contains both types of sporangia.

Within a developing megasporangium, all but one of the **megasporocytes** (potential spore-producing cells) degenerate. This remaining megasporocyte, nourished in part by

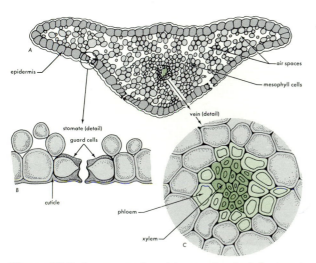

Figure 18.8 *A,* cross section of *Lycopodium* leaf. *B,* stomate detail. *C,* vein detail.

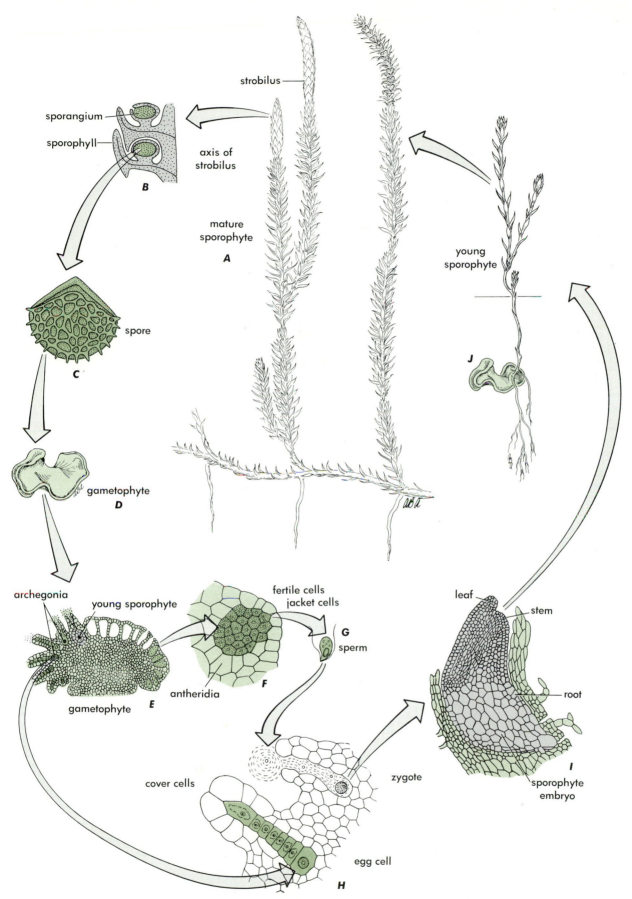

Figure 18.9 Stages in the life cycle of *Lycopodium annotinum*. *A*, mature sporophyte; *B*, section of sporophylls showing location of sporangia; *C*, spore; *D*, gametophyte; *E*, section of gametophyte showing location of gametangia; *F*, antheridium; *G*, sperm; *H*, archegonia before and after fertilization; *I*, section of embryo; *J*, underground gametophyte with young sporophyte.

fluid resulting from the degenerating spores and in part by cells surrounding the spore cavity, increases greatly in size. During meiosis, four large spores, called **megaspores,** are formed from the megasporocyte. Each megaspore germinates and gives rise to a **female gametophyte (megagametophyte;** Fig. 18.12G).

Only a few microsporocytes within the developing microsporangium degenerate. The approximately 250 microsporocytes that remain undergo meiosis, each forming four small spores, or **microspores** (Fig. 18.12E and 18.12H).

The production by a given species of two distinct types of meiospores—megaspores and microspores—is called **heterospory.** Thus, *Selaginella* is heterosporous. *Lycopodium*, which produces but one spore type, is said to be **homosporous.**

Female Gametophyte. Repeated cell divisions within the megaspore result in the female gametophyte, which is contained within the megaspore wall until it nears maturity. Its increase in size eventually ruptures the megaspore wall, and a small cushion of colorless gametophytic tissue protrudes from the megaspore along the lines of rupture (Fig. 18.12G). Archegonia develop on the protruding cushion.

The megaspore (containing the female gametophyte) may be shed from the cone or strobilus at almost any stage of the development of the gametophyte. In some species, it is retained in the strobilus until well after fertilization. In any event, fertilization occurs only when sufficient water, either from rain or dew, is present, allowing sperm to swim to archegonia.

Male Gametophyte. Upon germination, microspores divide into two cells. One of them, the **prothallial cell,** is small and does not divide further; it represents the vegetative portion of the male gametophyte. The other cell, by repeated divisions, develops into an antheridium composed of a jacket of sterile cells enclosing 128 or 256 biflagellate sperm (Fig. 18.12J, 18.12K, and 18.12L). This development takes place within the microspore wall. Microspores are shed from the microsporangium midway in the development of the male gametophyte and grow to maturity without direct connection with the parent sporophyte or the soil. Usually, microspores sift down to the bases of the megasporophylls below the microsporophylls. In this position, they are close to the developing female gametophytes. The sperm escape when the microspore wall ruptures.

Sperm swim to the archegonia, which grow on that portion of the female gametophyte protruding from the megaspore (Fig. 18.12M). Fertilization ensues, and the resulting zygote initiates the diploid or sporophyte generation.

Sporophyte Embryo. Of the two cells formed by the first division of the zygote, only one develops into an embryo. The other cell grows into an elongated structure, the **suspensor,** which pushes the developing embryo into gametophytic tissue, where there is a food supply (Fig. 18.12M).

The embryo is a structure with a foot, a root, two embryonic leaves, and a shoot (Fig. 18.12N). In certain species of *Selaginella*, the embryo is held by the megaspore and retained within the strobilus. It does not pass into a dormant state, as do embryos of seeds, but continues to grow. Young

Figure 18.10 Two species of *Selaginella*. A, *S. watsonii*, a montane species growing in crevices in granite; B, *S. emmeliana*, a native to tropical American and requiring a moist, warm climate.

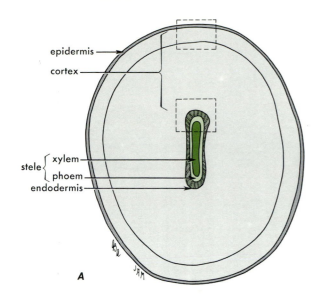

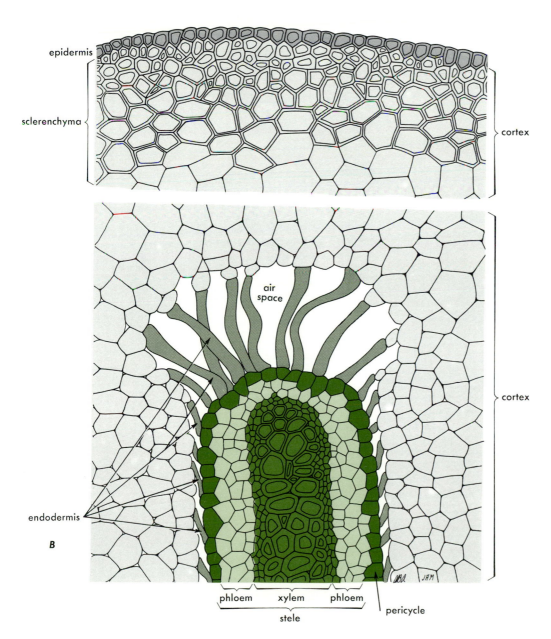

Figure 18.11 Cross section of a stem of *Selaginella*. *A*, diagram showing distribution of stem tissues. A single, flattened cylinder of vascular tissue is supported in an air space by filaments of endodermal cells, and the cortex is composed of parenchyma and sclerenchyma. *B*, detail showing cellular structure.

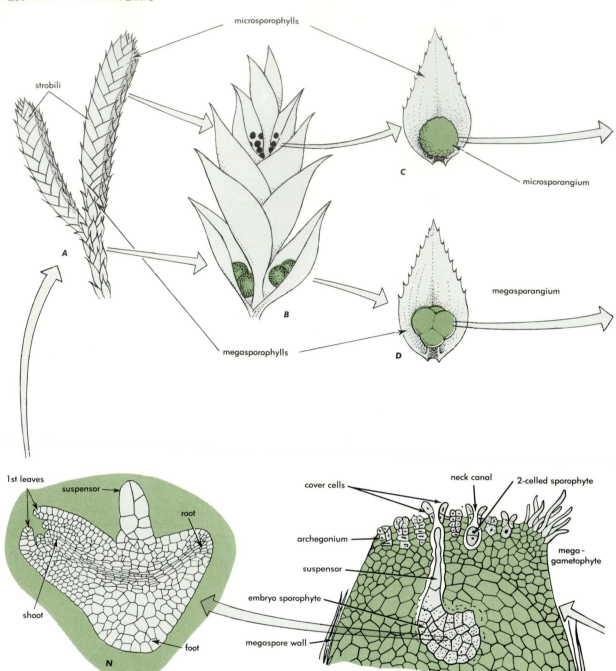

Figure 18.12 Stages in the life cycle of *Selaginella*. *A*, shoot of *S. watsonii* showing two strobili; *B*, enlarged portion of strobilus showing sporophylls with sporangia; *C*, microsporophyll with microsporangium; *D*, megasporophyll with megasporangium; *E*, microsporangium discharging microspores; *F*, megasporangium discharging megaspores; *G*, megagametophyte within megaspore; *H*, microspore; *I*, enlarged section through a microspore; *J*, microgametophyte within microspore; *K*, antheridium with developing sperm cells; *L*, mature sperm cells in microsporangium; *M*, enlarged section through megagametophyte showing archegonia and a young embryo pushed by its suspensor into the vegetative tissue of the gametophyte; *N*, embryo sporophyte which would normally be embedded in gametophyte tissue.

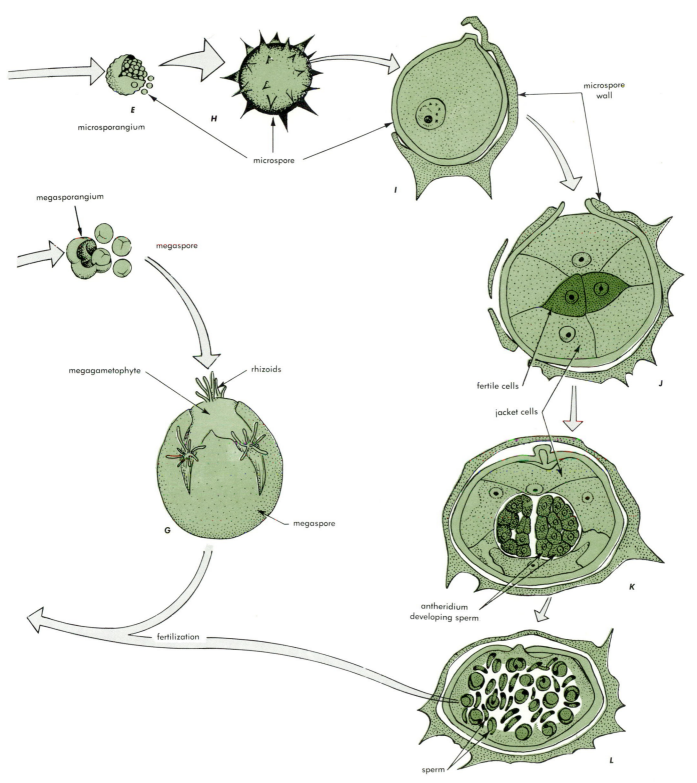

microsporangium

E

H

microspore

microspore wall

I

megasporangium

megaspore

megagametophyte

rhizoids

G

megaspore

fertile cells

J

jacket cells

antheridium
developing sperm

K

fertilization

sperm

L

Fig. 18.12 continued

sporophytes may be found extending from the strobilus of parent sporophytes. If the developing embryo were to pass into a period of dormancy while being held by the parent sporophyte, a condition would arise that approaches the seed habit.

ISOETES

The genus *Isoetes* (quillworts) includes more than 60 species. They are small, herbaceous plants that grow submerged, for at least part of the year, in fresh water (Fig. 18.13). The prominent part of the body is a group of slender leaves that bear sporangia at their bases. The leaves die back each year to a storage organ (a corm) just below the soil surface.

DIVISION SPHENOPHYTA

Members of the Sphenophyta are uncommon today, but they once grew very abundantly. In the Coal Age forests, only the Lycophyta were more prominent. *Calamites* (see Fig. 14.1) is among the most common extinct forms. *Calamites* grew 10 m tall and had a trunk 30 cm in diameter. Along

Figure 18.13 *Isoetes.*

the trunk, whorls of branches gave rise to smaller branches, and the latter developed whorls of leaves at the nodes. The upright stems arose from a horizontal rhizome.

Today, there is but a single genus *(Equisetum)*, with about 25 species (Fig. 18.14). Few members exceed 2 m in height. Many species inhabit cool, moist places, but *Equisetum arvense* also grows in dry habitats. Most species are characterized by the presence of silica in the epidermis of stems. Because of this trait, they were used in colonial days to scour pots and pans and thus were called scouring rushes. Members of the genus are also commonly known as the horsetails.

MATURE SPOROPHYTE

The sporophyte of *Equisetum* is the dominant phase of its life history. In one tropical species, the sporophyte is vinelike and may reach 7 m in length. Usually, 1 m represents the maximum upright growth. All species are perennials and have a branched rhizome from which upright stems arise. Stems, depending on the species, branch either profusely or sparingly. In either case they are straight and marked by vertical ridges and distinct nodes (Fig. 18.14*C*). Bases of nodes are sheathed by whorls of small simple leaves. Many branches may also form at each node. The microphyllous leaves are much reduced in size, nongreen, and in many species, short-lived. The stems are green and are, therefore, the organs of food manufacture. Roots occur at the nodes of the rhizomes, but they may also develop from stem nodes if the stem becomes procumbent on a moist surface.

Except at nodes, stems of *Equisetum* are hollow and the ribs make prominent markings on their outer circumferences. The ridges are formed of sclerenchyma tissue and not only project outward but extend inward almost to the small vascular bundles (Fig. 18.15). There are air spaces, or canals, between vascular bundles. Photosynthetic tissue lies between these air canals and the thin, outer layer of sclerenchyma tissue and epidermis. Vascular strands are also marked by a canal. Arms of xylem extend outward from this canal, and phloem tissue lying outward from the canal is present between radial arms of xylem. Some species contain vessels in the xylem. An endodermis is present, but its location varies from species to species; it may surround each vascular bundle or encircle a ring of vascular bundles. In some species, there is also an inner endodermis.

REPRODUCTION

In all species of *Equisetum,* the sporangium-bearing organs, **sporangiophores,** are specialized structures very different from ordinary leaves. They are grouped together in strobili, or cones (Fig. 18.14*B*) at the summit of main upright branches and occasionally on lateral branches. In most species, strobili are borne on ordinary vegetative shoots; in a few species, they are formed only on special fertile shoots.

The sporangiophores are stalked, shield-shaped structures borne at right angles to the main axis of the cone. The cone can be compared to a pole to which open umbrellas

Figure 18.14 Stems of *Equisetum telmateia*. *A*, fertile stems with strobili; note much-branched, adjacent sterile stems. *B*, strobilus with sporangiophores. *C*, enlarged view of the stem, showing three nodes, vertical ridges, and reduced leaves.

have been fastened, the handles of the umbrellas being at right angles to the pole. Sporangia are attached to the underside of the shield (umbrella), close to its edge. They extend horizontally inward, toward the axis of the cone. Four meiospores with the haploid number of chromosomes are formed from each sporocyte. All meiospores are morphologically alike. *Equisetum,* therefore, is homosporous. Upon maturity the meiospores are discharged from the sporangia. The spores are fragile and normally live only a few days.

Gametophytes

Gametophytes are small green bodies about the size of a pinhead and consist of a cushionlike base with many erect, delicate lobes (Fig. 18.16). They are easily cultured in the laboratory.

Gametangia are similar to those of *Lycopodium*. Fertilization takes place only when there is free water. Sperm are spiral shaped and multiciliate.

The zygote develops directly into an embryo. No suspensor is formed. The embryo is similar to those already described, except that the foot is small or lacking.

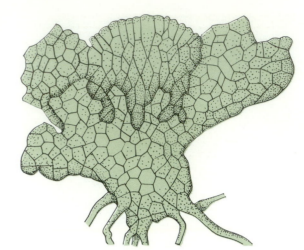

Figure 18.16 Gametophyte of *Equisetum*.

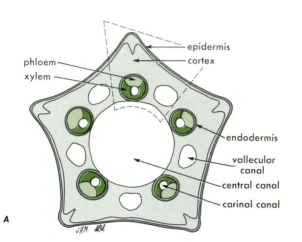

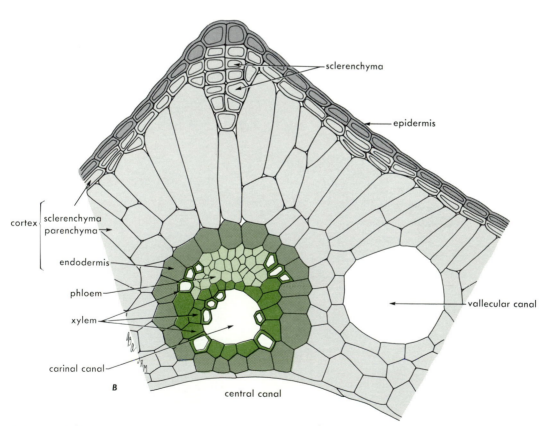

Figure 18.15 Cross section of *Equisetum* stem. *A*, diagram showing arrangement of tissues; *B*, enlarged view showing details of cellular structure.

DIVISION PTEROPHYTA

The division Pterophyta comprises the ferns, some 10,000 species, most of which are shade-loving plants of small size; their upright leaves, or **fronds,** are generally their most prominent feature (Fig. 18.17). Most species native to temperate zones have an underground rhizome with leaves and roots arising at nodes; a few are small, floating aquatics without rhizomes. Some ferns of tropical regions grow into fairly large trees. All Pterophyta have a definite alternation of generations, with both sporophyte and gametophyte being autotrophic plants.

The Pterophyta comprise four orders, the largest of which, the Filicales, includes the **true ferns.** The Filicales are divided into 11 or more families, of which the largest and best known in the United States is the Polypodiaceae. Most of the discussion below concerns representatives of this family.

MATURE SPOROPHYTE

The sporophyte, which is the dominant generation of all ferns, possesses an underground stem or rhizome from which leaves and adventitious roots arise (Fig. 18.17). Young fern leaves are rolled into tight spirals and consist chiefly of meristematic tissue. As the leaf matures, it unwinds from the base upwards. The upright, expanded portion is mature, but the leaf continues to grow by cell division at its coiled tip. In certain species, the apical meristem may remain active for years, resulting in leaves that are nearly 3 m long. The uncoiled, fully expanded fern leaf lacks any residual meristematic tissue.

Figure 18.17 *A,* typical, mature fern sporophyte, showing rhizome below ground and fronds above ground, lady fern, *Athyrium felix-femina. B-E,* aquatic ferns: *B, C, D, Salvinia* in the Salviniales; *E, Marsilia* in the Marsiliales.

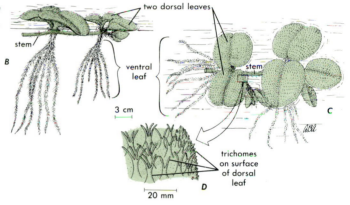

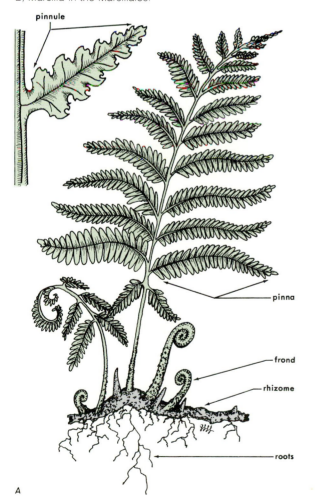

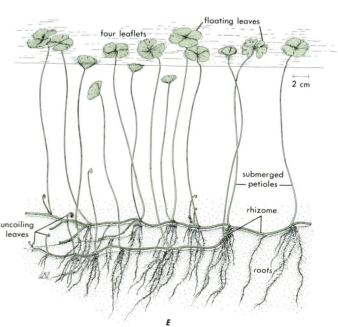

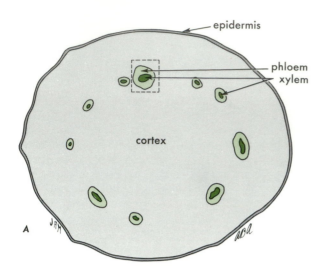

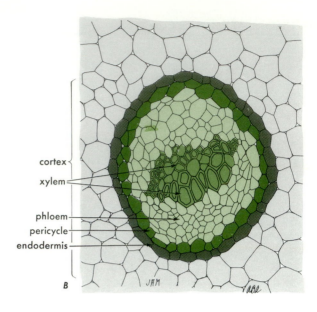

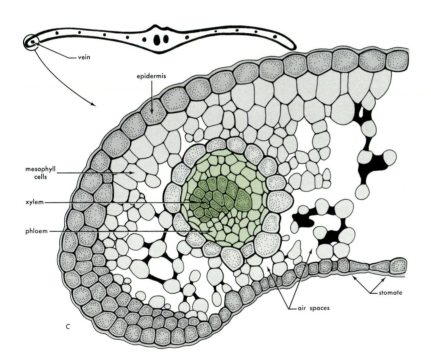

Figure 18.18 Fern anatomy. *A, Polypodium* stem cross section, showing relationship of tissues; *B,* a higher magnification of *A,* showing cellular detail; *C,* leaf cross section of *Polypodium.*

Leaves, the most prominent part of the fern plant, vary greatly in form (Fig. 18.18). Most fern leaves are compound, although simple types exist in all groups. The most common type of fern leaf has a stout or rigid petiole, which is prolonged to form a rachis from which leaflets (pinnae and pinnules) arise.

The vascular tissue of ferns is organized into one or more vascular strands, each having a core of xylem, surrounded by phloem and separated from the ground parenchyma by an endodermis (Fig. 18.19). These vascular bundles are interconnecting, so that, as in *Lycopodium,* the precise arrangement varies from section to section. However, there are arrangements characteristic of various gen-

era. For instance, in *Polypodium* there is a ring of small bundles placed well out from the center of the rhizome (Fig. 18.19*A*). Each strand is formed of a central core of xylem, surrounded by phloem, and the whole has a bounding endodermis (Fig. 18.19*B*). The xylem consists largely of tracheids, vessels being known in only two genera. Sieve cells occur in the phloem.

A typical fern leaf has a well-developed epidermis with chloroplasts in the epidermal cells and stomata on the lower surface (Fig. 18.19*C*). The mesophyll is either relatively undifferentiated or divided into palisade and spongy parenchyma layers.

REPRODUCTION
Vegetative Reproduction

Vegetative reproduction occurs in one of two ways: (1) by death and decay of the older portions of the rhizome and the subsequent separation of the younger growing ends; and (2) by the formation of deciduous leaf-borne "buds," which become detached and grow into new plants. Such buds occur in only a few genera.

Sexual Reproduction

In the sexual life cycle, independent sporophyte and gametophyte generations alternate with each other. The vegetative structure of the sporophyte has been described above. Spores are borne in sporangia, which ordinarily develop on the lower surface or margins of fronds (Fig. 18.20). Not all leaves are fertile (i.e., spore-producing), and the fertile leaves in some species are dissimilar in structure to the sterile (non-spore-producing) leaves. The distribution of sporangia on the leaf surface varies considerably in different genera and species (Fig. 18.20). Sporangia may (a) cover much of the lower surface, (b) be grouped in **sori** (singular, **sorus**) and grow in a definite relationship with veins, or (c) grow only along margins of the leaf.

When sporangia are grouped together in sori, a struc-ture called the **indusium** (Fig. 18.21*B*), which is sometimes umbrella-like, may be present, thereby protecting young and developing sporangia. Frequently, marginal sporangia are protected by the curled edge of the leaf itself, which forms a **false indusium.**

The sporangium is a delicate, watch-shaped case, consisting of a single layer of epidermal cells, only one row of which possesses heavy walls (Fig. 18.21*B*). This row, which nearly encircles the sporangium in the Polypodiaceae, is the **annulus;** it functions in opening dried mature sporangia and aids in dispersal of ripe spores.

The young sporangium is filled with sporocytes. Meiosis results, as always, in spores with a reduced number of chromosomes. Ferns, with the exception of two families, the Marsileaceae and the Salviniaceae, are homosporous. These two families have widely distributed species which are small, aquatic, and unfernlike in appearance.

Gametophyte. Gametophytes, or **prothallia,** of the Polypodiaceae are small, flat, green, heart-shaped structures with rhizoids on their lower surface (Figs. 18.21*D* and 18.22). In most species, they mature rapidly and are not long-lived. Antheridia and archegonia are borne on the same prothallus (Fig. 18.21*D*, 18.21*E*, 18.21*F*, 18.21*G*). Antheridia are formed

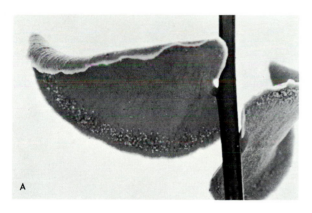

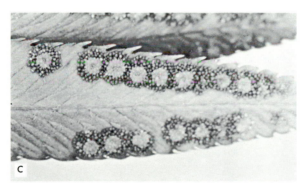

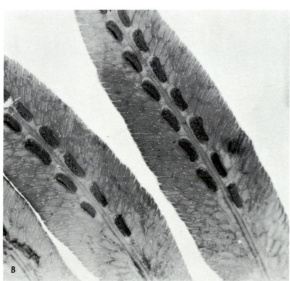

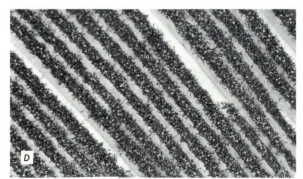

Figure 18.19 Diversity in sori. *A, Pellaea,* showing sporangia in a band close to leaflet margin; *B, Woodwardia,* showing sori parallel to midrib and covered with indusia; *C. Polystichum,* showing round sori with centrally attached indusia; *D, Asplenium,* showing sporangia in long rows on veins.

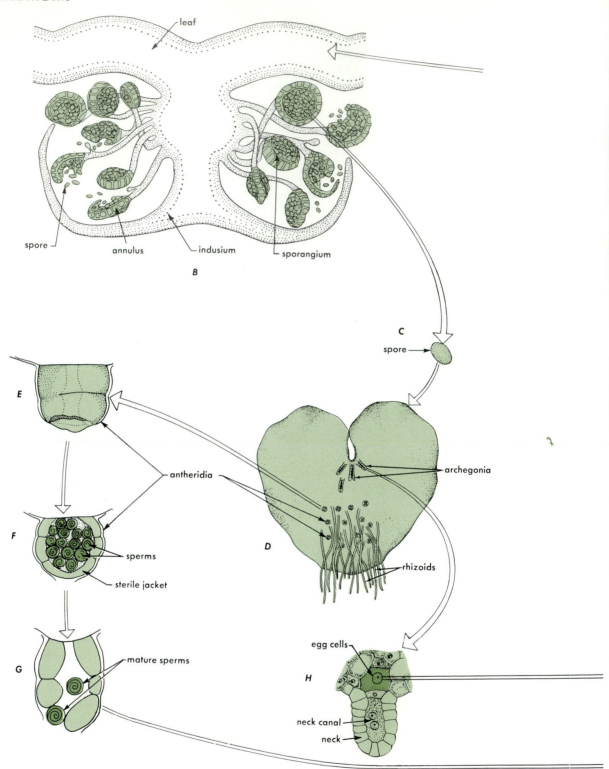

leaf

spore · annulus · indusium · sporangium

B

C

spore

E

antheridia

archegonia

F

sperms

sterile jacket

D

rhizoids

G

mature sperms

egg cells

H

neck canal

neck

Fig. 18.20

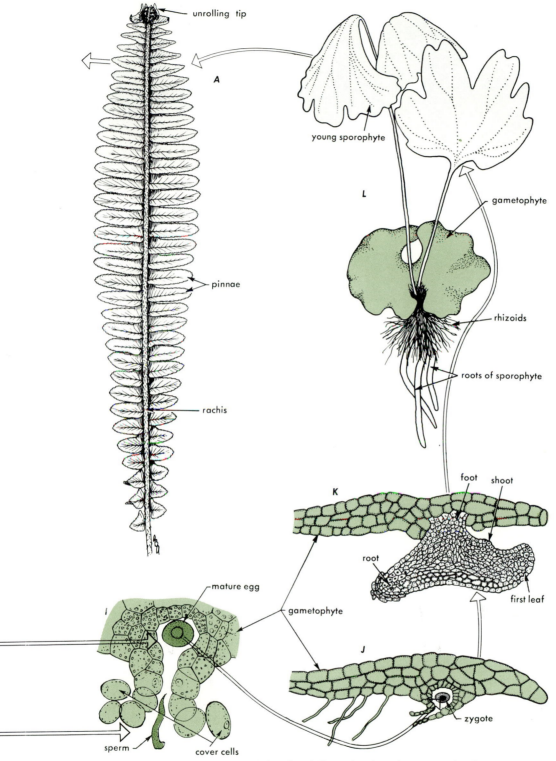

Figure 18.20 Stages in the life cycle of a fern. *A*, fern frond; *B*, section through a sorus, showing indusium, sporangia, and spores; *C*, spore; *D*, gametophyte; *E*, antheridium; *F*, section of antheridium showing sperm; *G*, open antheridium discharging sperm; *H*, archegonium; *I*, archegonium receptive to sperm; *J*, section of gametophyte showing a zygote with the venter of the archegonium; *K*, section of gametophyte and embryo; *L*, gametophyte with a young sporophyte still attached.

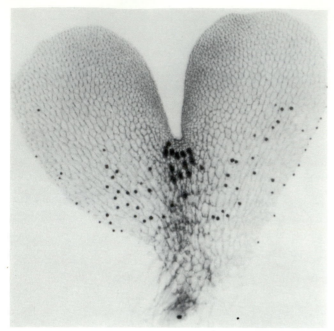

Figure 18.21 Photographs of fern gametophytes. *A,* a view from beneath, showing rhizoids, antheridia (dots), and archegonia (near notch); *B,* a gametophyte (G) with attached young sporophyte (S) emerging and overtopping the gametophyte.

when the prothallus is very young and are scattered over its lower surface. Normally, 32 sperm develop within each antheridium.

Archegonia form later than antheridia and are usually clustered close to the notch, also on the undersurface of the gametophyte (Fig. 18.21*D* and 18.21*H*).

Fertilization occurs when free water is present and sperm are thus able to swim to the neck of the archegonium (Fig. 18.21*I*). The resulting zygote is diploid and rapidly develops into an embryo sporophyte comprising a foot, root, stem, and leaf (Fig. 18.21*J* and 18.21*K*). The embryo develops directly into a young sporophyte, without a dormancy period.

SUMMARY

1. Lower vascular plants is a general name for vascular plants that do not produce seeds. This group consists of four divisions: Psilophyta, Lycophyta, Sphenophyta, and Pterophyta. The latter division, the ferns, contains the most species and is the most widespread.

2. The Psilophyta is represented by only two genera, *Psilotum* and *Tmesipteris.* Sporophytes of *Psilotum* consist of a branched, underground rhizome with rhizoids (no true roots) and an upright, green, branched stem with scalelike microphylls. It is homosporous and the meiospores germinate to produce nonphotosynthetic, small gametophytes. Free water is required for fertilization. The embryo sporophyte is at first attached to the gametophyte by an absorbing organ (the foot), and it depends for nourishment on the gametophyte.

3. The Lycophyta is represented by only three genera in this book: *Lycopodium, Selaginella,* and *Isoetes.* Two other genera, *Stylites* and *Phylloglossum,* were not discussed. Fossil forms, including trees, were prominent in Coal Age forests.

4. Sporophytes of *Lycopodium* (club mosses) are less than 20 cm tall and exhibit true roots, stems, and leaves. The stems are sheathed with small leaves that contain an epidermis with stomata, a mesophyll with intercellular air spaces, and a central vein. Sporangia are borne in the axils of leaves; they may be clustered near the tips of branches in a strobilus (cone). The genus is homosporous, and the meiospores produce small gametophytes that may or may not have chlorophyll. Free water is required for fertilization, and the embryo sporophyte is at first attached to the gametophyte by a foot, as in *Psilotum.*

5. Sporophytes of *Selaginella* are similar in appearance to those of *Lycopodium.* Vessels are present in the xylem of some species. Sporangia are borne in the axils of leaves and they are clustered into strobili. Sporangia and spores are of two kinds. Megasporangia produce megaspores that germinate and grow (within the ruptured spore wall) into female gametophytes. Microsporangia produce microspores that germinate and grow (within the spore wall) into male gametophytes. Microspores may sift down the strobilus to lie near megaspores, and then they rupture to release sperm; if free water is present, fertilization can be accomplished in the strobilus. The embryo is at first pushed deep within the female gametophyte by a suspensor, and nutrition is gained from the gametophyte through a foot. *Selaginella* is heterosporous.

6. Sporophytes of Sphenophyta (horsetails) are about 1 m tall and consist of an underground rhizome, with true roots, and an upright, jointed, silica-rich stem. Tree-sized members of the Sphenophyta, now extinct, were prominent in the Coal Age forests. Microphylls and branches arise in whorls at the nodes. Vessels are present in some species. Sporangia are grouped into complex strobili. The division is homosporous. Meiospores germinate and grow into small, photosynthetic gametophytes. Free water is required for fertilization. The embryo stage is similar to that of *Psilotum.*

7. Sporophytes of Pterophyta (ferns) are typically below 1 m in height, but some reach tree stature. In most cases, the stem is completely underground as a rhizome, and only the large leaves and petioles are above ground. A few ferns are small, floating aquatics. Vessels are present

in some species. Leaves may exhibit palisade and spongy parenchyma. Sporangia are produced on specialized leaves in some species, or on the undersides of all leaves in other species. The sporangia may be clustered into compact groups called sori and covered with an indu-sium. Most ferns are homosporous and, if so, the meio-spores germinate and grow into small, photosynthetic, heart-shaped gametophytes (prothallia). Free water is necessary for fertilization, and the embryo sporophyte is much like that of *Psilotum*.

Table 18.2. Comparison of Representative Lower Vascular Plants

Generation	*Psilotum*	*Lycopodium*	*Selaginella*	*Equisetum*	Ferns
Sporophyte	A branching stem, scale leaves, no roots	Prostrate branching stem with upright branches, roots, and leaves	Prostrate branching stem with upright branches, roots, and leaves, also scales	Rhizome, upright, jointed stem, small leaves	Rhizome, roots, and leaves
	Herbs or epiphytes	Herbs	Herbs	Herbs	Herbs or trees
	Simple vascular system	Simple vascular system	Simple vascular system, but vessels are present	Simple vascular system but vessels may be present	Stem: several strands of vascular tissue each with xylem surrounded by phloem. Vessels: only in two genera. Sieve cells: no companion cells in phloem
	One type of sporangium	One type of sporangium	Megasporangia and microsporangia	One type of sporangium	One type of sporangium (mostly)
	Homospory	Homospory	Heterospory	Homospory	Homospory (mostly)
	No sporophylls	Sporophylls present	Sporophylls present	Sporangiophores	Leaves bear spores in sporangia
	No cone	A cone in some species	Cone	Cone composed of sporangiophores	No cone
	Embryo develops attached to gametophyte	Embryo develops attached to gametophyte	Embryo develops within female gametophyte on sporophyll	Embryo develops attached to gametophyte	Embryo develops attached to gametophyte
Gametophyte	Irregular subterranean structure, associated with fungus	Irregular to tapered subterranean structure, associated with fungus	Male gametophyte: one prothallial cell, and an antheridium within microspore	A single green thallus resembling *Anthoceros*	Heart-shaped, small, green, completely independent thallus
	Gametangia embedded in thallus	Gametangia embedded in thallus	Female gametophyte: small amount of tissue within megaspore, cushion protrudes in which gametangia are embedded	Gametangia embedded in thallus, neck of archegonium protruding	Antheridium of approximately 6 to 10 cells, archegonia partially embedded in gametophyte
	Motile sperm	Motile sperm	Motile sperm	Motile sperm	Motile sperm

19 GYMNOSPERMS

Limited by the lack of vascular tissues, the first land plants were probably small forms that crowded the margins of ponds and waterways. Large land plants became possible with the evolution of woody vascular tissue. Vascular tissues also opened the way to the evolution of roots, which allowed plants to colonize areas with no surface water. Nevertheless, the lower vascular plants were still limited to regions that were at least seasonally very wet. The reason is that lower vascular plants have free-living gametophytes as part of their life cycle, and fertilization involves free-swimming sperm cells.

In the next stage of plant evolution, plants were able to break away from moist habitats by means of three new structures: the *pollen grain,* the *ovule,* and the *seed.* These innovations are first clearly seen in the **gymnosperms,** a mixed group of forms that are named for the trait of forming seeds on the surface of leaves (the term *gymnosperm* literally means "naked-seeded"). The conifers (firs, pines, and others) are familiar gymnosperms.

The gymnosperms form distinct male and female gametophytes. A pollen grain is a male gametophyte, with just two or three cells inside a waxy wall. Carried by water, wind, or animals, the pollen grain travels passively to the female structures. This eliminates the need for swimming sperm (but, as will be discussed, some gymnosperms still retain a much reduced swimming stage).

The female gametophyte remains attached to a leaf of the parent sporophyte throughout its development. In this way it receives shelter, water, and food; and it is never exposed to the rigors of the terrestrial environment. The female gametophyte is enclosed in a jacket of sterile cells called the **nucellus** (equivalent to a sporangium), and the nucellus in turn is enclosed in one or two layers of cells called the **integuments.** The whole unit is called an **ovule.** A stalk connects the ovule to the leaf. With this arrangement, the gametophyte has a much better chance of producing eggs than if it were free-living.

Given ovules, only minor advances were needed to evolve seeds. The zygote develops into an embryonic sporophyte inside the gametophyte. This involves no innovations, for the bryophytes and lower vascular plants also form embryos. But in lower forms, the embryo grows to maturity at its site of origin, on the parent gametophyte. Not so in the seed plant: here the embryo reaches a certain size and becomes dormant, while the integuments harden into a protective seed coat. The resulting unit—the seed—may remain dormant for years, allowing time for dispersal and the passage of adverse seasons.

With these new traits, the gymnosperms dominated world vegetation from about 225 million years ago until they were displaced by still more advanced plants, the angiosperms (Chapter 20). The gymnosperms and angiosperms are often called **higher vascular plants** because of their advanced reproductive traits.

The term *gymnosperm* is not a formal taxonomic grouping but is viewed as a stage of evolution. There are four divisions of living gymnosperms, encompassing about 600 species. By far the largest is the division **Coniferophyta,** whose members are a major source of lumber and paper pulp in North America. This group is represented by such common trees as pines *(Pinus),* spruces *(Picea),* firs *(Abies),* and true cedars *(Cedrus).* We will consider the Coniferophyta in detail. Two minor divisions that were widespread in the distant past are the **Cycadophyta** and **Ginkgophyta.** The **Gnetophyta** is an artificial fourth division made up of the three unusual genera *Gnetum, Ephedra,* and *Welwitschia.*

DIVISION CONIFEROPHYTA

CLASSIFICATION

Without exception, the well-known and economically important gymnosperms of temperate zones belong to the Coniferophyta. There are nine families and 550 species. They comprise pines, hemlocks, firs, spruces, junipers, yews, redwoods, and many others. Not all of them bear cones, yet the cone is such a conspicuous feature of many of them that the division has been named the Coniferophyta, or cone-bearers. Most conifers form true cones. Juniper berries are in reality cones with fleshy adhering scales. In yews, seeds are surrounded at the base by a more or less pulpy, berrylike body and are not borne in cones. In either case, the seeds are naked, not being surrounded by an ovary wall.

The following brief descriptions of several common genera enable us to introduce the more important conifers in some of the seven families.

Family Pinaceae

Pinus **(Pines).**   Pines are usually large trees. Some members of bristlecone pine *(Pinus longaeva)* are the oldest living things, over 5000 years old. The leaves are needlelike, two or more growing together (except singly in *Pinus monophylla*) in a **fascicle,** or group, which is sheathed at the base (Fig. 19.1). The cones vary greatly in size and shape and are very characteristic of the species to which they belong.

Figure 19.1 Pines. *Pinus jeffreyi* (Jeffrey pine), showing the trunk of one tree and the crown of a second tree. *B*, mature ovulate cones of *P. ponderosa* (ponderosa pine), a closely related species. *C*, fascicle of ponderosa pine needles.

***Abies* (Firs).** Firs are stately trees of a symmetrical, cylindrical, or pyramidal shape. The leaves are flat and linear; in cross section they are relatively broad, without marked angles. The cones are erect on the branches and shatter at maturity (Fig. 19.2).

***Picea* (Spruces).** These trees closely resemble the firs, but they can be distinguished by the position of the leaves on the branchlets, by the angular appearance of the leaves in cross section, and by the cones that are hanging and do not shatter at maturity (Fig. 19.3).

***Tsuga* (Hemlocks).** The trees of this genus are pyramidal with slender horizontal branches. The leaves are usually two-ranked, linear, flat, and with a short petiole. They resemble the leaves of firs but are much shorter. The cones are small.

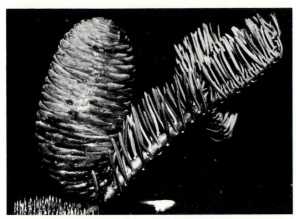

Figure 19.2 *Abies*. Ovulate cone and branch. Note arrangement of needles.

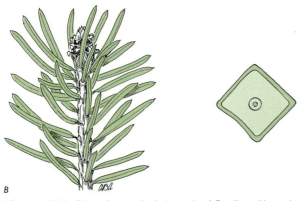

Figure 19.3 *Picea* (spruce). *A*, branch of *P. alba* with ovulate cones; *B*, leaves of *Picea*. For anatomical detail of the needle, see Fig. 19.10.

***Pseudotsuga* (Douglas Fir).** Only two species occur in the United States. One of them, the Douglas fir *(Pseudotsuga menziesii),* is the most important timber tree of the United States. It may grow to a height of 60 m, with a trunk diameter of 3 m. Its leaves are flat, like those of the true firs, but have white lines on either margin and a groove along the upper surface. The cones are 5 to 11 cm long, pendulous, and easily recognized by bracts extending outward below each scale.

***Larix* (Larches).** Trees belonging to this genus grow in the cooler portions of the Northern Hemisphere. They differ from most other members of the Pinaceae in that they are deciduous. The American larch, or tamarack, is a tall tree frequently found in bogs. The needles are short, linear, and grouped in crowded clusters on short spurs. On leading shoots, however, the needles are arranged spirally. The cones are small and persistent (Fig. 19.4).

Figure 19.4 *Larix lyallii* (larch). *A*, branch with needles; *B*, mature ovulate cones.

Figure 19.5 *Juniperus occidentalis* (juniper) growing at 2200 m elevation in the Sierra Nevada of California. *A*, tree; *B*, branch with staminate cones; *C*, branch with ovulate cones.

Family Cupressaceae

This family contains the junipers, cypresses, and false cedars (true cedars, *Cedrus* species, are in the Pinaceae). The leaves are borne singly but are usually small and scalelike and are crowded closely around the stems (Fig. 19.5). The cones are either woody, as in *Cupressus,* or fleshy, as in *Juniperus.* The cone of juniper is often called a "berry," but it is a modified cone, composed of fleshy scales that completely enclose the seeds (Fig. 19.5).

Family Taxodiaceae

This family contains the bald cypress, redwood, and Sierra bigtree. Redwood *(Sequoia sempervirens)* may be the tallest tree in the world: individuals over 60 m are common, and the record height is 112 m. Redwood has outstanding lumber qualities, not the least of which is its resistance to decay. Consequently, logging has removed 90% of all virgin redwood forest in the past 150 years. Only half of the remaining acreage is protected from future logging. Sierra bigtree *(Sequoiadendron giganteum)* is the most massive of any plant or animal species in the world; specimens have been measured with trunk diameters up to 9.8 m and heights above 82 m. The trunk alone of the General Sherman tree weighs 625,000 kg (over 680 tons).

Family Taxaceae

With two exceptions, ovulate cones are not borne by members of the yew family. The seed of this family so strongly resembles a drupe or a nut that it is usually called a fruit. The embryo is protected by an outer flesh called an **aril,** and an inner hard pit. Both these tissues are derived from the integuments of the ovule; hence, the structure is morphologically a naked seed. As shown in Fig. 19.6, the outer red aril of ground pine *(Taxus canadensis)* almost encloses the seed. It drops away when the seed is mature.

Native yews are common only in a few places in the United States. Possibly the best-known yew is the English yew, which, because of the excellent bows that were made from its wood, is closely linked with English history and folklore. Yews are now widely cultivated.

Figure 19.6 Branches of *Taxus* (yew) with seeds. Note berry-like aril surrounding seed.

LIFE HISTORY OF A CONIFER

The following outline of a gymnosperm life cycle is based largely on the genus *Pinus*. It differs from life cycles of other genera mainly in requiring three summers for completion. Significant stages of pollination, fertilization, and maturation of the seed occur within a decimeter of each other on a given branch, thus making it an ideal subject for class study.

Sporophyte

All species of pines are trees. They range in size and age from scrubby, 70 year-old fire-adapted species such as knobcone pine *(Pinus attenuata)* to large, straight-trunked ponderosa pine (*Pinus ponderosa;* Fig. 19.1) 500 years old, to twisted, slow-growing bristlecone pine (Fig. 19.7) 5000 years old.

The series of tree rings preserved in such long-lived conifers has value in estimating past climates. Each ring's width reflects temperature and soil moisture conditions. By comparing recent ring widths with known climatic data, it is possible to extrapolate back thousands of years. At the Laboratory of Tree-Ring Research in Tucson, Arizona, scientists have estimated climatic changes back to 3500 B.C. (Fig. 19.7*B*). These trees have also permitted anthropologists and archeologists to correct ages of material previously dated by the ^{14}C method. This method is hundreds of years in error for material older than 1000 B.C., and the corrected dates have revolutionized archeological thinking.

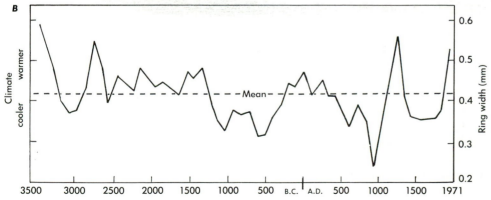

Figure 19.7 *A, Pinus longaeva* (bristlecone pine) in the White Mountains of California, growing near timberline. *B,* average ring widths for the past 5500 years. Rings wider than 0.4 mm indicate a warmer than average growing season (April-October), rings narrower than that suggest cooler conditions. (From LaMarche, Science 183: 1046 (1974).

Conifer wood has no vessels. A three-dimensional reconstruction from a small piece of *Sequoia* is diagramed in Fig. 19.8. It is extremely regular. Tracheids are elongated cells that appear in cross section as square, hollow cells. Those formed in the spring are largest in diameter; as the season progresses, newly formed tracheids are smaller in diameter with thicker walls. The heavy-walled cell is called a **fiber-tracheid.** Note that the narrow xylem rays are composed of brick-shaped parenchyma cells, sometimes with dark-staining contents. Other parenchyma cells run in vertical columns; these are called axial parenchyma.

Many gymnosperms produce resin. Typically, resin occurs in long **resin ducts** that are surrounded by parenchyma cells (Fig. 19.9). Resin plays a role in herbivore defense by inhibiting the activity of wood-boring insects. *Sequoia* does not produce resin.

In the phloem, companion cells are missing, but **sieve cells,** fibers, and parenchyma cells are present.

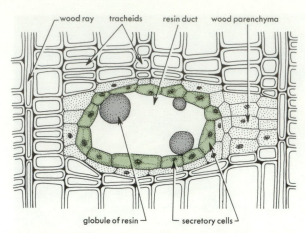

Figure 19.9 Resin duct in pine wood, as seen in cross section.

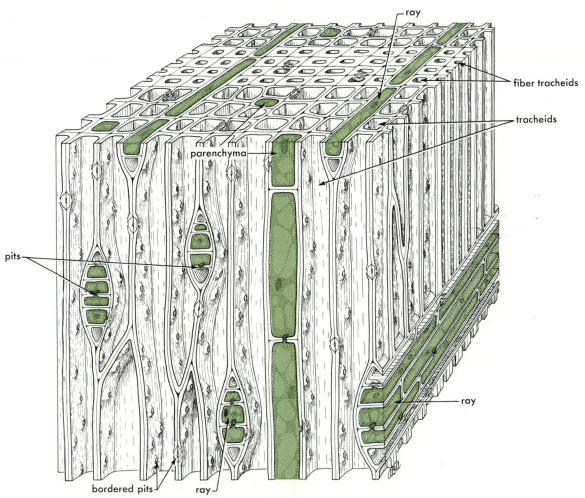

Figure 19.8 Block diagram of secondary xylem of redwood (*Sequoia sempervirens*).

The needle leaves of *Pinus* show considerable adaptation to drought (Fig. 19.10): sunken stomata, thick cuticle, fibrous epidermis, close-packed mesophyll, and veins only in the center of a thick leaf. Such modifications are important in winter, when soil moisture is frozen and the evergreen leaves are subjected to drying winds.

All conifers produce two kinds of spores—**microspores** and **megaspores**—borne in cones that are morphologically distinct. The two types of cones are known, respectively, as **staminate** (or pollen) and **ovulate** (or seed) **cones.**

Staminate (Pollen) Cone

Staminate cones average 10 mm or less in length, by 5 mm in diameter. They are borne in groups, usually on lower branches of trees (Fig. 19.11*A*). Each cone is composed of a large number of small scales (microsporophylls) arranged spirally on the axis of the cone. Two microsporangia develop on the under surface of each scale (Fig. 19.11*B*).

Stages of microspore development are quite similar to those of *Selaginella* and to spores of mosses and ferns. Microsporocytes, which are surrounded by a nutritive cell layer, the **tapetum,** undergo meiosis, and four microspores result. As usual, each microspore contains the haploid number of chromosomes. The nucleus within the newly formed microspore divides several times by mitosis, forming a **pollen grain** that contains two viable, haploid nuclei and vestiges of several vegetative cells (Fig. 19.12*H*). Enormous numbers of pollen grains are finally shed from the microsporangium. They are light in weight and bear two wings that may facilitate dispersal by wind.

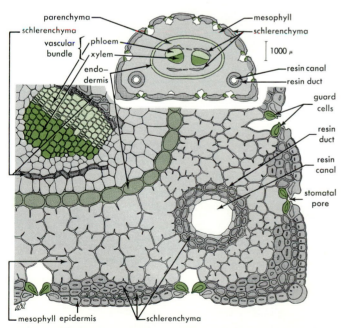

Figure 19.10 Cross section of a pine leaf. The inset (above) shows the entire cross section, with major tissues outlined; the large drawing shows cellular detail of a wedge from the epidermis into one of the vascular bundles at the center.

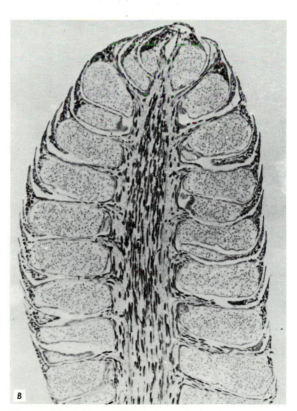

Figure 19.11 Staminate cones of *Pinus*. *A*, end of a branch with a cluster of staminate cones holding ripe pollen. *B*, longitudinal section of one staminate cone, showing microsporangia attached to the underside of each scale with pollen grains inside.

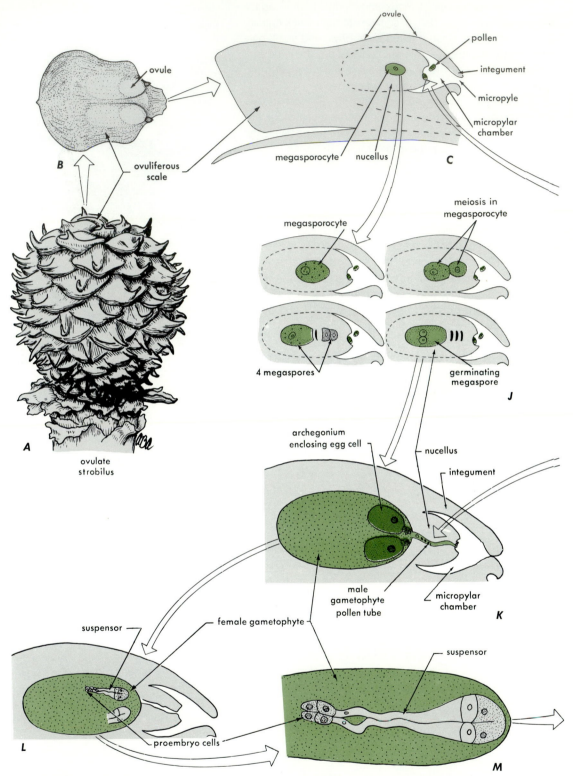

Figure 19.12 Stages in the life cycle of a pine. *A,* young ovulate strobilus, or cone; *B,* scale from ovulate cone showing two ovules on upper surface; *C,* section of ovuliferous scale showing megasporocyte cell and pollen in micropylar chamber; *D,* staminate cone; *E,* scale from staminate cone; *F,* cross section of staminate scale; *G,* winged pollen grain; *H,* development of male gametophyte in pollen grain; *I,* germinated pollen grain with pollen tube; *J,* formation of linear tetrad of megaspores; *K,* female gametophyte with two egg cells in archegonia and pollen tube; *L,* suspensor and proembryo sporophyte within female gametophyte; *M,* proembryo; *N,* section of seed showing seed coat, female gametophyte, and embryo sporophyte; *O,* winged seed; *P,* seedlings.

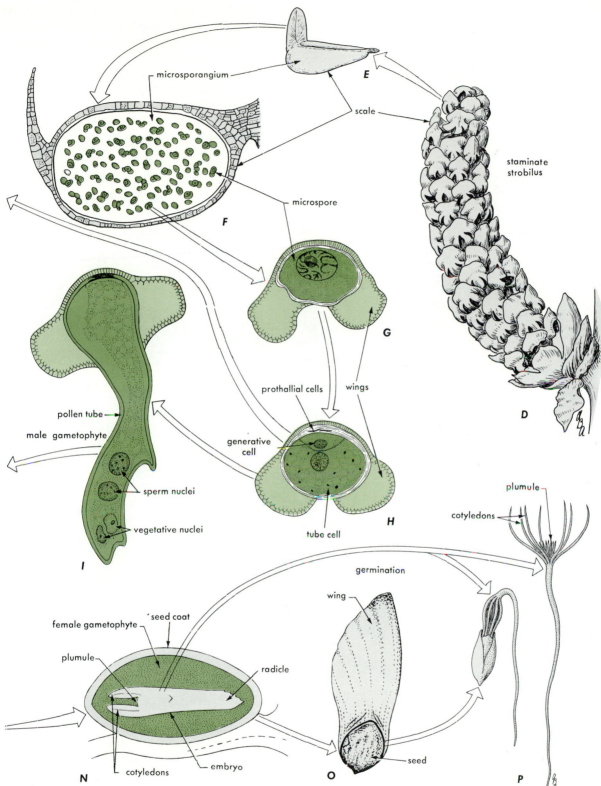

microsporangium

scale

staminate strobilus

E

microspore

F

D

prothallial cells

wings

G

pollen tube

male gametophyte

generative cell

sperm nuclei

vegetative nuclei

tube cell

H

I

plumule

cotyledons

germination

wing

female gametophyte

seed coat

plumule

radicle

cotyledons

embryo

seed

N

O

P

Fig. 19.12 continued

Ovulate (Seed) Cone

The ovulate cone, when mature, is the well-known cone of pines and other conifers. Each cone is composed of an axis upon which are borne several woody scales in a spiral fashion. Ovulate cones develop at tips of young branches in early spring. Two ovules develop on the upper surface of an **ovuliferous scale** (Figs. 19.12*B* and 19.13). Each ovuliferous scale holds two ovules, and each ovule encloses a single megasporangium.

The ovules first appear as small protuberances on the upper surface of this scale, close to the axis of the cone. A protective layer of cells, the **integument,** develops early on the outer surface of the ovule. In the end of each ovule near the axis of the cone, there is a small opening, the **micropyle,** through which pollen grains can enter.

In pine, one megasporocyte lies in the center of each ovule (Fig. 19.12*C*); several are contained in ovules of redwoods and cypresses. Like microsporocytes, the mega-

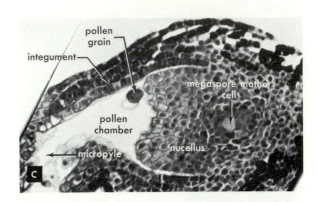

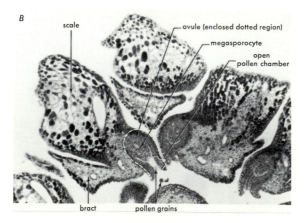

Figure 19.13 Ovulate cones of *Pinus. A,* longitudinal section of entire cone at the time of pollination, less than 1 cm long. *B,* close-up view of ovules from *A;* note the pollen chambers and micropyles. *C,* an ovule with a pollen grain in the pollen chamber. *D,* cone at the time of embryo development, scales closed. *E,* section through tip of ovule after pollination but before fertilization.

sporocyte is surrounded by a nutritive tissue called the **nucellus.** The nucellus is in reality a megasporangium because it surrounds the area where megaspores arise.

Female Gametophyte. The megasporocyte soon divides by meiosis. Four megaspores, usually arranged in a single row of four cells, result from each megasporocyte. The nucleus within each megaspore has the haploid number of chromosomes. Generally, only one of the four megaspores develops into a female gametophyte; the other three degenerate. Germination of the remaining megaspore and growth of the female gametophyte progresses very slowly. Several months are required in most conifers, and 13 months are required in pine.

The development of the female gametophyte takes place entirely within the ovule. There are approximately 11 mitotic divisions of the megaspore before cell walls begin to appear between the newly formed nuclei. The walls gradually form, resulting in a small mass of gametophytic tissue, completely enclosed by the diploid cells of the ovule. The adjacent sporophytic tissue of the ovule, the nucellus, is in part digested by the developing female gametophyte.

While cell walls are being laid down in the developing female gametophyte, two or more archegonia differentiate at its micropylar end. Figures 19.12K and 19.13E show that the ovule at this stage consists of integuments, nucellus, and a female gametophyte containing several archegonia, each with its enclosed egg. Directly beneath the micropyle is a space, the **micropylar chamber.** Nucellar tissue lies between the micropylar chamber and the archegonia.

Pollination

As discussed previously, **pollination** in flowers is the transfer of pollen from anther to stigma; in conifers, it is the transfer of pollen from staminate cone to ovulate cone. Conifer pollen is windblown. In many species, ovulate cones are borne on higher branches of the tree and staminate cones are concentrated on lower branches; since pollen does not blow directly upward, cross pollination is usual.

Pollination occurs in most conifers in early spring soon after the ovulate cone emerges from the dormant bud, about the time of meiosis. At this age, scales of young cones turn slightly away from the axis so that pollen grains can sift down to the axis of the cone. Here they come in contact with a sticky substance secreted by the ovule. As this material dries, it draws some pollen grains through the micropyle into the micropylar chamber, where they lodge close to the developing female gametophyte (Fig. 19.13). The pollen grain germinates and develops slowly into a tubular male gametophyte as it grows through the nucellus. Thus, male and female gametophytes of conifers develop to maturity in close proximity within the ovule. They are both dependent on nucellar tissue for nourishment and protection.

Several nuclear divisions occur in the tube, but no cell walls are formed. Two of the last-formed nuclei are **sperm nuclei.** This branched **pollen tube,** containing two sperm nuclei and several vegetative nuclei, is the mature male gametophyte (Fig. 19.12I).

Fertilization and Embryo Formation

Development of male and female gametophytes is coordinated so that the egg is formed and ready for fertilization when the pollen tube, containing two sperm nuclei, reaches the archegonium. In pine, this development requires about one year. At the time of fertilization, pine cones are generally green and the scales are tightly closed, showing a spiral pattern of arrangement (Fig. 19.13). The sperm nuclei, together with other protoplasmic contents of the pollen tube,

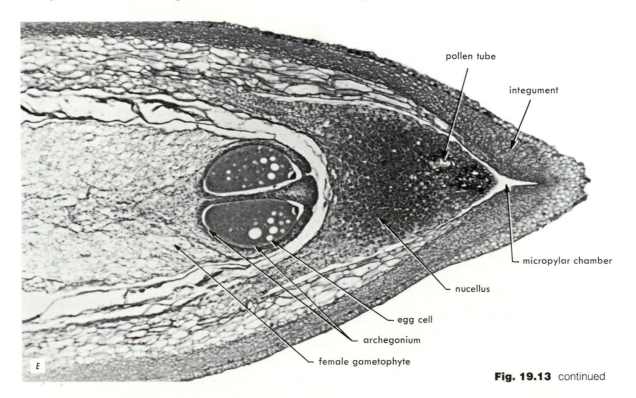

Fig. 19.13 continued

are discharged directly into the egg cell. Sperm nuclei do not possess cilia and hence are not actively motile. One sperm nucleus comes in contact with the egg nucleus and unites with it. The nonfunctioning sperm nucleus and the other protoplasmic material that are discharged into the egg cells soon undergo disorganization.

The formation of the embryo is preceded by the development of a relatively elaborate **proembryo.** This structure consists of four tiers of four cells each (Fig. 19.12*M*). The four cells farthest from the micropylar end of the proembryo may each develop into an embryo. The intermediate cells are the suspensor cells; they elongate greatly and push the embryo cells deep into the female gametophyte. While this development is taking place, the female gametophyte continues to grow, digesting most of the rest of the nucellus. It enlarges and becomes packed with food to be used not only for the growth of the embryo but also as a reserve in the seed.

As discussed, the female gametophyte usually contains two or more archegonia. Since the egg in each archegonium may be fertilized and since each of the four embryo-forming cells may give rise to an embryo, it is possible for eight or more embryos to develop in every seed. Normally, only one embryo survives, but seeds with two well-formed embryos are not rare.

The mature embryo consists of several **cotyledons** or seed leaves, an **epicotyl, hypocotyl,** and **radicle** or rudimentary root (Fig. 19.12*N*). The embryo is embedded, as previously mentioned, in the enlarged female gametophyte. Both embryo and female gametophyte are surrounded by a papery shell and a hard protective **seed coat,** both formed from the integument of the ovule. The whole structure is the **seed.** In many pines, the seeds are winged.

Normally, in pines, the seed matures some 12 months after fertilization; two years thus intervene between initiation of ovules and formation of seeds. A generalized life history of a pine is shown in Table 19.1.

Table 19.1. A Generalized Life History of a Pine

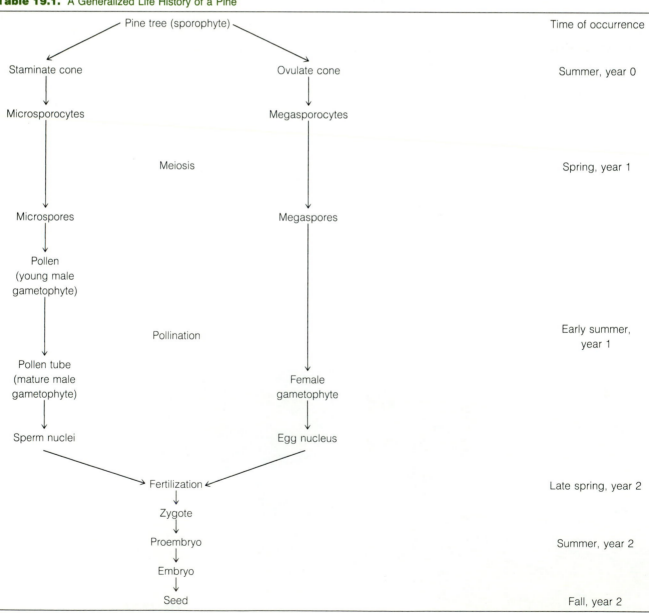

	Time of occurrence
Pine tree (sporophyte)	
Staminate cone → Ovulate cone	Summer, year 0
Microsporocytes → Megasporocytes	
Meiosis	Spring, year 1
Microspores → Megaspores	
Pollen (young male gametophyte)	
Pollination	Early summer, year 1
Pollen tube (mature male gametophyte) → Female gametophyte	
Sperm nuclei → Egg nucleus	
Fertilization	Late spring, year 2
Zygote	
Proembryo	Summer, year 2
Embryo	
Seed	Fall, year 2

Gymnosperm seeds may remain dormant for many years, and some may remain embedded in the old mature cones for six years or more. Heat causes the resin that coats the cones of some species to melt, allowing the cones to open and release the seeds. This behavior has considerable survival value, since large numbers of seeds are released after fires, and injured trees are thus replaced by the young ones. In many species, however, seeds are shed soon after they are mature.

DIVISION CYCADOPHYTA

Some 200 million years ago, members of the Cycadophyta formed an extensive portion of the earth's flora, probably constituting the food supply for at least some of the herbivorous dinosaurs. Today, there remain 9 to 10 genera of cycads with about 100 species growing in widely separated areas of the earth's surface but largely confined to the tropics. Only one genus, *Zamia,* occurs naturally in the continental United States; it is found in southern Florida. Various genera are, however, cultivated outdoors in warmer regions, and in greenhouses elsewhere.

Two cycads, *Cycas revoluta* and *Dioön spinulosum,* are shown in Fig. 19.14. Both were grown in the conservatory at Golden Gate Park in San Francisco. These trees are palm-like in appearance; in fact, *Cycas revoluta* is known as Sago palm. In *Zamia,* the genus native to Florida, the trunk is largely subterranean, but in *Cycas* it forms a single straight bole. The trees grow slowly, a 2 m high specimen being perhaps as much as 1000 years old.

It is readily evident that these trees are really gymnosperms and not palms because of their large cones (Fig. 19.16). A primitive feature of the division is the presence of flagellate sperm.

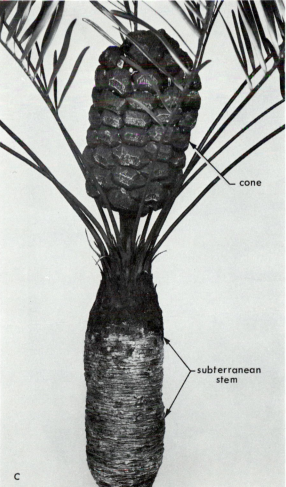

Figure 19.14 Cycads. *A, Cycas revoluta* showing a microsporangiate cone. *B, Dioön spinulosum* showing ovulate cone. *C, Zamia* showing ovulate cone and subterranean stem.

DIVISION GINKGOPHYTA

Only one living representative, the maidenhair tree *(Ginkgo biloba)*, remains of this very ancient division of plants. It has been reported as growing wild today in forests of remote western China. *Ginkgo* has, however, been grown for centuries on Chinese and Japanese temple grounds and is now cultivated in many countries. It is a large tree with characteristic small, fan-shaped leaves (Fig. 19.15) that are divided into two lobes. The sperm are flagellated. In the United States the *Ginkgo* is commonly planted as a city sidewalk tree.

DIVISION GNETOPHYTA

This division consists of only three genera, and in some respects their traits place them as intermediate between the gymnosperms and the angiosperms. For example, they contain vessels in the xylem, the ovules appear to be surrounded by two integuments, and the pollen-producing structures superficially resemble stamens. All these traits are shared by angiosperms and are absent in other gymnosperms, yet true flowers and fruit are absent—seeds are still borne naked.

Only one genus, *Ephedra* (joint-fir, Mormon tea), is found in North America. Its 40 species are distributed in warm, temperate parts of the Mediterranean region, India, China,

Figure 19.15 Branch with leaves and seeds of *Ginkgo biloba* (maidenhair tree).

Figure 19.16 *Ephedra viridis,* a member of the Gnetophyta. *A,* a portion of the woody shoot; *B,* close-up of *A,* showing two scalelike leaves attached to a young, green stem. *C,* (next page) female strobili. Note "pollination drop" at the cone tips.

the southwestern deserts of the United States, and mountainous parts of South America. It is a shrubby xerophyte, with whorls of small deciduous leaves at prominent joints; most photosynthesis is accomplished in the green stems (Fig. 19.16).

The other genera look quite different from *Ephedra* and exhibit important life cycle differences. Their inclusion into one division is a matter of convenience for classification, and the resulting division is quite artificial. *Welwitschia* is another xerophyte, distributed along the southwest coast of Africa. It has two long, straplike leaves that trail along the soil surface (Fig. 19.17). *Gnetum* is a tropical genus of 30 species, mainly climbing lianas. The leaves look very much like typical dicot leaves.

Figure 19.16 continued

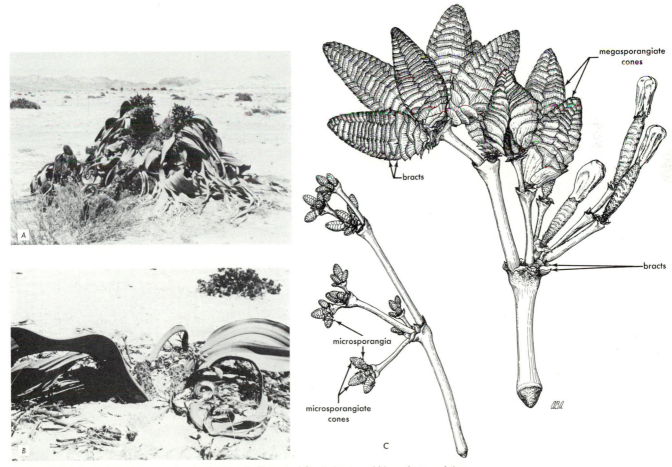

Figure 19.17 *Welwitschia,* growing in the Namib Desert of Southwestern Africa. *A,* one of the larger plants, about 1.5 m tall and carbon dated at 1500 years old; *B,* closer view of smaller plant, showing the two leaves emerging from the nearly buried stem; *C,* male (left) and female (right) cones borne near the growing base of the long, leathery leaves. The plants are unisexual.

SUMMARY

1. The gymnosperms are higher vascular plants and possess both vascular tissue and seeds. Seeds aid in plant dispersal, protection of the embryo, and establishment of the young sporophyte. A second advancement is pollination, which permits sperm to reach eggs while surrounded by protecting tissue and in the absence of free water.

2. Of the four divisions of gymnosperms—Cycadophyta, Ginkgophyta, Gnetophyta, and Coniferophyta—the latter contains the most species, is the most widespread, and contributes the most to modern vegetation types. The Coniferophyta contains seven families, several of which were discussed. The Pinaceae contains pines, firs, spruces, hemlocks, Douglas fir, larch, and true cedars (*Cedrus* spp.). The Cupressaceae contains junipers, cypresses, and false cedars. The Taxodiaceae contains bald cypress, redwood, and Sierra bigtree. The Taxaceae contains the yews. Other families are better represented in the Southern Hemisphere.

3. The *Pinus* (pine) life cycle was examined as representative of the Coniferophyta. The sporophyte is large and anatomically complex. Vessels are absent, but considerable secondary xylem is produced. Leaves are modified for arid conditions. Pine is heterosporous. The microspores are produced in small (staminate) cones on lower branches. Megaspores are produced in other (ovulate) cones on upper branches; these cones are at first small, but over a two year period they mature into large, woody cones.

4. The microspores begin dividing while still within the spore wall. After two divisions, and the elaboration of air-filled wings on the outside of the thickened spore wall, growth ceases and the immature male gametophytes are liberated to the wind. These are pollen grains. Some, by chance, sift down into young ovulate cones and are sealed in when the cone scales close.

5. Two ovules are present on each scale in the ovulate cone. An ovule consists of an outer integument, nucellar tissue, and the female gametophyte. One nucellar cell functions as a megasporocyte, and, of the four resulting megaspores, only one survives. It begins to divide to form the female gametophyte at the same time that pollination (the transfer of pollen from male to female cone) occurs. At maturity, the female gametophyte is a relatively undifferentiated, multicellular, nonphotosynthetic thallus embedded in the nucellus. At one end are two or more archegonia.

6. If the pollen grain is drawn through the micropyle (an opening in the integument), it comes to lodge against the nucellus. It germinates and a pollen tube begins to grow toward the female gametophyte through the nucellus. Finally, a third cell division produces two sperm, and the formation of the mature male gametophyte is complete. No antheridium is recognizable.

7. Fertilization is accomplished when the pollen tube ruptures near an archegonium and one sperm unites with an egg. The resulting zygote divides and differentiates into an embryo. A mature embryo lies surrounded by female gametophyte tissue, some remnants of the nucellus, and a seed coat (from the integument). The embryo is dormant until the seed is shed and comes to lie in an appropriate microenvironment.

8. In general, life cycles from thallophytes to conifers have shown a definite trend or pattern: the dominance and complexity of the sporophyte have increased and the dominance and complexity of the gametophyte have declined.

9. The Cycadophyta contain about 100 species of tropical, slow-growing, palmlike plants. The Ginkgophyta consists of a single species that may now be extinct in the wild but is widely cultivated. The Gnetophyta exhibit some traits that link them more closely to the flowering plants than to other gymnosperm groups. Only one of its three genera, *Ephedra,* is found in North America. This latter division is quite artificial in terms of classification.

20 ANGIOSPERMS

The angiosperms, or flowering plants, are the dominant plants of the world today. They include nearly all the crop plants of orchard, garden, and field. Hardwood forests, shrublands, grasslands, and deserts are composed chiefly of flowering plants. They show great variation in form, from simple stemless, free-floating duckweed *(Lemna)* through a series of herbaceous types to shrubs and trees such as oaks and beeches *(Fagus)*.

THE ORIGIN OF ANGIOSPERMS

How did the flowering plants come into being? This is a subject of much current research. The best evidence has come from recent studies of fossil pollen. The earliest angiosperm pollen grains have been found near the west coast of Africa, in rocks that are about 130 million years old. At that time the present African coast was in the heart of a supercontinent called Gondwanaland, which included Africa, South America, part of North America, Antarctica, and Australia (Fig. 20.1). By 120 million years ago the angiosperms were spreading rapidly, and they have dominated the world vegetation for the past 100 million years.

The angiosperms probably arose from gymnosperms, and they were probably successful because they had new vegetative and reproductive features that promote survival under conditions of drought and cold (the world climate turned cooler and dryer around the time of the great angiosperm radiation).

The chief vegetative advance found in angiosperms is an improved vascular system (Table 20.1). In the phloem, gymnosperms have distinct sieve cells, whereas angiosperms have sieve tubes made of many united sieve tube members. Compared to sieve cells, sieve tubes have a larger diameter and larger sieve pores. These features probably reduce the energy cost of food transport.

A still greater advance is found in the xylem. A few gymnosperms have xylem vessels, but in the great majority, tracheids serve as both conducting and stiffening elements. By contrast, all angiosperms have a well-developed division of labor in the xylem, with thick-walled stiffening fibers and large, relatively thin-walled vessels as conducting units. Both fibers and vessels arise from cells that would have become tracheids in an ancestral plant. The division of labor results in stronger wood and more efficient water transport.

The most important advance in angiosperms is the **flower,** which has many features that improve the efficiency of reproduction. Before considering the virtues of the flower, we will review its place in the life cycle of the flowering plant (see Chapters 12 and 13 for a more complete account).

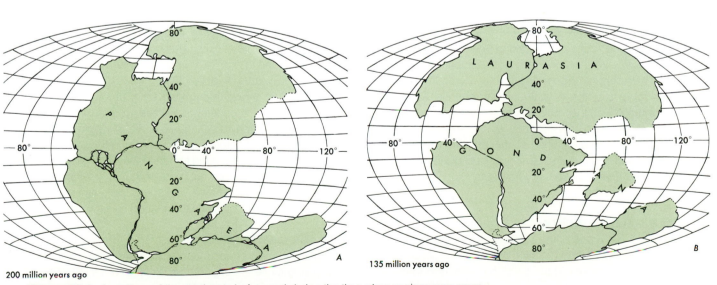

200 million years ago

135 million years ago

Figure 20.1. Locations of the continents before and during the time when angiosperms arose.

Table 20.1. Comparison of the Conifers and Flowering Plants

Generation	Gymnosperms	Angiosperms
	Roots, stems, and leaves present	Roots, stems, and leaves present
	Trees and shrubs, no herbs	Trees, shrubs, and herbs
	Stem: one core of xylem surrounded by phloem, vessels in one small division only	Stem: one core of xylem surrounded by phloem, or several vascular strands with phloem exterior
	Sieve cells without companion cells	Companion cells and sieve-tube members
	Pith present in stems, not in roots	Pith in stems, not in roots, of dicots; in some monocot roots
Sporophyte	Cambium develops in all species	Cambium in perennial forms; absent generally in annuals
	Staminate and ovulate cones, generally with staminate and ovulate scales	Flowers, stamens, and carpels
	Heterospory	Heterospory
	Microspores in microsporangia	Microspores in anther sacs (microsporangia)
	Megaspores in nucellus	Megaspores in nucellus
	Ovules present, exposed on scales	Ovules present, enclosed within carpels
	Embryo develops within a seed	Embryo develops within a seed
Fertilization	Double fertilization absent; no endosperm	Double fertilization, resulting in zygote and primary endosperm cell
	Pollen tube: male gametophyte	Pollen tube: male gametophyte
	Female gametophyte within ovule	Female gametophyte (embryo sac) within ovule
	Sperm cells formed in pollen tube	Sperm cells formed in pollen tube
Gametophyte	Motile sperm in several primitive genera, but free water not needed for fertilization	
	Nonmotile sperm in all other forms	Nonmotile sperm
	A greatly reduced archegonium completely embedded in the female gametophyte	Archegonium only suggested by synergid cells of the embryo sac
	Embryo within female gametophyte and seed coat	Embryo within seed coats, remains of nucellus, and endosperm

REVIEW OF THE ANGIOSPERM LIFE CYCLE

The flower is a shoot bearing floral leaves (Fig. 20.2). In a complete flower, the floral leaves are sepals, petals, stamens, and carpels. Some flowers have only the essential reproductive structures—stamens and carpels. Other flowers are unisexual: they have either stamens or carpels.

All flowering plants produce seeds, except some that human beings have modified so that they are now seedless (e.g., seedless grapes and bananas). In angiosperms, seeds are borne within a closed structure, the ovary, which eventually becomes the fruit.

MALE GAMETOPHYTE

The anther is the part of the stamen responsible for production of pollen. Microsporocytes form within the pollen sac or chamber (Fig. 20.3A). They divide by meiosis to form haploid (1n) microspores, which in turn divide mitotically to form a generative nucleus and a tube nucleus (Fig. 20.3B). A heavy, sculptured wall forms about the microspore, and a mature pollen grain results. The pollen is then shed and conveyed in one fashion or another (Chapter 12) to a stigma, where it germinates. A pollen tube (Fig. 20.3C) penetrates the stigma and grows through the style, down to the ovule within the ovary. The generative nucleus usually divides within the pollen tube into two sperm nuclei. Each sperm nucleus, with its associated cytoplasm, is a sperm cell.

Free water is not needed for fertilization in angiosperms; motile, ciliated sperm are not produced by any members of this division. The pollen tube with sperm cells and the tube nucleus, if present, constitute the mature male gametophyte.

FEMALE GAMETOPHYTE

An enlarged cell, the megasporocyte, within nucellar tissue of a young ovule undergoes meiosis and forms four megaspores arranged in a row. While this process is occurring,

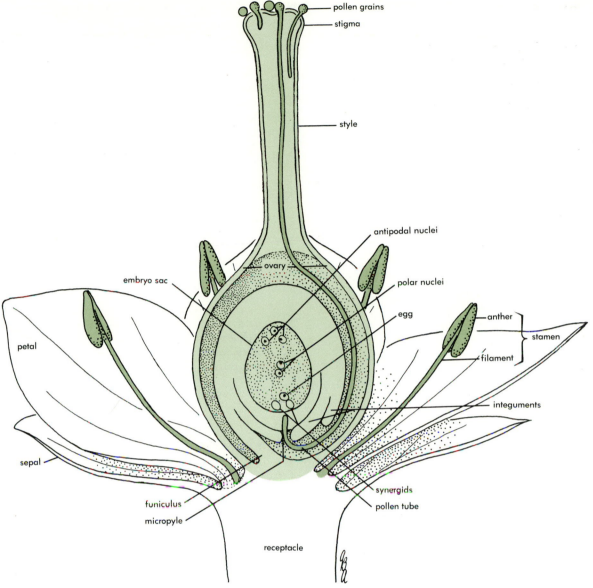

Figure 20.2. Hypothetical flower and its parts.

integuments form about the nucellar tissue, resulting in the formation of an ovule. The three megaspores closest to the micropyle generally disintegrate. The remaining megaspore undergoes three successive mitotic divisions, resulting in seven cells of the embryo sac, or female gametophyte: one egg cell, two synergid cells, one central cell with two polar nuclei, and three antipodal cells (Fig. 20.2).

FERTILIZATION AND SEED DEVELOPMENT

With penetration of the pollen tube into the mature embryo sac, the stage is set for fertilization. In angiosperms, fertilization involves not only the union of the egg cell with the sperm cell but, in addition, the union of a second sperm cell with the central cell to form the primary endosperm cell.

Since there are two nuclei in the central cell, the nucleus of the primary endosperm cell will have three sets of chromosomes. The fusions of two sperm cells, one with the egg, the second with the central cell, is **double fertilization.**

The resulting zygote generally develops directly into a small proembryo, from one end of which a typical embryo develops. The fate of the primary endosperm cell depends on the species. This cell gives rise to an endosperm that plays some role in nourishing the developing embryo. In a few genera, notably the grasses, the endosperm enlarges and persists in the seed to form the source of nourishment for the young seedling. A seed always includes an embryo and a food supply to enable the young seedling to establish itself, except in rare cases like orchids where the embryo

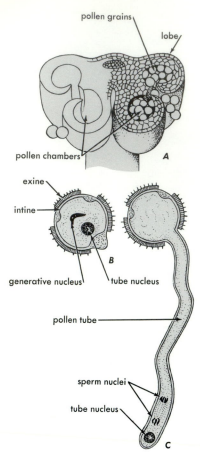

Figure 20.3. Development of the pollen tube. *A,* cross section of anther at the time of pollen release; *B,* mature pollen grain showing emergence of the pollen tube; *C,* pollen tube containing two sperm.

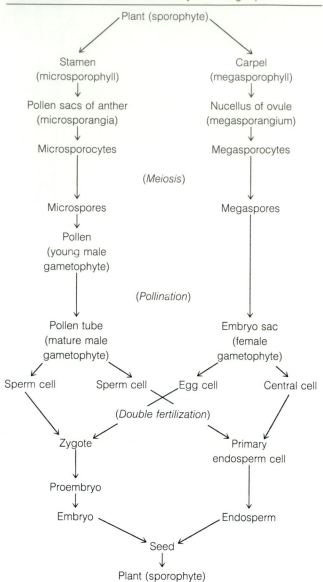

Table 20.2. Generalized Life History of an Angiosperm

Plant (sporophyte)

Stamen (microsporophyll)			Carpel (megasporophyll)
Pollen sacs of anther (microsporangia)			Nucellus of ovule (megasporangium)
Microsporocytes			Megasporocytes

(Meiosis)

Microspores			Megaspores
Pollen (young male gametophyte)			

(Pollination)

Pollen tube (mature male gametophyte)			Embryo sac (female gametophyte)
Sperm cell	Sperm cell	Egg cell	Central cell

(Double fertilization)

Zygote		Primary endosperm cell
Proembryo		
Embryo		Endosperm

Seed

Plant (sporophyte)

remains quite rudimentary. Food is generally stored either in cotyledons of the embryo or in endosperm.

With the germination of a seed and the development of the seedling into a flowering plant, the life cycle of an angiosperm is completed. The essential steps are shown in Table 20.2.

ADVANCES IN REPRODUCTION AND THE LIFE CYCLE

In the evolution of terrestrial plants, the life cycle has changed in many ways that promote survival in dry-land habitats. The flowering plants show several advances over their immediate ancestors, the gymnosperms (Table 20.1). The chief advances in the life cycle and reproduction are described below.

REDUCTION OF THE GAMETOPHYTE PHASE

In moving from the most primitive vascular plants to the angiosperms, there is a progressive reduction in the size and importance of the gametophyte, and a matching in-

crease in the sporophyte phase of the life cycle. The gametophyte of the fern is a free-living plant. In gymnosperms, the gametophytes have been reduced to microscopic forms that depend on food derived from the parental sporophyte. Nevertheless, in gymnosperms the female gametophyte still has many cells. Angiosperms carry the reduction one step further; the female gametophyte is reduced to a few (usually seven) cells. This reduction probably leads to more efficient reproduction, in that less energy is lost when an egg is not fertilized (a gametophyte is of little use unless it leads to a new sporophyte).

ENCLOSURE OF THE OVULES

In gymnosperms, the ovules are borne on the exposed surface of fertile leaves. In angiosperms, the fertile leaves are folded and joined at the margins to form carpels, with the ovules sealed inside. This change has important benefits: (a) it shelters the ovules against drying and attack by predators or parasites; (b) it allows the tissue of the ovule to control the process of fertilization, since pollen tubes must pass through the stigma and style to reach the ovules; and (c) the wall of the ovule matures into a structure (the fruit wall) that aids in the dispersal of seeds and that often helps to control seed germination so that seedlings emerge when conditions are favorable for survival.

DOUBLE FERTILIZATION AND EMBRYO NUTRITION

Double fertilization is a unique trait of flowering plants. One sperm nucleus joins with the egg to form a zygote; another joins with the polar nuclei to initiate the endosperm. This arrangement probably helps to make good use of energy reserves, for the ovule does not accumulate food until fertilization has occurred. By contrast, in gymnosperms the female gametophyte matures and stores much food before fertilization. In this case an unfertilized ovule represents a sizable waste of food energy.

ANIMAL POLLINATION

A chief benefit of the flower is that it attracts animals, which are usually more efficient carriers of pollen than the wind. (Gymnosperms are wind-pollinated.) The sterile parts of the flower (bracts, petals, sepals, nectaries, and scent glands) are especially important in attracting pollinators. Close symbiotic associations occur between many plants and their pollinators, and much of the diversity in both insects and flowering plants is related to pollination. The rapid spread of flowering plants 120 million years ago was probably due in part to pollen transfer by insects.

PHYLOGENY AND CLASSIFICATION OF FLOWERING PLANTS

The flowering plants are classified as the division **Anthophyta,** with the two subdivisions **Monocotyledonae** and **Dicotyledonae.** Many systems have been proposed for dividing the members into classes and orders. A system that has found much favor with American botanists was developed early in the twentieth century by Charles E. Bessey. His approach was phylogenetic and was based on an effort to distinguish between primitive and advanced characters (Fig. 20.4). While Bessey's classification is much debated, it provides a reasonable introduction to the orders and families of flowering plants.

THE BESSEYAN SYSTEM

Bessey regarded the order Ranales as being the most primitive. In this order, the family Magnoliaceae is especially primitive, while the Ranunculaceae are somewhat more advanced. The Christmas rose (*Helleborus*) and buttercups (*Ranunculus*) are representatives of these families.

If we consider the arrangement of flower parts in magnolias and buttercups to be most primitive, then it is possible to compare them to other families to determine their apparent relationships. The principal tendencies in the evolution of the flower, according to the Besseyan system, are as follows:

1. From an elongated to a shortened floral axis
2. From a spiral to a whorled condition of floral parts
3. From numerous and separate stamens and carpels to few and connate stamens and carpels
4. From numerous and separate sepals and petals to few and connate sepals and petals
5. From complete and perfect flowers to incomplete and imperfect flowers
6. From regular to irregular flowers
7. From hypogyny to epigyny

According to the Besseyan view, there were at least three main lines of advance from the primitive Ranalian type of flower. These lines culminated in mints (Lamiaceae), asters (Asteraceae), and orchids (Orchidaceae).

SELECTED FAMILIES OF ANGIOSPERMS

We will examine selected families of angiosperms with two objectives in mind: (1) to learn the characteristics of some important angiosperm families, and (2) to discover how these families fit into the Besseyan system of angiosperm classification. In our discussion of these matters so far, floral characteristics have been emphasized. The emphasis is placed on floral parts because their form and structure are little influenced by the external environment. They also are structurally more constant from generation to generation than are vegetative parts. Because of this constancy in structure, floral parts have a much greater value in both plant identification and in tracing evolutionary relationships than do vegetative characters. However, the latter cannot be neglected.

Since the magnolia and buttercups are thought to be primitive forms, our discussion will begin with these families and will then proceed to follow through a line of ascent ending in the mint family. We will consider the line culminating in the Asteraceae (Compositae)[1] and will complete our discussion with the line of monocotyledonous families. It should be further noted that lines of development are branched and treelike, not straight. The arrangement of the families discussed here as they occur in the Besseyan system is shown in Fig. 20.4.

[1]According to rules of plant nomenclature, family names should end in -ceae. Traditionally, the names of certain families such as Cruciferae and the Compositae have not ended this way. In this chapter the currently more acceptable synonym names ending in -ceae will be used.

MAGNOLIACEAE TO LAMIACEAE

In this line of ascent (Fig. 20.4), all flowers in all families are hypogynous. Syncarpy (the fusion of carpels) occurred very early, as did a reduction in number of floral parts. Other types of connation and the change to irregular flowers appeared somewhat later but still early in the line of ascent. Based on their floral characteristics, primarily, it is possible to show an evolutionary connection between the following families: Magnoliaceae, Ranunculaceae, Papaveraceae, Malvaceae, Brassicaceae, Solanaceae, Salicaceae, and Lamiaceae (Fig. 20.4).

The Malvaceae is a clearly marked family with relatively primitive characters as shown by its regular, perfect, and hypogynous flowers. There are numerous stamens with connate anthers, and the five or more carpels are connate. In Solanaceae, the flowers, while perfect and regular, show connation of sepals, petals, and carpels, with stamens adnate to the corolla tube. The Salicaceae is a family in which reduction in both number and size of floral parts has resulted in its advanced condition. The flowers are imperfect and grouped in catkins, and both calyx and corolla are lacking. The pistillate flowers are hypogynous and, in staminate flowers, the number of stamens has been reduced to one or two. The Lamiaceae, or mints, represent the most advanced family in this line of ascent. The flowers are hypogynous, with an irregular symmetry; there is a reduction in number of parts to four or two; and sepals, petals, and carpels are connate. The stamens are not only adnate to the corolla tube but are also highly specialized for insect pollination. We will consider some of the general characteristics of these families.

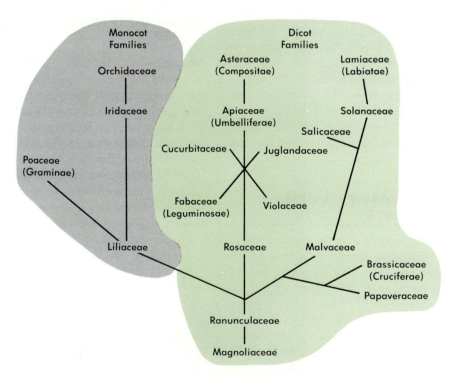

Figure 20.4. The evolutionary tendencies of the Angiospermae according to the Besseyan system. (Courtesy of H. E. McMinn.)

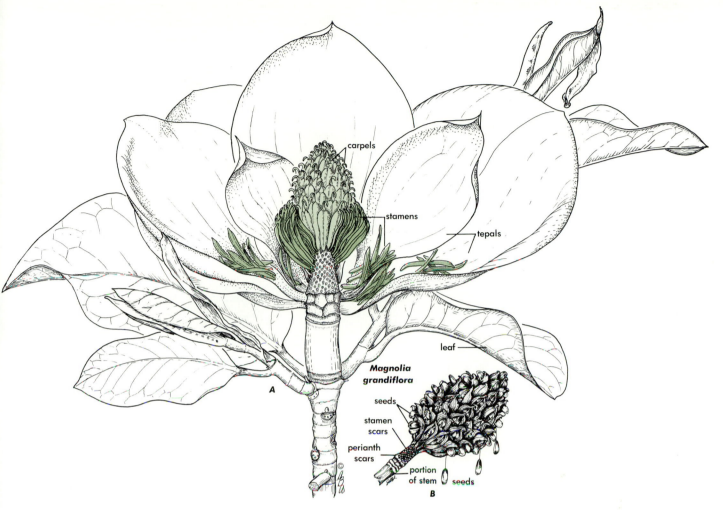

Figure 20.5. *Magnolia grandiflora* (magnolia) flower (*A*) and fruit (*B*) with hanging seeds.

Magnoliaceae

This group and the Ranunculaceae share the distinction of being very ancient. The tulip tree, *Liriodendron,* as shown by fossil remains, once had a very wide distribution. Most species of the Magnoliaceae are trees or fairly large shrubs. Magnolias themselves are magnificent trees; *Magnolia* *grandiflora* grows to 30 m or more with stiff, large, evergreen leaves and large white blossoms that are 18 to 20 cm across. The magnolias are largely warm-climate species, but tulip trees can tolerate northern winters (Fig. 20.5). There are 12 genera and 230 species[2] in the family.

[2]The estimates of genus-species numbers were taken from J. C. Willis, *A Dictionary of the Flowering Plants and Ferns*, 8th ed., revised by H. K. Airy Shaw (Cambridge: Cambridge University Press, 1973). Note that species numbers from other authors may differ depending on the system used to calculate their estimates.

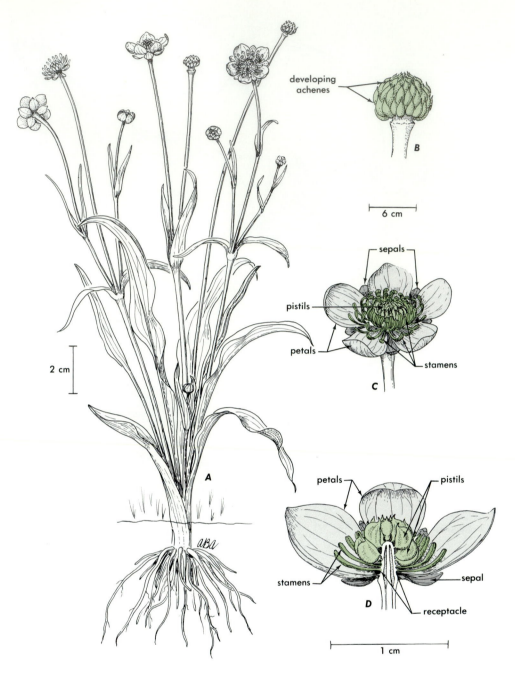

Figure 20.6. *Ranunculus alismaefolius* (water plantain). *A*, entire plant; *B*, fruit head; *C*, flower; *D*, longitudinal section of flower.

Ranunculaceae

Most of the plants in this family are herbs, but there are a few small shrubs and woody climbers. Leaves are frequently compound, a feature that has led to the common name of crowfoot family. There are about 50 genera and approximately 1900 species growing largely in the north temperate and arctic regions. Many of our common garden flowers belong to this family, such as *Delphinium, Aquilegia, Paeonia, Ranunculus* (Fig. 20.6), *Anemone,* and *Clematis.* The marsh marigold *(Caltha)* and *Hepatica,* common spring flowers, are members of this family. One species, *Aconitum napellus,* yields the drug aconite, a cardiac and respiratory sedative; and some species of *Delphinium* are poisonous to livestock.

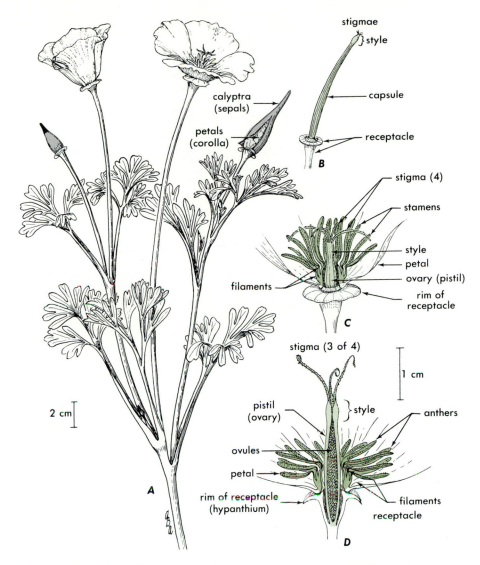

Figure 20.7. *Eschscholzia californica* (California poppy). *A*, plant habit; *B*, fruit; *C*, flower with exposed pistil; *D*, section of flower showing internal view of ovary.

Papaveraceae

This is a small family of only 250 species, but it is of great economic importance because it contains the source of opium. The opium poppy *(Papaver somniferum)* is native to the Middle East, but it is now widely cultivated in India, Southeast Asia, and the Mediterranean region. Incisions are made in the fruit of the opium poppy when it is still immature, and the milky latex (opium) is collected. It contains a mixture of many alkaloids, including morphine, codeine, and papaverine. Perhaps only 5% of the opium crop is sold legally. Most species of this family are annual or perennial herbs. The flowers are typically solitary, with 4 to 12 free, often crumpled petals, numerous stamens, and a single large pistil that lacks a style (Fig. 20.7). Many species have ornamental value.

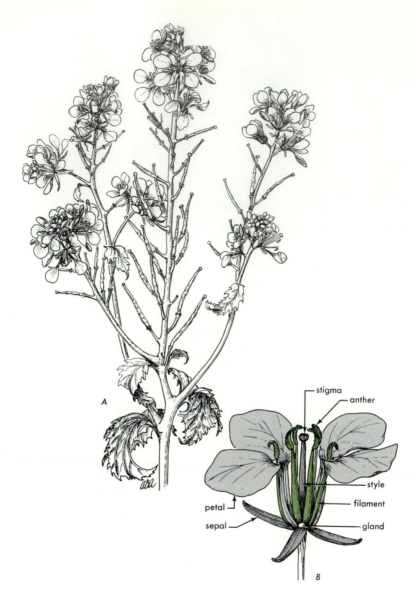

Figure 20.8. *Brassica kaber* (charlock) branch (*A*) and flower (*B*).

Brassicaceae (Cruciferae)

This is a large, distinct family of many cultivated forms, as well as some that are noxious weeds. Cabbage, cauliflower, broccoli, brussels sprouts, kohlrabi, and kale are all horticultural varieties of a single species, *Brassica oleracea*. Wild mustard, another *Brassica* species, is a sometimes noxious weed in grain fields, though it is sometimes used as cover crop in orchards (Fig. 20.8). Radishes, turnips, and stocks are other members of this family, as are many garden herbs. The family has a worldwide distribution in temperate and subarctic zones; all are herbs. There are about 375 genera and 3200 species.

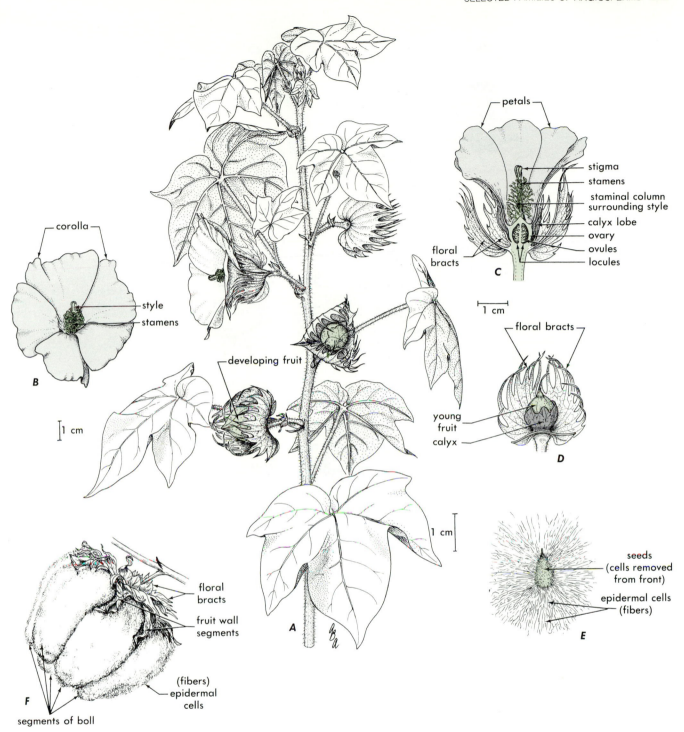

Figure 20.9. *Gossypium hirsutum* (cotton). *A*, plant habit; *B*, flower; *C*, section of flower; *D*, young fruit; *E*, single seed showing seed-coat fibers; *F*, cotton boll.

Malvaceae

This is a large family of 95 genera and 1000 species. Cotton, taken from seed coats of various members of the genus *Gossypium,* makes this family important from the viewpoint of politics, agriculture, and industry (Fig. 20.9). The cotton of commerce occurs as long hair or fuzz on seeds, which are borne in large capsules. Herbs, shrubs, and trees, and such ornamentals as *Hibiscus,* okra, hollyhocks, and numerous weeds are members of this family.

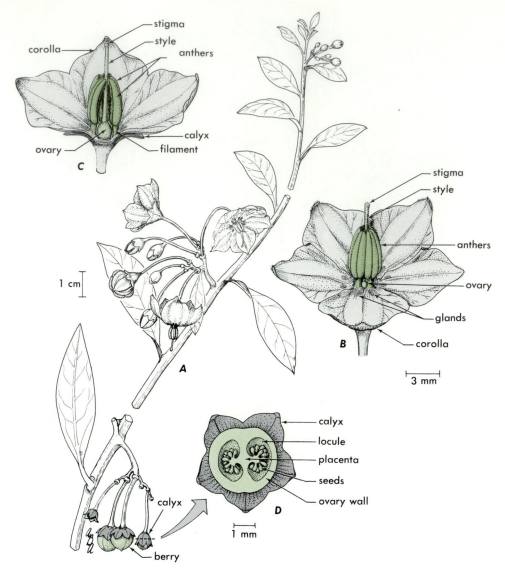

Figure 20.10. *Solanum parishii. A,* plant habit; *B,* flower; *C,* section of flower; *D,* cross section of ovary.

Solanaceae

This large family has many tropical forms, and it is also well represented in temperate regions. There are about 90 genera and over 2000 species, some 1700 of the latter being in a single genus, *Solanum* (Fig. 20.10). To this family belong tomatoes, potatoes, tobacco, eggplant, peppers, *Petunias,* and *Salpiglossis.* Although the family is of worldwide distribution, most of the cultivated forms were brought under domestication in the Western Hemisphere. There are many poisonous and drug plants in the family, such as belladonna and tobacco. Even such a common plant as the tomato was long supposed to be poisonous, as its scientific name *Lycopersicon,* which means "wolf peach," seems to indicate. There are many erect and climbing herbs in the family, as well as some shrubs and small trees.

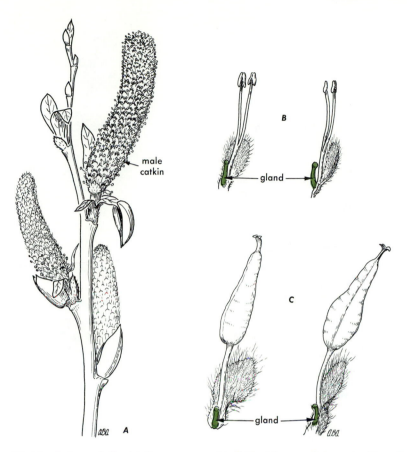

Figure 20.11. *Salix* sp. (willow) inflorescence and individual flowers. *A*, branch with male catkins; *B*, single male flowers. *C*, single female flowers.

Salicaceae

The willow family is comprised of three genera (*Salix*, willows; *Populus*, poplars; *Chosenia*, an asiatic shrub), with about 530 species. They are mostly trees, but some shrubby forms are known. They are very abundant in the Northern Hemisphere, mostly in temperate zones. Baskets are woven from branches of the basket willow, and paper pulp is obtained from trunks of one species. Willows are common along water courses (Fig. 20.11), frequently overhanging or even choking mountain streams, to the ill comfort of fishermen. Pussy willow is a familiar example of this family.

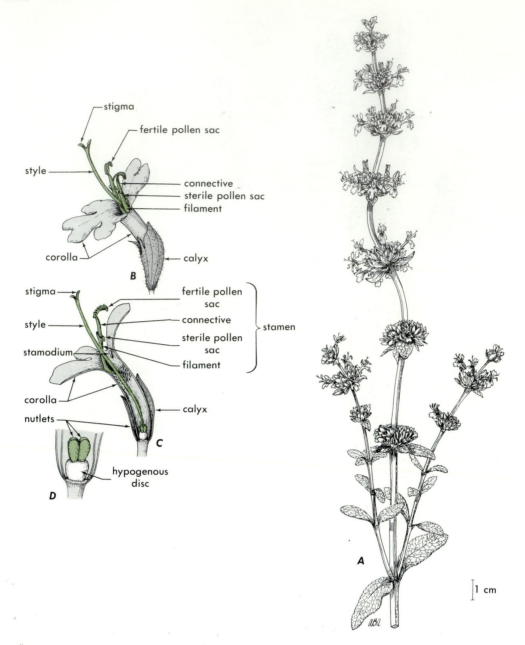

Figure 20.12. *Salvia mellifera* (sage). *A*, plant habit; *B*, flower; *C*, longitudinal section of flower; *D*, developing fruit.

Lamiaceae (Labiatae)

Mints represent the supposed highest advance of one of the three lines of angiosperm evolution. It is a large family of considerable economic importance, largely because of volatile oils produced by certain of its members. Peppermint, spearmint, thyme, sage (*Salvia;* Fig. 20.12), and lavender are examples. There are about 180 genera with 3500 species that are well distributed over the surface of the earth. There are herbs and shrubs in the family, generally with characteristic square stems and opposite leaves.

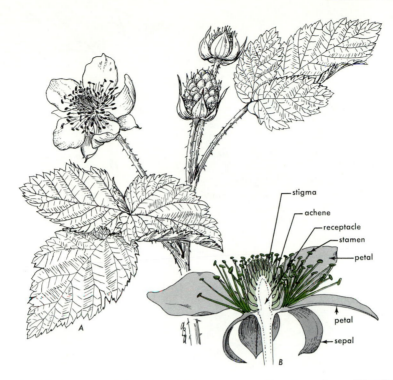

stigma
achene
receptacle
stamen
petal

petal

sepal

A

B

Figure 20.13. . *Rubus* sp. (boysenberry) flowering branch (*A*) with one flower (*B*) split open to show the fleshy receptacle.

MAGNOLIACEAE TO ASTERACEAE

The second line of ascent involves the Rosaceae, Fabaceae, Apiaceae, and Asteraceae, with an offshoot family, the Cucurbitaceae. This large group of families is characterized by an early change from hypogyny to epigyny. Following this there is a division into two sublines, one being marked by a lack of connation and a retention of regular flowers. In the other subline, connate and irregular flowers both occur in advanced forms.

The Rosaceae (rose family) is one of the more primitive families in this line of evolutionary development. There exists a considerable variety of flower structure within the family. The more primitive members are regular, perfect, with numerous parts, and with such a slight degree of perigyny as to be recognizable only with careful observation. This is exemplified by the boysenberry (Fig. 20.13). Changes within the family involve a reduction in the number of floral parts and a shift from slight perigyny to true perigyny and to epigyny. Connation also occurs. Members of the family generally have regular flowers.

Fabaceae (pea family) are considered to be a more advanced family than Rosaceae, even though their flowers are hypogynous. There is a reduction in the number of parts, the gynoecium having only one carpel; connation of stamens and of some petals occurs, and the corolla parts are irregular. Apiaceae represent a further advance over Rosaceae in that the flowers are epigynous and exhibit syncarpy (Fig. 20.13). This line of ascent finds its climax in the Asteraceae, whose flowers all possess advanced characters, such as reduction in number of parts, modification of parts (pappus), irregular flowers, and some imperfect flowers, connation in each whorl, and adnation.

The Cucurbitaceae culminate a branch line from the rose order. Flowers of Cucurbitaceae have advanced characters such as loss of floral parts, imperfect flowers, epigyny, and connation.

Rosaceae

Whereas grasses and legumes supply bread and vegetables of basic importance, the rose family supplies fruit for dessert and roses for decoration. There are more than 2000 species in this family and over 100 genera, not counting the almost numberless cultivated forms of roses, peaches, apples, cherries, boysenberries (Fig. 20.13), almonds, and so on. The family is of worldwide distribution and is somewhat heterogeneous. Many of the principles of angiosperm evolution can be demonstrated with its members. There are trees, shrubs, and herbs in the family. Leaves are usually alternate and bear stipules.

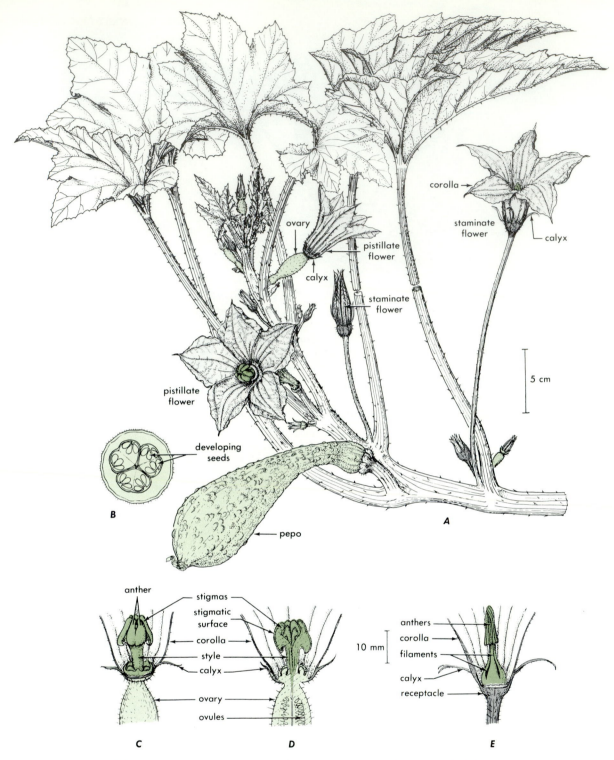

Figure 20.14. *Cucurbita pepo* (squash). *A*, branch with leaves, flowers and fruit; *B*, cross section of ovary; *C*, female flower; *D*, section of female flower; *E*, male flower.

Cucurbitaceae

The gourd or melon family is of worldwide distribution in warmer regions of the world, and various peoples have selected different forms for domestication (Fig. 20.14). It is also probably the only group in which fruits are highly prized for ornamental purposes and for use as containers of various types. Pumpkins and squashes are of Western Hemisphere origin. Cucumbers and melons have probably come from

Africa and central Asia. Other species appear to have been first cultivated in the tropics of Asia, Polynesia, and India. They frequently use tendrils to climb and are annual or perennial vines. Most are rapid-growing and frost-tender. There are about 110 genera with 640 species. Stems are usually soft and hairy or prickly. The generally simple leaves are large and sometimes deeply cut. Flowers are usually unisexual.

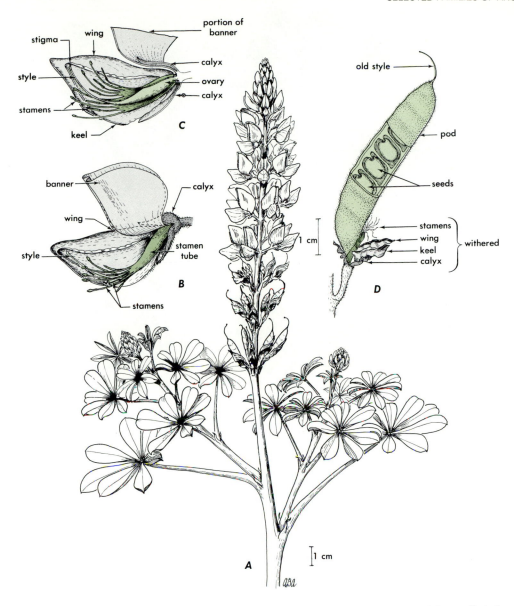

Figure 20.15. *Lupinus succulentus* (lupine). *A*, plant habit; *B*, flower; *C*, section of flower; *D*, fruit pod.

Fabaceae (Leguminosae)

The legume family has such distinctive characteristics that its members can frequently be recognized with only little experience (Fig. 20.15). It is one of the three largest families, with 600 genera and 12,000 species. All major growth forms are represented: herbs, both annuals and perennials, shrubs, vines, and trees. It is of worldwide distribution. While less heterogeneous than the rose family, considerable variation is found among its various members. It is a family of considerable importance in supplying food for humans and their animals. Many legumes are used for ornaments, from shade trees to cut flowers. Some lumber is obtained from the black locust, and some of the tropical species furnish wood for fine cabinet work. Association of the nitrogen-fixing bacteria with roots of legumes places this family in a unique position relative to maintenance of soil fertility. Peas, beans, peanuts, clovers, and lupines are common herbaceous legumes; wisteria is a vine; brooms and redbuds are shrubs or low trees; and locusts and acacias are trees. The *Mimosa* genus alone has some 450 species, including the sensitive plant of the florist shop; there are others of varying habit from tall trees to low herbs. Leaves of legumes are prevailingly pinnately compound and generally have stipules. Sometimes a leaflet is modified to a tendril.

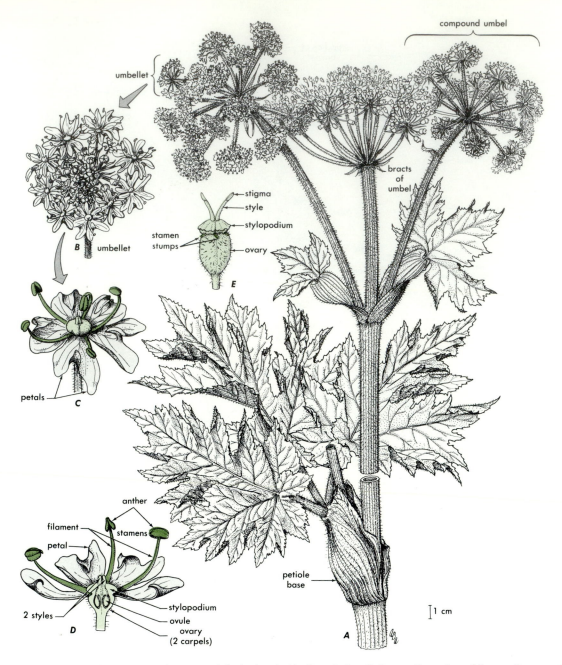

Figure 20.16. *Heracleum lanatum* (cow-parsnip). *A,* plant habit; *B,* umbellet; *C,* flower; *D,* section of flower; *E,* young fruit.

Apiaceae (Umbelliferae)

The parsley family forms a very distinctive group of plants, many of which can be easily recognized. There are over 200 genera and 3000 species growing in all regions of the world, though they are confined to mountains in the tropic zone. The name is derived from the typical umbel type of inflorescence (Fig. 20.16). The family contains many crop plants, as well as some that produce drugs and a number of kitchen herbs. Poison hemlock, a member of this family, is famous because of its use by the Greeks as a means of carrying out the death sentence; it was given as such to Socrates. The genus is widespread, constituting a hazard to livestock not only in pastures of Greece but also on open ranges of California. Carrots, celery, and parsnips, as well as parsley, are members of this family. They are mostly herbs, rarely small shrubs, with generally hollow stems. Leaves are alternate, mostly compound, and are frequently much dissected. The petioles expand at the base and may somewhat sheath the stem.

Flowers are small, being borne in umbels, which are in turn grouped in umbels, so a large inflorescence is formed. Frequently, umbels at all levels are subtended by bracts, forming a characteristic involucre. Flowers are regular, or the outer flowers are irregular, always epigynous and perfect, although they may not always be complete. There are five stamens, alternating with the same number of petals. The inferior gynoecium is composed of two carpels, each with but a single seed. Fruits are indehiscent, although carpels separate from each other when mature.

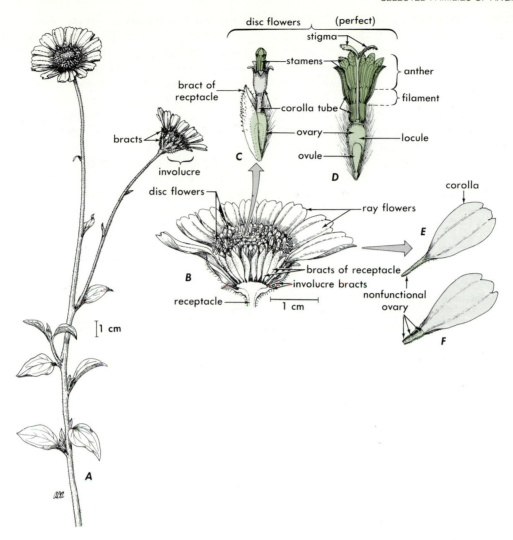

Figure 20.17. *Encelia californica* (tarweed). *A*, plant habit; *B*, section of inflorescence; *C* and *D*, perfect disk flowers; *E* and *F*, sterile ray flowers.

Asteraceae (Compositae)

The Asteraceae is the largest family of angiosperms (Fig. 20.17). Only a few of these plants are woody. In this family of herbaceous plants there are around 900 genera and about 13,000 known species. The family is not only of worldwide distribution but in most places is relatively abundant. The family is not noted for its food plants: endive *(Cichorium),* artichokes *(Cynara),* chicory, lettuce, and sunflower form most of its contribution in this respect. Nor is it famous as a producer of drugs or other commercial products. One species, safflower *(Carthamus),* is now being grown for oil, and members of the genera *Taraxacum* and *Parthenium* are being considered as possible sources of latex for rubber. Seed catalogs and florist shops will show members of this family in their full grandeur: *Dahlia, Chrysanthemum, Aster, Zinnia, Tagetes, Gaillardia, Ageratum,* and many others. Dandelions color lawns, and various wild species are abundant in meadows and fields. There are 400 species of the genus *Artemisia* that cover arid portions of the world and supply, among other things, tarragon for fancy vinegar and nectar for desert honey. Leaves are of various shapes.

MONOCOTYLEDONOUS LINE FROM MAGNOLIACEAE TO ORCHIDACEAE

There are many obvious differences between the dicotyledonous plants such as the Magnolias and the monocotyledonous line of evolution. These have been discussed and are briefly reviewed in Table 20.3.

However, some primitive monocot plants, particularly certain water weeds such as arrowheads and water plantains, have much in common with buttercups and marsh marigolds. The families selected to represent the monocots are Liliaceae, Poaceae, Iridaceae, and Orchidaceae. Lilies differ from magnolias in all distinctive monocot characteristics. In addition, they show a reduction in the number of floral parts and a connation of carpels. Irises show a single advance over lilies in that they are epigynous. Many irises are regular, perfect, and have separate perianth parts; stamens show no connation; syncarpy occurs as it does in lilies. Orchids show evolutionary advance in that they have irregular flowers that are highly specialized for insect pollination. Stamens have been reduced in number to one or two.

The order to which grasses belong may have arisen from the order to which lilies belong. The primitive characters of grasses include hypogyny, regular flowers, and connation only of carpels. The loss of perianth parts and the reduction in number of stamens and carpels mark them as a more advanced family than the Liliaceae.

Table 20.3. Tabulation of Differences between Monocotyledonous and Dicotyledonous plants

Monocot Plants	Dicot Plants
1. One cotyledon or seed leaf	**1.** Two cotyledons or seed leaves
2. Generally marked parallel leaf venation	**2.** Generally marked netted venation of leaves
3. Flower parts typically in groups of three or multiples of three	**3.** Flower parts typically in groups of four or five
4. Vascular bundles of stems scattered throughout a cylindrical mass of ground tissue	**4.** Vascular bundles of stems usually arranged in the form of a cylinder
5. Vascular cambium lacking in most forms	**5.** Vascular cambium present in forms having secondary growth

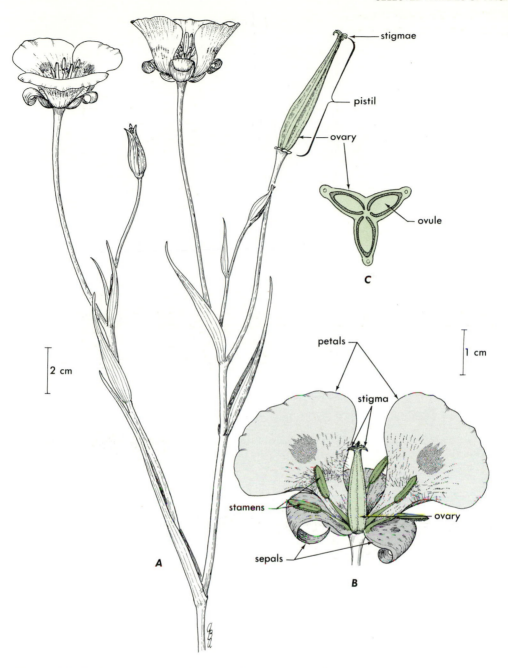

Figure 20.18. *Calochortus luteus* (Mariposa lily). *A,* flowering branch; *B,* flower with one stamen and petal removed; *C,* section through ovary.

Liliaceae

Unlike grasses, lilies are grown largely for ornamental purposes, although two genera, *Allium* (e.g., onions, garlic, and leeks) and *Asparagus,* are grown extensively for food. Some species yield drugs, and others poison cattle in pastures or on ranges of western states (Fig. 20.18). There are about 4100 species in some 280 genera. Most lilies grow from a bulb or a bulblike organ and flower in a single growing season, after which the shoot dies down. A few, such as Joshua trees, are woody perennials. Tulips, hyacinths, day lilies, and aloes are other examples of the lily genera.

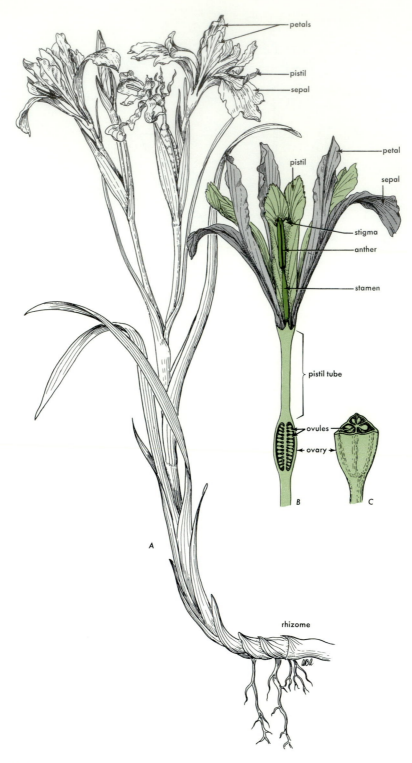

Figure 20.19. *Iris douglasiana* (iris). *A*, entire flowering plant; *B*, flower showing inside of inferior ovary; *C*, section through ovary.

Iridaceae

The iris family contains about 60 genera and 800 species. They are all herbaceous, largely perennial forms, usually with rhizomes, bulbs, or corms, as in lilies. Leaves and flowering stalks last for only one season. Some of the choicest florists' plants occur in this family—*Iris* (Fig. 20.19), *Gladiolus, Freesia, Crocus, Watsonia,* and so forth.

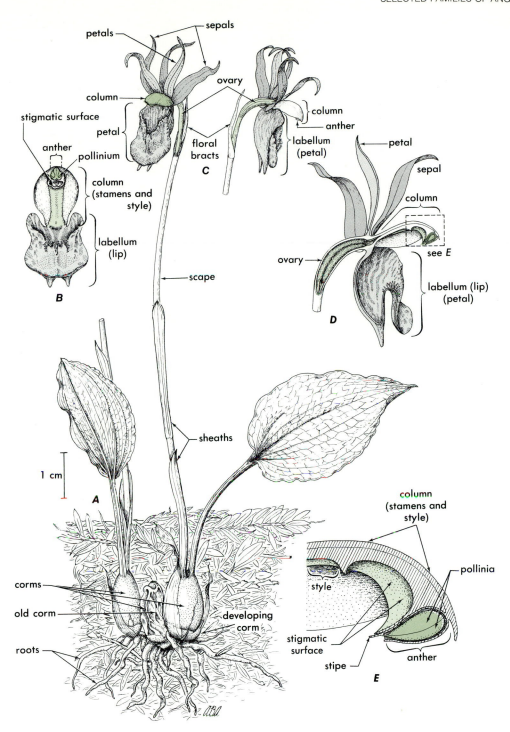

Figure 20.20. *Calypso bulbosa* (orchid). *A,* plant habit showing leaves, stem, corms, and flowers; *B,* face view of flower; *C,* side view of flower; *D,* section of flower; *E,* detail of column.

Orchidaceae

It is agreed by all that orchids represent the most advanced family in the Monocotyledoneae (Fig. 20.20). They are all herbaceous, occurring throughout the world largely in the tropics, but with a few genera extending into colder temperate regions. There are some 17,000 recognized species in several hundred genera, making this one of the three largest families of angiosperms. All forms are perennial, with tuberous, bulbous, or otherwise thickened roots. They may be erect, prostrate, or climbing. A few are saprophytic and lack chlorophyll; others are epiphytes, growing on trees without benefit of soil.

Poaceae (Gramineae)

Humans and other mammals have been associated with the grass family for far more years than those of recorded history. It is probably not an overstatement to say that without the grass family, human civilization, as we know it, could not have developed. Different grasses were domesticated by the three centers of civilization: wheat, rye, barley, and oats by peoples of the Mediterranean region and Southwest Asia; corn by natives of the Western Hemisphere; and rice and millet *(Eleusine)* by peoples in the Orient. Grasses are grown for human and animal consumption, for ornament, and for uses in arts and industry. There are over 620 genera and about 10,000 species. Most grasses are herbaceous, either annuals or perennials; a few bamboos are climbers and some are woody and grow up to 20 m tall. Grasses of one kind or another grow in all kinds of soil and situations. Grasses can be conveniently divided into six groups according to their use: (1) Bamboos, mostly evergreen, stout perennials, are used for construction in many parts of the Orient, and some have been introduced into warmer parts of the United States as ornamentals. (2) Cereals and some other annual grasses supply grain and forage for both human beings and animals. (3) Sugar-producing species, such as sugar cane, are strong, upright perennials growing only in the tropics. (4) Sod-forming grasses, perennials that cover many square miles of the earth's surface, are used in lawns and meadows and for forage. (5) Bunch grasses, common in semi-arid regions, are perennials that do not form rhizomes and do not produce a turf, as sod grasses do. (6) There is a large group of grasses grown for ornamental purposes, such as pampas grass. Flowers and vegetative characteristics of grasses have been described in some detail in Chapter 12.

SUMMARY OF ANGIOSPERM FAMILIES

1. Angiosperm families, according to the Besseyan system of classification, are all derived from the primitive order Ranales, to which buttercups and magnolias belong.

2. Important changes in floral evolution, according to the Besseyan system, are as follows: (a) from a spiral to a whorled arrangement of floral parts; (b) from many parts to few or even to a loss of parts; (c) from separate floral parts to connate parts; (d) from a regular flower to an irregular flower; (e) from hypogyny to epigyny.

3. According to the Besseyan system of classification there are three branched lines of ascent from the Ranales.

4. The first line of ascent following the Ranales includes: Malvaceae (mallow family), Papaveraceae (poppy family), Brassicaceae (mustard family), Solanaceae (potato family), Salicaceae (willow family), and Lamiaceae (mint family).

5. The second line comprises: Rosaceae (rose family), Fabaceae (pea family), Apiaceae (carrot family), and Asteraceae (sunflower family), with Cucurbitaceae (melon family) as an offshoot from Rosaceae.

6. The third line comprises the Monocotyledoneae in the following order: Liliaceae (lily family), Iridaceae (iris family), and Orchidaceae (orchid family). The Poaceae (grass family) is an offshoot from the Liliaceae.

APPENDIX
BASIC IDEAS OF CHEMISTRY

The living body is built of very small units of matter called **molecules.** There are thousands of kinds of molecules, which differ in their size, shape, behavior, and role in the body. The largest molecules carry hereditary information and are slender threads that may reach 1 mm in length. Molecules of table sugar (sucrose) are closer to the average size: about three million molecules of sucrose placed end to end would span the printed word *cube*. The living body builds nearly all of its own molecules by rearranging the parts of simpler molecules taken from the environment. The systems in the body that build new molecules are themselves collections of molecules.

WHAT ARE MOLECULES MADE OF?

A molecule is built of still smaller units of matter called **atoms.** An atom in turn contains three kinds of material particles: **electrons, protons,** and **neutrons.** At the center of the atom is a unit called the **atomic nucleus,** which is made of protons and neutrons. Electrons move separately around the nucleus. As will be explained below, the electrons can bind atoms together into molecules. However, the *nucleus*—more precisely, the *number of protons* in the nucleus—determines the chemical identity of the atom. Nuclei occur in nature with anywhere from one to 92 protons.

A molecule that contains nuclei of differing kinds is called a **chemical compound.** In contrast, a unit of matter is called an **element** if all of its nuclei contain the same number of protons. Several elemental substances were obtained and given common names before the discovery of atomic structure, such as silver, gold, iron, sulfur, and oxygen. The nuclei of an element may differ in the number of neutrons they contain, but this difference does not influence the chemical behavior of the nucleus. The neutrons seem to be important chiefly in holding the nucleus together as a unit. Nuclei with the same number of protons but differing numbers of neutrons are called **isotopes.** An isotope with a number of neutrons far from the average may be unstable and may spontaneously decompose, a process known as **radioactive decay.** Such a nucleus splits into parts, releasing much energy. Decay events are rare in the molecules of life and will not be considered further.

Some of the elements most prominent in life are listed below, along with the letter symbols that chemists use to indicate their presence in a molecule:

Name of Element	Symbol	Number of Protons
*Hydrogen	H	1
*Carbon	C	6
*Nitrogen	N	7
*Oxygen	O	8
Sodium	Na	11
Magnesium	Mg	12
*Phosphorus	P	15
Sulfur	S	16
Chlorine	Cl	17
Potassium	K	19
Calcium	Ca	20
Iron	Fe	26

The elements that are starred provide almost all the nuclei of biological molecules. These nuclei combine in many different combinations, along with electrons, to form thousands of compounds.

Chemists indicate the composition of a molecule by using the elemental symbols to show the kinds of nuclei present. Subscript numbers indicate how many nuclei there are of each kind. Thus, the symbol CH_4 represents a molecule that has four hydrogen atoms and one carbon atom (the subscript "1" is assumed if no other number is supplied).

WHAT HOLDS THE PARTS OF THE MOLECULE TOGETHER?

Protons, neutrons, and electrons are far smaller than the atom as a whole. Protons and neutrons are about equal in weight, and both are about 1840 times heavier than an electron. Thus almost all the weight or mass of an atom is in the nucleus. Nevertheless, the nucleus is extremely tiny. If an atom were magnified to the size of a house, the nucleus would be about as large as a pinhead. Since the electrons are also very small, this means that the atom (or the molecule) is mostly empty space.

The forces that hold the nuclei and electrons together in a molecule are electrical. Each electron carries a unit of negative electrical charge, and each proton has an equal-sized unit of positive electrical charge. Neutrons carry no charge. Two particles attract one another if they carry opposite charges and repel or push one another away if they carry charges of the same sign. These forces grow stronger as the particles approach closer. (But the attraction between an electron and a nucleus turns into a repulsion if they approach too closely. This keeps the electrons and nuclei from colliding.)

Because of these electrical forces, each electron and nucleus in the molecule experiences a combination of pulls and pushes from all the other particles at all times. The molecule holds together because the attraction between nuclei and electrons is enough to overcome the forces of repulsion.

A nucleus can hold in its vicinity about as many electrons as it has protons. Therefore, most molecules have an equal number of protons and electrons. Such a molecule is said to be **electrically neutral,** because any force that its electrons exert on a distant object is countered by an opposite force exerted by the protons.

Though most molecules are electrically neutral, there are many kinds of molecules that have more protons than electrons or vice-versa. These molecules are called **ions.** In symbolizing an ion, chemists indicate the number and sign of the excess charge with a superscript: for example, K^+ indicates the potassium ion, with one more proton than the number of electrons; SO_4^{-2} indicates the sulfate ion, with two more electrons than protons. Ions with a positive charge are **cations;** those with a net negative charge are **anions.** Ions that have the same charge repel one another at a distance; those with opposite signs tend to approach and stay close together. When two opposite ions are in contact, chemists often say they are joined by an **ionic bond.**

WHAT DETERMINES THE SHAPE OF A MOLECULE?

Nuclei are heavy and take up definite positions in the molecule. Their locations can be shown accurately in drawings or models. But the electrons continually move at high speed around and between the nuclei. Chemists describe this situation by saying that the nuclei are embedded in a cloud of electrons. Thus molecules do not have sharp boundaries. But two molecules repel one another if they get too close, because the two electron clouds repel one another. The size of a molecule is the distance at which this repulsion becomes great enough to prevent an oncoming molecule from moving any closer.

Though electrons do not settle in definite positions, they have regions in which they spend most of their time. The region in which a given electron can usually be found is called an **orbital.** The orbitals do not have sharp boundaries, but they can be shown in diagrams by means of shaded areas.

The nuclei act as centers about which the orbitals are

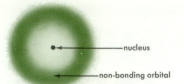

Figure A.1.

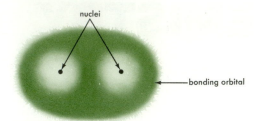

Figure A.2.

oriented. Some of the electrons spend all their time near a single nucleus (Fig. A.1). These electrons are said to be in a **nonbonding orbital.** Other electrons travel between two or more nuclei. These are **bonding electrons.** The region of space they use is called a **bonding orbital** (Fig. A.2). The two nuclei that share the bonding electrons are said to be joined by a **covalent bond.** The sharing of electrons results in a powerful force holding the two nuclei at a stable distance from one another.

Covalent bonds within a molecule are shown by **structural formulas,** in which the nuclei are represented by the elemental symbols and a covalent bonding orbital is shown by a line between the nuclei; for example, C—N. Electrons that are in nonbonding orbitals are usually not shown in the structural formula.

It is important to know how orbitals form around the nuclei of hydrogen (H), carbon (C), nitrogen (N), and oxygen (O), since these make up the bulk of the molecules of life. H is the simplest nucleus, consisting of just one proton. The charge on the proton is so weak that only one orbital forms around the H nucleus. C, N, and O nuclei have six, seven, and eight protons, respectively, and they exert a much stronger attraction for electrons. Each of these nuclei is the center for five orbitals. Since these three elements are almost alike with respect to the kinds of orbitals that form around them, they are considered as a group below.

One of the five orbitals is a spherical region that lies very close to the nucleus (Fig. A.3). The electrons in this orbital are nonbonding electrons and are usually omitted from diagrams of molecules. But in studying this nonbond-

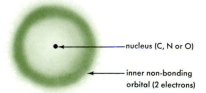

Figure A.3.

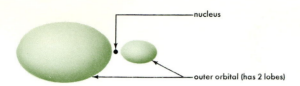

nucleus

outer orbital (has 2 lobes)

Figure A.4.

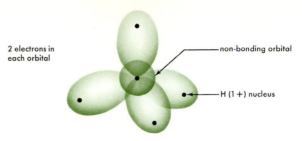

2 electrons in each orbital

non-bonding orbital

H (1 +) nucleus

Figure A.7.

ing orbital we encounter a general rule that applies to all orbitals: *each orbital contains exactly two electrons* when the nucleus is part of a stable molecule. An orbital cannot hold more than two electrons, and if for some reason an orbital loses one of its electrons, there is a powerful tendency for the molecule to rearrange itself or to react with other molecules so that each orbital again has two electrons.

Most commonly, the four remaining orbitals all have the same shape: each is approximately dumbbell-shaped, with the nucleus at the narrow point and with one lobe much larger than the other (Fig. A.4). The smaller lobe can be ignored. These four equal orbitals tend to be oriented in space so that their axes are as far from one another as possible (Fig. A.5). This happens because the electrons of each orbital push those of neighboring orbitals away. (But actually the outer orbitals overlap more than the diagrams indicate, so that electrons shield the nucleus from external contact equally in all directions.)

In the case of carbon, all four of the outer orbitals are bonding orbitals. This is seen most simply with CH_4, or methane, which is shown in Fig. A.6 with the structural formula and as it would appear in a ball-and-stick model. The C nucleus with its inner electrons forms the center of the molecule (Fig. A.7). Each of the outer, oblong orbitals has an H nucleus embedded in one end, with the C nucleus at the

other end. Therefore, C forms four covalent bonds. Every H nucleus shares two electrons. These shared electrons might be considered as half-time electrons from the point of view of the hydrogen nucleus. Together, they add up to placing one full unit of negative charge near the H nucleus, to balance the proton's unit of positive charge. The C nucleus has the two electrons of the inner orbital to itself and also shares four pairs of half-time electrons in the bonding orbitals. Therefore, six electrons are likely to be near the C nucleus at any moment, balancing the six positive charges of the nucleus. Overall, the molecule is neutral; it has 10 electrons and 10 protons.

In summary, methane illustrates two important rules of molecular structure: (1) *each nucleus tends to control enough full-time (unshared) and half-time (shared) electrons to equal its own charge;* and (2) *this is achieved in a way that places two electrons in each orbital.* The same rule applies to all other cases in which C occurs in biological molecules, and it applies equally to H, N, and O.

Notice also that the four bonds made by a C atom tend to be rigidly spaced; the angle between any two of the bonds is about 109°. This is due to orbital repulsion, and it does much to fix the internal geometry of the molecule. No drawing of a molecule on a flat page can adequately show these angles, so chemists frequently resort to physical models by using parts such as balls to represent nuclei and springs or sticks to show covalent bonds.

The structural rules seen above can explain why nitrogen (N) nuclei commonly form three covalent bonds, as in ammonia (NH_3) (Fig. A.8). The N nucleus has 4 electrons to itself: 2 in the inner nonbonding orbital and 2 in an outer orbital that are not shared with other nuclei (this is also a nonbonding orbital) (Fig. A.9). Also, the N nucleus shares 6 electrons in the three bonding orbitals. Counting these half-time electrons, there are a total of 7 negative charges near the N nucleus at any moment, matching the 7 protons of the nucleus. The N nucleus forms only three covalent bonds rather than four because the extra proton in the N nucleus

Figure A.5.

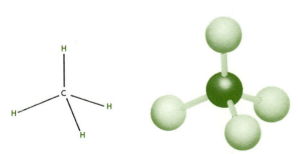

H

C

H

H

H

Figure A.6.

N

H

H

H

Figure A.8.

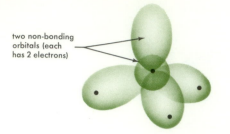

two non-bonding
orbitals (each
has 2 electrons)

Figure A.9.

Figure A.10.

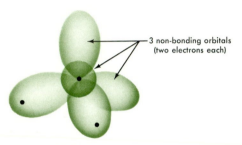

3 non-bonding orbitals
(two electrons each)

Figure A.11.

Figure A.13.

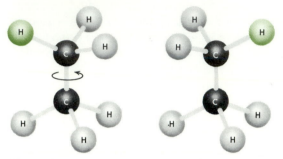

Figure A.14. The substance ethane C_2H_6, as it would appear in a ball-and-stick model. One of the H atoms has been shaded to show how rotation around the axis of the central (C-C) bond can affect the internal geometry of the molecule.

is sufficient to hold 2 electrons in one of the outer orbitals without sharing with another atom. As in methane, there are 10 protons and 10 electrons in the ammonia molecule. Notice that the usual repulsion between the orbitals occurs, so that the three bonds between N and H form a pyramidal arrangement.

Similar rules apply to the behavior of oxygen (O), as illustrated by water (H_2O) (Fig. A.10). Oxygen forms only two bonds rather than four because the O nucleus has enough protons to hold pairs of electrons in two of the outer orbitals without having to share with other nuclei (Fig. A.11). The 8 protons of the O nucleus are balanced by 6 full-time electrons in the three nonbonding orbitals, plus 4 half-time electrons that are shared in the bonds between O and H. Again, there are 10 protons and 10 electrons in the molecule.

A less common but still important orbital arrangement is seen in molecules like ethylene, C_2H_4 (Fig. A.12). Each C nucleus has the usual pair of inner, nonbonding electrons

and shares two pairs of electrons in bonds with H nuclei. But also, *four* electrons are shared between the two C nuclei. The sharing of four electrons is known as a **double bond.** Chemists represent a double bond by two lines between the nuclei. Double bonds such as C≡N, C≡C, and C≡O are common in the molecules of life.

Since an orbital can contain only two electrons, the double bond consists of two orbitals (Fig. A.13). One of the orbitals has three parts: an oval region between the two C nuclei plus a small lobe on either end. The other orbital consists of two sausage-shaped regions. The two electrons in this orbital spend part of their time in one sausage-shaped region and part of their time in the other. Remember that the diagram of an orbital represents only the places where the electrons spend *most* of their time; they can pass between two lobes, but they spend little time in the intervening space.

When two nuclei are joined by a double bond, all their other bonds lie in a single plane. This is a rigid arrangement that is maintained because the orbitals in the double bond resist being twisted. By contrast, a single bond forms an axis about which the joined parts of a molecule may freely rotate. Figure A.14 illustrates this arrangement with the substance ethane, C_2H_6. Rotation does not change the angle

Figure A.12.

between two neighboring bonds, but rotation lends a good deal of variability to the shape of a molecule if there are several nuclei joined into chains by single bonds.

HOW MOLECULES BEHAVE: POLARITY, HYDROGEN BONDS, AND SOLUBILITY

Molecules attract or repel one another because of their electrical charges. This is important in establishing how molecules will behave as a group.

Even neutral molecules may have local regions of positive and negative charge. This happens because electrons that are shared between two unlike nuclei usually spend more of their time close to one nucleus than the other. A measure called **electronegativity** expresses the tendency of a nucleus to dominate in sharing electrons: a higher value means a stronger tendency to hoard the electrons. Electronegativities for some common elements are as follows:

Element	Electronegativity
O	3.5
N	3.0
C	2.5
S	2.5
H	2.1
P	2.1

Because of these differences in electronegativity, a molecule that contains a mixture of nuclei tends to be negative near the N and O nuclei, and positive near the C, H, P, and S nuclei. A molecule that has such internal charge separations is said to be **polar.** There is a degree of polarity in every bond between two unlike nuclei, but between C and H the polarity is insignificant, whereas in bonds such as O—H, C—O, C—N, and N—H the polarity is high enough to affect the behavior of the molecule toward other molecules. Water is a strongly polar molecule. The local charge concentrations are less than the value of a whole electron and are shown in diagrams by the symbols $\delta(-)$ and $\delta(+)$:

$$\delta(-)$$
$$O$$
$$H \qquad H$$
$$\delta(+) \qquad \delta(+)$$

The H of one water molecule is attracted to the O of a neighboring water molecule because of their opposite charges (Fig. A.15). This attraction represents a weak bond called a **hydrogen bond.** It is only about 1/16 as strong as the covalent bond between the H and O within a water molecule. A hydrogen bond is indicated in diagrams by a dotted line.

Each of the two outer nonbonding orbitals in a water molecule can become involved in a hydrogen bond, and each H of the molecule can form another hydrogen bond. Therefore, each water molecule can, and does, form as many as four hydrogen bonds with neighboring water molecules or other polar molecules that may be present. The force exerted by these bonds allows water molecules to move about only by breaking and re-forming hydrogen bonds one at a time.

Other molecules besides water can form hydrogen bonds. This happens when a hydrogen nucleus that is sharing electrons with O or N comes close to another O or N, with the three nuclei approximately in a straight line. Since most biological molecules contain O—H or N—H, they can make hydrogen bonds with one another or with water.

A mass of pure substance consists of many molecules of the same kind packed regularly (a crystal) or irregularly (an amorphous solid or liquid). On contact with water, the molecules may leave such a mass to become surrounded by water molecules. The mass is said to **dissolve;** the molecules that become embedded in the water are called **solute** molecules; the water is said to act as a **solvent;** and the resulting mixture of molecules is a **solution.** The number of solute molecules in a given volume of solvent is the **solute concentration.** Concentrations are usually expressed in terms of **molarity.** For example, a solution may be 0.01 molar, abbreviated as 0.01 M. A 1-M solution contains, by definition, 6.023×10^{23} molecules of solute (one mole) in one liter of solution. A 0.01-M solution contains 1/100 as much solute as a 1-M solution, and so on.

For most solutes, there is a limit to the concentration of solute molecules that a given solvent will accept. Excess solute molecules spontaneously aggregate to form masses such as crystals. A solution with the limiting solute concentration is said to be **saturated.** The **solubility** of a substance is the concentration at the point of saturation. Solubility tends to be high when the solute molecules are strongly attracted to solvent molecules.

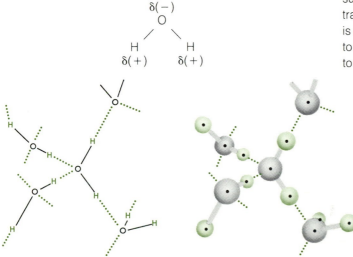

Figure A.15. Several water molecules joined by hydrogen bonds. (Dotted lines represent hydrogen bonds). On the left, structural formulas; on the right, the arrangement as it would look using ball-and-stick models.

A general rule governing solubility is that *water will accept polar but not nonpolar molecules.* Polarity is a factor because the water molecules form a network, joined by hydrogen bonds between molecules. A solute can move into this network if it can form hydrogen bonds with water; that is, if it is polar. If not, the network of water molecules resists being parted to make room for the solute molecules. Ions often are particularly soluble: solids composed of positive and negative ions, such as NaCl, often readily break up in contact with water because each ion acquires a blanket of oriented water molecules.

Nonpolar molecules dissolve freely in nonpolar solvents such as oil or gasoline. By contrast, polar substances attract one another and tend not to dissolve in nonpolar solvents. These facts can be summarized by the dictum "like dissolves like." But molecules differ widely in their degree of polarity, and correspondingly, they differ widely in the degree of solubility in various solvents.

HOW MOLECULES BEHAVE: CHEMICAL REACTIONS

A **chemical reaction** is an event in which new kinds of molecules are built by rearranging the parts of old molecules. Chemists symbolize a reaction by means of an arrow, with the initial molecules **(reactants)** at the tail and the final molecules **(products)** at the head.

Some common types of reactions are illustrated in Fig. A.16. The simplest are **rearrangements,** or **isomerizations,** in which a single molecule undergoes an internal reorganization. There are also **addition** reactions, in which two molecules join to form a larger molecule. The reverse is a **decomposition** reaction. The most common kind of reaction is the **transfer,** in which two molecules meet and exchange, or transfer, parts to form two new molecules.

What governs the occurrence of chemical reactions? Most biological molecules are stable and resist change. This

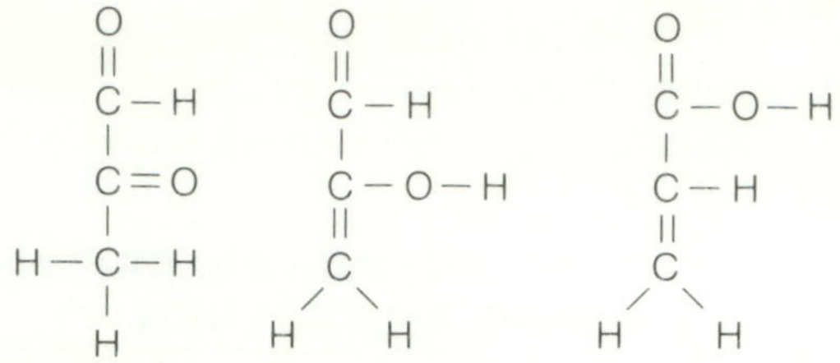

Figure A.17. Three possible structures that could occur with the formula $C_3H_4O_2$.

is because the nuclei and electrons fall into an arrangement that is the best compromise between the forces of attraction and repulsion. Any small change from this condition will create an imbalance of forces that tends to restore the initial condition. As illustrated in Fig. A.17, there may be several different stable arrangements that a given collection of nuclei and electrons fall into, given the chance. Some of these arrangements are more stable than others, but all have a degree of stability for the reasons given above. To go from one arrangement to another, the molecule has to pass through an intermediate arrangement that can be reached only by working against electrical forces. The intermediate condition is a highly unstable arrangement that spontaneously gives way to any other arrangement in which the forces are again balanced. A molecule that is in the process of changing is called an **activated intermediate.**

A collision between two molecules may set the stage for an exchange of parts, as illustrated in Fig. A.18. But the exchange first requires the two molecules to form an unstable **activated complex.**

Activated complexes form only when outside forces push the molecules out of their stable arrangements. The required force is most commonly provided by the force of collision between two molecules. Momentum carries the molecules together until their shapes become distorted, just as a basketball is flattened on one side at the moment of impact with

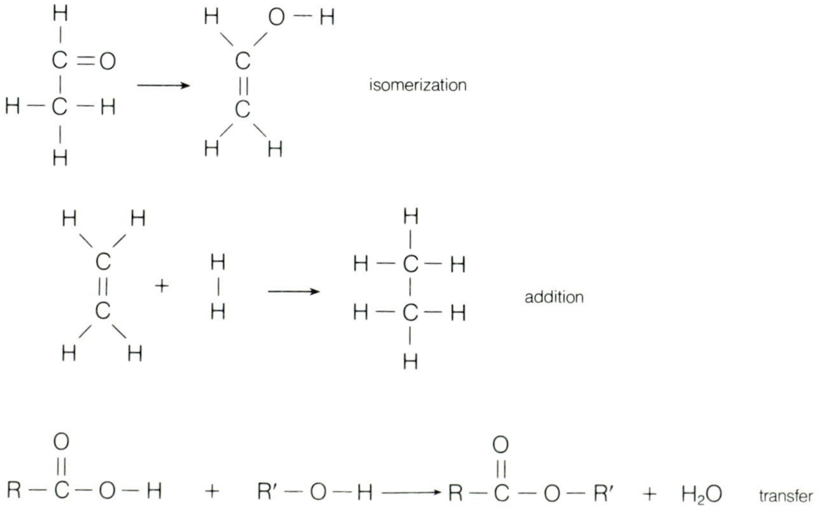

Figure A.16. Examples showing three general types of reactions.

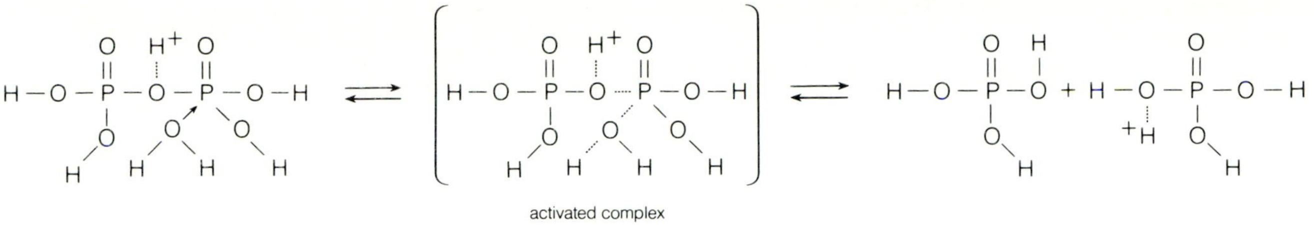

Figure A.18. The reaction between pyrophosphoric acid and water, illustrating an activated complex. The bonds shown with dotted lines in the activated complex are those that will be affected by the reaction. As written, the reaction is a hydrolysis, one of several kinds of transfer reactions.

the floor. Sufficient distortion brings the molecules into the activated state, enabling a chemical reaction to follow. A lesser degree of distortion allows the colliding molecules to rebound without reacting.

Chemists routinely use energy concepts to discuss chemical reactions. A typical graph of the energy relations in a reaction is shown in Fig. A.19. The molecules contain a certain amount of **stored,** or **potential, energy** in their stable condition. The activated complex contains more potential energy. This reflects the fact that force must be applied (work must be done) to bring the stable molecule into the activated condition. Moving molecules have **kinetic energy,** or energy of motion, in addition to their potential energy. A collision between two molecules converts this kinetic energy into potential energy: the colliding molecules are momentarily at rest in a distorted state. When the molecules rebound from the collision, some of the potential energy is converted back into kinetic energy. In this process the total amount of energy remains constant. Energy cannot be created or destroyed by the reactions in life, though the form of the energy can be changed.

The products of a reaction contain, between them, either more or less potential energy than the original reactants; the difference is made up by a compensating change in the amount of kinetic energy carried by the molecules. In Fig. A.19 the products have less potential energy and therefore more kinetic energy than the reactants. The kinetic energy of the molecules is known as **heat.** Thus heat is produced by the reaction in Fig. A.19.

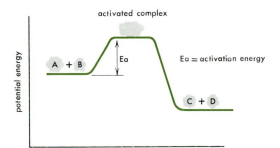

Figure A.19. Potential energy changes that occur during a reaction event in which molecules A and B collide, form an activated complex, and react to form C and D. Graph shows the potential energy of (A + B), the activated complex, and (C + D).

The difference in energy between the activated intermediate and the separate molecules is termed the **activation energy.** This is the amount of energy that must be supplied in the collision if there is to be a reaction event. The higher the hill, the less likely it is that the collision will produce a reaction.

Whether the collision causes a reaction will depend on the nature of the molecules, their orientation when they hit, and how hard they hit. Most collisions at the temperatures of life fail to bring about chemical reactions because of deficiencies in one of these factors.

In principle, all chemical reactions are **reversible.** Using Fig. A.19 as an example, if A and B can collide to form an activated complex, C and D can also collide and form the same kind of activated complex. The latter can break apart to form either A and B, or C and D. Reversibility is shown in reaction diagrams by putting two opposed arrows at each step.

Which of the two opposed reactions will run faster? The rate of a reaction depends on the frequency of collisions, hence on the concentrations of the reacting molecules. If we refer again to Fig. A.19, the reaction from left to right depletes the supply of A and B so that the reaction rate declines with time. The products C and D accumulate and collide with one another increasingly often, raising the rate of the back reaction. Thus the overall rate, the difference between forward and back reactions, depends on the prevailing balance of concentrations. **Equilibrium** occurs when the concentrations of all the molecules are just sufficient for the forward and back reactions to run at the same rate. At equilibrium the overall rate is zero; the opposed reactions cancel.

What determines the balance of concentrations at equilibrium? In most reactions (Fig. A.19) there is a difference in potential energy between the reactant and product molecules. This gives the back reaction a different activation energy requirement than the forward reaction: in the reaction of Fig. A.6, the back reaction has a higher energy hill to climb. A higher activation energy requirement means that fewer collisions will have enough energy to form the activated complex. Therefore more collisions must occur per unit time if the back reaction rate is to equal the forward reaction rate. The rates will only be equal when there are more of the low-energy molecules than the high-energy molecules. In Fig. A.19, C and D will be more abundant than A and B at equilibrium.

These ideas suggest that reactions should have the overall effect of replacing molecules that are rich in potential energy with molecules that are lower in potential energy. Nevertheless, living bodies do form many molecules that are higher in potential energy than the raw materials from which they are made. This is possible because the energy relations discussed above refer only to the *sum* of potential energy, taking both product molecules together. As long as this sum is greater for A and B than for C and D, the latter will tend to be the more abundant at equilibrium. If C is very low in potential energy, D may be richer in potential energy than A or B.

In biological systems, reactions rarely come to equilibrium because various materials are continually being withdrawn or supplied by other reactions. Also, back reactions are often insignificant because their activation energy requirements are too high.

SPECIAL REACTIONS: ACIDS AND BASES

Acidity results from a special type of transfer reaction involving water and a hydrogen nucleus (proton). When two water molecules are joined by a hydrogen bond, occasionally the hydrogen nucleus jumps from one molecule to the other:

$$H-O\overset{H}{\underset{H-O}{}} \quad \overset{H}{\longrightarrow} \quad H-O_{+}\overset{H}{\underset{H}{}} \quad + \quad \overset{H}{\underset{-O}{}}$$

hydronium   hydroxyl

The products are H_3O^+, known as **hydronium,** and OH^-, called the **hydroxyl** ion. A meeting between hydronium and a hydroxyl ion usually results in a return to the two water molecules: the ionization of water strongly favors the two un-ionized molecules.

The concentration of hydronium ions determines the acidity of the solution. More hydronium means a more acidic solution. Chemists express the hydronium concentration by a measure known as the **pH.** The scale runs from 0 to 14. A change of 1 unit on the pH scale is a 10-fold change in hydronium concentration. The lower the pH, the higher is the hydronium concentration. Thus, a solution with pH 2 has 10 times as much hydronium as a solution with pH 3. Pure water has a pH of 7. Most living materials have a pH value near 7. Stomach juice in human beings is very acidic, with a pH around 1. Even at this strongly acidic pH, though, there are many more neutral water molecules than hydronium ions in the solution: pH 1 represents about 1 hydronium for every 180 neutral water molecules.

An **acid** is a compound that *raises* the hydronium concentration in water. Any compound that can donate a proton

to a water molecule is an acid. If we use the symbol HA for a generalized acid, the ionization is:

$$HA + H_2O \longrightarrow H_3O^+ + A^-$$

Acids vary widely in strength. The very strong acid HCl (hydrochloric) completely breaks up into Cl^- and H_3O^+ when it contacts water. A weak acid remains mostly un-ionized, as HA. Water and acetic acid (which causes the sourness in vinegar) are both weak acids.

A **base** is a molecule that **reduces** the concentration of hydronium in water. For example, OH^- is a base. There are several ways in which a substance can decrease the hydronium concentration. Thus B^- can be an ion that reacts as:

$$B^- + H_3O^+ \longrightarrow BH + H_2O$$

The OH^- ion acts in this way. As another example, BOH may be a compound that ionizes in water to release OH^-. The OH^- then reacts as shown above.

Instead of talking about hydronium, chemists often abbreviate by limiting their attention to the H^+ ion itself, ignoring the part played by H_2O in carrying the proton (as H_3O^+). From this viewpoint, pH refers to the H^+ concentration. But free H^+ never occurs in the solution.

FUNCTIONAL GROUPS

The molecules of life are often large and complex. But they are easier to understand when we use the concept of **functional groups**. A functional group is a part of the molecule that is especially likely to participate in chemical reactions.

The functional groups **carboxyl**
$$(-\overset{\overset{\textstyle O}{\|}}{C}-O-H)$$

and **phosphoryl**
$$(-O-\overset{\overset{\textstyle O}{\|}}{\underset{\underset{\textstyle O-H}{|}}{P}}-O-H)$$

act as acids. They can donate H^+ to water:

$$R-\overset{\overset{\textstyle O}{\|}}{C}-O-H + H_2O \longrightarrow R-\overset{\overset{\textstyle O}{\|}}{C}-O^- + H_3O^+$$

$$R-O-\overset{\overset{\textstyle O}{\|}}{\underset{\underset{\textstyle O-H}{|}}{P}}-O-H + 2H_2O \longrightarrow R-O-\overset{\overset{\textstyle O}{\|}}{\underset{\underset{\textstyle O^-}{|}}{P}}-O^- + 2H_3O^+$$

The carbon of carboxyl and the phosphorus of phosphoryl tend to have a partial positive charge; the attached oxygens have partially drawn away the shared electrons, leaving the C and P nuclei relatively exposed to attack by other molecules. Therefore, these nuclei tend to be involved often in chemical reactions. By contrast, C nuclei that are only bonded to other C or H nuclei are less subject to chemical attack.

The **amino** group ($-N\overset{H}{\underset{H}{<}}$) acts as a base. The nonbonding orbital of the N atom can accept a proton from water or hydronium:

$$R-N\overset{H}{\underset{H}{<}} + H_3O^+ \longrightarrow R-\overset{H}{\underset{H}{\overset{|}{N}}}\overset{+}{-}H + H_2O$$

Hydroxyl ($-O-H$) and **carbonyl** ($>C=O$) are also functional groups. They do not appreciably form ions unless they are joined together as carboxyl, but the oxygen atoms in these groups pull electrons from adjoining C nuclei, creating polarity and exposing the C nuclei to attack. A common reaction is that of a carboxyl with a hydroxyl group, resulting in a product called an **ester** plus a water molecule (see Fig. A.16). Such reactions, which give off a water molecule, are called **condensations.** The opposite, cleavage of a molecule by adding water, is a **hydrolysis** reaction. An example of a hydrolysis is shown in Fig. A.18. Both condensations and hydrolysis reactions are transfer reactions.

OXIDATION AND REDUCTION

An **oxidation-reduction** reaction is one in which electrons are transferred from one group or molecule to another. The entity that loses the electrons is said to be **oxidized;** the one that gains the electrons is **reduced.** Often such a reaction also involves a transfer of atomic nuclei.

A simple example is the ionization of chlorophyll, an important event in photosynthesis. After absorbing a unit of light energy, the chlorophyll molecule passes an electron to a compound that acts as an electron acceptor:

chlorophyll + acceptor $\rightarrow$ chlorophyll$^+$ + acceptor$^-$

Here chlorophyll is oxidized and the acceptor is reduced.

Somewhat more complex is the formation of the compound NADPH from NADP$^+$, another event in photosynthesis:

$$2Z^- + NADP^+ + H_3O^+ \longrightarrow 2Z + NADPH + H_2O$$

In this reaction, which probably proceeds by several steps, NADP$^+$ receives two electrons, with Z as the electron donor. NADP$^+$ also gains a proton. The donor Z$^-$ is oxidized; NADP$^+$ becomes reduced.

On binding to oxygen, most atoms lose electrons—they become oxidized (hence the term, *oxidation*). This is because O attracts electrons very strongly. A molecule is said to be in a reduced state if it has C-H and C-C bonds. Sugar is a reduced compound:

$$H-\overset{\overset{\displaystyle H}{|}}{\underset{\underset{\displaystyle OH}{|}}{C}}-\overset{\overset{\displaystyle H}{|}}{\underset{\underset{\displaystyle OH}{|}}{C}}-\overset{\overset{\displaystyle H}{|}}{\underset{\underset{\displaystyle OH}{|}}{C}}-\overset{\overset{\displaystyle OH}{|}}{\underset{\underset{\displaystyle H}{|}}{C}}-\overset{\overset{\displaystyle H}{|}}{\underset{\underset{\displaystyle OH}{|}}{C}}-\overset{\overset{\displaystyle H}{\|}}{\underset{\underset{\displaystyle H}{}}{C}}{=}O$$

Reduced compounds spontaneously react with O_2 (with a suitable catalyst) to give products in which the C and H are more fully combined with O and therefore are more highly oxidized. This occurs because it transfers electrons from C and H to O. The reverse, the removal of O, is unfavorable unless energy is supplied by the environment. Therefore, in the presence of O_2, reduced compounds are rich in energy as compared to oxidized products. The stored energy is often called **redox energy** and can be defined as energy associated with an unfavorable distribution of electrons between the nuclei. This redox energy is released when a reaction allows the electrons to assume a more favorable arrangement.

GLOSSARY

Abbreviation	Meaning
A.S.	Anglo-Saxon
D.	Dutch
dim.	diminutive
F.	French
Gr.	Greek
It.	Italian
L.	Latin
Lapp.	Lappland
M.E.	Medieval English
M.L.	Medieval Latin
N.L.	New Latin
O.F.	Old French
O.E.	Old English
R.	Russian
Sp.	Spanish

A. Abbreviation for **angstrom;** a unit of length. There are 10,000,000,000 angstroms in a meter.

Abscisic acid (ABA). A type of plant hormone that affects stomata and influences many aspects of development.

Abscission (L. *abscissus,* to cut off). A process in which an organ is detached from a plant by means of physiological mechanisms.

Absorption (L. *ab,* away + *sorbere,* to suck in). The process of taking in material or energy.

Absorption spectrum. A graph relating the ability of a substance to absorb light of various wavelengths.

Accessory (M.L. *accessorius,* an additional appendage). Something aiding or contributing in a secondary way, such as buds in addition to the main axillary bud.

Achene. Simple, dry, one-seeded indehiscent fruit, with seed attached to ovary wall at one point only.

Acid. A substance that can donate a hydrogen ion.

Actinomorphic (Gr. *aktis,* ray + *morphe,* form). Having radial symmetry; consisting of equal parts arranged uniformly around a common center.

Action spectrum (F. *acte,* a thing done). A graph relating the degree of physiological response (e.g., phototropism, photosynthesis) caused by different wavelengths of light.

Active transport. The transport of solutes across a cellular membrane by mechanisms that expend metabolic energy; can lead to transport against an electrochemical potential gradient.

Adaptation (L. *ad,* to + *aptere,* to fit). Adjustment of an organism to the environment. Also, a feature of an organism that has arisen through natural selection in response to a particular environment and that promotes survival in that environment.

Adaptive radiation. Expansion of a taxonomic group, with the formation of many diversified species, in response to an ecological opportunity.

Adenine. A purine base present in nucleic acids and nucleotides.

Adenosine triphosphate (ATP). A high-energy organic phosphate of great importance in energy transfer in cellular reactions.

Adhesion (L. *adhaerere,* to stick to). A sticking together of unlike things or materials.

Adnation (L. *adnasci,* to grow to). In flowers, the growing together of two or more whorls to a greater or lesser extent; compare with **connation.**

ADP. Adenosine diphosphate.

Adsorption (L. *ad,* to + *sorbere,* to suck in). The concentration of molecules or ions of a substance at a surface or an interface (boundary) between two substances.

Advanced. Recent in origin.

Adventitious (L. *adventicius,* not properly belonging to). Referring to a structure arising from an unusual place: buds at other places than leaf axils, roots growing from stems or leaves.

Aeciospore (Gr. *aikia,* injury + spore). One of the dikaryotic asexual spores of rust fungi.

Aecium (plural, **aecia**) (Gr. *aikia,* injury). In rust, a sorus that produces aeciospores.

Aerate. To supply or impregnate with common air, such as by bubbling air through a culture solution.

Aerobic (Gr. *aer,* air + *bios,* life). Using or requiring molecular oxygen.

Aerobic respiration. Respiration involving molecular oxygen.

Agar (Malay *agaragar*). A gelatinous substance obtained mainly from certain species of red algae.

Aggregate fruit (L. *ad,* to + *gregare,* to collect; to bring together). A fruit developing from the several separate carpels of a single flower; e.g., a strawberry; compare with **multiple fruit.**

Alcohol. Any of a class of hydroxyl compounds, of which

ethanol (C_2H_5OH), or grain alcohol, is the most common example; names of these compounds end in -ol.

Aleurone (Gr. *aleuron*, flour). The outermost cell layers of the endosperm of wheat and other cereal grains.

Alfisol. Modified podsol soil, typical of the northern part of the deciduous forest.

Alga (plural, **algae**) (L. *alga*, seaweed). Photosynthetic, eucaryotic organism that lacks water-conducting tissue, does not form an embryo, and does not form reproductive cells inside a jacket of sterile cells.

Algin. A long-chain polymer of mannuronic acid found in the cell walls of the brown algae.

Alkali (Arabic *alqili*, the ashes of the plant saltwort). A substance with marked basic properties.

Alleles (Gr. *allos*, another). Variant forms of a gene.

Alpine (L. *Alpes*, the Alps Mountains). Referring to high regions of a mountain that are above the tree line.

Alternate. Referring to bud or leaf arrangement in which there is one bud or one leaf at a node.

Alternation of generations. The alternation of haploid (gametophytic) and diploid (sporophytic) phases in the life cycle of many organisms; the phases (generations) may be morphologically quite similar or very distinct, depending on the organism.

Amensalism (L. *a*, not + *mensa*, table). An ecological relationship in which an organism of one species inhibits organisms of another species by adding something to the environment.

Amino acid. One of the building blocks of a protein.

Amyloplast (L. *amylum*, starch + *plastos*, formed). Cytoplasmic organelle specialized to store starch. Abundant in roots in storage organs such as tubers.

Anabolism (Gr. *ana*, up + metabolism). The constructive phase of metabolism, in which more complex molecules are built from simpler substances.

Anaerobe (Gr. *a*, without + *aer*, air + *bios*, life). An organism able to live in the absence of free oxygen or in greatly reduced concentrations of free oxygen.

Anaphase (Gr. *ana*, up + *phais*, appearance). The stage in mitosis in which half chromosomes, or sister chromatids, move to opposite poles of the cell.

Androecium (Gr. *andros*, man + *oikos*, house). All the stamens in a flower.

Angiosperm (Gr. *angion*, a vessel + *sperma* from *speirein*, to sow). A plant that forms seeds inside an ovary; a flowering plant.

Anion. A negatively charged ion.

Anisogamy (Gr. *an*, prefix meaning not + *isos*, equal + *gamete,* spouse). The condition in which the gametes, though similar in appearance, are not identical. Compare with **heterogamy.**

Annual (L. *annualis*, within a year). A plant that completes its life cycle within one year and then dies.

Annual ring. Ring of xylem in wood, which indicates an annual increment of growth in trees growing in temperate regions.

Annular vessels (L. *annularis*, a ring). Vessels with rings of secondary wall material.

Annulus (L. *anulus* or *annulus*, a ring). In ferns, a row of specialized cells in a sporangium, of importance in opening of the sporangium; in mosses, thick-walled cells along the rim of the sporangium to which the peristome teeth are attached.

Anther (M.L. *anthera*, from the Gr. *anthos*, meaning flower). Pollen-bearing portion of stamen.

Antheridiophore (Gr. *anthos*, flower + *phoros*, to bear). In some liverworts, an elongate structure, bearing antheridia.

Antheridium (anther + *-idion*, a Gr. dim. ending, thus, a little anther). A type of male gametangium, consisting of a sterile jacket of cells enclosing one or more sperm-producing cells.

Anthocyanin (Gr. *anthos*, a flower + *kyanos*, dark blue). A blue, purple, or red vacuolar pigment.

Anticlinal cell division. Cell division in which the newly formed cell wall is perpendicular to the nearest surface of an organ.

Antipodal (Gr. *anti*, opposite + *pous*, foot). Referring to cells at the end of the embryo sac opposite that of the egg apparatus.

Apex (L. *apex*, a tip, point, or extremity). The tip, point, or angular summit of anything: the tip of a leaf; that portion of a root or shoot containing apical and primary meristems.

Apical dominance. The inhibition of lateral buds or meristems by the shoot apex.

Apical meristem. A mass of meristematic cells at the very tip of a shoot or root.

Aplanospore (Gr. *a*, not + *planetes*, wanderer + *sporos*, seed, spore). A nonmotile spore, one that is carried passively by wind, water, or other organisms.

Apomixis (Gr. *apo*, away from + *mixis*, a mingling). Asexual reproduction, the formation of offspring without a stage in which gametes fuse. Some authorities restrict the term to cases in which maternal cells substitute for the zygote in forming seed.

Apothecium (Gr. *apotheke*, a storehouse). A cup-shaped or saucer-shaped open ascocarp.

Arboretum (L. *arbor*, tree; also *arboretum*, a place grown with trees). A place, often outdoors, set aside for the display of living plants, including herbs and shrubs as well as trees; contrasts with an herbarium, which displays dead, preserved remains of plants.

Archegoniophore (Gr. *archegonos*, founder of a race + *phoros*, to bear). Elongate structure that bears archegonia; found on some liverworts.

Archegonium (Gr. *archegonos*, literally a little founder of a race). Female gametangium or egg-bearing organ, in which the egg is protected by a jacket of sterile cells.

Aril (M.L. *arillus*, a wrapper for a seed). An accessory seed covering formed by an outgrowth at the base of the ovule in *Taxus*.

Ascus (plural, **asci**) (Gr. *askos*, a bag). A cell in which nuclei fuse and go through meiosis, after which the protoplast divides and forms meiospores. The original cell wall remains as a sac enclosing the spores. The ascus is the defining characteristic of the ascomycete fungi.

-ase. A chemical suffix indicating an enzyme.

Asexual (Gr. *a*, without + *sexualis*, sexual). Any type of reproduction that does not involve the union of nuclei and meiosis.

Aspect (L. *aspectus*, appearance). The direction of slope of a surface, as in a hillside with a south-facing aspect.

Assimilation (L. *assimilare*, to make like). The transformation of food into protoplasm.

Atom (Gr. *atomos*, indivisible). A unit of matter consisting of an atomic nucleus and one or more electrons. The nucleus consists of one or more protons and from zero to many neutrons.

ATP. See **adenosine triphosphate.**

Auricles (L. *auricula*, dim. of *auris*, ear). In grasses, small projections that grow out from the opposite side of the leaf sheath at its upper end where it joins the blade.

Autotrophic (Gr. *auto*, self + *trephein*, to nourish with food). Pertaining to an organism that is able to manufacture its own food.

Auxin (Gr. *auxein*, to increase). A type of hormone that regulates many aspects of plant growth and development.

Axial system (L. *axis*, axle). System of cells in secondary tissues that are oriented parallel to the long axis of the stem; forms from fusiform initials of the vascular cambium.

Axil (L. *axilla*, armpit). The upper angle between a petiole of a leaf and the stem from which it grows.

Axile placentation. Condition in which ovules arise on the axis of an ovary with several locules.

Axillary bud. A bud formed in the axil of a leaf.

Banner (M.L. *bandum*, a standard). Large, broad, and conspicuous petal of a legume type of flower.

Bark (Swedish *bark*, rind). The external tissues of a woody stem or root, from the vascular cambium outward.

Base. A substance that can accept a proton (H^+). Also, the purine and pyrimidine groups in nucleic acids and nucleotides are collectively called bases; they act as bases.

Basidiospore (M.L. *basidium*, a little pedestal + spore). Type of meiospore borne by basidia in the basidiomycetes.

Basidium (plural, **basidia**) (M.L. *basidium*, a little pedestal). A cell that bears meiospores on its surface, characteristic of the basidiomycete fungi.

Benthon (Gr. *benthos*, the depths of the sea). Attached aquatic plants and animals, collectively.

Berry. A simple fleshy fruit, the ovary wall fleshy and including one or more carpels and seeds.

Biennial (L. *biennium*, a period of two years). A plant that requires two years to complete its life cycle. Flowering is normally delayed until the second year.

Bifacial leaf (L. *bis*, twice + *facies*, face). A leaf having structurally distinct upper and lower surfaces.

Binomial (L. *binominis*, two names). The pair of names that make up the taxonomic name of a species; a genus name followed by a specific epithet.

Bioassay (Gr. *bios*, life + L. *exagere*, to weigh or test). To test for the presence or quantity of a substance by using an organism's response as an indicator.

Biology (Gr. *bios*, life + *logos*, word, speech, discourse). The science that deals with living things.

Biotic (Gr. *biotikos*, relating to life). Relating to life.

Bladder (O.E. *blaedre*, a blister). A gas-filled sac whose buoyancy keeps some aquatic plants upright.

Bloom (Gr. *blume*, flower). A flower; a waxy coating on a fruit; the coloring of a body of water by a high density of planktonic organisms.

Bordered pit. A pit in a tracheid or vessel member having a distinct rim of the cell wall overarching the pit membrane.

Botanic garden. See **arboretum.**

Botany (Gr. *botane*, plant, herb). The science dealing with plant life.

Bract (L. *bractea*, a thin plate of precious metal). A modified leaf, from the axil of which arises a flower or an inflorescence.

Bud (M.E. *budde*, bud). An undeveloped shoot, largely meristematic tissue, generally protected by modified scale leaves. Also, a swelling on a yeast cell that will become a new yeast cell when released.

Bud scale. A modified protective leaf of a bud.

Bud scar. A scar left on a twig when the bud or bud scales fall away.

Bulb (L. *bulbus*, a modified bud). A short underground stem enclosed by many fleshy leaves filled with stored food.

Bundle scar. Scar left where conducting strands passing out of the stem into the leaf stalk were broken off when the leaf fell.

Bundle sheath. Sheath of parenchyma cells that surround the vascular bundles of leaves, sometimes called border parenchyma.

C₃ cycle. The part of the photosynthetic process in which sugar is made from CO_2. Named for the fact that compounds with three carbon atoms are the first stable products formed after CO_2 fixation. Also called the Calvin–Benson cycle. Occurs in all plants.

C₃ plant. A plant that has the C_3 cycle but not the C_4 cycle of photosynthesis.

C₄ cycle. A pathway of photosynthesis in which compounds with four carbon atoms are the immediate products of CO_2 fixation.

Callose (L. *callum*, thick skin + *ose*, a suffix indicating a carbohydrate). An amorphous polysaccharide deposited around pores in sieve-tube members.

Callus (L. *callum*, thick skin). Mass of thin-walled cells, usually developed as the result of wounding or in tissue cultures.

Calorie (L. *calor*, heat). The amount of heat needed to raise the temperature of 1 g of water 1°C (usually from 14.5° to 15.5°C); also called gram-calorie; 1000 calories = 1 kilocalorie.

Calvin cycle. See **C₃ cycle.**

Calyptra (Gr. *kalyptra*, a veil, covering). In bryophytes, an envelope covering the developing sporophyte, formed by growth of the venter of the archegonium.

Calyx (Gr. *kalyx*, a husk, cup). Sepals collectively; outermost flower whorl.

Cambium (L. *cambium*, one of the alimentary body fluids supposed to nourish the body organs). A layer, usually regarded as one or two cells thick, of persistently meristematic tissue, giving rise to secondary tissues, resulting in growth in diameter.

Canopy (Gr. *kanopeion*, a cover over a bed to keep off gnats). The leafy portion of a tree or shrub.

Capillaries (L. *capillus*, hair). Very small spaces, or very fine bores in a tube.

Capsule (L. *capsula*, dim. of *capsa*, a case). A simple, dry, dehiscent fruit, with two or more carpels.

Carbohydrate (chemical combining forms, *carbo*, carbon + *hydrate*, containing water). Compounds with the general formula $C_n(H_2O)_n$ or $C_nH_{2n}O_n$.

Carbon fixation. The enzymatic reaction in which CO_2 is attached to a receiver compound such as ribulose diphosphate, thereby adding to the supply of organic carbon. Occurs chiefly in photosynthesis.

Carotene (L. *carota*, carrot). A reddish-orange plastid pigment.

Carotenoids. A class of fat-soluble compounds (lipids) that includes carotenes as well as xanthophylls; most of them absorb light and appear yellow, orange, or red.

Carpel (Gr. *karpos*, fruit). A floral leaf bearing ovules along the margins.

Carpogonium (Gr. *karpos*, fruit + *gonos*, producing). Female gametangium (in red algae).

Carpospore (Gr. *karpos*, fruit + spore). One of the spores produced in a carpogonium.

Carposporophyte (Gr. *karpos*, fruit + *sporos*, seed, spore + *phuton*, plant). One of two sporophyte generations in certain red algae; grows attached to the gametophyte generation, in contrast to the tetrasporophyte.

Caruncle (L. *caruncula*, dim. of *caro*, flesh, wart). A spongy outgrowth of the seed coat, especially prominent in the castor bean seed.

Caryopsis (Gr. *karyon*, a nut + *opsis*, appearance). A simple, dry, one-seeded, indehiscent fruit, with the pericarp fused to the seed coat. Also called a grain.

Casparian strip. A band around each endodermal cell, in which the radial and transverse cell walls are impregnated with suberin.

Catabolism (Gr. *kata*, down + metabolism). The phase of metabolism in which complex substances are broken down into simpler molecules, the chief role being to provide chemically reactive materials for use in anabolism and to provide ATP for cellular work.

Catalyst (Gr. *kateluein*, to dissolve). A substance that accelerates a chemical reaction but that is not used up in the reaction.

Cation. An ion with a positive electrical charge.

Catkin (literally a kitten, apparently first used in 1578 to describe the inflorescence of the pussy willow). A type of inflorescence, really a spike, generally bearing only pistillate flowers or only staminate flowers, which eventually fall intact from the plant.

Caulescent (Gr. *kaulos*, a plant stem). A plant whose stem bears leaves separated by visibly elongated internodes, as opposed to a rosette plant.

Cell (L. *cella*, small room). The smallest structural unit in the organism that is capable of self-reproduction. It is surrounded by a plasma membrane and contains a store of DNA together with a metabolic system.

Cell cycle. Sequence of metabolic steps during which a cell prepares for and conducts DNA synthesis and mitosis; consists of G_1 (pre-DNA synthesis phase), S (DNA synthesis), G_2 (premitotic phase), and M (mitosis).

Cellulose (cell + -ose, a suffix indicating a carbohydrate). An unbranched polymer composed of glucose molecules; the chief fibrous component of plant cell walls.

Cell wall. A layer of material, chiefly elongated polymers, that is laid down outside the plasma membrane of most plant cells and that serves to protect the protoplast and to limit its expansion.

Cenozoic (Gr. *kainos*, recent + *zoe*, life). The geologic era extending from 65 million years ago to the present.

Central cell. Central portion of an embryo sac consisting of the polar nuclei (or secondary nucleus if they have fused) and surrounding cytoplasm. Called the primary endosperm cell after fertilization.

Central placentation. Ovules attached to the central axis of a one-loculed ovary.

Centromere (L. *centrum*, center + Gr. *meros*, part). Specialized part of chromosomes where spindle fibers attach and where two chromatids connect. Two **kinetochores,** one on each chromatid, compose one centromere.

Chalaza (Gr. *chalaza*, small tubercle). The region on a seed at the upper end of the raphe where the funiculus spreads out and unites with the base of the ovule.

Chaparral (Sp. *chaparro*, an evergreen oak). A vegetation type characterized by small-leaved, evergreen shrubs growing together, forming a nearly impenetrable scrub; shrubby oaks are found in chaparral of California and the Mediterranean region, but other genera are typical of chaparral in Chile, South Africa, and Australia.

Chemotropism (Gr. *khemia*, alchemy + *tropos*, turning). A response in which the direction of growth is controlled by the concentration of a chemical substance in the environment.

Chernozem (R. *Cherny*, black + *zem*, earth). See **mollisol.**

Chitin (Gr. *chiton*, a coat of mail). A polymer in which the monomer unit is the modified sugar *N*-acetyl glucosamine; it is the principal stiffening material in the cell walls of most fungi and in the exoskeletons of insects and crustaceans.

Chlamydospore (Gr. *chlamys*, a horseman's or young man's coat + spore). A heavy-walled resting asexual spore.

Chlorenchyma (Gr. *chloros*, green + *-enchyma*, a suffix meaning tissue). Parenchyma tissue possessing chloroplasts.

Chloro- (Gr. *chloros*, green). A prefix meaning "green."

Chlorophyll (chloro- + *phyllon*, leaf). The green pigment found in the chloroplast, important in the absorption of light energy in photosynthesis.

Chloroplast (chloro- + Gr. *plastos*, formed). The organelle that performs photosynthesis in eukaryotic cells.

Chlorosis (chloro- + Gr. *osis*, diseased state). A diseased condition in which leaves are yellow due to an infection or mineral deficiency that prevents chlorophyll formation.

Chroma-, chromo- (Gr. *khroma*, color). A prefix meaning "color."

Chromatid (chroma- + Gr. *-id*, suffix meaning daughters of). Either of two daughter strands of a duplicated chromosome that are still joined at the centromeres.

Chromatin. The material that carries hereditary information in eukaryotic cells, composed of DNA and protein.

Chromoplast (chroma- + *plastos*, formed). Specialized plastid containing yellow or orange pigments.

Chromosome (chroma- + Gr. *soma*, body). A body that carries genes in a cell, composed of chromatin in eukaryotes; composed of naked DNA in prokaryotes.

Cilium (plural, **cilia**) (Fr. *cil*, an eyelash). Extension of a eukaryotic cell containing microtubules in a characteristic pattern, which undulates and imparts motion to the cell.

Cisterna (plural, **cisternae**) (L. *cisterna*, a reservoir). A flattened sac, composed of a continuous surrounding membrane, together with the enclosed space.

Citric acid cycle. A system of reactions that contributes to the catabolic breakdown of fuels in respiration and that provides building materials for a number of important anabolic pathways. Also called the **Krebs cycle** and the **tricarboxylic acid (TCA) cycle.**

Cladode (Gr. *kladodes*, having many shoots). A branch resembling a foliage leaf.

Class (L. *classis*, a division of Roman people). The taxonomic category below the division and above the order.

Clay. Soil particles less than 2 μm in diameter, composed mainly of aluminum (Al), oxygen (O), and silicon (S).

Cleistothecium (plural, **cleistothecia**) (Gr. *kleistos*, closed + *thekion*, a small receptacle). The closed, spherical ascocarp of the powdery mildews.

Climax community. The last stage of a natural succession; a community capable of maintaining itself as long as the climate does not change.

Clone (Gr. *klon*, a twig or slip). The aggregate of individual organisms produced asexually from one individual.

Closed bundle. A vascular bundle lacking residual procambium.

Coal Age. The Carboniferous period, beginning 345 million years ago and ending 280 million years ago.

Coalescence (L. *coalescere*, to grow together). A condition in which there is union of separate parts of any one whorl of flower parts; synonyms are **connation** and **cohesion.**

Codon. A sequence of three nucleotides in DNA or RNA, which specifies a particular amino acid.

Coenocyte (Gr. *koinos*, shared in common + *kytos*, a vessel). A plant or filament whose protoplasm is continuous and multinucleate and without any division by walls into separate protoplasts.

Coenzyme. A substance, usually nonprotein and of low molecular weight, necessary for the action of some enzymes.

Coevolution (L. *co*, together + evolution). The process of two or more interacting species evolving through time as a unit.

Cohesion (L. *cohaerere*, to stick together). Union or holding together of parts of the same materials; the union of floral parts of the same whorl, as petals to petals.

Coleoptile (Gr. *koleos*, sheath + *ptilon*, down, feather). The first leaf in germination of grasses, which sheaths the succeeding leaves.

Coleorhiza (Gr. *koleos,* sheath + *rhiza,* root). Sheath that surrounds the radicle of the grass embryo and through which the young root bursts.

Collenchyma (Gr. *kolla,* glue + *-enchyma,* a suffix denoting a kind of cell or tissue). A flexible supporting tissue of stems and leaves, composed of closely packed cells with thickened primary walls.

Colloid (Gr. *kolla,* glue + *eidos,* form). Referring to matter composed of particles, ranging in size from 0.0001 to 0.000001 mm, dispersed in some medium, as clay particles in soil.

Colony (L. *colonia,* a settlement). A growth form characterized by a group of closely associated but poorly differentiated cells; sometimes filaments can be associated together in a colony (as in *Nostoc*), but more typically, unicells are associated in a colony.

Community (L. *communitas,* a fellowship). All the populations within a given habitat; usually the populations are thought of as being interdependent.

Companion cells. Cell associated with sieve-tube members.

Compensation depth (L. *compensare,* to counterbalance). The depth in a body of water at which the light intensity supports just enough photosynthesis to equal respiration.

Competition (L. *competere,* to strive together). An ecological interaction in which two organisms require the same resource, of which there is an insufficient supply.

Complete flower. A flower having sepals, petals, stamens, and carpels.

Compound leaf. A leaf whose blade is divided into several distinct leaflets.

Conceptacle (L. *conceptaculum,* a receptacle). A cavity or chamber of a frond (of *Fucus,* for example) in which gametangia are borne.

Conduction (L. *conducere,* to bring together). Act of moving or conveying water through the xylem in plant organs.

Cone (Gr. *konos,* a pine cone). See **strobilus.**

Conidiophore (conidium + Gr. *phoros,* bearing). A hypha that bears conidia.

Conidium (plural, **conidia**) (Gr. *konis,* dust). Asexual spores of certain fungi, produced by cutting off the tip of a hypha or by extruding a bud from the tip of a hypha.

Conifer (cone + L. *ferre,* to carry). A cone-bearing tree in the division Coniferophyta.

Conjugation (L. *conjugatus,* united). Process of sexual reproduction involving the fusion of isogametes or of specialized cell extensions.

Connation (L. *connatus,* to be born together). Condition in the flower in which there is a union of similar parts of any one whorl of appendages; synonym of **coalescence.**

Convergence. Evolution of similar traits in organisms that are descended from unlike ancestors, resulting from exposure to similar environmental selective pressures.

Cork. An external, secondary tissue impermeable to water and gases.

Cork cambium. The cambium from which cork develops.

Corm (Gr. *kormos,* a trunk). A short, solid, vertical, enlarged underground stem in which food is stored.

Corolla (L. *corolla,* dim. of *corona,* a wreath, crown). Petals, collectively; usually the conspicuous colored flower whorl.

Cortex (L. *cortex,* bark). A region of primary tissues in a stem or root, bounded externally by the epidermis and internally by the vascular tissue (in stems) or the pericycle (in roots).

Cotyledons (Gr. *kotyledon,* a cup-shaped hollow). Seed leaves; the first leaves formed in the embryo, specialized for absorbing and sometimes for storing reserve foods.

Covalent bond. A binding force that holds two atoms together, due to the sharing of electrons.

Crista (plural, **cristae**) (L. *crista,* a crest). A fold in the inner membrane of the mitochondrion.

Cross. Sexual reproduction in which the two parents differ.

Crossing-over. An exchange of corresponding segments of homologous chromatids; occurs during meiosis.

Cross-pollination. The transfer of pollen from a stamen to the stigma of a flower on another plant, except in clones.

Cultural eutrophication. Eutrophication that results from pollution of bodies of water by human activities. See **eutrophication.**

Cuticle (L. *cuticula,* dim. of *cutis,* the skin). Waxy layer on outer wall of epidermal cells.

Cutin (L. *cutis,* the skin). Waxy substance that is but slightly permeable to water, water vapor, and gases; a major part of the cuticle.

Cutinization. Impregnation of cell wall with a substance called cutin.

Cyme (Gr. *kyma,* a wave, a swelling). A type of inflorescence in which the apex of the main stalk or the axis of the inflorescence ceases to grow quite early, relative to the laterals.

Cystocarp (Gr. *kystos,* bladder + *karpos,* fruit). A peculiar diploid spore-bearing structure formed after fertilization in certain red algae.

Cyto- (Gr. *kytos,* a hollow vessel). A prefix meaning ''cellular.''

Cytochrome (cyto- + Gr. *chroma,* color). A class of several electron-transport proteins serving as carriers in mitochondrial oxidations and in photosynthetic electron transport.

Cytokinesis (cyto- + Gr. *kinesis,* motion). Division of the cytoplasm into two by the formation of a new cell wall. Usually occurs during telophase of mitosis or meiosis.

Cytokinin. A class of hormones that participate in controlling many developmental processes in plants.

Cytology (cyto- + Gr. *logos*, word, speech, discourse). The science dealing with the cell.

Cytoplasm (cyto- + Gr. *plasma*, form). All the protoplasm of a protoplast outside the nucleus.

Cytosine. A pyrimidine base found in DNA and RNA.

Deciduous (L. *deciduus*, falling). Referring to trees and shrubs that lose their leaves in the fall.

Decomposer (L. *de*, from + *componere*, to put together). An organism that obtains food by breaking down dead organic matter into simpler molecules.

Dehiscent (L. *dehiscere*, to split open). Opening spontaneously when ripe; splitting into definite parts.

Deletion (L. *deletus*, to destroy, to wipe out). Used here to designate an area, or region, lacking from a chromosome.

Demography (Gr. *demos*, people + *graphos*, writing). The science of vital statistics of populations.

Denaturation. A major change in the folding of a protein, usually irreversible, caused by environmental factors.

Dendrogram (Gr. *dendron*, tree + *gramme*, what is written or drawn). A graph showing the relationship between things at different levels of similarity; the graph resembles the limbs of a tree.

Denitrification (L. *de*, to denote an act undone + *nitrum*, *nitro*, a combining form indicating the presence of nitrogen + *facere*, to make). Conversion of nitrates into nitrites or into gaseous oxides of nitrogen, or even into free nitrogen.

Deoxyribonucleic acid (DNA). A polymer composed of nucleotides in which the sugar is the compound deoxyribose. DNA carries the hereditary information of the cell.

Desert scrub (M.E. *schrubbe*, shrub). A vegetation type characterized by evergreen or drought-deciduous shrubs growing together rather openly, generally in an area with annual precipitation below 25 cm.

Detritus (L. *detritus*, worn away). Particulate organic matter released in the process of decomposition of dead organisms or parts of organisms (such as plant litter).

Development (F. *developper*, to unfold). Changes in the plant body that result from controlled processes of cell division, cell growth, and cell differentiation.

Diatom (Gr. *diatomos*, cut in two). Member of a group of golden-brown algae with silicious cell walls fitting together much as do the halves of a pill box.

Diatomite. Diatomaceous earth; that is, sedimentary deposits made up of the silica wall remains of diatoms.

Dicotyledon (Gr. *dis*, twice + *kotyledon*, a cup-shaped hollow). A plant whose embryo has two cotyledons.

Dictyosome (Gr. *diktyon*, a net + *soma*, body). One of the component parts of the Golgi apparatus; in plant cells a complex of flattened double lamellae.

Differentially permeable. See **selectively permeable.**

Differentiation (L. *differere*, to carry different ways). Developmental change of a cell, tissue, or organ, leading to features that support the performance of specialized functions.

Diffuse porous. Wood with an equal and random distribution of large xylem vessel members throughout the growth season.

Diffuse secondary growth. Secondary growth, such as in palm trees, that is caused by a proliferation of parenchyma cells and not by vascular cambium.

Diffusion (L. *diffusus*, spread out). The random migration of ions or molecules, resulting from thermal motion.

Dikaryon (Gr. *di*, two + *karyon*, nut). A hypha or mycelium in which each cell contains two haploid nuclei, the two usually derived from different parent organisms. The dikaryotic condition is often abbreviated as the $n + n$ condition.

Dinoflagellate (Gr. *dinein*, to whirl + L. *flagellum*, a whip). The common name for members of the algal division Pyrrhophyta; the organisms are typically unicellular and motile, with cell walls made up of overlapping plates.

Dioecious (Gr. *dis*, twice + *oikos*, house). Unisexual; having the male and female elements in different individuals.

Diploid (Gr. *diploos*, double + -*oid*). Having two sets of chromosomes in each cell.

Divergence. Evolutionary change in which the descendants of similar ancestors become increasingly dissimilar as a result of isolation and different histories of mutation and natural selection.

Division. The taxonomic category below the kingdom and above the class, equivalent to phylum.

DNA. See **deoxyribonucleic acid.**

Dominant (L. *dominari*, to rule). *Genetics:* Said of an allele that generates the same phenotype whether heterozygous or homozygous. *Ecology:* Referring to the species of a community that receives the full force of the macroenvironment; usually the most abundant of such species.

Dormant (L. *dormire*, to sleep). Being in a state of reduced physiological activity such as occurs in seeds, buds, etc.

Dorsiventral (L. *dorsum*, the back + *venter*, the belly). Having upper and lower surfaces distinctly different, as many leaves do.

Double bond. A covalent bond that involves four electrons.

Drupe (L. *drupa*, an overripe olive). A simple, fleshy fruit, derived from a single carpel, usually one-seeded, in which the exocarp is thin, the mesocarp fleshy, and the endocarp stony.

Early wood. The portion of an annual ring formed during spring, characterized by large cells and thin walls; also called **spring wood.**

Ecology (Gr. *oikos*, home + *logos*, discourse). The study of life in relation to environment.

Ecosystem (Gr. *oikos*, house + *synistanai*, to place together). An inclusive term for a living community and all the factors of its nonliving environment.

Ecotype (Gr. *oikos*, house + *typos*, the mark of a blow). Genetic variant within a species that is adapted to a particular environment yet remains interfertile with all other members of the species.

Edaphic (Gr. *edaphos*, soil). Pertaining to soil conditions that influence plant growth.

Egg (A.S. *aeg*, egg). The nonmotile gamete in species that have distinct motile and nonmotile gametes.

Elater (Gr. *elater*, driver). An elongated, spindle-shaped, sterile, hygroscopic cell in the sporangium of some Bryophyta sporophytes.

Electron. An elementary particle of matter that bears a unit of negative electrical charge. See **atom.**

Electron microscope. A microscope that uses a beam of electrons rather than light to produce a magnified image.

Element (L. *elementa*, the first principles). In modern chemistry, a substance that cannot be divided or reduced by any known chemical means to a simpler substance; 92 natural elements are known, of which gold, carbon, oxygen, and iron are examples; several, including plutonium, have been formed in atomic piles.

Embryo (Gr. *en, in* + *bryein,* to swell). A young sporophytic plant, while still retained in the gametophyte or in the seed.

Embryo sac. The female gametophyte of the angiosperms; generally composed of an egg cell, two synergid cells, three antipodal cells, and a central cell.

Endocarp (Gr. *endon*, within + *karpos*, fruit). Inner layer of fruit wall (pericarp).

Endodermis (Gr. *endo*, within + *derma*, skin). The layer of living cells, bound by a Casparian strip, that surrounds the vascular tissue in nearly all roots and certain stems and leaves.

Endogenous (Gr. *endon*, within + *genes*, born). Produced within the cell or organism.

Endoplasmic reticulum (Gr. *endon*, within + *plasma*, anything formed or molded; L. *reticulum*, a small net). A system of membrane-bound cisternae found in the cytoplasm.

Endosperm (Gr. *endon*, within + *sperma*, seed). The nutritive tissue formed within the embryo sac of seed plants; it is often consumed as the seed matures but remains in the seeds of corn and other cereals.

Energy (Gr. *en*, at + *ergon*, work). A measure of the capacity of a system to undergo spontaneous change, or to do work, or to cause change in another system.

Environment (O.F. *environ*, around). The living and nonliving factors that surround a given organism or community of organisms.

Enzyme (Gr. *en*, in + *zyme*, yeast). A protein that acts as a catalyst to speed chemical reactions.

Epicotyl (Gr. *epi*, upon + *kotyledon*, a cup-shaped hollow). The upper portion of the axis of the embryo or seedling above the cotyledons.

Epidermis (Gr. *epi*, upon + *derma*, skin). A superficial layer of cells occurring on all parts of the primary plant body except the root cap and the apical meristems.

Epigyny (Gr. *epi*, upon + *gyne*, woman). The arrangement of floral parts in which the ovary is embedded in the receptacle so that the other parts appear to arise from the top of the ovary.

Epiphyte (L. *epi*, upon + *phyton*, a plant). A plant that grows upon another plant but is not parasitic.

Epithet. A descriptive word. In taxonomy, a word that distinguishes one species from others in the same genus; often chosen to emphasize a characteristic feature.

Equilibrium. In chemistry, the condition in which forward and backward reactions cancel so that the concentrations of materials do not change with time. More broadly, a state in which all sources of variation cancel, resulting in constancy.

ER. Endoplasmic reticulum.

Ethylene. C_2H_4, a hormone that participates in the control of many developmental processes in plants.

Etiolation (F. *etioler*, to blanch). A condition involving increased stem elongation, poor leaf development, and lack of chlorophyll found in plants growing in the absence, or in a greatly reduced amount, of light.

Eukaryote (L. *eu*, true + *karyon*, a nut, referring in modern biology to the nucleus). Any organism characterized by having cellular organelles, including a nucleus, bounded by membranes. Also spelled **eucaryote.**

Eutrophication (Gr. *eu*, good, well + *trephein*, to nourish). Pollution of bodies of water resulting from slow, natural, geological, or biological processes such as siltation or encroachment of vegetation or accumulation of detritus; also called natural eutrophication.

Evapotranspiration. See **transpiration.**

Evolution (L. *evolutio*, an unrolling). A progressive change in the genetic makeup of a population, with accompanying changes in characteristics of the organisms.

Exine (L. *exterus*, outside). Outer coat of pollen.

Exocarp (Gr. *exo*, without, outside + *karpos*, fruit). Outermost layer of fruit wall (pericarp).

Exogenous (Gr. *exe*, out, beyond + *genos*, race, kind). Produced outside of, originating from, or due to external causes.

Facultative (L. *facultas*, capability). Able to adapt to varied conditions; e.g., a facultative anaerobe can survive either with or without oxygen. Contrast to **obligate.**

Family. The taxonomic category below the order and above the genus.

Fascicle (L. *fasciculus*, a small bundle). A cluster of leaves or other organs.

Fascicular cambium. Vascular cambium that arises within a vascular bundle.

Fermentation (L. *fermentum*, a drink made from fermented barley, beer). Catabolic breakdown of fuels by a process that does not involve molecular oxygen.

Fern (O.E. *fearn*, fern). Common name for members of the division Pterophyta, part of the lower vascular plants.

Fertilization (L. *fertilis*, capable of producing fruit). The union of gametes. Synonymous with **syngamy.**

Fiber (L. *fibra*, a fiber or filament). An elongated, tapering, thick-walled strengthening cell occurring in various parts of plant bodies.

Fiber-tracheids. Xylem elements found in pine that are structurally intermediate between tracheids and fibers.

Field capacity. The amount of water retained in a soil (generally expressed as percent by weight) after large capillary spaces have been drained by gravity.

Filament (L. *filum*, a thread). Stalk of stamen bearing the anther at its tip; also, a slender row of cells (certain algae).

Flagellum (plural, **flagella**) (L. *flagellum*, a whip). A long cilium.

Flora (L. *floris*, a flower). An enumeration of all the species that grow in a region; also, the collective term for all the species that grow in a region.

Floret (F. *fleurette*, a dim. of *fleur*, flower). One of the small flowers that make up the composite inflorescence (head) or the spike of the grasses.

Flower (F. *fleur*, L. *flos*, a flower). Floral leaves grouped together on a stem and adapted for sexual reproduction in the angiosperms.

Follicle (L. *folliculus*, dim. of *follis*, bag). A simple, dry, dehiscent fruit, with one carpel, splitting along one suture.

Food (A.S. *fōda*). Any organic substance that furnishes energy and building materials directly for vital processes.

Food chain. The path along which caloric energy is transferred within a community (from producers to consumers to decomposers).

Foot. The portion of the sporophyte in bryophytes and lower vascular plants that is sunk in gametophyte tissue and absorbs food from the gametophyte.

Fossil (L. *fossilis*, dug up). Any impression, impregnated remains, or other trace of an animal or plant of past geological ages that has been preserved in the earth's crust.

Fossil fuel. Hydrocarbon deposits, currently mined and refined for use as fuel (coal, gas, oil), that were originally deposits of detritus of now-extinct plants.

Fret (O.F. *frette*, latticework). Regions of thylakoids that extend between the grana in a chloroplast.

Frond (L. *frons*, branch, leaf). A synonym for a large divided leaf, especially for a fern leaf.

Fruit (L. *fructus*, that which is enjoyed, hence the product of the soil, trees, cattle, etc.). A matured ovary; in some seed plants, other parts of the flower may be included; also applied, as **fruiting body,** to reproductive structures of some fungi.

Frustule (L. *frustulum*, little piece). A diatom cell, composed of two overlapping halves (valves).

Fucoxanthin (Gr. *phykos*, seawood + *xanthos*, yellowish brown). A brown pigment found in brown algae.

Fungus (plural, **fungi**) (L. *fungus*, a mushroom). A eukaryotic organism that lacks plastids and that reproduces by means of spores.

Funiculus (L. *funiculus*, dim. of *funis*, rope or small cord). A stalk of the ovule, containing vascular tissue.

Fusiform initials (L. *fusus*, spindle + form). Meristematic cells in the vascular cambium that develop into xylem and phloem cells comprising an axial system.

Gametangium (gamete + Gr. *angeion*, a vessel). A structure that bears gametes.

Gamete (Gr. *gamos*, marriage). A cell that is specialized for syngamy.

Gametic life cycle. A life cycle in which the haploid phase is limited to gametes, as in some diatoms and many animals.

Gametophyte (gamete + Gr. *-phyte*, a plant). In species with a sporic life cycle, the gamete-producing plant, which is haploid.

Gel (L. *gelare*, to freeze). Jellylike colloidal mass.

Gemma (plural, **gemmae**) (L. *gemma*, a bud). A multicellular unit that serves for asexual reproduction in some fungi and bryophytes.

Gene (Gr. *genos*, race, offspring). A group of base pairs in the DNA molecule that determines one or more hereditary characters.

Generation (L. *genus*, birth, race, kind). Any stage in a life cycle characterized by a particular chromosome number, as the gametophyte and sporophyte generations. Also called a **phase.**

Gene recombination. The appearance of gene combinations in the progeny different from the combinations present in the parents.

Genet (L. *generatus*, to beget). A genetically unique individual; also, a group of genetically identical plants; synonymous with a **clone.**

Genetic code. The set of correspondences between amino acids and nucleotides that relates a gene to the formation of a protein.

Genetics. The study of biological information and heredity.

Genotype. The specific set of alleles in an organism; the genetic makeup of an organism.

Genus (plural, **genera**) (Gr. *genos*, race, stock). The taxonomic category below family and above the species.

Geotropism (Gr. *ge*, earth + *tropos*, turning). A growth curvature induced by gravity.

Germination (L. *germinare*, to sprout). The beginning of growth of a seed, spore, or other once-dormant structure.

Gibberellins. A class of hormones that participate in controlling many developmental processes in plants.

Girdle (O.E. *gyrdel*, enclosure, girdle). The region of a frustule where the two valves overlap.

Glucose (Gr. *glykys*, sweet + *ose*, a suffix indicating a carbohydrate). A common hexose sugar.

Glume (L. *gluma*, husk). An outer and lowermost bract of a grass spikelet.

Glycogen (Gr. *glykys*, sweet + *genes*, born). A polysaccharide built of glucose, akin to starch, serving as a food reserve in animals and fungi and some prokaryotes.

Glycolysis (Gr. *glykys*, sweet + *lysis*, a loosening). Early steps in the catabolic breakdown of glucose, in which glucose is converted to pyruvic acid with a formation of ATP and NADH.

Golgi body (Italian cytologist Camillo Golgi, 1844–1926, who first described the organelle). In animal cells, a complex perinuclear region thought to be associated with secretion; in plant cells a series of flattened plates, more properly called **dictyosomes.**

Granum (plural, **grana**) (L. *granum*, a seed). A stack of tightly appressed thylakoids in the chloroplast.

Ground cover. The area of ground covered by a plant when its canopy edge is projected perpendicularly down.

Ground meristem (Gr. *meristos*, divisible). A primary meristem that gives rise to cortex and pith.

Ground tissues. Category of primary tissues (parenchyma, collenchyma, and sclerenchyma) that provide such basic functions as storage, support, and secretion.

Growth (A.S. *growan*, probably from Old Teutonic *gro*, from which grass is also derived). An irreversible increase in size.

Growth retardant. A chemical (such as cycocel, CCC) that selectively interferes with normal hormonal promotion of growth, but without appreciable toxic effects.

Guanine. A purine base found in DNA and RNA.

Guttation (L. *gutta*, drop, exudation of drops). Exudation of water from plants, in liquid form.

Gymnosperm (Gr. *gymnos*, naked + *sperma*, seed, sperm). The common name for a group of divisions, including the Coniferophyta, which bear exposed seeds, in contrast to the flowering plants that bear the seeds enclosed in a fruit (mature ovary).

Gynoecium (Gr. *gyne*, woman + *oikos*, house). All the carpels in a flower.

Haploid (Gr. *haploos*, single + -oid). Having a single set of chromosomes.

Hapteron (Gr. *haptein*, to fasten). An individual branch of the holdfast organ of a kelp; also called haptere (plural, hapteres).

Haustorium (plural, **haustoria**) (M.L. *haustrum*, a pump). A projection that acts as a penetrating and absorbing organ.

Head. An inflorescence, typical of the composite family, in which flowers are sessile and grouped closely on a receptacle.

Heartwood. Wood in the center of old secondary stems that is plugged with resins and tyloses and is not active.

Helix (Gr. *helix*, anything twisted). A spiral that is drawn out along an axis.

Hemicellulose. A class of polysaccharides of the cell wall, built of several different kinds of simple sugars linked in various combinations.

Herb (L. *herba*, grass, green blades). A seed plant that does not develop woody tissues.

Herbaceous (L. *herbaceus*, grassy). Lacking woody tissues.

Herbal (L. *herba*, grass). A book that contains the names and descriptions of plants, especially those that are thought to have medicinal uses.

Herbarium (L. *herba*, grass). A collection of dried and pressed plant specimens.

Herbicide (L. *herba*, grass or herb + *cidere*, to kill). A chemical used to kill plants, frequently chemically related to a hormone.

Herbivore (L. *herba*, grass + *vorare*, to devour). An organism that eats plants, generally applied to animals.

Heredity (L. *hereditas*, being an heir). The transmission of parental traits to the offspring.

Hermaphroditic (Gr. god Hermes + goddess Aphrodite). Having both male and female reproductive structures in the same individual.

Hetero- (Gr. *heteros*, other, different). A prefix meaning "unlike" or "other."

Heterocyst (hetero- + Gr. *cytis*, a bag). An enlarged cell in the filaments of certain cyanobacteria, associated with nitrogen fixation.

Heteroecious (hetero- + Gr. *oikos*, house). Referring to fungi that cannot carry through their complete life cycle unless two different host species are present.

Heterogametes (hetero- + gamete). Gametes dissimilar from each other in size and behavior, like egg and sperm.

Heterogamy (hetero- + Gr. *gamos*, union or reproduction). Reproduction involving two types of gametes.

Heterophylly (hetero- + Gr. *phyllon*, leaf). Condition where a plant has different leaf forms at different times in its life cycle. Often induced by environmental factors.

Heterospory. The condition of producing two kinds of spores (megaspores and microspores); hence two kinds of gametophytes.

Heterothallic. Incapable of self-fertilization. Not applied to seed plants.

Heterotrichy (hetero- + Gr. *trichos*, a hair). In the algae, the occurrence of two types of filaments, erect and prostrate.

Heterotrophic (hetero- + Gr. *trephein*, to nourish). Requiring nourishment from outside sources.

Heterozygous (hetero- + Gr. *zygon*, yoke). Having two different alleles for a given gene in the diploid cell.

Hexose (Gr. *hexa*, six + -*ose*, suffix indicating carbohydrate). A carbohydrate with six carbon atoms.

Higher vascular plant. Collective term for all seed-producing plants, both gymnosperms and angiosperms.

Hilum (L. *hilum*, a trifle). Scar on seed that marks the place where the seed broke from the stalk.

Homo- (Gr. *homo*, same). A prefix meaning ''same.''

Homologous (Gr. *homologos*, the same). Corresponding in structure and evolutionary origin; e.g., the flower petal and the foliage leaf are lateral appendages that evolved from the same primitive structure.

Homologous chromosomes. Chromosomes that have the same set of genes.

Homospory. The condition of producing only one kind of spore, hence one kind of gametophyte.

Homothallic. Capable of self-fertilization. Not applied to seed plants.

Homozygous (Gr. *homos*, one and the same + *zygon*, yoke). Having two identical alleles for a given gene in the diploid cell.

Hormogonia (singular, **hormogonium**) (Gr. *hormos*, necklace + *gonos*, offspring). Short filaments, the result of a breaking apart of filaments of certain blue-green algae at the heterocysts.

Hormone (Gr. *hormaein*, to excite). A compound that is normally produced by a plant and whose sole function is to act as a signal in controlling development.

Host (L. *hospis*, host). An organism that harbors a parasite.

Humidity, relative (L. *humidus*, moist). The ratio of the weight of water vapor in a given quantity of air to the total weight of water vapor that quantity of air is capable of holding at the temperature in question, expressed as a percentage.

Humus (L. *humus*, the ground). Decomposing organic matter in the soil.

Hybrid (L. *hybrida*, offspring of a tame sow and a wild boar, a mongrel). The offspring of two plants or animals differing in at least one Mendelian character; or the offspring formed by mating two plants or animals that differ genetically.

Hydathode (Gr. *hydro*, water + O.E. *thoden*, stem, or *thyd-dan*, to thrust). A structure, usually on leaves, that releases liquid water during guttation.

Hydrogen acceptor. A substance capable of accepting hydrogen atoms or electrons in the oxidation–reduction reactions of metabolism.

Hydrogen bond. A weak bond that occurs between a hydrogen atom and an oxygen or nitrogen atom when the hydrogen is already covalently bonded to another oxygen or nitrogen atom. The hydrogen bond is not covalent.

Hydrolysis (Gr. *hydro*, water + *lysis*, loosening). Reaction of a compound with water, attended by decomposition into less complex compounds.

Hydrophyte (Gr. *hydro*, water + *phyton*, plant). A plant that normally grows in a wet habitat.

Hymenium (Gr. *hymen*, a membrane). Spore-bearing tissue in various fungi.

Hypanthium (L. *hypo*, under + Gr. *anthos*, flower). Fusion of calyx and corolla part way up their length to form a cup, as in many members of the rose family.

Hypertrophy (Gr. *hyper*, over + *trephein*, to nourish with food). A condition of overgrowth or excessive development of an organ or part.

Hypha (plural, **hyphae**) (Gr. *hyphe*, a web). A slender, elongated, threadlike cell or filament of cells of a fungus.

Hypocotyl (Gr. *hypo*, under + *kotyledon*, a cup-shaped hollow). The portion of an embryo or seedling between the cotyledons and the radicle or young root.

Hypogyny (Gr. *hypo*, under + *gyne*, female). A condition in which the receptacle is convex or conical, and the flower parts are situated one above another in the following order, beginning with the lowest: sepals, petals, stamens, carpels.

Hypothesis (Gr. *hypothesis*, foundation). A tentative theory or supposition provisionally adopted to explain certain facts and to guide in the investigation of other facts.

Imbibition (L. *imbibere*, to drink). The absorption of liquids or vapors into the ultramicroscopic spaces or pores found in such materials as cellulose or a block of gelatin; an adsorption phenomenon.

Imperfect flower. A flower lacking either stamens or pistils.

Imperfect fungi. Fungi reproducing only by asexual means.

Incomplete flower. A flower lacking one or more of the four kinds of flower parts.

Indehiscent (L. *in*, not + *dehiscere*, to divide). Not opening by valves or along regular lines.

Indicator species. A species that has a narrow range of tolerance for one or more environmental factors so that, from its occurrence at a site, one can predict these factors at that site (e.g., nutrient availability or summer temperatures).

Indoleacetic acid (IAA). A natural auxin.

Indusium. Membranous growth of the epidermis of a fern leaf that covers a sorus.

Inferior ovary. An ovary that is more or less (sometimes completely) attached to the calyx and corolla.

Inflorescence (L. *inflorescere*, to begin to bloom). A flower cluster.

Inheritance (O.F. *enheritance*, inheritance). The reception or acquisition of characters or qualities by transmission from parent to offspring.

Inorganic. Referring to compounds that do not contain *both* carbon and hydrogen.

Integuments (L. *integumentum*, covering). The outermost layers of cells of the ovule, surrounding the female gametophyte (nucellus) and giving rise to the seed coat.

Intercalary (L. *intercalare*, to insert). Referring to primary meristematic tissue or primary growth in regions that are separated from the apex of the organ by mature tissue.

Intercellular (L. *inter*, between + cells). Lying between cells.

Interfascicular cambium (L. *inter*, between + *fasciculus*, small bundle). Cambium that develops between vascular bundles.

Internode (L. *inter*, between + *nodus*, a knot). The region of a stem between two successive nodes.

Interphase (L. *inter*, between + phase). Period between mitotic divisions, consists of G1, pre-DNA synthesis phase; S, DNA synthesis; and G2, post-DNA synthesis phase.

Intine (L. *intus*, within). The innermost coat of a pollen grain.

Intracellular (L. *intra*, within + cell). Lying within cells.

Introgressive hybridization (L. *intro*, to the inside + *gress*, walk + *hybrida*, half-breed). Backcrossing between complete or partial hybrids and the original parental stock.

Involucre (L. *involucrum*, a wrapper). A whorl or rosette of bracts surrounding an inflorescence.

Ion. An atom or molecule that has a net negative or positive electrical charge because the number of electrons does not equal the number of protons.

Ionic compound. A substance whose molecules readily break up into ions when placed in contact with water.

Irregular flower. A flower in which one or more members of at least one whorl are of a different form from other members of the same whorl; zygomorphic flower.

Isobilateral leaf (Gr. *isos*, equal + L. *bis*, twice, twofold + *lateralis*, pertaining to the side). A leaf having the upper and lower surfaces and anatomy essentially similar.

Isodiametric (Gr. *isos*, equal + diameter). Having diameters equal in all directions, as a ball.

Isogametes (Gr. *isos*, equal + gametes). Gametes similar in size and behavior.

Isogamy (Gr. *isos*, equal + *gamete*, spouse). The condition in which the gametes are identical.

Isolating barriers. Any barrier to the crossing (hybridization) of plants; includes biological, physiological, and ecological categories.

Isomers (Gr. *isos*, equal + *meros*, part). Two or more compounds having the same molecular formula but different internal structure; for example, glucose and fructose, both having the formula $C_6H_{12}O_6$.

Karyogamy (Gr. *karyon*, nut + *gamos*, marriage). The fusion of two nuclei.

Keel (A.S. *ceol*, ship). A structure of the legume type of flower, made up of two petals loosely united along their edges.

Kelp. (M.E. *culp*, seaweed). A collective name for any of the large brown marine algae.

Kinetochore (Gr. *kinein*, to move + *chorein*, to move apart). A microtobule-organizing center that occurs beside the centromere of a chromosome, forming microtubules that bind the chromosome to the spindle apparatus.

Kingdom. The largest formal taxonomic category.

Krebs cycle. See **citric acid cycle.**

K selection. Natural selection that favors long-lived, late-maturing individuals that devote a small fraction of their resources for reproduction; tree species are K strategists.

Lamella (plural, **lamellae**) (Gr. *lamin*, a thin blade). Parallel, closely spaced cellular membranes, as in thylakoids and cisternae.

Lamina (L. *lamina*, a thin plate). Blade or expanded part of a leaf.

Lateral bud. A bud that grows out of the side of a stem.

Laterite (L. *later*, a brick). See **oxisol.**

Late wood. The portion of an annual ring that formed during summer and fall, characterized by cells with a small diameter and thick walls; also called **summer wood.**

Latex (L. *latex*, juice). A milky secretion.

Leach (O.E. *leccan*, to moisten). To extract a soluble or movable substance (as ions or clay particles or bits of organic matter) by causing water to filter down through a material (as a soil horizon).

Leaf. Lateral organ formed by a shoot apical meristem, containing vascular tissue; usually photosynthetic; distinguished from a stem by the absence of a persistent meristem at its tip, and the absence of nodes, internodes, and buds along its axis.

Leaf axil. The point where the upper surface of the base of a leaf meets the stem.

Leaflet. Separate part of the blade of a compound leaf.

Leaf primordium (L. *primordium*, a beginning). A lateral outgrowth from the apical meristem, which will become a leaf.

Leaf scar. Characteristic scar on stem axis made after leaf abscission.

Legume (L. *legumen*, any leguminous plant, particularly bean). A simple, dry dehiscent fruit with one carpel, splitting along two sutures.

Lemma (Gr. *lemma*, a husk). Lower bract that subtends a grass flower but is located above the glumes.

Lenticel (M.L. *lenticella*, a small lens). A structure of the bark that permits the passage of gas inward and outward.

Leucoplast (Gr. *leukos*, white + *plastos*, formed). A colorless plastid.

Liana (F. *liane* from *lier*, to bind). A plant that climbs upon other plants, depending upon them for mechanical support; a plant with climbing shoots.

Lichen (Gr. *leichen*, thallus plants growing on rocks and trees). A composite plant consisting of a fungus living symbiotically with an alga.

Lignification (L. *lignum*, wood + *facere*, to make). Formation of lignin in the spaces between the cellulose microfibrils of a cell wall.

Lignin (L. *lignum*, wood). A polymeric substance, formed by linking together molecules derived from coniferyl alcohol.

Ligule (L. *ligula*, dim. of *lingua*, tongue). In grass leaves, an outgrowth from the upper and inner side of the leaf blade where it joins the sheath.

Line (L. *linea*, line). A taxonomic group and all of its ancestors.

Linkage. The grouping of genes on the same chromosome.

Linked characters. Characters of a plant or animal controlled by genes grouped together on the same chromosome.

Lipase (Gr. *lipos*, fat + *-ase*, suffix indicating an enzyme). Any enzyme that breaks fats into glycerin and fatty acids.

Lipids (Gr. *lipos*, fat + L. *ides*, suffix meaning son of; now used in the sense of having the quality of). An artificial grouping of compounds consisting of substances that are insoluble in water and soluble in fat solvents.

Liverwort (liver + M.E. *wort*, a plant, literally, a liver plant, so named in medieval times because of its fancied resemblance to the lobes of the liver). Common name for the class Hepaticae of the Bryophyta.

Loam (O.E. *lam* or Old Teutonic *lai*, to be sticky, clayey). A particular soil texture class, referring to a soil having 30–50% sand, 30–40% silt, and 10–25% clay.

Lobed leaf (Gr. *lobos*, lower part of the ear). A leaf divided by clefts or sinuses.

Locule (L. *loculus*, dim. of *locus*, a place). A cavity of the ovary in which ovules occur.

Lodicules (L. *lodicula*, a small coverlet). Two scalelike structures that lie at the base of the ovary of a grass flower.

Lower vascular plants. Vascular plants that reproduce by means of spores rather than seeds; the Psilophyta, Lycophyta, Sphenophyta, and Pterophyta.

Lumen (L. *lumen*, light, an opening for light). The cavity of the cell within the cell walls.

Lysis (Gr. *lysis*, a loosening). A process of disintegration or cell destruction.

Macroenvironment (Gr. *makros*, large + O.F. *environ*, about). The environment due to the general, regional climate; traditionally measured some 1.5 m above the ground and away from large obstructions.

Macronutrient (Gr. *makros*, large + L. *nutrire*, to nourish). An essential element required by plants in relatively large quantities.

Mating type. A term used to indicate the mating compatibilities in lower organisms that do not show a male–female sex differentiation. Individuals must be of different mating types (usually designated + and −) to cross successfully.

Matrix (L. *mater*, mother). Surrounding substance; also, the fluid contained within a mitochondrion.

Medulla (L. *medulla*, marrow). The filamentous center of certain lichens and kelp blades and stipes.

Mega- (Gr. *megas*, great). A prefix indicating large size.

Megaphyll (mega + *phyllon*, leaf). A leaf that is joined to the vascular system of the stem by more than one vein, or leaf trace; a leaf gap usually occurs in the vascular tissue of the stem above the point where the leaf is attached.

Megasporangiate cone. In gymnosperms, a cone that produces megaspores and, ultimately, seeds; synonyms include female cone, ovulate cone, seed cone.

Megasporangium (Gr. *megas*, large + sporangium). Sporangium that bears megaspores.

Megaspore (mega + *spora*, seed). A spore that gives rise to a female gametophyte.

Megasporocyte. A diploid cell in which meiosis occurs, forming four megaspores. Synonymous with **megaspore mother cell.**

Megasporophyll (Gr. *megas*, large + spore + *phyllon*, leaf). A leaf bearing megasporangia.

Meiocyte (meiosis + Gr. *kytos*, currently meaning a cell). Any cell in which meiosis occurs.

Meiosis (Gr. *meioun*, to make smaller). A type of nuclear division in which four cells are produced, each with half the chromosome number of the original cell. Two rounds of cell division, meiosis I and meiosis II, are involved. Also called **reduction division.**

Meiospore (meiosis + spore). Any spore resulting from the meiotic divisions.

Membrane (L. *membrana*, skin covering an organ or member). An organized sheet of material, composed of protein and lipid, that encloses a protoplast or an organelle.

Meristem (Gr. *meristos*, divisible). Undifferentiated tissue, the cells of which are capable of active cell division and differentiation into specialized tissues.

Meristoderm (meristem + epidermis). The outer meristematic cell layer (epidermis) of some Phaeophyta.

Mesocarp (Gr. *mesos*, middle + *karpos*, fruit). Middle layer of fruit wall (pericarp).

Mesophyll (Gr. *mesos*, middle + *phyllon*, leaf). Parenchyma tissue of leaf between epidermal layers.

Mesophyte (Gr. *mesos*, middle + *phyton*, plant). A plant that normally grows in moist habitats.

Mesozoic (Gr. *mesos*, middle + *zoe*, life). A geologic era beginning 225 million years ago and ending 65 million years ago.

Messenger RNA. A type of RNA that carries information for making specific proteins.

Metabolism (M.L. from Gr. *metabolos*, to change). The overall set of chemical reactions occurring in an organism or cell.

Metabolite (Gr. *metabolos*, changeable + *ites*, one of a group). A chemical that is a normal cell constituent capable of entering into the biochemical transformations within living cells.

Metaphase (Gr. *meta*, after + *phasis*, appearance). Stage of nuclear division at which the centromeres of the chromosomes lie in the central plane or equator of the spindle.

Metaxylem. Last formed primary xylem.

Micro- (Gr. *mikros*, small). A prefix meaning "small"; in the metric system, the prefix indicates 1/1,000,000th of a unit.

Microbody (micro- + body). A cellular organelle, always bound by a single membrane, frequently spherical, from 20 to 60 nm in diameter, containing a variety of enzymes.

Microcapillary space. Exceedingly small spaces, such as those found between microfibrils of cellulose.

Microenvironment (micro- + O.F. *environ*, about). The environment close enough to the surface of a living or nonliving object to be influenced by it.

Microfibrils (micro- + Gr. *fibrils*, dim. of fiber). Very small fibers of the cell wall.

Microfilament (micro- + L. *filum*, a thread). A proteinaceous filament 4–7 nm in diameter involved in moving particles in the cytoplasm.

Microfossil (micro- + L. *fossilis*, dug up). Fossils of microscopic organisms, visible only when thin sections of rock are examined.

Micrometer (micro- + Gr. *metron*, measure). One millionth (10^{-6}) of a meter, or 0.001 mm; also called a micron and abbreviation μm.

Micronutrient (micro- + L. *nutrire*, to nourish). An essential element required by plants in relatively small quantities.

Microphyll (micro- + Gr. *phyllon*, leaf). A leaf that is connected to the vascular system of the stem by a single vein, or leaf trace; no leaf gap occurs.

Micropylar chamber. The space between the micropyle and the nucellus; sealed off from the outside when the micropyle closes after pollination.

Micropyle (micro- + Gr. *pulon*, orifice, gate). A pore leading from the outer surface of the ovule between the edges of the integuments down to the surface of the nucellus.

Microsporangiate cone. In gymnosperms, a cone that produces microspores and, ultimately, pollen; synonyms include male cone, pollen cone, and staminate cone.

Microsporangium (plural, **microsporangia**) (micro- + sporangium). A sporangium that bears microspores.

Microspore (micro- + Gr. *spora*, seed). A spore that gives rise to a male gametophyte.

Microsporocyte (micro- + spore + L. *cyta*, vessel). A diploid cell in which meiosis occurs, resulting in four microspores; synonymous with **microspore mother cell.**

Microsporophyll (micro- + spore + *phyllon*, leaf). A leaf bearing microsporangia.

Microtubule organizing center (MTOC). A small region in the cell that can initiate the assembly of protein molecules into microtubules. MTOCs occur at the spindle poles, kinetochores, phragmoplast, and adjacent to the plasma membrane.

Middle lamella (L. *lamella*, a thin plate or scale). Thin layer separating two adjacent protoplasts and remaining as a distinct cementing layer between adjacent cell walls.

Milli- (L. *mille*, thousand). 1/1000th of a unit.

Millimeter. 1/1000th of a meter; 0.0394 in.

Mineral. A substance formed in nature by nonbiological processes.

Mitochondrion (plural, **mitochondria**) (Gr. *mitos*, thread + *chondrion*, a grain). A membrane-bounded organelle associated with intracellular respiration.

Mitosis (plural, **mitoses**) (Gr. *mitos*, a thread). Nuclear division, involving coiling and equal distribution of duplicate chromosomes into derivative nuclei; the chromosome number remains constant. Compare with **meiosis.**

Mitospore (mitosis + spore). A spore forming after mitosis.

Mixed bud. A bud containing both rudimentary leaves and flowers.

Molecular biology. A field of biology that emphasizes the interaction of biochemistry and genetics in the life of an organism.

Molecule (F. *môle* + *clue*, a dim., literally, a little mass). A unit of matter, the smallest portion of an element or a compound that retains chemical identity with the substance in mass; the molecule usually consists of a union of two or more atoms, some organic molecules containing a very large number of atoms.

Mollisol (L. *mollis*, soft + *solum*, soil, solid). One of the 10 world soil orders, characterized by containing more than 1% organic matter in the top 17.5 cm and associated with grassland vegetation; synonymous with **chernozem.**

Mono- (Gr. *monos*, solitary). A prefix meaning "single."

Monocotyledon (mono- + *kotyledon,* a cup-shaped hollow). A plant whose embryo has one cotyledon.

Monocot. A common term for a **monocotyledon.**

Monoecious (mono- + Gr. *oîkos,* house). Having male and female reproductive structures on the same individual, but in different organs. Monoecious flowering plants have stamens and carpels in separate flowers on the same plant.

Monophyletic (mono- + Gr. *phyle,* tribe). Having a single evolutionary origin. Refers to a taxonomic group in which all the members derive from one ancestral line that is also included in the group. Compare with **polyphyletic.**

Morphogenesis (Gr. *morphe,* form + L. *genitus,* to produce). The changes in body form that occur during development of an organism.

Morphology (Gr. *morphe,* form + *logos,* discourse). The study of form and its development. Also used loosely in referring to the form of an organism.

Moss (L. *muscus,* moss). A plant of the class Musci, in the division Bryophyta.

mRNA. A synonym for **messenger RNA.**

Multi- (L. *multus,* many). A prefix meaning "many."

Multiple fruit. (A cluster of matured ovaries produced by separate flowers, e.g., a pineapple.

Mutation (L. *mutare,* to change). A change in the DNA of a cell, resulting from chromosome breakage or chemical modification of one or more bases; the event that gives rise to a new allele.

Mutualism (L. *mutuus,* reciprocal). An interaction with advantage to all organisms involved.

Mycelium (Gr. *mykes,* mushroom). The vegetative body of a true fungus, composed of branching hyphae.

Myco- (Gr. *mykes,* fungi). A prefix meaning "fungal."

Mycology (myco- + *logos,* discourse). The branch of botany dealing with fungi.

Mycorrhiza (myco- + *riza,* root). A symbiotic association between a fungus and usually the root of a higher plant.

NAD, NAD⁺. Nicotinamide adenine dinucleotide, a coenzyme that can accept two electrons and a proton; an important electron/hydrogen carrier in catabolism.

NADH. Reduced NAD, which differs from NAD in having two additional electrons and one additional proton.

NADP, NADP⁺. Nicotinamide adenine dinucleotide phosphate, a coenzyme resembling NAD but with an additional phosphate group; an important electron/hydrogen carrier in photosynthesis and anabolism.

NADPH. Reduced NADP. See NADH.

Naked bud. A bud not protected by bud scales.

Nano-. A prefix meaning 1/1,000,000,000th of a unit.

Nanometer. 1/1,000,000,000th of a meter.

Natural classification. A classification scheme that is based on the phylogenetic nature of the organisms classified; contrasts with an artificial classification, which separates organisms on the basis of convenient traits but fails to show the evolutionary relationships among the organisms.

Natural selection. The effect of the environment on the reproductive success of variant phenotypes within a species, allowing some phenotypes to reproduce more abundantly than others.

Nectar (Gr. *nektar,* drink of the gods). A fluid rich in sugars secreted by nectaries, which are often located near or in flowers.

Nectar guide. A mark of contrasting color or texture that guides pollinators to nectaries in a flower.

Nectary (Gr. *nektar,* the drink of the gods). A nectar-secreting gland.

Net productivity. The arithmetic difference between calories produced in photosynthesis and calories lost in respiration.

Net radiation. The arithmetic difference between incoming solar radiation and outgoing terrestrial radiation.

Net venation. Veins of leaf blade visible to the unaided eye, branching frequently and joining again, forming a network.

Neutron (L. *neuter,* neither). An uncharged particle found in the atomic nucleus of all elements except hydrogen.

Niche (It. *nicchia,* a recess in a wall). The functional relationship of an organism to its ecosystem.

Nitrification (L. *nitrum,* nitro, a combining form indicating the presence of nitrogen + *facere,* to make). Change of ammonium salts into nitrates through the activities of certain bacteria.

Nitrogen fixation. The process of reducing the nitrogen of N_2, in which N becomes bound to C and H.

nm. See **nanometer.**

n + n. See **dikaryon.**

Node (L. *nodus,* a knot). The point on the stem where leaves, axillary buds, and branches arise; often marked by a slight swelling.

Nonseptate. Descriptive of hyphae or algal filaments lacking cross-walls.

Nucellus (L. *nucella,* small nut). Tissue composing the chief part of the young ovule, in which the embryo sac develops; megasporangium.

Nucleic acid. A polymeric molecule consisting of subunits called **nucleotides,** linked together in a chain.

Nucleolus (L. *nucleolus,* a small nucleus). A dense body in the nucleus, containing parts of ribosomes plus the DNA that codes for the production of those parts.

Nucleoside. A component of nucleic acids, consisting of a sugar (ribose or deoxyribose) linked to a base (adenine, cytosine, guanine, thymine, uracil, or their derivatives).

Nucleotide. A nucleoside to which a phosphate unit is attached.

Nucleus (L. *nucleus*, kernel of a nut). A membrane-bound organelle that contains most of the DNA in eukaryotic cells.

Numerical taxonomy. A field of taxonomy that does not place subjective weight on any particular type of evidence that shows relationships between taxa.

Nut (L. *nux*, nut). A dry, indehiscent, hard, one-seeded fruit, generally produced from a compound ovary.

Obligate (L. *obligare*, to oblige). Requiring certain conditions for survival. Obligate anaerobes cannot survive in the presence of oxygen; obligate parasites can only live as parasites.

-oid (Gr. *oides*, like). A suffix meaning ''like.''

Oogamy (Gr. *oion*, egg + gamete). Sexual reproduction involving eggs and sperm. Contrast **isogamy, anisogamy,** and **heterogamy.**

Oogonium (L. dim. of Gr. *oogonos*, a little egg layer). Egg-producing gametangium that is not surrounded by a jacket of sterile cells.

Oospore (Gr. *oion*, an egg + spore). A resistant spore that develops from a zygote, resulting from the fusion of heterogametes.

Open bundle. A vascular bundle with residual procambium.

Operculum (L. *operculum*, a lid). In mosses, the cap of a sporangium.

Opposite. Referring to a leaf arrangement in which there are two leaves opposite each other at a node.

Order. A taxonomic category below the class and above the family.

Organ (L. *organum*, an instrument or engine of any kind). A part or member of a plant body adapted to a particular function.

Organelle. An organized unit within the cell, composed of many molecules and performing a specific function.

Organic. Referring to compounds that contain both carbon and hydrogen; also generally referring to the material products of living organisms.

Organic evolution. See **evolution.**

Organism. An individual living body.

-ose. A chemical suffix indicating a carbohydrate.

Osmosis (Gr. *osmos*, a pushing). Diffusion of water through a selectively permeable membrane.

Ovary (L. *ovum*, an egg). Enlarged basal portion of the pistil, which becomes the fruit.

Ovulate. Referring to a cone, scale, or other structure bearing ovules.

Ovulate cone. See **megasporangiate cone.**

Ovule (F. *ovule*, from L. *ovulum*, dim. of *ovum*, egg). The structure in seed plants that contains the egg, consisting of the female gametophyte with egg cell, and one or two integuments; after fertilization, develops into the seed.

Ovuliferous (ovule + L. *ferre*, to bear). Referring to a scale or sporophyll bearing ovules.

Oxidation. The removal of electrons or hydrogen, or the addition of oxygen to a compound.

Oxisol (Gr. *oxus*, sharp, sour + L. *solum*, soil, solid). One of the 10 world soil orders, characterized by a soil horizon at least 30 cm thick that is highly weathered, rich in clay, but low in nutrients, red in color, and associated with tropical vegetation, synonymous with **laterite.**

P_{fr} and P_r. abbreviations for the far-red (FR) or red (R) absorbing forms of phytochrome (P).

Palea (L. *palea*, chaff). Upper bract that subtends a grass flower.

Paleo- (Gr. *palaios*, ancient). Prefix meaning ''ancient.''

Paleobotany. The study of ancient plants, based on fossil evidence.

Paleoecology (Gr. *palaios*, ancient). A field of ecology that reconstructs past vegetation and climate from fossil evidence.

Paleozoic (paleo- + *zoe*, life). A geologic era beginning 570 million years ago and ending 225 million years ago.

Palisade parenchyma. Elongated cells, containing many chloroplasts, found just beneath the upper epidermis of leaves.

Palmate (L. *palma*, palm of the hand). Referring to an arrangement in which similar parts arise from a common point.

Palmate venation. A pattern of leaf venation in which several principal veins spread out from the upper end of the petiole.

Panicle (L. *panicula*, a tuft). An inflorescence, the main axis of which is branched, and whose branches bear loose racemose flower clusters.

Pappus (L. *pappus*, woolly, hairy seed or fruit of certain plants). Scales or bristles representing a reduced calyx in composite flowers.

Parallel venation. Type of venation in which veins of a leaf blade that are clearly visible to the unaided eye are parallel to each other.

Paraphysis (plural, **paraphyses**) (Gr. *para*, beside + *physis*, growth). A sterile hair or hypha growing beside reproductive cells or hyphae in the reproductive structures of certain fungi, algae, and bryophytes.

Parasitism. An ecological relationship in which one organism derives nourishment from a living organism of another species without benefitting the other organism.

Parenchyma (Gr. *parenkhein*, to pour in beside). A tissue composed of living cells with thin primary walls and no secondary walls; usually with large vacuoles.

Parietal (F. *pariétal*, attached to the wall, from L. *paries*, wall). Belonging to, connected with, or attached to the wall of a hollow organ or structure, especially of the ovary or cell.

Parietal placentation. A type of placentation in which placentae are on the ovary wall.

Parthenocarpy (Gr. *parthenos*, virgin + *karpos*, fruit). The development of fruit without fertilization.

Parthenogenesis (Gr. *parthenos*, virgin + *genesis*, origin). The development of a gamete into a new individual without fertilization.

Passive solute absorption. Absorption due only to forces of simple diffusion.

Pathogen (Gr. *pathos*, suffering + *genesis*, beginning). An organism that causes a disease.

Pathology (Gr. *pathos*, suffering + *logos*, account). The study of diseases, their effects on plants or animals, and their treatment.

Pathway. A sequence of chemical reactions, each governed by an enzyme, that gradually transform a starting molecule into some final product.

Peat (M.E. *pete*, of Celtic origin, a piece of turf used as fuel). Any mass of semicarbonized vegetable tissue, such as *Sphagnum*, formed by partial decomposition in water.

Pectin (Gr. *pektos*, congealed). A class of polymers of the cell wall that are built chiefly of partially oxidized sugars.

Pedicel (L. *pediculus*, a little foot). Stalk or stem of the individual flowers of an inflorescence.

Peduncle (L. *pedunculus*, a late form of *pediculus*, a little foot). Stalk or stem of a flower that is borne singly; or the main stem of an inflorescence.

Penicillin. An antibiotic derived from the mold *Penicillium*.

Pentose (Gr. *pente*, five + -ose). A five-carbon sugar.

Perennial (L. *perennis*, lasting the whole year through). A plant that lives from year to year.

Perfect flower. A flower having both stamens and pistils.

Peri- (Gr. *peri*, around). A prefix meaning "around."

Perianth (peri- + Gr. *anthos*, flower). The petals and sepals taken together.

Pericarp (peri- + Gr. *karpos*, fruit). Fruit wall, developed from ovary wall.

Periclinal division. Cell division in which the newly formed cell wall is parallel to the nearest surface of the organ.

Pericycle (peri- + Gr. *kyklos*, circle). A meristematic tissue found in roots and sometimes other organs, that is bounded externally by the endodermis and internally by the vascular tissue. In roots, pericycle gives rise to branch roots and parts of the vascular cambium and cork cambium.

Periderm (peri- + Gr. *derma*, skin). Protective tissue that replaces the epidermis after secondary growth is initiated. Consists of cork, cork cambium, and phelloderm.

Peridium (plural, **peridia**) (Gr. *peridion*, a little pouch). External covering of the hymenium of certain fungi; in Myxomycetes, the hardened envelope that covers the sporangium.

Perigyny (peri- + Gr. *gyne*, a female). A condition in which the receptacle is more or less concave, at the margin of which the sepals, petals, and stamens have their origin, so that these parts seem to be attached around the ovary; also called half-inferior.

Perisperm. Nutritive tissue in some seeds that forms from the nucellus.

Peristome (peri- + Gr. *stoma*, a mouth). In mosses, a fringe of teeth about the opening of the sporangium.

Perithecium (peri- + Gr. *theke*, a box). A spherical or flask-shaped ascocarp having a small opening.

Permafrost (L. *permanere*, to remain + A.S. *freosan*, to freeze). Soil that is permanently frozen; usually found some distance below a surface layer that thaws during warm weather.

Permeable (L. *permeabilis*, that which can be penetrated). Said of a membrane, cell, or cell system through which substances may diffuse.

Peroxysome. An organelle of the microbody class that contains enzymes capable of destroying peroxides.

Petals (Gr. *petalon*, a leaf or plate). Whorl of flower parts, usually colored, encircling the androecium.

Phase (Gr. *phainein*, to show). General term for any subdivision of a process, sequence, or system.

Petiole (L. *petiolus*, a little foot or leg). A stalk of a leaf.

Phelloderm (Gr. *phellos*, cork + *derma*, skin). A layer of cells formed in the stems of some plants from the inner cells of the cork cambium.

Phellogen (Gr. *phellos*, cork + *genesis*, birth). Cork cambium, a cambium giving rise externally to cork and in some plants internally to phelloderm.

Phenotype (Gr. *phaneros*, showing + type). The bodily characteristics of an organism.

Phloem (Gr. *phloos*, bark). Food-conducting tissue, consisting of sieve tube members or sieve cells, companion cells, parenchyma, and fibers.

Phosphorylation. A reaction in which phosphate is added to a compound, e.g., the formation of ATP from ADP and inorganic phosphate.

Phospholipid. A molecule found in cellular membranes, made of a glycerol molecule to which two fatty acids and a phosphate-containing group are linked.

Photo- (Gr. *photos*, light). Prefix meaning "light."

Photon. A particle of light, having zero mass, carrying a specific quantity (a quantum) of energy, associated with a specific wavelength.

Photoperiodism. A response in which developmental events are timed by reference to the lengths of days or nights.

Photophosphorylation. A reaction in which light energy is converted into chemical energy in the form of ATP produced from ADP and inorganic phosphate.

Photoreceptor (photo- + L. *receptor*, a receiver). A light-absorbing molecule involved in converting light into some metabolic (chemical energy) form, e.g., chlorophyll and phytochrome.

Photosynthesis (photo- + Gr. *syn*, together + *tithenai*, to place). A process in which light energy is used to drive the formation of organic compounds.

Phototropism (photo- + Gr. *tropos*, a turning). Influence of light on the direction of plant growth.

Phyco- (Gr. *phykos*, seaweed). A prefix referring to algae.

Phycobilins. Red and blue accessory pigments characteristic of red algae (Rhodophyta) and blue-green algae (Cyanobacteria).

Phycocyanin (phyco- + Gr. *kyanos*, blue). A blue phycobilin pigment.

Phycoerythrin (phyco- + Gr. *erythros*, red). A red phycobilin pigment.

Phylogenetic classification. See **natural classification.**

Phylogeny (Gr. *phylon*, race or tribe + *genesis*, beginning). The evolution of a group of related individuals.

Phylum. See **division.**

Physiology (Gr. *physis*, nature + *logos*, discourse). The science of the functions and activities of living organisms.

Phytobenthon (Gr. *phyton*, a plant + *benthos*, depths of the sea). Attached aquatic plants, collectively.

Phytochrome. A pigment found in green plants; it is associated with the control of development in response to light stimuli.

Phytoplankton (Gr. *phyton*, a plant + *planktos*, wandering). Free-floating plants, collectively.

Pigment. A substance that absorbs visible light.

Pileus (L. *pileus*, a cap). Umbrella-shaped cap of fleshy fungi.

Pinna (plural **pinnae**) (L. *pinna*, a feather). Leaflet or division of a compound leaf (frond).

Pinnate (L. *pinna*, a feather). Having similar parts arranged in sequence along both sides of an axis. Compare with **palmate.**

Pinnate venation. A pattern of leaf veins in which lateral veins depart at regular intervals from a midrib.

Pioneer community. The first stage of a succession.

Pistil (L. *pistillum*, a pestle). Seed-bearing organ of a flower, consisting of a single carpel or two or more carpels fused together; typically subdivided into ovary, style, and stigma.

Pistillate flower. A flower having pistils but no stamens.

Pit. A thin area of a secondary cell wall.

Pith. The parenchymatous tissue occupying the central portion of a stem.

Placenta (plural, **placentae**) (L. *placenta*, a cake). The tissue within the ovary to which the ovules are attached.

Placentation (L. *placenta*, a cake + *-tion*, state of). Manner in which the placentae are distributed in the ovary.

Plankton (Gr. *planktos*, wandering). Free-floating aquatic plants and animals, collectively; generally microscopic.

Plasmalemma (Gr. *plasma*, anything formed + *lemma*, a husk of a fruit). A synonym for **plasma membrane.**

Plasma membrane. The membrane that separates the living protoplast from the external environment; found in all cells.

Plasmid. A small, circular unit of DNA that carries genes; distinct from the chromosomes.

Plasmodesma (plural, **plasmodesmata**) (Gr. *plasma*, something formed + *desmos*, a bond, a band). Fine protoplasmic thread passing through the wall that separates two protoplasts.

Plasmodium (Gr. *plasma*, something formed + mod. L. *odium*, something of the nature of). In myxomycetes, a coenocytic mass of protoplasm, with no surrounding wall.

Plasmogamy (Gr. *plasma*, anything molded or formed + *gamos*, marriage). The fusion of protoplasts, not accompanied by nuclear fusion.

Plasmolysis (Gr. *plasma*, something formed + *lysis*, a loosening). The separation of the cytoplasm from the cell wall due to removal of water from the protoplast.

Plastid (Gr. *plastis*, a builder). A class of organelles, including the chloroplast and several related kinds of bodies; the latter kinds are associated with storage of food materials and some of them (chromoplasts) are highly pigmented.

Plumule (L. *plumula*, a small feather). The first bud of an embryo or that portion of the young shoot above the cotyledons.

Podzol (R. *pod*, ground + *zola*, ashes). A soil type; see **spodosol.**

Polar (Gr. *pol*, an axis). *Chemistry:* Having distinct regions of negative and positive electrical charge. *Physiology:* Differing in activity along an axis.

Polar transport. The directed movement within plants of compounds (usually hormones) predominantly in one direction; polar transport overcomes the tendency for diffusion in all directions.

Pollen (L. *pollen*, fine flour). The germinated microspores or partially developed male gametophytes of seed plants.

Pollen mother cell. See **microsporocyte.**

Pollen tube. The male gametophyte of seed plants, which grows through the nucellus of a gymnosperm ovule or through the pistil of an angiosperm.

Pollination. The transfer of pollen from a stamen or staminate cone to a stigma or ovulate cone.

Pollinium (L. *pollentis*, powerful, or *pollinis*, fine flour + *ium*, group). A mass of pollen that sticks together and is transported by pollinators as a mass; present in orchids and milkweeds.

Pollutant (L. *polluere*, to dirty + *lutum*, mud). An unnatural, human-related substance that is introduced to the environment; may be gas, liquid, or solid, easily broken down or long-lasting.

Poly- (Gr. *polys*, many). A prefix meaning "many."

Polymer. A molecule that is made by coupling together many small molecules (monomers) that are similar to one another.

Polymerization. The chemical union of monomers to produce a polymer.

Polypeptide. See **protein.**

Polyphyletic (poly- + Gr. *phyle*, tribe). Of many origins. Said of a taxonomic group containing organisms from two or more lines of descent, there being no common ancestor within the group.

Polyploid (poly- + Gr. *ploos*, fold). Having more than two sets of chromosomes per cell.

Polysaccharide (poly- + Gr. *sakkharon*, sugar). A polymer in which the subunits are sugars.

Polysome. Several ribosomes that are translating the same mRNA molecule, hence are linked together in a chain. Also called a **polyribosome.**

Pome (Gr. *pomme*, apple). A simple fleshy fruit, the outer portion of which is formed by the floral parts that surround the ovary.

Population (L. *populus*, people). A group of closely related, interbreeding organisms.

Pore spaces. Spaces between soil particles that may be filled with air or water and into which root hairs may penetrate; the larger the soil particles, the larger the pore space.

Prairie (L. *pratum*, meadow). Grassland vegetation, with trees essentially absent; often considered to have more rainfall than does the steppe.

Predation (L. *predatio*, plundering). A form of biological interaction in which one organism is destroyed (by ingestion); parasitism is a form of predation.

Pre-prophase band. A band of microtubules that forms near the plasma membrane before mitosis and that positions the plane of cell division.

Primary endosperm cell. A cell of the embryo sac after fertilization, generally containing a nucleus resulting from fusion of the two polar nuclei with a sperm nucleus; the endosperm develops from this cell.

Primary meristems. Meristems of the shoot or root tip giving rise to the tissues of the primary plant body.

Primary pitfield. Thin area in a primary cell wall, traversed by many plasmodesmata.

Primary thickening meristem. A meristem at the shoot tip of monocot plants that contributes to elongation and lateral growth.

Primary tissues. Tissues that derived from primary meristems, including epidermis, vascular, and ground tissues.

Primitive (L. *primus*, first). Referring to a taxonomic trait thought to have evolved early in time.

Primordium (L. *primus*, first + *ordiri*, to begin to weave; literally beginning to weave, or to put things in order). The beginning or origin of any part of an organ.

Pro- (Gr. *pro*, before, forward). A prefix meaning "first."

Procambium. A primary meristem that gives rise to primary vascular tissues and, in most woody plants, to the vascular cambium.

Procaryote. See **prokaryote.**

Producers. Organisms that form organic matter by means of photosynthesis, thus forming the base of the ecological food chain.

Proembryo (pro- + L. *embryon*, embryo). A group of cells arising from the division of the fertilized egg cell before the cells that are to become the embryo are recognizable.

Prokaryote (pro- + Gr. *karyon*, a nut, referring in modern biology to the nucleus). An organism in which the DNA is not separated from the cytoplasm by an envelope. The Cyanobacteria and other bacteria are prokaryotes.

Prophase (pro- + Gr. *phasis*, appearance). An early stage in nuclear division, characterized by coiling of the chromosomes and formation of the mitotic spindle.

Proplastid. An undifferentiated plastid, which can develop into any one of the specialized types of plastids (such as the chloroplast and chromoplast).

Protease. An enzyme that breaks down proteins.

Protein (Gr. *proteios*, primary). A polymer in which the subunits are amino acids, joined by peptide bonds. Also called a **polypeptide.**

Proterozoic (Gr. *protero*, before in time + *zoe*, life). The earliest geologic era, beginning about 4.5–5 billion years ago and ending 570 million years ago; also called the Precambrian era.

Prothallial cells. Sterile cells of the male gametophyte in some lower plants, considered to be remnants of a vegetative body.

Prothallus. The gametophyte of a fern. Also called a **prothallium.**

Proto- (Gr. *protos*, first). A prefix signifying earliness in origin or rank.

Protochlorophyll (proto- + Gr. *chloros*, green + *phyllos*, leaf). One of the precursors of chlorophyll.

Protoderm (proto- + Gr. *derma*, skin). A primary meristem that gives rise to epidermis.

Proton. An elementary particle that bears a unit of positive electrical charge, found in the nucleus of the atom. The nucleus of a hydrogen atom, symbolized H$^+$, is a single proton.

Protonema (plural, **protonemata**) (proto- + Gr. *nema*, a thread). A filamentous stage in the development of the gametophyte of a moss.

Protoplasm (proto- + Gr. *plasma*, something formed). The substance of the living cell, excluding the cell wall and the contents of the vacuole.

Protoplast. In cell culture, a living cell that has been liberated from its wall. The term is used more generally as a synonym for "cell" when there is a need to emphasize the distinction between a cell and its wall; e.g., "solutes can move either within the protoplast, or within the wall."

Protoxylem. First-formed primary xylem.

Pseudopodium (Gr. *pseudes*, false + *podion*, a foot). In myxomycetes, an armlike projection from the body by which the plant creeps over the surface.

Purines. A class of compounds, some members of which (e.g., adenine and guanine) serve as bases in nucleotides.

Pyrenoid (Gr. *pyren*, the stone of a fruit + L. *oïdes*, like). A denser body occurring within the chloroplasts of certain algae and liverworts and apparently associated with starch deposition.

Pyrimidines. A class of compounds, some members of which (cytosine, thymine, uracil) serve as bases in nucleotides.

Quantum (L. *quantum*, how much). An elemental unit of energy.

Quiescent center (L. *quiescere*, to rest). Disk-shaped region of root apex containing slowly dividing cells.

Raceme (L. *racemus*, a bunch of grapes). An inflorescence in which the main axis is elongated and the flowers are born on pedicels that are about equal in length.

Rachilla (Gr. *rhachis*, a backbone + L. dim, ending *-illa*). Shortened axis of spikelet.

Rachis (Gr. *rhachis*, a backbone). Main axis of spike; axis of fern leaf (frond) from which pinnae arise; in compound leaves, the extension of the petiole corresponding to the midrib of an entire leaf.

Radicle (L. *radix*, root). Portion of the plant embryo that develops into the primary or seed root.

Ramet (L. *ramus*, a branch). A plant produced by asexual reproduction, hence one member of a genet or clone.

Random plant distribution. A distribution of a plant species within an area such that the probability of finding an individual at one point is the same for all points.

Raphe (Gr. *rhaphe*, seam). Ridge on seeds, formed by the stalk of the ovule, in those seeds in which the funiculus is sharply bent at the base of the ovule.

Raphides (Gr. *rhaphis*, a needle). Fine, sharp, needlelike crystals.

Ray initials. Meristematic cells in the vascular cambium that develop into xylem and phloem cells comprising the ray system.

Ray system (L. *radius*, a beam or ray). System of cells in secondary tissues that are oriented perpendicular to the long axis of the stem, form from ray initials of the vascular cambium.

Receptacle (L. *receptaculum*, a reservoir). Enlarged end of the pedicel or peduncle to which other flower parts are attached.

Recombination (L. *re*, repeatedly + *combinatus*, joined). The mixing of genotypes that results from sexual reproduction.

Red tide. The coloring of near-shore, marine water by a high density of microscopic algal organisms that may additionally release toxic by-products into the water.

Reduction. (F. *reduction*, L. *reductio*, a bringing back). Any chemical reaction involving the removal of oxygen from or the addition of hydrogen or an electron to a substance.

Reduction division. See **meiosis.**

Regular flower. An actinomorphic flower.

Reproductive isolation. The separation of populations in time or space so that genetic flow between them is cut off.

Residual meristem. Cylindrical region near the shoot tip in which the cells, still resembling those of the apex, are made visible by the differentiation of nearby pith and cortex.

Resin duct. A tube lined with resin-secreting cells, containing resins; occurs in the leaves and wood of conifers.

Respiration (L. *re*, repeatedly, + *spirare*, to breathe). In the cell, the catabolic process by which sugars and other fuels are oxidized and broken down, with some of the energy captured in the formation of ATP.

Rhizoid (Gr. *rhiza*, root + -oid). A nonvascular, filamentous structure that performs functions comparable to those of a root (anchorage and absorption).

Rhizome (Gr. *rhiza*, root). An elongated, underground, horizontal stem.

Rhizophores (Gr. *rhiza*, root + *phoros*, bearing). Leafless branches that grow downward from the leafy stems of certain Lycophyta and give rise to roots when they come into contact with the soil.

Ribonucleic acid (RNA). A class of nucleic acid polymer in which the nucleotides contain the sugar ribose.

Ribose. A pentose sugar.

Ribosomes (ribo, from RNA + Gr. *somatos*, body). Small particles 10–20 nm in diameter, containing RNA and protein; active in protein synthesis.

Ring porous. Wood with large xylem vessel members mostly in early wood; compare with **diffuse porous.**

Ripening (A.S. *rifi,* perhaps related to reap). Changes in a fruit that follow seed maturation and that prepare the fruit for its function of seed dispersal.

RNA. See **ribonucleic acid.**

Root (A.S. *rōt*). The descending axis of a plant, normally below ground, serving to anchor the plant and absorb and conduct water and mineral nutrients.

Root cap. A thimblelike mass of living cells covering and protecting the apical meristems of a root; site of perception of gravity in geotropism.

Root hair. A cylindrical outgrowth of a cell of the root epidermis, serving to reach and absorb water in regions lateral to the axis of the root.

Root pressure. Pressure in the xylem arising as a result of osmosis in the root.

Rosette. A shoot with a very short stem, composed of several unelongated internodes with fully expanded leaves.

r selection. Natural selection that favors short-lived, early-maturing individuals that devote a large fraction of their resources to reproduction; annual herbs are r strategists.

Runner. A stem that grows horizontally along the ground surface.

Samara (L. *samara,* the fruit of the elm). Simple, dry, one- or two-seeded indehiscent fruit with pericarp bearing a winglike outgrowth.

Sand. Soil particles between 50 and 2000 μm in diameter.

Saprophyte (Gr. *sapros,* rotten + *phyton,* a plant). An organism deriving its food from the dead body or the nonliving products of another plant or animal.

Sapwood. Peripheral wood that actively transports.

Savannah (Sp. *sabana,* a large plain). Vegetation of scattered trees in a grassland matrix.

Schizocarp (Gr. *schizein,* to split + *karpos,* fruit). Dry fruit with two or more united carpels that split apart at maturity.

Sclereids (Gr. *skleros,* hard). Sclerenchyma cells having variably shaped, heavily lignified cell walls.

Sclerenchyma (Gr. *skleros,* hard + *-enchyma,* a suffix denoting tissue). A strengthening tissue composed of cells with heavily lignified cell walls.

Scrub (A.S. *scrob,* a shrub). Vegetation dominated by shrubs; described as thorn forest in areas with moderate rainfall, or as chaparral or desert in areas with low rainfall.

Scutellum (L. *scutella,* a dim. of *scutum,* shield). Single cotyledon of grass embryo.

Seaweed. Any large, marine alga; usually a red or brown alga; includes the kelps.

Secondary tissues. Those tissues, xylem, phloem, and periderm, that form from secondary meristems.

Secretory structures. Any of a number of specialized plant structures such as nectaries, glands, and hydathodes that secrete secondary plant substances.

Seed. A matured ovule, containing an embryo within a seed coat. Nutritive tissue may also be present within the seed coat.

Seed coat. A hardened, outer layer of the seed, derived from the integument(s) of the ovule, and which functions to prevent mechanical disturbance to and water loss from the embryo; it may also regulate germination in several ways.

Seed plant. Common name for members of the gymnosperm and angiosperm groups and for extinct members of other groups that produced seeds.

Selectively permeable. Referring to a membrane that permits some kinds of molecules to pass through while not allowing other kinds of molecules to pass through.

Self-pollination. Transfer of pollen from the stamens to the stigma of either the same flower or flowers on the same plant.

Seminal root. The root or roots forming from primordia present in the seed.

Sepals (M.L. *sepalum,* a covering). Whorl of sterile, leaflike structures that usually enclose the other flower parts.

Septate (L. *septum,* fence). Divided by cross-walls into cells or compartments.

Septum (L. *septum,* fence). Any dividing wall or partition; frequently a cross-wall in a fungal or algal filament.

Serpentine (L. *serpens,* a serpent). Referring to soil derived from metamorphic parent material characterized (among other things) by low calcium (Ca), high magnesium (Mg), and a greenish-gray color.

Sessile (L. *sessilis,* low, dwarf, from *sedere,* to sit). Sitting, referring to a leaf lacking a petiole or a flower or fruit lacking a pedicel.

Seta (plural, **setae**) (L. *seta,* a bristle). In bryophytes, a short stalk of the sporophyte, which connects the foot and the capsule.

Sexual reproduction. Reproduction that requires meiosis and fertilization for a complete life cycle.

Shade tolerance. The ability to grow slowly in the shade of an overstory canopy; an essential characteristic for a climax species.

Sheath. Part of leaf that wraps around the stem, as in grasses.

Shoot (derivation uncertain). A young stem and its attached leaves.

Shoot tip. Terminal portion of the shoot containing apical and primary meristems and cells in early stages of differentiation.

Sibling species. Species morphologically nearly identical but incapable of producing fertile hybrids.

Side chain. A part of a polymer that extends laterally from the main chain.

Sieve cell. A long and slender sieve element without a companion cell, with relatively unspecialized sieve areas, and with tapering end walls that lack sieve plates; found in gymnosperms and ferns.

Sieve plate. Perforated wall area in a sieve-tube member through which pass strands connecting sieve-tube protoplasts.

Sieve tube. A series of sieve-tube members forming a long cellular tube specialized for the conduction of food materials.

Sieve-tube member. An enucleate phloem cell primarily responsible for photosynthate transport; separated from other sieve-tube members by sieve plates.

Silique (L. *siliqua*, pod). The fruit characteristic of Brassicaceae (mustards); two-celled, the valves splitting from the bottom and leaving the placentae with a partition stretched between.

Silt. Soil particle between 2 and 50 μm in diameter.

Simple pit. Pit not surrounded by an overarching border; in contrast to **bordered pit.**

Single bond. A covalent bond that involves the sharing of two electrons.

Siphonous (Gr. *siphon*, a tube). A synonym for **coenocytic.**

Soil (L. *solum*, soil, solid). The uppermost stratum of the earth's crust, which has been modified by weathering and organic activity into (typically) three horizons: an upper A horizon that is leached, a middle B horizon in which the leached material accumulates, and a lower C horizon that is unweathered parent material.

Soil order. In soil taxonomy, the highest level of classification. There are 10 soil orders in the world.

Soil series. In soil taxonomy, equivalent to the genus. Soil series are named for the locations where they were first identified, as Miami or Mendocino.

Soil texture. Refers to the amounts of sand, silt, and clay in a soil, as a sandy loam, loam, or clay texture.

Soil type. In soil taxonomy, the smallest unit recognized. It has a binomial, the first being the soil series, the second reflecting the texture of the surface horizon, e.g., Miami silt loam or Mendocino sandy loam.

Solute (L. *solutus*, from *solvere*, to loosen). A dissolved substance.

Solution. A homogeneous mixture, the molecules of the dissolved substance (e.g., sugar), the solute, being dispersed between the molecules of the solvent (e.g., water).

Solvent. The most abundant type of molecule in a given solution. Also, a substance that can dissolve other substances.

Soredium (plural, **soredia**) (Gr. *soros*, a heap). A sexual reproductive body of lichens, consisting of a few algal cells surrounded by fungal hyphae.

Sorus (plural, **sori**) (Gr. *soros*, a heap). A cluster of sporangia in ferns.

Species (L. *species*, kind). The taxonomic category below the genus and above the variety; the largest taxonomic group in which the members can freely interbreed; A group of very similar organisms, usually capable of interbreeding, that are related by recent common ancestry.

Sperm (Gr. *sperma*, the generative substance or seed of a male animal). A male gamete.

Spermagonium (plural, **spermagonia**) (Gr. *sperma*, sperm + *gonos*, offspring). Flask-shaped structure characteristic of the sexual phase of the rust fungi; bearing receptive hyphae and spermatia.

Spermatophyte (Gr. *sperma*, seed + *phyton*, plant). A seed plant.

Spike (L. *spica*, an ear of grain). An inflorescence in which the main axis is elongated and the flowers are sessile.

Spikelet (L. *spica*, an ear of grain + dim. ending -*let*). The unit of inflorescence in grasses; a small group of grass flowers.

Spindle (A.S. *spinel*, an instrument employed in spinning thread by hand). Referring in mitosis and meiosis to the spindle-shaped intracellular aggregate of microtubules involved in chromosome movement.

Spine (L. *spina*, thorn). Modified leaf in the form of a hardened, pointed lateral projection.

Spodosol (Gr. *spodos*, wood ashes; R. *pod*, under + *zola*, ashes). One of the 10 world soil orders, characterized by an ashy, sandy, bleached, acidic A_2 horizon and associated mainly with coniferous forest vegetation; synonymous with **podzol.**

Sporangiophore (sporangium + Gr. -*phore*, a root of *pherein*, to bear). A branch bearing one or more sporangia.

Sporangium (spore + Gr. *angeion*, a vessel). Spore case.

Spore (Gr. *spora*, seed). A reproductive cell that develops into a plant without union with other cells.

Sporic life cycle. A life cycle that includes alternation of generations, the sporophyte and gametophyte being more than zygote and gametes, respectively.

Sporocyte (spore + L. *cyta*, vessel). A diploid or haploid cell that will undergo mitosis or meiosis to produce spores.

Sporophore (spore + Gr. *pherein*, to bear). The fruiting body of fleshy and woody fungi, which produces spores.

Sporophyll (spore + Gr. *phyllon*, leaf). A spore-bearing leaf.

Sporophyte. In alternation of generations, the plant in which meiosis occurs, producing meiospores; hence a plant with an even number of chromosome sets, usually the diploid number.

Spring wood. See **early wood.**

Stamen (L. *stamen,* the standing-up things or a tuft of thready things). Flower structure made up of an anther (pollen-bearing portion) and a filament.

Staminate cone. See **microsporangiate cone.**

Staminate flower. A flower having stamens but no pistils.

Starch (M.E. *sterchen,* to stiffen). A polysaccharide composed of glucose; the chief food storage material of many plants.

Statolith (Gr. *statos,* standing + *lithos,* stone). An organelle that moves to its position in a cell as a result of gravity, thus providing an initial sensing of, or orientation to, gravity by a cell.

Steady state. A conditioin in which some feature of a system remains constant while material or energy from the environment flows through the system.

Stele (Gr. *stele,* a post). The central cylinder, inside the cortex, of roots and stems of vascular plants.

Stem (O.E. *stemn*). The portion of the vascular plant that bears nodes, internodes, and leaves.

Steppe (R. *step,* a lowland). An arid grassland vegetation.

Sterigma (plural, **sterigmata**) (Gr. *sterigma,* a prop). A slender, pointed protuberance at the end of a basidium, which bears a basidospore.

Stigma (L. *stigma,* a prick, a spot, a mark). Receptive portion of the style to which pollen adheres.

Stipe (L. *stipes,* post, tree trunk). A stemlike portion of an alga or fungus.

Stipule (L. *stipula,* dim of *stipes,* a stock or trunk). A leaflike structure from either side of the leaf base.

Stolon (L. *stolo,* a shoot). A stem that grows horizontally along the ground surface.

Stoma (plural, **stomata**) (Gr. *stoma,* mouth). Epidermal structure on stems and leaves composed of two guard cells plus the small pore between them, through which gases pass.

Strain. A very uniform group of organisms within a species.

Strobilus (Gr. *strobilos,* a cone). A number of modified, spore-producing scales or leaves (sporophylls) grouped together on an axis.

Stroma (Gr. *stroma,* a bed or covering). The fluid material within the chloroplast, in which thylakoids are embedded. Also, a mass of protective vegetative filaments.

Style (Gr. *stylos,* a column). Slender column of tissue that arises from the top of the ovary and through which the pollen tube grows.

Suberin (L. *suber,* the cork oak). A waxy material found in cell walls or cork tissue and endodermis.

Substrate. A molecule that engages in a reaction catalyzed by an enzyme. Also, the solid medium on which a plant grows.

Succession (L. *successio,* a coming into the place of another). A sequence of changes in time of the species that inhabit an area, from an initial pioneer community to a final climax community.

Succulent. A plant with fleshy, water-storing parts.

Sucrose. Table sugar, a disaccharide made of a molecule of glucose linked to a molecule of fructose.

Sugar. A simple carbohydrate such as glucose.

Summer wood. See **late wood.**

Superior ovary. An ovary completely separate and free from the calyx.

Suspensor (L. *suspendere,* to hang). A cell or chain of cells developed from a zygote whose function is to place the embryo cells in an advantageous position to receive food.

Suture (L. *sutura,* a sewing together; originally the sewing together of flesh or bone wounds). The junction, or line of junction, of contiguous parts.

Symbiosis (Gr. *syn,* with + *bios,* life). An association of two different kinds of living organisms.

Sympetaly (Gr. *syn,* with + *petalon,* leaf). A condition in which petals are united.

Synandry (Gr. *syn,* with + *andros,* a man). A condition in which stamens are united.

Syncarpy (Gr. *syn,* with + *karpos,* fruit). A condition in which carpels are united.

Synergids (Gr. *synergos,* toiling together). The two cells at one end of the embryo sac, which, with the third (the egg), constitute the egg apparatus.

Syngamy. See **Fertilization.**

Synsepaly (Gr. *syn,* with + sepals). A condition in which sepals are united.

Taiga (Teleut *taiga,* rocky mountainous terrain). A broad northern belt of vegetation dominated by conifers; also, a similar belt in mountains just below alpine vegetation.

Tannin. A substance that has an astringent, bitter taste.

Tapetum (L. *tapete,* carpet). The tissue that lines developing pollen sacs (microsporangia) of seed plants; it degenerates and provides nutrition to the tissue within.

Taxon (plural, **taxa**) (Gr. *taxis,* order). A general term for any taxonomic rank, from subspecific to divisional.

Taxonomy (Gr. *taxis,* arrangement + *nomos,* law). Systematic botany; the science dealing with the describing, naming, and classifying of plants.

TCA cycle. See **citric acid cycle.**

Teliospore (Gr. *telos,* completion + spore). Resistant spore characteristic of the Heterobasidiomycete fungi, in which karyogamy and meiosis occur and from which a basidium develops.

Telium (plural, **telia**) (Gr. *telos*, completion). A sorus of teliospores.

Telophase (Gr. *telos*, completion + phase). The last stage of mitosis, in which daughter nuclei are reorganized.

Tendril (L. *tendere*, to stretch out, to extend). A slender coiling organ that aids in the support of stems.

Terminal bud. A bud at the end of a stem.

Testa (L. *testa*, brick, shell). The outer coat of the seed.

Tetrad (Gr. *tetradeion*, a set of four). A group of four chromatids occurring in prophase I of meiosis; also, the group of four cells that result from meiotic cell division.

Tetraploid (Gr. *tetra*, four + *ploos*, fold). Having four sets of chromosomes per nucleus.

Tetraspores (Gr. *tetra*, four + spores). Four spores formed by division of the sporocyte.

Tetrasporophyte (Gr. *tetra*, four + *sporos*, seed, spore + *phyton*, plant). One of two sporophyte generations in certain red algae; produces meiospores in tetrad clusters.

Thallophytes (Gr. *thallos*, a sprout + *phyton*, plant). A division of plants whose body is a thallus, i.e., lacking roots, stems, and leaves.

Thallus (Gr. *thallos*, a sprout). Plant body without true roots, stems, or leaves.

Thermoperiod (Gr. *therme*, heat + *period*, a cycle). A difference in temperature between day and night.

Thorn. Hard, sharp lateral appendage of a stem; a type of modified stem.

Thylakoid (Gr. *thylakos*, sac + -oid). A flattened, membrane-bound sac in which are embedded the light-capturing systems of photosynthesis.

Thymidine. A nucleoside incorporated in DNA but not in RNA.

Thymidine ³H. Tritiated or radioactive thymidine.

Thymine. A pyrimidine occurring in DNA but not in RNA.

Tiller (O.E. *telga*, a branch). A grass stem arising from a lateral bud at a basal node; tillering is the process of tiller formation.

Tissue. A group of cells that perform a collective function.

Tonoplast (Gr. *tonos*, stretching tension + *plastos*, molded, formed). The cytoplasmic membrane bordering the vacuole; so-called by de Vries, as he thought it regulated the pressure exerted by the cell sap.

Toxin (L. *toxicum*, poison). A poisonous secretion of a plant or animal.

Tracheid (Gr. *tracheia*, windpipe). An elongated, tapering xylem cell, with lignified pitted walls, adapted for conduction and support.

Tracheophytes (Gr. *tracheia*, windpipe + *phyton*, plant). Vascular plants.

Trait. A distinctive definable characteristic; a mark of individuality.

Transfer cells. Specialized cells modified by their cell wall projections for efficient short distance transport.

Transfusion tissue (L. *trans*, across + *fundere*, to pour, to melt). In pine needles, the tissue surrounding the central veins, which may serve to transfer water, nutrients, and food between the vascular tissue and the mesophyll.

Translocation (L. *trans*, across + *locare*, to place). The transfer of food materials or products of metabolism.

Transpiration (F. *transpirer*, to perspire). The giving off of water vapor from the surface of leaves.

Trichogyne (Gr. *trichos*, a hair + *gyne*, female). Receptive hairlike extension of the female gametangium in the Rhodophyta and ascomycetes.

Trichome (Gr. *trichoma*, a growth of hair). A short filament of cells.

Triose (Gr. *treis*, three + -ose). Any three-carbon sugar.

Tritium. A hydrogen atom, the nucleus of which contains one proton and two neutrons; it is written as ³H; the more common hydrogen nucleus consists only of a proton.

Tropical rain forest. Vegetation with several tree strata, characteristic of tropical lowland regions.

Tropism (Gr. *trope*, a turning). An orientation of the direction of growth in an organ, guided by an external stimulus such as light or gravity.

Tuber (L. *tuber*, a bump, swelling). A much-enlarged, short, fleshy underground stem tip.

Tundra (Lapp. *tundar*, hill). Meadowlike vegetation at low elevation in cold regions that do not experience a single month with average daily maximum temperatures above 50°F.

Turgid (L. *turgidus*, swollen, inflated). Swollen, distended; referring to a cell that is firm due to water uptake.

Turgor pressure (L. *turgor*, a swelling). The pressure within the cell resulting from the absorption of water into the vacuole and the imbibition of water by the protoplasm.

Tylosis (plural, **tyloses**) (Gr. *tylos*, a lump or knot). A growth of one cell into the cavity of another.

Type specimen. The herbarium specimen selected by a taxonomist to serve as a basis for the naming and descriptions of a new species.

Ultisol. A modified podsol soil, with red or yellow B horizon, representative of the southern deciduous forest.

Umbel (L. *umbella*, a sunshade). An inflorescence, the individual pedicels of which all arise from the apex of the peduncle.

Unavailable water. Water held by the soil so strongly that root hairs cannot readily absorb it.

Unicell (L. *unus,* one + cell). An organism consisting of a single cell; generally used in describing algae.

Uniseriate (L. *unus,* one + M.L. *seriatus,* to arrange in a series). Said of a filament having a single row of cells.

Uracil. A pyrimidine found in RNA but not in DNA.

Uredium (plural **uredia**) (L. *uredo,* a blight). A sorus of uredospores.

Uredospore (L. *uredo,* a blight + spore). A red, one-celled summer spore in the life cycle of the rust fungi.

Vacuole (L. *vasculum,* a small vessel). An organelle consisting of a single membrane that encloses a watery solution, serving chiefly for water and solute storage.

Variety. Differences among members of a group. In taxonomy, a group below the level of the species.

Vascular (L. *vasculum,* a small vessel). Referring to any plant tissue or region consisting of or giving rise to conducting tissue, e.g., bundle, cambium, ray.

Vascular bundle. A strand of tissue containing primary xylem and primary phloem (and procambium if present) and sometimes enclosed by a bundle sheath of parenchyma or fibers.

Vascular cambium. Cambium giving rise to secondary phloem and secondary xylem.

Vascular plant. Any plant that has both xylem and phloem.

Vegetation (L. *vegetare,* to quicken). The plant cover that clothes a region; it is formed of the species that make up the flora but is characterized by the abundance and life form (tree, shrub, herb, evergreen, deciduous plant, etc.) of certain of them.

Venation (L. *vena,* a vein). Arrangement of veins in leaf blade.

Venter (L. *venter,* the belly). Enlarged basal portion of an archegonium in which the egg cell is borne.

Ventral suture (L. *ventralis,* pertaining to the belly). The line of union of the two edges of a carpel.

Vernalization (L. *vernalis,* belonging to spring + *izare,* to make). The promotion of flowering by naturally or artificially applied periods of extended low temperature; seeds, bulbs, or entire plants may be so treated.

Vessel (L. *vasculum,* a small vessel). A series of xylem elements whose function it is to conduct water and mineral nutrients.

Vessel member. One of the cells composing a vessel. Also called a vessel element.

Vitamins (L. *vita,* life + amine). Naturally occurring organic substances, akin to enzymes, necessary in small amounts for the normal metabolism of plants and animals.

Water potential. A measure of the free energy associated with the water in a system; useful in predicting the direction of water movement.

Weathering. Physical and chemical change in parent material that leads to soil formation.

Weed (A.S. *wēod,* used at least since 888 in its present meaning). Generally an herbaceous plant or shrub not valued for use or beauty, growing where unwanted, and regarded as using ground or hindering the growth of more desirable plants.

Whorl. A circle of three or more flower parts or leaves.

Whorled. Referring to a bud or leaf arrangement in which there are three or more buds or three or more leaves at a node.

Wings. Lateral petals of a legume type of flower.

Wood (M.E. *wode, wude,* a tree). Secondary xylem.

Xanthophyll (Gr. *xanthos,* yellowish brown + *phyllon,* leaf). A yellow chloroplast pigment.

Xerophyte (Gr. *xeros,* dry + *phyton,* plant). A plant that normally grows in dry habitats.

Xylem (Gr. *xylon,* wood). The principal water- and mineral-conducting tissue in a vascular plant, composed of tracheids and/or vessels, fibers, and parenchyma.

Zoosporangium (Gr. *zoon,* an animal + sporangium). A sporangium bearing zoospores.

Zoospore (Gr. *zoon,* an animal + spore). A motile spore.

Zygomorphic (Gr. *zugon,* a yoke + *morphe,* form). Referring to bilateral symmetry; said of organisms or a flower, capable of being divided into two symmetrical halves only by a single longitudinal plane passing through the axis.

Zygospore (Gr. *zugon,* a yoke + spore). A thick-walled resistant spore developing from a zygote resulting from the fusion of isogametes.

Zygote (Gr. *zugon,* a yoke). A protoplast resulting from the fusion of gametes.

Zygotic life cycle. A life cycle in which the diploid phase is represented only by the zygote.

ILLUSTRATION CREDITS

Chapter 1: Introduction
1.1 Courtesy of R. M. Liebaert.
1.2 Courtesy of N. J. Lang.
1.3 Courtesy of N. J. Lang.
1.4 Courtesy of D. Dreyfus.
1.5 Courtesy of R. M. Liebaert.
1.6 Courtesy of R. M. Liebaert.
1.7 Courtesy of D. Brown.

Chapter 2: Metabolism
2.8 From *The Atlas of Protein Sequence and Structure*, Vol. 5, Suppl. 2, 1976, © The National Biomedical Research Foundation; redrawn from R. M. Sweet, H. T. Wright, J. Janin, C. H. Chothia, and D. M. Blow, *Biochemistry* 13:4220, © 1974, American Chemical Society.
2.11 From V. A. Greulach and J. E. Adams, *Plants: An Introduction to Modern Botany*, 3rd ed., © 1976, John Wiley & Sons, New York. Used with permission.

Chapter 3: The Plant Cell
3.1 From Hooke, *Micrographia* (1664). The Council of the Royal Society of London for Improving Natural Knowledge.
3.5 Courtesy of C. H. Lin.
3.6 Courtesy of H. T. Horner, Jr.
3.9 Courtesy of D. Branton.
3.10 Courtesy of H. T. Horner, Jr.
3.12 Courtesy of E. G. Cutter.
3.13 *A* and *B*, from E. H. Newcomb, *Protoplasma* 84:3,11, © Springer-Verlag, New York, 1975.
3.14 Courtesy of J. P. Braselton.
3.22 Courtesy of A. Bajer, *Chromosoma* 25:249, © Springer-Verlag, New York, 1971.
3.23 Courtesy of A. Bajer.
3.24 Courtesy of P. K. Hepler and W. T. Jackson, *J. Cell Biol.* 38:437, 1968.
3.26 Courtesy of M. C. Ledbetter.
3.18 Courtesy of R. Seagull.
3.28 Redrawn from K. Esau and J. Cronshaw, *Protoplasma* 65:1, © 1966 by Springer-Verlag. Reprinted with permission of the publisher.
3.29 Courtesy of S. Wick. Reproduced from the *Journal of Cell Biology*, 81:688, 1981. By copyright permission of The Rockefeller University Press.

Chapter 4: The Structure and Development of the Plant Body, and the Anatomy of Stems
4.6 *A* and *B*, redrawn from C. C. Forsaith, *The Technology of New York Timbers*, New York State College of Forestry Publication 18.
4.8 From J. S. Pate and B.E.S. Gunning, *Protoplasma* 68:140, 1969. Used with permission.
4.9 Courtesy of A. S. Crafts.

4.10 *A*, courtesy of J. Cronshaw and K. Esau, *J. Cell Biol.* 38:25, 1968. Used with permission.
4.11 *A* through *D*, redrawn from C. C. Forsaith, *The Technology of New York Timbers*, New York State College of Forestry Publication 18.
4.12 *B* and *C*, courtesy of D. Hess.
4.14 Courtesy of J. Lin.
4.16 Redrawn with permission of C. B. Beck.
4.35 *B*, redrawn with permission from K. Esau, *Plant Anatomy*, 2nd ed., © 1965, John Wiley & Sons, New York.
4.36 *A*, courtesy of R. Parker, National Audubon Society; *B*, courtesy of D. T. Smithers, National Audubon Society.

Chapter 5: The Structure of Leaves
5.5 *A* and *B*, courtesy of S. Lynch.
5.6 Courtesy of G. Breckon.
5.7 *A*, from R. M. Holman and W. W. Robbins, *A Textbook of General Botany*, © 1939, John Wiley & Sons, New York; *B*, courtesy of C. H. Lin; *C*, courtesy of W. W. Thomson.
5.13 Redrawn after G. S. Avery, *Amer. J. Bot.* 20:565, 1933.

Chapter 6: The Structure of Roots
6.2 *E*, from L. Kutschera, *Wurselatlas*, Geschaftfuhrer des DLG-Verlages, Frankfurt, 1960.
6.10 Courtesy of P. Barlow.
6.13 *A* through *D*, redrawn from K. Esau, *Plant Anatomy*, 2nd ed., © 1965, John Wiley & Sons, New York.

Chapter 7: The Absorption and Transport Systems
7.9 Courtesy of E. L. Proebsting.
7.10 Courtesy of D. R. Hoagland.
7.11 Courtesy of D. R. Hoagland.
7.13 Courtesy of D. R. Hoagland.
7.14 Courtesy of U. S. Forest Service.

Chapter 9: The Control of Growth and Development
9.10 Courtesy of L. Rappaport.
9.13 Courtesy of J. Goeschl and H. Pratt.
9.14 Courtesy of U.S. Department of Agriculture.
9.15 Courtesy of Plant Industry Station, Crops Research Division, Agricultural Research Service, U.S. Department of Agriculture.
9.16 Courtesy of Plant Industry Station, Crops Research Division, Agricultural Research Service, U.S. Department of Agriculture.
9.18 Courtesy of A. Lang.

Chapter 10: Plant Ecology
10.1 From *Silvics of Forest Trees of the United States*, U.S. Department of Agriculture Handbook No. 271, Washington, D.C., 1965.

16.16 Courtesy of K. Schmitz.
16.17 *A* and *C,* redrawn after G. M. Smith, *Freshwater Algae of the United States,* McGraw-Hill, New York, 1933. Used with permission of McGraw-Hill Book Co.
16.18 Redrawn after F. Moewns.
16.22 Redrawn from R. F. Scagel, *et al., An Evolutionary Survey of the Plant Kingdom,* 1965, Wadsworth Publishing Company, Belmont, California. Reprinted by permission of the publisher.
16.24 *F* through *P,* redrawn from R. F. Scagel, *et al., An Evolutionary Survey of the Plant Kingdom,* 1965, Wadsworth Publishing Company, Belmont, California. Reprinted by permission of the publisher.

Chapter 17: Bryophytes
17.9 Redrawn from G. M. Smith, F. M. Gilbert, G. S. Bryan, R. I. Evans, and J. F. Stauffer, *A Textbook of General Botany,* Macmillan, New York. Reprinted with permission of Macmillan Publishing Company, Inc. Copyright © 1949, by L. H. Bailey, renewed © 1977, by E. Z. Bailey.
17.11 Redrawn from W. P. Schimper, *Recherches sur les mousses,* Strasbourg, 1948.
17.12 Redrawn from R. F. Scagel *et al., An Evolutionary Survey of the Plant Kingdom,* 1965, Wadsworth Publishing Company, Belmont, California. Reprinted by permission of the publisher.
17.13 *B,* courtesy of C. Laning.
17.14 From C. Hebant, *Naturalia Monspeliensia Ser. Bot.* 18:301, 1967.
17.18 Courtesy of W. Russell and D. Hess.

Chapter 18. Lower Vascular Plants
18.1 Courtesy of Field Museum of Natural History.
18.2 Redrawn from H. P. Banks, *Evolution and Plants of the Past,* Wadsworth Publishing Company, Belmont, California. Reprinted by permission of the publisher.
18.5 Courtesy of D. Brandon.

18.6 Redrawn from M. Hirmer, *Handbuch der Paläobotanik* 1:182–232, 1927.
18.9 *B,* redrawn from M. G. Sykes, *Annals of Botany* (London) 22:63, 1909; *C,* redrawn from E. Pritzel in A. Engler and K. Prantl, *Die näturlichen pflanzenfamilien.* 1897–1915, 2nd ed.; *E, G,* and *H,* redrawn from H. Bruchmann, *Flora* 101:220, 1912, with permission of Gustav Fischer Verlag, New York; *J,* redrawn from material supplied by A. J. Eames.
18.12 *E, H, I, J, K,* and *L,* redrawn from R. A. Slagg, *American Journal of Botany* 19:106–107, 1932; *I, J,* and *O* redrawn from H. Bruchmann, *Flora* 104:180, 1915. With permission of Gustav Fischer Verlag, New York.
18.16 Redrawn from E. R. Walker, *The Botanical Gazette* 92:7, © 1940, by the University of Chicago Press.
18.20 *E, F,* and *G,* redrawn from M. E. Hartmann, *The Botanical Gazette* 91:259, © 1939, by the University of Chicago Press; *H* and *I,* redrawn from D. H. Campbell, *Mosses and Ferns,* Macmillan, New York, 1905; *J* and *K,* redrawn from R. M. Holman and W. W. Robbins, *A Textbook of General Botany,* © 1939, John Wiley & Sons, New York.
18.21 *A,* courtesy of D. Brandon.

Chapter 19: Gymnosperms
19.3 *A,* courtesy of the American Museum of Natural History; *B,* redrawn from L. H. Bailey, *The Cultivated Conifers in North America,* Macmillan, New York, 1933.
19.12 *G, H,* and *I,* redrawn from Coulter and Chamberlain, *Morphology of the Gymnosperms,* University of Chicago Press, 1917. With permission from the University of Chicago Press.
19.13E Slide courtesy of Triarch Products.
19.14 Courtesy of the Field Museum of Natural History.
19.16 Courtesy of A. Addicott.

Chapter 20: Angiosperms
20.3 Redrawn from Bonnier and Sablon, *Cours de Botanique,* Librairie Generale de l'Enseignement, 1923.

INDEX TO GENERA

This index includes all the genera mentioned in the text. Species are not shown, thus a given set of entries for a genus may refer to more than one species. In parentheses following the genus is the family (for genera in the Anthophyta) or the class (for genera in the Mycota) or the division (for all other genera). Common names are listed, but they give only cross-references to the scientific name, and all page entries appear after the scientific name. Common names selected are those preferred by Bailey (*Manual of Cultivated Plants*) or Munz and Keck (*A California Flora*).

INDEX OF SUBJECTS

This index includes terms, topics, place names, persons, and plant names above the genus level. The generic index should be consulted for other plant names. Not all persons mentioned in the text are listed here. Pages on which terms have been defined are shown in bold face. The Glossary (not included in this index) should be consulted for additional definitions.